Soil Chemistry and Ecosystem Health

Related Society Publications

Chemical Equilibrium and Reaction Models

Chemical Mobility and Reactivity in Soil Systems

Interactions of Soil Minerals with Natural Organics and Microbes

Rates of Soil Chemical Processes

Reactions and Movement of Organic Chemicals in Soils

Replenishing Soil Fertility in Africa

Selenium in Agriculture and the Environment

Sorption and Degradation of Pesticides and Organic Chemicals in Soil

Soil Chemistry and Ecosystem Health

Proceedings of a workshop sponsored by Division S-2 Soil Chemistry of the Soil Science Society of America in St. Louis, MO, 28 Oct. 1995. Cosponsors of the workshop include: International Union of Pure and Applied Chemistry; International Society for Ecosystem Health; International Society of Soil Science; Society for Environmental Geochemistry and Health; Society of Environmental Toxicology and Chemistry; Division of Environmental Chemistry, American Chemical Society; Soil Ecology Section of the Ecological Society of America; and Divisions S-3 and S-11 of the Soil Science Society of America

Editor

P.M. Huang

Coeditors

D.C. Adriano
T.J. Logan
R.T. Checkai

Organizing Committee

P.M. Huang, G.W. Bailey, J. Berthelin, J. Buffle, and A.L. Page

Editor-in-Chief SSSA

Jerry M. Bigham

Managing Editor

David M. Kral

Associate Editor

Marian K. Viney

SSSA Special Publication Number 52

Published by
Soil Science Society of America, Inc.
Madison, Wisconsin, USA

1998

Cover Design: Patricia Scullion

Soil Science Society of America, Inc.
677 South Segoe Road, Madison, WI 53711 USA

Library of Congress Registration Number: 98-60385

Printed in the United States of America

CONTENTS

FOREWORD

The Soil Science Society of America (SSSA) has a long and distinguished history of developing programs and publications dealing with the fundamental properties of soils. These publications, slide sets, and more recently CDS, have provided essential reference materials for a variety of audiences. Topics such as irrigation and drainage, liming of acid soils, methods of soil analysis, soil mineralogy, and waste management are examples of topics emphasized in SSSA books. The subject matter of these publications has emphasized the traditional strengths of soil science—soil chemistry, soil physics, soil microbiology, soil fertility, pedology, soil mineralogy, and so forth. Research on basic soil processes has been facilitated by the material synthesized in these publications. In addition, they have provided the insight needed to tackle practical problems in the use and management of soils for agriculture, engineering, and environmental applications.

This book entitled *Soil Chemistry and Ecosystem Health* deals with soils and their role in ecosystem health. Soils are an integral component of terrestrial ecosystems but it may not be as apparent that soils also play a key role in maintaining the health of air and water ecosystems as well. This volume emphasizes the key role of soil chemistry in understanding the functioning of ecosystems. The authors represent several subdisciplines within soil science in addition to the field of ecology and ecosystem science. Information has been summarized for scales ranging from the microscopic to landscapes. Importantly, this work demonstrates the key interrelationships between scientific disciplines studying ecosystems and the necessity for interdisciplinary efforts to understand the increasingly complex issues facing science and society.

Lee E. Sommers
President, SSSA

PREFACE

Soil is an integral part of the environment. The pedosphere, hydrosphere, atmosphere, and biosphere are environmental compartments that overlap and are intimately associated in the ecosystem. Therefore, what happens in soil should have a profound impact on not only soil quality but also ecosystem health, which is defined in terms of ecosystem sustainability as a function of activity, organization, and resilience. An ecological system is healthy and free from distress syndrome if it is stable and sustainable—that is, if it is active and maintains its organization and autonomy over time and is resilient to stress.

Among soil processes, chemical and biogeochemical reactions occurring in soil environments play a vital role in governing mechanisms of transformations, speciation, bioavailability, toxicity, dynamics, and transport processes of radionuclides, metals, and organics of environmental concern. Therefore, soil chemical and biogeochemical processes have a tremendous impact on soil quality and the stability and sustainability of the ecosystem. In this context, soil chemists have to share an enormous responsibility in sustaining ecosystem health. It is essential for soil chemists to interact with pure chemists, physicists, biologists, and scientists in other disciplines to develop and apply basic analytical tools to ecotoxicological risk assessment procedures, and to seek understanding of pertinent chemical and biogeochemical reactions pertaining to soil quality and ecosystem health. Furthermore, it is equally important for soil chemists to interact with scientists in other pertinent disciplines such as ecology, toxicology, and health science to investigate the impact of soil chemical and biogeochemical processes on the health of the ecosystem.

The workshop "Soil Chemistry and Ecosystem Health" was sponsored by Division S-2 Soil Chemistry of the Soil Science Society of America (SSSA) and cosponsored by Division S-3 Soil Biology and Biochemistry and Division S-11 Soil and Environmental Quality of SSSA, International Union of Pure and Applied Chemistry, International Society for Ecosystem Health, International Society of Soil Science, the Society for Environmental Geochemistry and Health, Society of Environmental Toxicology and Chemistry, Division of Environmental Chemistry of American Chemical Society, and the Soil Ecology Section of the Ecological Society of America. The workshop, which was supported by SSSA Outreach Funds, was held immediately prior to the 1995 SSSA Annual Meeting in St. Louis, MO. The Executive Committee of the SSSA gave its approval for the proposed special publication and appointed the editorial committee as indicated in a letter of 27 Apr. 1995 by then President David E. Kissel. The objectives of this special publication are: (i) to critically assess the impact of soil chemical and biogeochemical processes on ecosystem health and (ii) to build a linkage between soil chemistry and other disciplines pertaining to ecosystem health on a global scale. It is hoped that this publication will provide useful information on the prospects of outreaching activities of soil chemists pertaining to ecosystem health on a global scale, foster communications of scientific contributions of soil chemists to users of soil science information in other scientific disciplines, promote the soil science profession, and bring positive recognition to soil science as a whole.

P.M. Huang, Editor
D.C. Adriano, Coeditor
T.J. Logan, Coeditor
R.T. Checkai, Coeditor

CONTRIBUTORS

Domy C. Adriano — Professor of Environmental Soil Science, University of Georgia, Savannah River Ecology Laboratory, Drawer E, Aiken, SC 29802

C.C. Ainsworth — Environmental Process and Ecotoxicology Group, Battle Pacific Northwest Laboratory, P.O. Box 999, Battelle Boulevard, Richland, WA 99352

J.S. Angle — Department of Plant Science, University of Maryland, College Park, MD 20742

S.L. Brown — Environmental Chemistry Laboratory, USDA-ARS, BARC-West, Beltsville, MD 20705

Mark L. Brusseau — Soil, Water and Environmental Science Department, Hydrology and Water Resources Department, University of Arizona, Tucson, AZ 85721

D.A. Cataldo — Environmental Process and Ecotoxicology Group, Battelle Pacific Northwest Laboratory, P.O. Box 999, Battelle Boulevard, Richland, WA 99352

R.L. Chaney — Environmental Chemistry Laboratory, USDA-ARS, BARC-West, Beltsville, MD 20705

Anna Chlopecka — Soil Scientist, University of Georgia, Savannah River Ecology Laboratory, Drawer E, Aiken, SC 29802

David C. Coleman — Research Professor of Ecology, Institute of Ecology, University of Georgia, Athens, GA 30602-2360

C.J. Driver — Environmental Process and Ecotoxicology Group, Battelle Pacific Northwest Laboratory, P.O. Box 999, Battelle Boulevard, Richland, WA 99352

R.J. Fellows — Environmental Process and Ecotoxicology Group, Battelle Pacific Northwest Laboratory, P.O. Box 999, Battelle Boulevard, Richland, WA 99352

Connie Gaudet — Head, Soil and Sediment Quality Section, Environment Canada, 8th Floor, Place Vincent Massey, Hull, QC, Canada K1A 0H3

Kim Ford Hayes — Professor of Environmental Engineering, Environmental Engineering and Water Resources, Department of Civil and Environmental Engineering, University of Michigan, Ann Arbor, MI 48109-2125

Paul F. Hendrix — Associate Professor, Institute of Ecology, University of Georgia, Athens, GA 30602

Daniel I. Kaplan — Senior Research Scientist, Pacific Northwest National Laboratory, Battelle Boulevard, Richland, WA 99352

H. Marschner Professor of Plant Nutrition (Deceased). Institute of Plant Nutrition, University Hohenheim, 70593, Stuttgart, Germany

John McCullum Study Director, West Kitikmeot/Slave Study, 703 Northwest Tower, 5201 50th Ave., Yellow Knife, NT X1A 3S9, Canada

M.J. McLaughlin Principal Research Scientist, Land and Water-Waite Campus, CSIRO, Private Bag No. 2, Glen Osmond, SA 5064, Australia

R. Merckx Professor of Soil Science, Laboratory for Soil Fertility and Soil Biology, Katholieke Universität, 3001 Heverlee, Belgium

Paul G. Mezey Director, Mathematical Chemistry Research Unit, Department of Chemistry and Department of Mathematics and Statistics, University of Saskatchewan, 110 Science Place, Saskatoon, SK, Canada S7N 5C9

Murray Miller Professor Emeritus, Department of Land Resource Science, University of Guelph, Guelph, ON, Canada N1G 2Hl

Eugene P. Odum Professor Emeritus of Ecology and Director Emeritus, Institute of Ecology, University of Georgia, Athens, GA 30602

David J. Rapport Professor, Faculty of Environmental Sciences, University of Guelph, Ecosystem Health, Blackwood Hall, Guelph, ON, Canada NlG 2W1

Ramon J. Seidler Research Microbiologist, US Environmental Protection Agency, National Health and Ecological Effects Research Laboratory, Corvallis, OR 97333

Erik Smolders Associate Professor of Soil Science, Laboratory for Soil Fertility and Soil Biology, Katholieke Universität, 3001 Heverlee, Belgium

Jean M. Smolen Research Associate, US Environmental Protection Agency, 960 College Station Road, Athens, GA 30605

Alan T. Stone Professor of Environmental Chemistry, Department of Geography and Environmental Engineering, the Johns Hopkins University, Baltimore, MD 21218

Samuel Justin Traina Professor of Soil Physical Chemistry, School of Natural Resources, Ohio State University, 2021 Coffey Road, Columbus, OH 43210

Dharni Vasudevan Assistant Professor, Nichols School of the Environment, Duke University, Durham, NC 27708-0328

Lidia S. Watrud Research Ecologist, U.S. Environmental Protection Agency, National Health and Ecological Effects Research Laboratory, Corvallis, OR 97333

Conversion Factors for SI and non-SI Units

Conversion Factors for SI and non-SI Units

To convert Column 1 into Column 2, multiply by	Column 1 SI Unit	Column 2 non-SI Units	To convert Column 2 into Column 1, multiply by
		Length	
0.621	kilometer, km (10^3 m)	mile, mi	1.609
1.094	meter, m	yard, yd	0.914
3.28	meter, m	foot, ft	0.304
1.0	micrometer, μm (10^{-6} m)	micron, μ	1.0
3.94×10^{-2}	millimeter, mm (10^{-3} m)	inch, in	25.4
10	nanometer, nm (10^{-9} m)	Angstrom, Å	0.1
		Area	
2.47	hectare, ha	acre	0.405
247	square kilometer, km^2 (10^3 m)^2	acre	4.05×10^{-3}
0.386	square kilometer, km^2 (10^3 m)^2	square mile, mi^2	2.590
2.47×10^{-4}	square meter, m^2	acre	4.05×10^3
10.76	square meter, m^2	square foot, ft^2	9.29×10^{-2}
1.55×10^{-3}	square millimeter, mm^2 (10^{-3} m)^2	square inch, in^2	645
		Volume	
9.73×10^{-3}	cubic meter, m^3	acre-inch	102.8
35.3	cubic meter, m^3	cubic foot, ft^3	2.83×10^{-2}
6.10×10^4	cubic meter, m^3	cubic inch, in^3	1.64×10^{-5}
2.84×10^{-2}	liter, L (10^{-3} m^3)	bushel, bu	35.24
1.057	liter, L (10^{-3} m^3)	quart (liquid), qt	0.946
3.53×10^{-2}	liter, L (10^{-3} m^3)	cubic foot, ft^3	28.3
0.265	liter, L (10^{-3} m^3)	gallon	3.78
33.78	liter, L (10^{-3} m^3)	ounce (fluid), oz	2.96×10^{-2}
2.11	liter, L (10^{-3} m^3)	pint (fluid), pt	0.473

	Mass		
2.20×10^{-3}	gram, g (10^{-3} kg)	pound, lb	454
3.52×10^{-2}	gram, g (10^{-3} kg)	ounce (avdp), oz	28.4
2.205	kilogram, kg	pound, lb	0.454
0.01	kilogram, kg	quintal (metric), q	100
1.10×10^{-3}	kilogram, kg	ton (2000 lb), ton	907
1.102	megagram, Mg (tonne)	ton (U.S.), ton	0.907
1.102	tonne, t	ton (U.S.), ton	0.907
	Yield and Rate		
0.893	kilogram per hectare, kg ha^{-1}	pound per acre, lb $acre^{-1}$	1.12
7.77×10^{-2}	kilogram per cubic meter, kg m^{-3}	pound per bushel, lb bu^{-1}	12.87
1.49×10^{-2}	kilogram per hectare, kg ha^{-1}	bushel per acre, 60 lb	67.19
1.59×10^{-2}	kilogram per hectare, kg ha^{-1}	bushel per acre, 56 lb	62.71
1.86×10^{-2}	kilogram per hectare, kg ha^{-1}	bushel per acre, 48 lb	53.75
0.107	liter per hectare, L ha^{-1}	gallon per acre	9.35
893	tonnes per hectare, t ha^{-1}	pound per acre, lb $acre^{-1}$	1.12×10^{-3}
893	megagram per hectare, Mg ha^{-1}	pound per acre, lb $acre^{-1}$	1.12×10^{-3}
0.446	megagram per hectare, Mg ha^{-1}	ton (2000 lb) per acre, ton $acre^{-1}$	2.24
2.24	meter per second, m s^{-1}	mile per hour	0.447
	Specific Surface		
10	square meter per kilogram, m^2 kg^{-1}	square centimeter per gram, cm^2 g^{-1}	0.1
1000	square meter per kilogram, m^2 kg^{-1}	square millimeter per gram, mm^2 g^{-1}	0.001
	Pressure		
9.90	megapascal, MPa (10^6 Pa)	atmosphere	0.101
10	megapascal, MPa (10^6 Pa)	bar	0.1
1.00	megagram, per cubic meter, Mg m^{-3}	gram per cubic centimeter, g cm^{-3}	1.00
2.09×10^{-2}	pascal, Pa	pound per square foot, lb ft^{-2}	47.9
1.45×10^{-4}	pascal, Pa	pound per square inch, lb in^{-2}	6.90×10^3

(continued on next page)

Conversion Factors for SI and non-SI Units

To convert Column 1 into Column 2, multiply by	Column 1 SI Unit	Column 2 non-SI Units	To convert Column 2 into Column 1, multiply by
		Temperature	
1.00 (K – 273)	Kelvin, K	Celsius, °C	1.00 (°C + 273)
(9/5 °C) + 32	Celsius, °C	Fahrenheit, °F	5/9 (°F – 32)
		Energy, Work, Quantity of Heat	
9.52×10^{-4}	joule, J	British thermal unit, Btu	1.05×10^{3}
0.239	joule, J	calorie, cal	4.19
10^{7}	joule, J	erg	10^{-7}
0.735	joule, J	foot-pound	1.36
2.387×10^{-5}	joule per square meter, J m^{-2}	calorie per square centimeter (langley)	4.19×10^{4}
10^{5}	newton, N	dyne	10^{-5}
1.43×10^{-3}	watt per square meter, W m^{-2}	calorie per square centimeter minute (irradiance), cal cm^{-2} min^{-1}	698
		Transpiration and Photosynthesis	
3.60×10^{-2}	milligram per square meter second, mg m^{-2} s^{-1}	gram per square decimeter hour, g dm^{-2} h^{-1}	27.8
5.56×10^{-3}	milligram (H_2O) per square meter second, mg m^{-2} s^{-1}	micromole (H_2O) per square centimeter second, μmol cm^{-2} s^{-1}	180
10^{-4}	milligram per square meter second, mg m^{-2} s^{-1}	milligram per square centimeter second, mg cm^{-2} s^{-1}	10^{4}
35.97	milligram per square meter second, mg m^{-2} s^{-1}	milligram per square decimeter hour, mg dm^{-2} h^{-1}	2.78×10^{-2}
		Plane Angle	
57.3	radian, rad	degrees (angle), °	1.75×10^{-2}

Electrical Conductivity, Electricity, and Magnetism			
10	siemen per meter, S m^{-1}	millimho per centimeter, mmho cm^{-1}	0.1
10^4	tesla, T	gauss, G	10^{-4}
Water Measurement			
9.73×10^{-3}	cubic meter, m^3	acre-inches, acre-in	102.8
9.81×10^{-3}	cubic meter per hour, m^3 h^{-1}	cubic feet per second, ft^3 s^{-1}	101.9
4.40	cubic meter per hour, m^3 h−1	U.S. gallons per minute, gal min^{-1}	0.227
8.11	hectare-meters, ha-m	acre-feet, acre-ft	0.123
97.28	hectare-meters, ha-m	acre-inches, acre-in	1.03×10^{-2}
8.1×10^{-2}	hectare-centimeters, ha-cm	acre-feet, acre-ft	12.33
Concentrations			
1	centimole per kilogram, cmol kg^{-1}	milliequivalents per 100 grams, meq 100 g^{-1}	1
0.1	gram per kilogram, g kg^{-1}	percent, %	10
1	milligram per kilogram, mg kg^{-1}	parts per million, ppm	1
Radioactivity			
2.7×10^{-11}	becquerel, Bq	curie, Ci	3.7×10^{10}
2.7×10^{-2}	becquerel per kilogram, Bq kg^{-1}	picocurie per gram, pCi g^{-1}	37
100	gray, Gy (absorbed dose)	rad, rd	0.01
100	sievert, Sv (equivalent dose)	rem (roentgen equivalent man)	0.01
Plant Nutrient Conversion			
	Elemental	***Oxide***	
2.29	P	P_2O_5	0.437
1.20	K	K_2O	0.830
1.39	Ca	CaO	0.715
1.66	Mg	MgO	0.602

1 Ecosystem Health: An Overview

David C. Coleman, Paul F. Hendrix, and Eugene P. Odum

Institute of Ecology
University of Georgia
Athens, Georgia

1–1 OVERVIEW OF THE ECOSYSTEM CONCEPT

Ecosystems consist of all organisms in a given space or volume, interacting with all of the abiotic factors, such as energy, inorganic nutrients, light, and others. The system is open, and freely exchanges matter and energy across real or perceived boundaries (Odum, 1989). The ecosystem is often considered to be a watershed or other geographically defined unit, most fruitfully studied at that level of resolution (Swank & Crossley, 1988; Bormann & Likens, 1979; Fig. 1–1; Coleman et al., 1992). Being a functional rather than structural concept, an ecosystem can be considered at any spatial scale, ranging from a soil crumb or an acorn, to the biosphere (Fig. 1–1). The ecosystem concept focuses on processes that occur in nature, with emphasis on processes of organic matter accumulation via net primary production, and then a large series of catabolic, or breakdown processes, involved in consumption and decomposition, which release much of the C bound in reduced forms back to CO_2, and dissipated energy. In the processes of production and decomposition, there is considerable immobilization of nutrients by living organisms, and then subsequently mineralization upon the consumption or death of these organisms (Fig. 1–2; Coleman et al., 1983).

As soil scientists, we are most interested in processes occurring in particular types of ecosystems, such as agricultural, forest, and grassland systems. The relative amounts of nutrients, such as N or P, and their availability in inorganic or organic pools, are of considerable importance. Much progress has been made in ecosystem-level studies, using conceptual and simulation models, which emphasize the fact that the *labile* fraction, or microbial biomass pool, is of central importance in assessing soil fertility and nutrient availabilities for both short-term and long-term studies (Fig. 1–3; Parton et al., 1987, 1989; Hendrix et al., 1992). Other pools, with intermediate (decades) and long-term (across several centuries) recalcitrant compounds are slower to turn over, but play important roles in ensuring ecosystem stability.

The principal focus in this volume is the interrelationships of soil chemistry and ecosystem health. We suggest that any ecosystem that is functioning well in all of the major processes of primary production, decomposition, and nutrient cycling, and one that is capable of changing or evolving over the long term

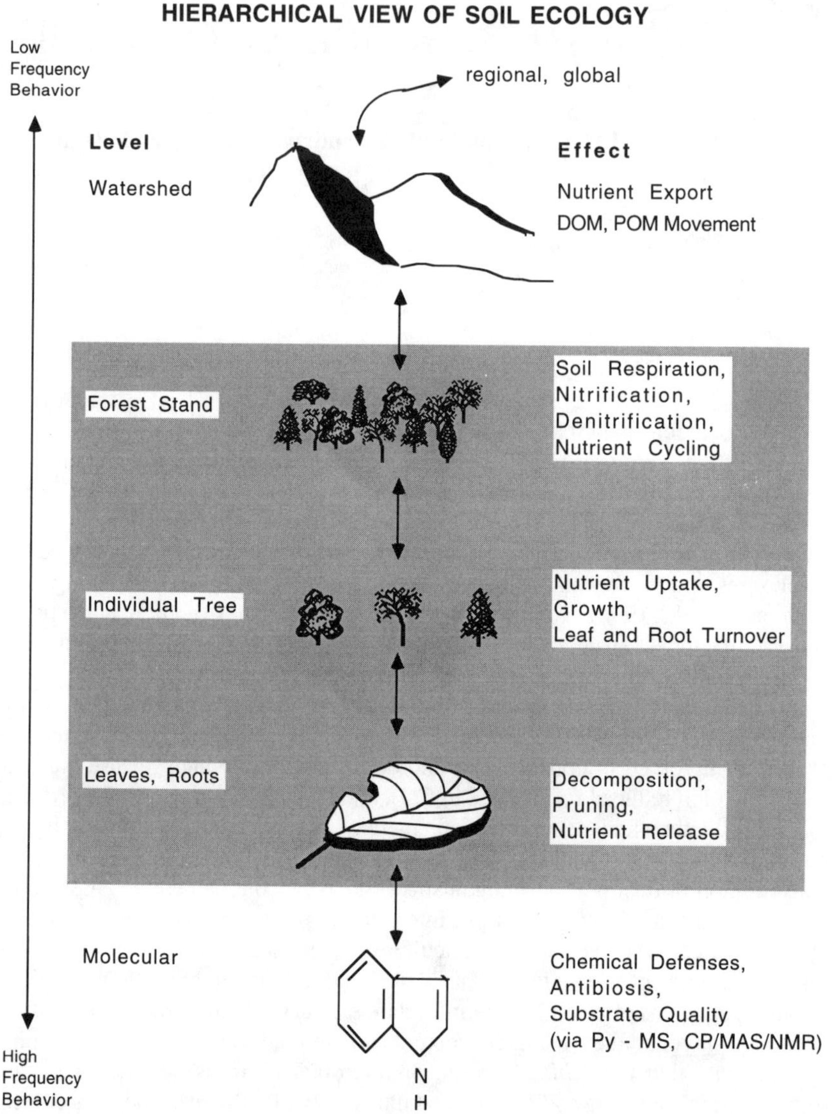

Fig. 1–1. Hierarchical view of soil ecological processes embedded in a landscape. Soil ecological phenomena span sizes from molecules to tissues, organs, organisms, communities, and watersheds in a region (Coleman et al., 1992).

(Gallopin, 1995), also can be considered as a *healthy ecosystem*. Health can be considered subjectively, and in a fashion analogous to the usage in medicine: what are our system's *vital signs*? It should thus be possible to measure some key variables and processes, and calculate indexes of *soil health*. There has been much discussion recently concerning *soil quality*. Several of the articles in the publication edited by Doran et al. (1994), relating to soil physical, chemical, and

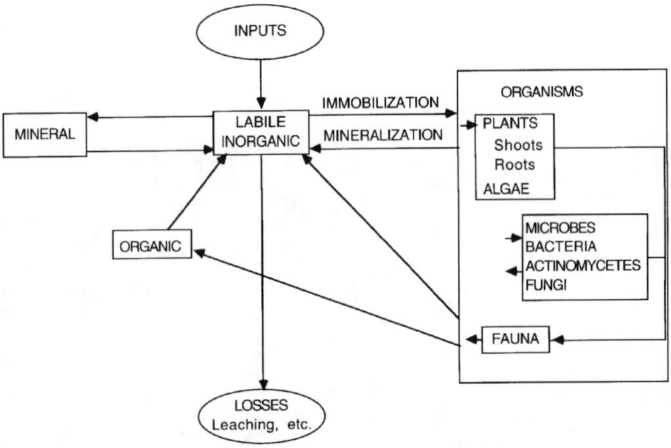

Fig. 1–2. Immobilization and mineralization processes in soils and the soil solution, with the mediating and facilitating effects of the biota indicated (Coleman et al., 1983).

biological attributes (e.g., Doran & Parkin, 1994; Turco et al., 1994; Linden et al., 1994) are germane to this topic, and are discussed later.

1–2 CENTRAL ROLES OF SOILS AS ORGANIZERS OF TERRESTRIAL ECOSYSTEMS

From a global or biospheric perspective, the amounts of C and major nutrients, such as N and P that are contained in soils are impressive when compared with any other portions of the biosphere. The standing stocks that are contained in a wide range of ecosystems within various biomes are quite variable across a range of habitat types Thus C amounts range from 156 Pg (10^{15} g = Pg) in tropical forests, to 213 Pg in grasslands and 59 Pg in deciduous forests (Anderson, 1992).

A factor of central importance to soil ecological studies is that soils have a range of textures and structural characteristics that enables an enormous array of plants, animals, and microbes to coexist and thrive. Humans have exploited the ability for soils to provide massive amounts of food, which can then be exported for use elsewhere within a country or elsewhere on the planet.

Another factor of great urgency in consideration of soils and their centrality to human well-being is the following: humans are directly exploiting >40% of the net primary production on the globe (Vitousek et al., 1986). That exploitation figure is probably increasing with the addition of 87.6 million people added to the global population every year, with the rate of addition steadily increasing (Brown & Flavin, 1996).

1–2.1 Fields and Watersheds—Foci of Activity

Several of the participants in this workshop consider processes that occur in various foci of activity, which have been termed *hot spots*, by Beare et al.

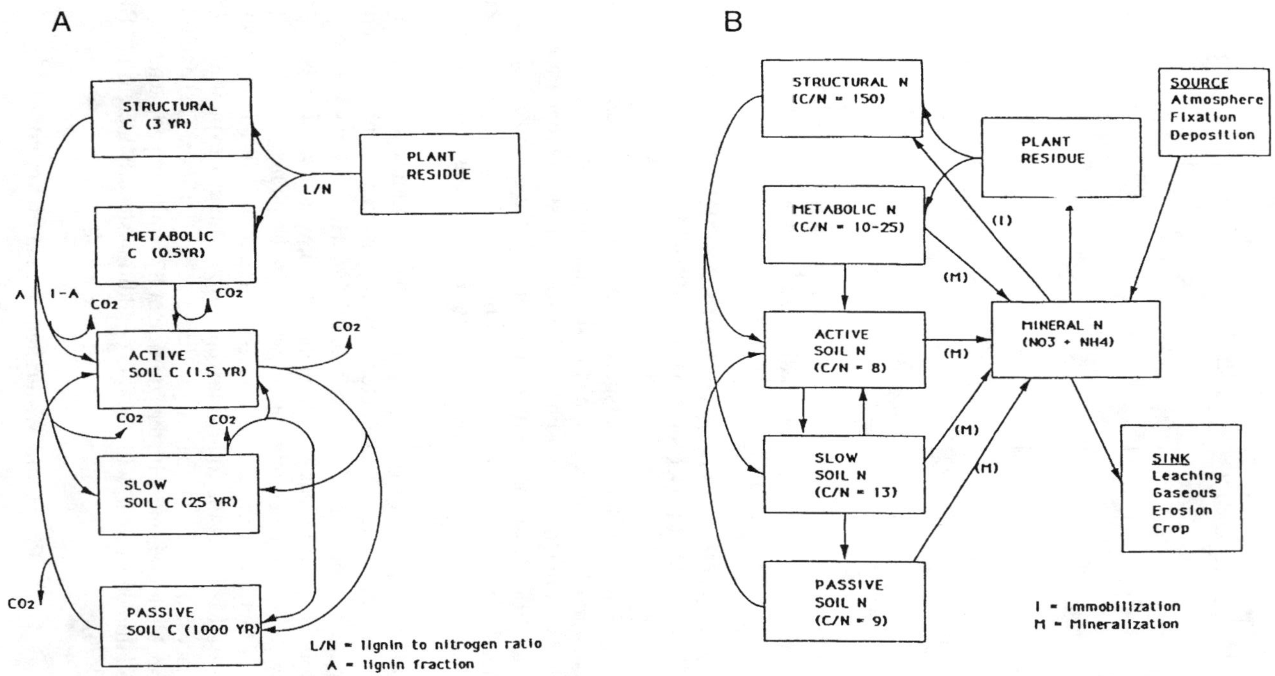

Fig. 1–3. Century model of soil organic matter turnover, showing flows of C and N in pools that are active, slow, and passive, depending on turnover rates (from Parton et al., 1987, 1989).

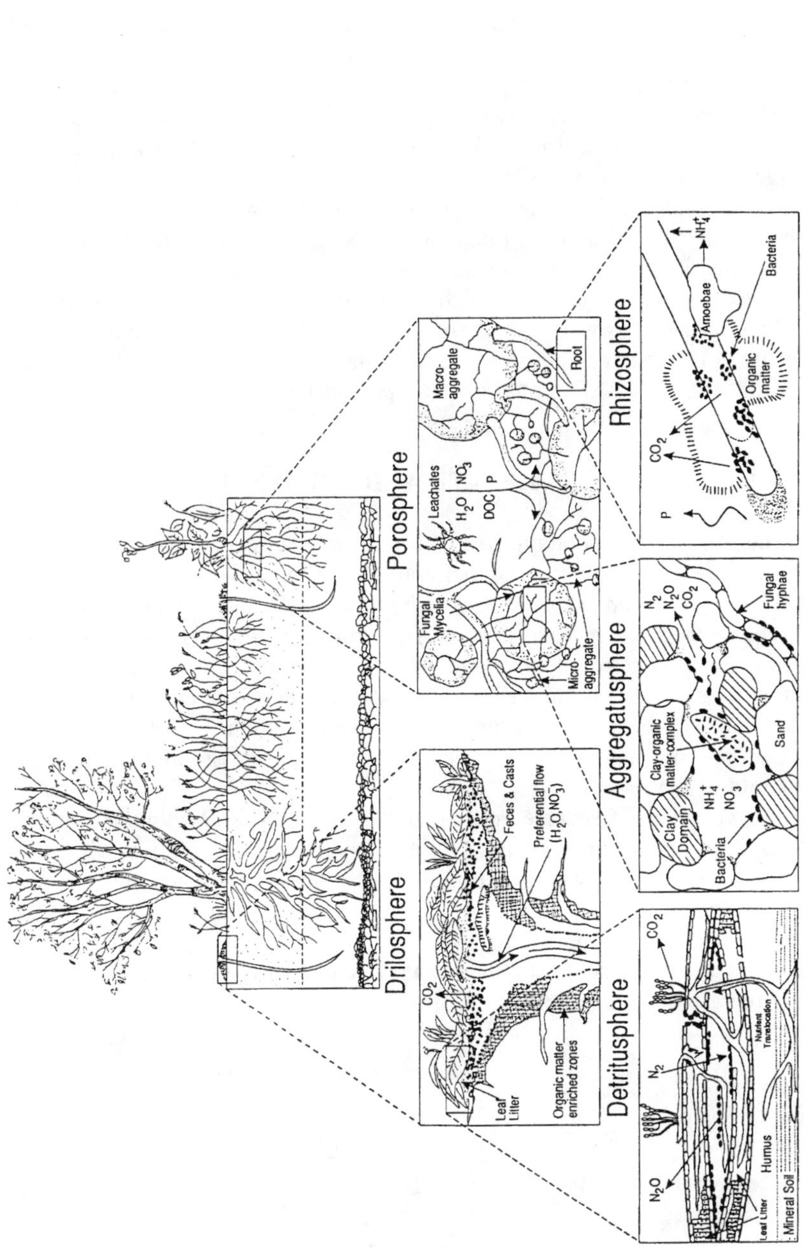

Fig. 1–4. Hot spots of activity within soils, ranging in size from aggregates to rhizosphere, detritus, drilosphere (earthworm burrows and macropores in general (porosphere); from Beare et al., 1995).

(1995; Fig. 1–4). These foci include, in ascending order of sizes, micro- and macroaggregates, rhizospheres, and macropores. The latter may result from root growth, or tunneling by various macrofauna, such as earthworms (e.g., *Lumbricus terrestris*). Particularly in tropical ecosystems, the influence of termite nests, the so- called *termitosphere* (Lavelle et al., 1992) can be very important, as well (Fig. 1–4; Beare et al., 1995).

As soil scientists working within ecosystems we should be concerned with how best to manage a complete range of ecosystem functions, within or encompassing many of these hot spots of interest. This is of central importance to successful ecosystem management, which is an area of emphasis for foresters and other land managers as well. It also should be noted that humans are an integral part of these ecosystems and probably have been for the last several millennia (Bormann et al., 1994; J.F. Franklin, 1995, personal communication).

1–3 ECOSYSTEM HEALTH AND THERMODYNAMICS

There are three general types of thermodynamic systems in existence: (i) an adiabatic system can exchange neither energy nor matter with its surroundings, (ii) a closed system can exchange energy but not matter, and (iii) an open system can exchange both energy and matter. All portions of the biosphere, including soils, are open systems (Johnson & Watson-Stegner, 1987). Open systems, including soils, can be best described by nonequilibrium thermodynamics, which tend toward a steady state characterized by minimum production of entropy. The soil–plant–microbe–fauna system is one that tends toward the minimum production of entropy, or disorder.

The following discourse relates to the Principle of Minimum Entropy Production for sustainability in these systems, with a view toward intensification of land use (Addiscott, 1995). Thermodynamic work is performed when energy in the form of heat is transferred from a source at a high temperature (e.g., the sun) to a sink at a low temperature (water or soils in the biosphere, or the infinite sink known as outer space). Continuous work requires infinite sources and sinks that are represented diagrammatically in Fig. 1–5 (Addiscott, 1995). During the

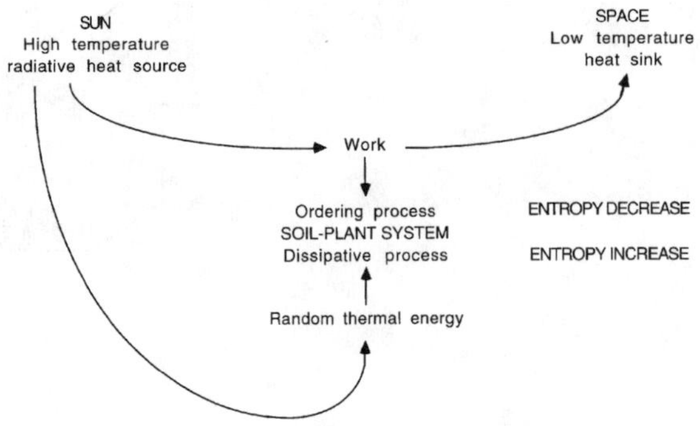

Fig. 1–5. The action of soil–plant–organism systems involved in ordering and dissipative processes, and their placement between a source (solar energy) and sink (outer space; from Addiscott, 1995).

processes of work done on earth, there may be considerable increases in order and thus, decreases in entropy at local scales. By definition, there are always dissipative processes that produce entropy at a local scale, the cost of doing work on any scale, following the Second Law of Thermodynamics. Photosynthesis and associated processes build complex ordered structures, namely carbohydrates and peptides, and others, from small molecules such as CO_2, H_2O, and NH_3, with dissipative processes involved in catabolism degrade these structures back to the small molecules. These aggrading and dissipative processes are summarized in Table 1–1, with various biological and physical processes considered separately. The great cycles of the biosphere are driven by such ordering and dissipative processes (Morowitz, 1970; Hutchinson, 1970). The soil–plant–microbial–faunal system is certainly a key component of many of the great biogeochemical cycles (Schlesinger, 1991), and the sustainability of our soil–plant systems depends on maintaining a reasonable balance, over the long run, between order and disorder.

If the ecosystem of interest, our soil–plant–microbial–faunal assemblage is markedly disturbed, will there be forces acting counter to the disturbance, i.e., acting homeostatically, to regain some sort of steady state or equilibrium condition? If so, how long will the ecosystem take to restore itself to a steady state following perturbation? Are there instances where a catastrophic perturbation occurs, and the system is unable to redirect itself toward any sort of steady state? The ecological literature is replete with many examples and case histories, and, if one is operating within certain boundary conditions, the system usually can approach some sort of steady state again. There are many examples of irreversible damage after ecosystem perturbation, such as after wholescale deforestation and subsequent degradation of the landscape, as in Amazonia. Usually, this is accompanied by massive losses of soil, so the ecosystem is truly physically altered. Interestingly, loss of the soil fauna in west African forests usually spelled a much greater likelihood of irreversible change, than if fauna were present and able to assist in the soil recuperative processes (P.H. Nye in Addiscott, 1995, personal communication). In numerous other instances, following occasional fires or flooding, ecosystems usually return to some sort of steady-state condition over

Table 1–1. Ordering and dissipative processes in the soil–plant systems categorized as biological or physical. The pairs are not necessarily exact opposites (from Addiscott, 1995).

Ordering processes Entropy decreases	Dissipative processes Entropy increases
Biological	
Photosynthesis	Respiration
Growth	Senescence
Formation of humus	Decomposition of humus
Physical	
Water flow (profile development)	Water flow (erosion, leaching)
Flocculation	Dispersion
Aggregation	Disaggregation
Development of structure	Breakdown of Structure
Larger units	Smaller units
Fewer of them	More of them
More ordered	Less ordered

years to decades following the perturbation. The key point is that the system retains its capacity for self-organization, which means that its biological potential is maintained.

Within the field of agricultural research, there are several classical examples of long-term perturbation experiments that have been ongoing for >150 yr. The Broadbalk, Grassland, and other plots at Rothamsted are world-renowned and only need be mentioned briefly here. The behavior of fertilizers and changes in organic matter over time are of great interest to us as ecologists and soil scientists. The experimental results may give us some insights into the concept of sustainability in agriculture and of resilience of terrestrial ecosystems in general.

Sustainability, whether for nonmanaged ecosystems, or managed ecosystems such as agricultural systems, is a topic that is the center of much debate, and deservedly so, given increasing population pressures on our biosphere. We have seen that natural systems that reach some asymptotic, mature phase are able to reach this point as the result of a minimum system-level production of entropy. We think of these systems as being in balance between system level production and respiration, with the P/R ratio being equal to one. One sign of entropy production is, as noted before, the degradation of substances with ordered structures and large molecular weights, resulting in production of small molecules such as CO_2, CH_4, N_2, and NO_3^-, all of which are environmentally undesirable when produced in excess. Therefore, it would behoove us to seek the establishment and operation of steady-state systems that have a long-term viability and operate in a sustainable fashion. To achieve this, Addiscott (1995) suggests four provisos, namely (i) the system must achieve a steady state; (ii) the flows within the system, particularly N, must be of appropriate magnitude; (iii) the capacity for self-organization must be retained in the system, thus ensuring that the biological potential is maintained; (iv) the maintenance of steady state must not involve an excessive expenditure of energy, e.g., fertilizer production, as entropy production from outside the system. The bottom line is that leaky systems, with many small molecules escaping from them, are inherently unstable, and becoming more disordered. Such a situation should be avoided.

1–4 ECOSYSTEM HEALTH AND SOIL QUALITY

A general definition of soil quality is "the capacity of a soil to function within ecosystem boundaries to sustain biological productivity, maintain environmental quality, and promote plant and animal health (Doran & Parkin, 1994). This general definition emphasizes what occurs within a bounded system, which might be a watershed unit, from which any particulate wind- or waterborne materials can be measured. It considers only plant and animal health, but the healthy activity of all groups of organisms, including microorganisms, while implied, should be stated explicitly. It is very difficult to measure the entire range of microbial species in a given ecosystem; indeed, it is virtually impossible, even with the use of various sophisticated molecular biological techniques (Paul & Clark, 1989; Torsvik et al., 1994; Coleman et al., 1994). Therefore, it becomes essential to measure indicator organisms and processes that can reveal a great deal about ecosystem health.

One of the major outputs of this workshop might well be an array of workable techniques for determining soil health and soil quality. One of the key approaches that we suggest is the measurement of keystone processes, and thus some sort of overall measurement of the health status of a given soil. This draws upon the usage, in aquatic ecology, from the last 20 yr, of such indices as the P/R, or net primary production to respiration ratio, measured as grams of C fixed to grams of C respired over a specified time span. Aquatic and terrestrial ecosystems are generally considered to be healthy if the P/R ratio is greater than one, more or less in steady state at a ratio of one, and a ratio less than one, with respiration being predominant; indicates that there is considerable input of organic materials that are then being oxidized, with considerable disruption to the system. This is a highly simplified version of a much more complex story. For example, small first- and second-order streams high up in mountains are principally dependent on inputs from outside the system, such as wind-blown leaves and other organic debris, hence the P/R ratio is <1. Also, if one were to analyze only the effluent waters from a coastal salt marsh, one might consider it to be very unhealthy. If one includes the primary production functions of the upland *Spartina* grasses, however, the overall productivity is generally positive, and indeed helps to maintain a very flourishing shrimp fishery in the coastal waters off many continental shelf areas of the world.

1–4.1 Biological Indicators of Soil Health: Development of Indices

For a microbe–soil–fauna approach to soil health, it is appropriate to consider various indicator indices. For example, the ratio of microbial biomass C to soil organic C (C_{mic}/C_{org}) relates to soil C availability and the tendency for a soil to accumulate or lose organic matter. This index was more useful in evaluating the status of a restored ecosystem, e.g., restored coal-mine land, than were single components, such as microbial C or soil organic C, separately (Insam & Domsch, 1988). Another index, more widely used, relates to soil temperature, soil management, ecosystem succession and heavy metal stress. The specific respiration rate, q_{CO2} is the ratio of basal respiration rate (as CO_2–C) per unit of microbial biomass (Insam & Haselwandter, 1989; Anderson & Domsch, 1990, 1993; Anderson, 1994). These are just the sorts of indices that can and should be used in a wide range of ecosystem types (Fig. 1–6).

A more recent approach to indices of soil quality has been a direct assessment of the available C pool, using a kinetic parameter (*AC*) derived from a modified Michaelis-Menten equation (Bradley & Fyles, 1995). This index was found to be preferable because the q_{CO2} ratio may confound energy availability and environmental stress, either one giving rather skewed values of q_{CO2}.

One needs specific measurements of labile C to know what is available to the microbial populations in soils. This is particularly important in the context of the hot spots noted above. For example, the impacts of living roots on soil organic matter decomposition can be considerable (Cheng & Coleman, 1990; Bottner et al., 1991). Several of our participants will be presenting extensive comments

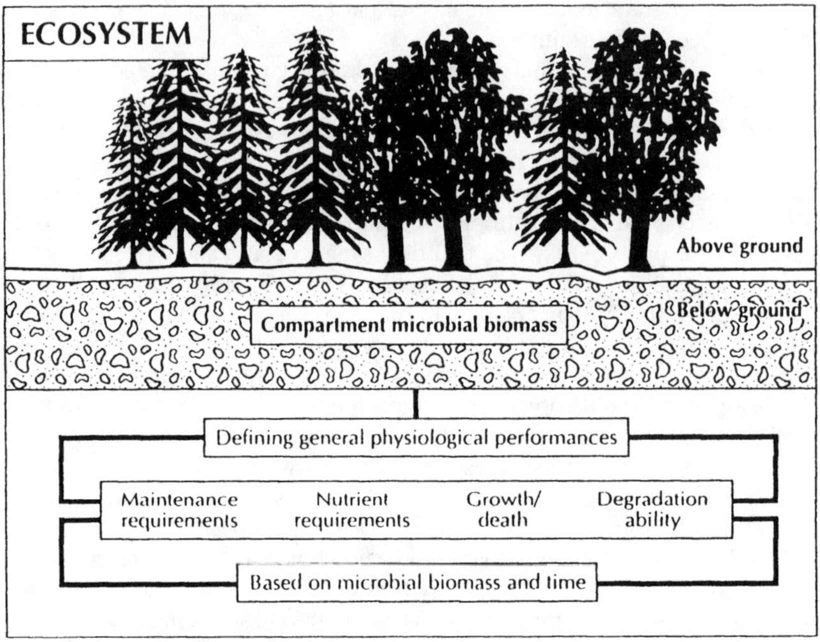

The prevailing assumption is that terrestrial ecosystems in a quasi–steady state are characterised by a constant flow of nutrients and energy, entering and leaving the system on a yearly basis, as well as entering and leaving the microbial biomass compartment. Microbial biomass communities adapt to the flow rate specific to the system.

C_{mic}/C_{org} ratio = percentage of microbial carbon to total soil carbon

Metabolic quotient for CO_2 (qCO_2) = μg CO_2-C/mg C_{mic}/h

Fig. 1–6. Specific respiration rates and ratios of microbial C to total soil organic C as useful indicators of ecosystem health (from Anderson, 1994).

about activities of roots, and specific rhizosphere-based phenomena, such as enzymatic or mycorrhiza-root interactions, for example, via microbial vs. mycorrhizal-based siderophores (e.g., Buyer et al., 1994; George et al., 1994; Crowley & Gries, 1994). These latter authors note the intricate interactions that occur between pathogens and mycorrhiza, competing in the mycorrhizosphere.

1–4.2 Nematodes as Key Indicator Organisms in Soil Systems

There are increasing demands for the use of indicator organisms as signs of ecosystem health, effectively integrating across several classes of environmental stressors, including chemical, physical, and biological (Linden et al., 1994; Rapport, 1995). Because numerous nematode species and functional groups (Freckman, 1982) are responsive to many chemical contaminants, nematodes can serve as sensitive indicators of ecosystem change that results from soil pollution

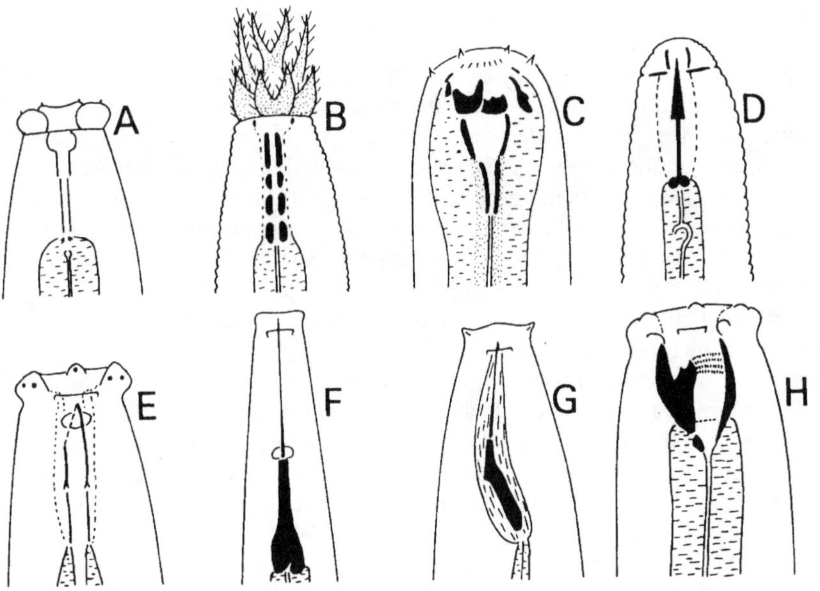

Fig. 1–7. Nematode anterior views, showing gradations from simple mouth opening (A and B), to stylet bearing, indicating plant root or fungal filament feeding (D through G), and predators (C and H; from Yeates & Coleman, 1982).

(Bongers, 1990; Niles & Freckman, 1998). Nematodes are aquatic organisms, inhabiting the water-films in soil pores and pore necks. They are bilaterally symmetric and unsegmented, with an acellular cuticle supported by a hydrostatic skeleton (Freckman & Baldwin, 1990). Nematodes typically go through five life-history stages, with four juvenile and one adult stage. They are very responsive to favorable conditions of moisture and food availability, completing their life cycle in as few as 8 d (at an average temperature of 20°C) for bacterial feeders. The range of functional groups based on their principal food sources includes bacterial-feeders, fungal-feeders, plant parasites (receiving most attention so far), omnivores, and predators (Yeates & Coleman, 1982; Yeates et al., 1993; Fig. 1–7).

In the process of ecosystem development known as succession, bacterial-feeding nematodes in the family Rhabditidae are dominant in the early stages, when there is a surplus of labile substrates, such as those in manure and fresh crop residues. The latter have a somewhat narrower ratio of C to N than mature residues (Griffiths et al., 1994; Niles & Freckman, 1998). As time passes, species in other families such as the Cephalobidae dominate; these nematodes are more tolerant of drought and persist through the later stages of succession (Ettema & Bongers, 1993). Fungal feeders become more prominent as more resistant substrates (e.g., lignin and cellulose, both with a very high C/N) accumulate in the ecosystem (Bouwman et al., 1993). In general, secondary succession in nematode communities parallels plant succession and soil development (de Goede et al., 1993). Relatively long-lived, so-called *K-selected* nematodes, those that have a relatively low egg output and slower population growth curves, are more numerous in mature communities than in earlier stages of succession (Hanel, 1995; Wasilewska, 1994).

Heavy metals and several other environmental stressors may cause the trophic structure to shift towards dominance by bacterial-feeding species of nematodes, e.g., Rhabditidae and Cephalobidae as in the early successional stages. In a number of cases, there can be an induction of metal-responsive genes, which produce metal-binding proteins such as metallothionein. This metallothionein induction allows the nematodes, as well as other invertebrates, to sequester metal ions and perhaps more readily tolerate contaminated habitats (Niles & Freckman, 1998). It is thus possible for nematode taxa to shift in dominance as a function of differential fitness to toxic chemicals, or differential responses to changing microbial populations, which are in turn showing responses to environmental insults such as acid precipitation. In summary, there is a growing amount of literature on the usage of indicator genera, such as *Caenorhabditis* sp. to monitor soil and aquatic habitats (Donkin & Dusenbery, 1993). Some of the key responsive species are termed *sentinel species* and display responses that are genetic, physiologic, morphologic, or behavioral indicators of contaminant exposure (Niles & Freckman, 1998).

One of the key indices of soil health has been the maturity index (MI; Bongers, 1990; Yeates, 1994), which applies different weightings to *r*-selected, rapidly growing genera, *colonizers*, and other types, which are *K*-selected, called *persisters*, ranked on a scale of from 1 to 5 (Table 1–2). The MI is calculated as the weighted mean of the individual colonizer–persister (*c–p*) values where $v(i)$ is the *c–p* value of Taxon i as given in Table 1–2 and $f(i)$ the frequency of that taxon in a sample. This index seems to be a more sensitive measure of environmental conditions than are ecological indices based on nematode feeding groups (Neher et al., 1995). The MI may be most sensitive when assessing the overall health of eutrophic habitats, which are characterized by high levels of microbial activity, or when monitoring polluted habitats, in which toxic chemicals kill omnivorous, persister nematodes (Niles & Freckman, 1998; Ruess & Funke, 1995).

1–5 INDEXES OF SOIL QUALITY

We now consider the features of a soil quality index, which was developed with regard to five specific soil functions (Doran & Parkin, 1994): Function 1—the ability to hold, accept, and release water to plants, streams, and subsoils (water flux); Function 2—the ability to hold, accept, and release nutrients and other chemicals (nutrient and chemical fluxes); Function 3—to promote and sustain root growth; Function 4—to maintain suitable soil biotic habitat; Function 5—to respond to management and resist degradation. These indexes are suitable, but rather general in scope. There is further need of more specifically targeted indexes of soil quality, and these are considered next.

More specific chemically oriented indexes include measurement of the relative abundances or levels of lignin, polyphenols, and C/N. Tian et al. (1993, 1995) developed a plant residue quality index (PRQI) that uses the C/N and lignin

Table 1–2. Colonizer–persister (*c–p*) values for terrestrial and aquatic nematode families: 1 = colonizer, 5 = persister (from Bongers, 1990).

Nematode family	*c–p* value	Nematode family	*c–p* value
Neotylenchidae	2	Achromadoridae	3
Angiunidae	2	Ethmalaimidae	3
Aphelenchidae	2	Cyatholaimidea	3
Aphelenchoididae	2	Desmodoridae	3
Rhabditidea	1	Microlaimidae	3
Alloionematidae	1	Odontolaimidae	3
Deploscapteridae	1	Aulolaimidae	3
Bunonematidae	1	Bastianiidae	3
Cephalobidae	2	Prismatolaimidae	3
Ostellidae	2	Ironidae	4
Panagrolaimidae	1	Tobrilidae	3
Myolaimidae	2	Onchulidae	3
Teratocephalidae	3	Tripylidae	3
Diplogasteridae	1	Alaimidae	4
Neodiplogasteridae	1	Bathyodontidae	4
Deplogasteroididae	1	Monochidae	4
Tylopharyngidae	1	Anatonchidae	4
Odontopharyngidae	1	Nygolaimidae	5
Monhysteridae	1	Dorylaimidae	4
Xyalidae	2	Chrysonematidae	5
Linhomoeidae	3	Thornenematidae	5
Plectidae	2	Nordiidae	4
Leptolaimidae	3	Thornenematidae	5
Halaphanolaimidae	3	Aporcelaimidae	5
Deplopeltidae	3	Belondiridae	5
Rhabdolaimidae	3	Actinolaimidae	5
Chromadoridae	3	Discolaimidae	5
Hypodontolaimidae	3	Leptonchidae	4
Choanolaimidae	4	Diphtherophoridae	3

and polyphenol concentration of plant residues. There was a large variation in PRQI among the 18 species of plant residues tested. The PRQI was defined as:

$$\text{PRQI} = [1/(a\ \text{C/N} + b\ \text{lignin} + c\ \text{polyphenols})] \times 100$$

where *a, b,* and *c* are coefficients of relative contribution of C/N ratio, lignin content (%) and polyphenol content (%) to plant residue quality, respectively. Tian et al. (1995) noted that with lower values of all three variables, there was a marked mulching effect, leading to enhanced macrofaunal activities, e.g., increased termite and earthworm activity, allowing greater aeration and nutrient mineralization.

A soil quality index could be extended usefully to include soil organic matter constituents. For example, in a study of the influences of soil biota on C dynamics and nutrient cycling in agroecosystems, a ratio of the two prevalent carbohydrates, mannose–xylose, was used to evaluate the relative contributions of carbohydrates of plant vs. microbial origin. The ratio was highest in the subplots where roots had been excluded and followed a descending gradient to a control plot, which included all biota (Hu et al., 1995; Table 1–3).

Table 1–3. Monosaccharides (g kg^{-1} soil), mannose-to-xylose ratios (M/X), and (mannose + galactose) to (arabinose + xylose) ratios ([M + G]/[A + X]) in 5- to 15-cm soils under four biotic treatments and control (from Hu et al., 1995).

	Biotic treatments†				
Measurement	Root exclusion	Earthworm addition	Fungicide	Arthropod inhibitor	Control
Arabinose	49.8 (12)b	106.6 (15)a	80.2 (14)ab	74.4 (17)ab	102.9 (20)a
Xylose	5.8 (1.5)b	12.1 (2.6)ab	10.1 (0.7)ab	11.1 (2.3)ab	18.8 (5.0)a
Mannose	27.6 (3.0)a	46.8 (6.4)a	36.2 (3.8)a	43.3 (9.2)a	45.4 (9.9)a
Galactose	15.0 (3.0)a	32.5 (6.5)ab	25.1 (2.0)b	19.3 (6.0)b	49.0 (10)a
Glucose	68.2 (19)a	135.9 (30)a	100.1 (8.6)a	121.2 (36)a	152.1 (39)a
M/X	4.8 (0.47)a	4.1 (0.92)a	3.6 (0.42)a	3.5 (0.45)a	2.8 (0.74)a
[M + G]/[A + X]	0.84 (0.12)a	0.68 (0.08)a	0.73 (0.12)a	0.89 (0.05)a	0.81 (0.11)a

† Values are averages of the analyses from four field replicates ($n = r$), with standard erors in parentheses. Means in the same row with the same letters are not significantly different at $P = 0.05$.

Several studies have been made of organic matter dynamics in agroecosystems under different management regimes (Phillips & Phillips, 1984; Hendrix et al., 1986; Andrén et al., 1990; de Ruiter et al., 1993) There is general agreement that the physicochemical and biological milieus in zero-tillage plots are significantly different than in heavily plowed treatments. Many constituents of the soil biota tend to be more numerous and more diverse in the zero-tillage environments (House et al., 1982; Table 1–4). This can have significant implications on C and

Table 1–4. Comparison of agroecosystem components and associated agroecosystem processes from conventional tillage (CT) and no tillage (NT) systems (from House, 1982).

Component or process	CT vs. NT
Crop yields	NT = CT (except during drought)
Crop biomass	Decreasing in both CT and NT
Weed biomass	NT > CT
Plant N dynamics	CT > NT (N flux)
Shoot to root ratios	CT > NT
N fixation	NTCT (?)
Surface crop and weed residues	NT >> CT
Litter decomposition rates	CT > NT
Surface litter (%N)	NT > CT
Soil total N	NT > CT in upper soil layer
Nitrification activity	NT > CT in upper soil layer NT > CT in middle soil layer (?)
Soil organic matter	NT > CT
Soil moisture	NT > CT
Ground water leaching (NO_3–N)	CT > NT (?)
Foliage arthropods	CT = NT
Crop herbivory by insects	CT > NT
N content of crop foliage	CT > NT
Arthropods species diversity	NT > CT
Soil arthropods (no. of individuals)	NT >> CT
N contained in arthropods:	
soil	NT > CT
foliage	CT > NT
Ecosystem N turnover time	NT > CT
Ecosystem N efficiency	NT > CT (?)

N turnover, as reviewed by Beare et al. (1992). In recent studies, we investigated the influences of soil biota (specifically plant roots, soil fungi, microarthropods and earthworms) on the dynamics of soil organic matter, in particular the size of soil carbohydrate and microbial biomass pools in an Ultisol from the southeastern USA (Hu et al., 1995).

At the initiation of the experiment, a 40-yr-old mixed grass sod (8.27 g C kg^{-1} soil and 0.84 g N kg^{-1} soil; 0–15-cm depth) was moldboard-plowed, disk-harrowed, and rotary-tilled to a depth of 15 cm. Five subplots (1.2 by 1.66 m) were established on each of four replicate plots and each subplot was enclosed with plastic walls to a 25-cm depth. The subplots were planted with grain sorghum [*Sorghum bicolor* (L.) Moench] in summer and fallowed in winter following tillage. Five biotic manipulation treatments were as follows: (1) root exclusion by a microscreen (Versapor-3000, 3 micrometer pore size, Gelman Sciences, Ann Arbor, MI), (2) inhibition of fungi with the fungicide Captan (cis-N-trichloromethylthio-4-cyclohexene-1,2-dicarboximide), (3) earthworm additions during the winter and spring months, (4) repulsion of arthropods with the arthropodicide naphthalene, and (5) untreated control. Further details of plot layout and biocides are presented by Hu et al. (1995).

Samples of soil were taken just prior to fall harvesting of the grain sorghum in November 1990 and 1991. Soil carbohydrates were measured following the methods of Cheshire et al. (1990), using ground soil samples (200 mg) hydrolyzed in H_2SO_4, filtering the suspension and neutralizing in Ba $(OH)_2$. The hydrolysates, after reduction with $NaBH_4$, were acetylated with acetic anhydride, and the resulting alditol acetates in acetone were determined in a Shimadzu gas chromatograph (Shimadzu, Tokyo), equipped with a flame ionization detector. Further chemical extraction details are given in Hu et al. (1995). The ratio of mannose to xylose (milligram of carbohydrates per kilogram of soil; M/X) was calculated and used as an indicator of the origin of the carbohydrates. Microbial biomass was determined by using a chloroform fumigation-extraction method adapted from Vance et al. (1987). Extracts (0.5 *M* K_2SO_4) were analyzed using a Shimadzu total organic C analyzer. Microbial biomass C was calculated using a K_{ec} factor of 0.33 (Sparling & West, 1988). Microbial biomass N was calculated on total microbial N determined by a persulfate oxidation method (Cabrera & Beare, 1993) using a K_{en} factor of 0.45 (Jenkinson, 1988).

Fungal densities were significantly lower after the fungicide treatment in both surface (0–5 cm) and deeper (5–15) cm soils, with fungal lengths less than one-half (187 000 km kg^{-1} soil) compared with the control (409 000 km kg^{-1} soil). Fungal densities also were lower in the root exclusion treatment at both depths, compared with the control. In the surface (0–5 cm) soils, the effect of root exclusion on the total carbohydrate content decreased to 139 mg carbohydrates per kg^{-1} soil, which was only 30% of the control (456 mg carbohydrates kg^{-1} soil). All of the other biotic treatments were more moderate in response. Thus, in the surface soils, all the major simple sugars were significantly lower in the root exclusion treatment than in the control, and the M/X was higher (Table 1–3), indicating a greater contribution from microbial-derived carbohydrates compared with those from plants. No such difference was found in the (galactose + man-

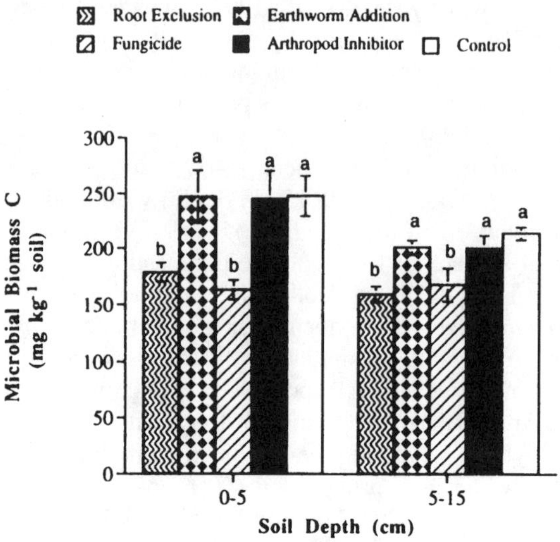

Fig. 1–8. Microbial biomass C in soils of biotically manipulated treatments in experimental plots at Horseshoe Bend, Athens, GA (see text for details; from Hu et al., 1995).

nose)/(arabinose + xylose) ratio, although it has been used as an indicator of the origin of carbohydrates (Oades, 1984).

For the surface (0–5 cm) soils, microbial biomass C was significantly lower in the root exclusion and fungicide treatment than in the control, earthworm addition, and arthropod inhibition treatments (Fig. 1–8). Microbial biomass N followed a similar pattern to that of the microbial biomass C (Fig. 1–9). There was a close linear relationship ($R = 0.64$) between total carbohydrates and microbial

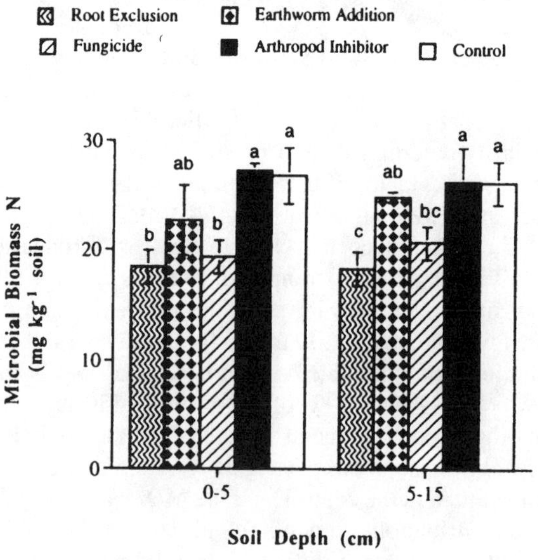

Fig. 1–9. Microbial biomass N in soils of biotically manipulated treatments in experimental plots at Horseshoe Bend, Athens, GA (see text for details; from Hu et al., 1995).

biomass C ($P < 0.01$), with the regression accounting for 41.5% of the variance (data not shown).

The foregoing analysis has been presented in detail to give some indication of the biotic and chemical interactions possible in agroecosystem field studies, and also is offered as an example of research leading to some indications of soil quality, which is often the result of microbial and faunal activity.

1–6 SUMMARY AND CONCLUSIONS

There are several take-home messages to be gleaned from a consideration of soils and ecosystem health. We have seen that soils serve as organizing centers in ecosystems. Soils are open systems, and evolve or develop over time. Soils that are healthy have homeostatic properties, within broadly set limits. Soils can be managed for long-term viability, and can even be suitable subjects for bioremediation (as noted in several chapters of this publication). There are numerous and active interfaces between chemistry, physics, and biology of soils, and this multitude of interfaces will be increasingly explored in the coming years ahead. One facet not covered in any detail in this chapter is the phenomenon of soil biodiversity. The diverse array of microbes and fauna serve as major causative agents in *natural bioremediation*, as noted earlier in this chapter, and provide buffering against changes, caused by humans or other agents.

REFERENCES

Addiscott, T.M. 1995. Entropy and sustainability. Eur. J. Soil Sci. 46:161–168.

Anderson, J.M. 1992. Responses of soils to climate change. Adv. Ecol. Res. 22:163–210.

Anderson, T.-H. 1994. Physiological analysis of microbial communities in soil: Applications and limitations. p. 67–76 *In* K. Ritz et al. (ed.) Beyond the biomass. Wiley-Sayce, Chichester, England.

Anderson, T.-H., and K.H. Domsch. 1990. Application of eco-physiological quotients (qCO_2 and qD) on microbial biomasses from soils of different cropping histories. Soil Biol. Biochem. 22:251–255.

Anderson, T.-H., and K.H. Domsch. 1993. The metabolic quotient for CO_2 (qCO_2) as a specific activity parameter to assess the effects of environmental conditions, such as pH, on the microbial biomass of forest soils. Soil Biol. Biochem. 25:393–395.

Andrén, O., T. Lindberg, K. Paustian, and T. Rosswall (ed.). 1990. Ecology of arable land. Organisms, carbon and nitrogen cycling. Munksgaard Int., Copenhagen.

Beare, M.H., D.C. Coleman, D.A. Crossley, Jr., P.F. Hendrix, and E.P. Odum. 1995. A hierarchical approach to evaluating the significance of soil biodiversity to biogeochemical cycling. Plant Soil 170:5–22.

Beare, M.H., R.W. Parmelee, P.F. Hendrix, W. Cheng, D.C. Coleman, and D.A. Crossley, Jr. 1992. Microbial and faunal interactions and effects on litter nitrogen and decomposition in agroecosystems. Ecol. Monogr. 62:569–591.

Bongers, T. 1990. The maturity index: An ecological measure of environmental disturbance based on nematode species composition. Oecologia 83:14–19.

Bormann, B.T., M.H. Brookes, E.D. Ford, A.R. Kiester, C. Oliver, and J.F. Weigand. 1994. A Framework for sustainable ecosystem management. Gen. Tech. Rep. 331. U.S. For. Serv., Pacific North West, Portland, OR.

Bormann, F.H., and G.E. Likens. 1979. Pattern and process in a forested ecosystem. Springer-Verlag, New York.

Bottner, P., J. Cortez, and Z. Sallih. 1991. Effect of living roots on carbon and nitrogen of the soil microbial biomass. p. 201–210. *In* D. Atkinson (ed.) Plant root growth. Blackwell, London.

Bouwman, L.A., G.H.J. Hoenderboom, A.C. Van Klinken, and P.C. de Ruiter. 1993. Effect of growing crops and crop residues in arable fields on nematode production. p. 127–131 *In* H.J.P. Eijsackers and T. Hamers (ed.) Integrated soil and sediment research: A basis for proper protection. Kluwer Acad. Publ., Dordrecht, the Netherlands.

Bradley, R.L., and J.W. Fyles. 1995. A kinetic parameter describing soil available carbon and its relationship to rate increase in C mineralization. Soil Biol. Biochem. 27:167–172.

Brown, L.R., and C. Flavin. 1996. State of the world 1996. W.W. Norton & Co., New York.

Buyer, J.S., M.G. Kratzke, and L.J. Sikora. 1994. Microbial siderophores and rhizosphere ecology. p. 67–80. *In* J.A. Manthey et al. (ed.) Biochemistry of metal micronutrients in the rhizosphere. Lewis Publ., Boca Raton, FL.

Cabrera, M.L., and M.H. Beare. 1993. Alkaline persulfate oxidation for determining total nitrogen in microbial biomass extracts. Soil Sci. Soc. Am. J. 57:1007–1013.

Cheng, W., and D.C. Coleman. 1990. Effect of living roots on soil organic matter decomposition. Soil Biol. Biochem. 22:781–787.

Cheshire, M.V., B.T. Christensen, and L.H. Sorensen. 1990. Labeled and native sugars in particle-size fractions from soils incubated with ^{14}C straw for 6 to 18 years. J. Soil Sci. 41:29–39.

Coleman, D.C., J. Dighton, K. Ritz, and K.E. Giller. 1994. Perspectives on the compositional and functional analysis of soil communities. p. 261–271 *In* K. Ritz et al. (ed.) Beyond the biomass. Wiley-Sayce, Chichester, England.

Coleman, D.C., E.P. Odum, and D.A. Crossley Jr. 1992. Soil biology, soil ecology, and global change. Biol. Fert. Soil 14:104–111.

Coleman, D.C., C.P.P. Reid, and C.V. Cole. 1983. Biological strategies of nutrient cycling in soil systems. Adv. Ecol. Res. 13:1–55.

Crowley, D.E., and D. Gries. 1994. Modeling of iron availability in the plant rhizosphere. p. 199–223 *In* J.A. Manthey et al. (ed.) Biochemistry of metal micronutrients in the rhizosphere. Lewis Publ., Boca Raton, FL.

de Goede, R.G.M., S.S. Georgieva, B.C. Verschoor, and J.-W. Kamerman. 1993. Changes in nematode community structure in a primary succession of blown-out areas in a drift sand landscape. Fund. Appl. Nematol. 16:501–513.

de Ruiter, P.C., J.C. Moore, K.B. Zwart, L.A. Bouwman, J. Hassink, J. Bloem, J.A. De Vos, J.C.Y. Marinissen, W.A.M. Didden, G. Lebbink, and L. Brussaard. 1993. Simulation of nitrogen mineralization in the below-ground food webs of two winter wheat fields. J. Appl. Ecol. 30:95–106.

de Ruiter, P.C., A. Neutel, and J.C. Moore. 1995. Energetics, patterns of interaction strengths, and stability in real ecosystems. Science (Washington, DC) 269:1257–1260.

Donkin, S.G., and D.B. Dusenbery. 1993. A soil toxicity test using the nematode *Caenorhabditis elegans* and an effective method of recovery. Arch. Environ. Contam. Toxicol. 25:141–151.

Doran, J.W., D.C. Coleman, D.F. Bezdicek, and B.A. Stewart (ed.). 1994. Defining soil quality for a sustainable environment. SSSA Spec. Publ. 35. SSSA and ASA, Madison, WI.

Doran, J.W., and T.B. Parkin. 1994. Defining and assessing soil quality. p. 3–21 *In* J.W. Doran et al. (ed.) Defining soil quality for a sustainable environment. SSSA Spec. Publ. 35. SSSA and ASA, Madison, WI.

Ettema, C.H., and T. Bongers. 1993. Characterization of nematode colonization and succession in disturbed soil using the maturity index. Biol. Fert. Soil 16:79–85.

Freckman, D.W. 1982. Parameters of nematode contribution to ecosystems. p. 81–97. *In* D.W. Freckman (ed.) Nematodes in soil ecosystems. Univ. Texas Press, Austin.

Freckman, D.W., and J.G. Baldwin. 1990. Nematoda. p. 155–200. *In* D.L. Dindal (ed.) Soil biology guide. John Wiley & Sons, New York.

Gallopin, G.C. 1995. The potential of agroecosystem health as a guiding concept for agricultural research. Ecosyst. Health 1:129–140.

George, E., V. Roemheld, and H. Marschner. 1994. Contribution of mycorrhizal fungi to micronutrient uptake by plants. p. 93–109. *In* J.A. Manthey et al. (ed.) Biochemistry of metal micronutrients in the rhizosphere. Lewis, Boca Raton, FL.

Griffiths, B.S., K. Ritz, and R.E. Wheatley. 1994. Nematodes as indicators of enhanced microbiological activity in a Scottish organic farming system. Soil Use Manage. 10:20–24.

Hanel, I. 1995. Secondary successional stages of soil nematodes in cambisols of south Bohemia. Nematologica 28:197–218.

Hendrix, P.F., D.C. Coleman, and D.A. Crossley, Jr. 1992. Using knowledge of soil nutrient cycling processes to design sustainable agriculture. J. Sust. Agric. 2:63–82.

Hendrix, P.F., R.W. Parmelee, D.A. Crossley, Jr., D.C. Coleman, E.P. Odum, and P. Groffman. 1986. Detritus food webs in conventional and no-tillage agroecosystems. Bioscience 36:374–380.

House, G.J. 1982. Nitrogen cycling in conventional and no-tillage agroecosystems: Analysis of pathways and processes. Ph.D. diss. Univ. of Georgia, Athens.

Hu, S., D.C. Coleman, P.F. Hendrix, and M.H. Beare. 1995. Biotic manipulations effects on soil carbohydrates and microbial biomass in a cultivated soil. Soil Biol. Biochem. 27:1127–1135.

Hutchinson, G.E. 1970. The biosphere. Sci. Am. 208:3–11.

Insam, H., and K.H. Domsch. 1988. Relationship between soil organic carbon and microbial biomass on chronosequences of reclamation sites. Microb. Ecol. 15:177–188.

Insam, H., and K. Haselwandter. 1989. Metabolic quotient of the soil microflora in relation to plant succession. Oecologia 79:174–178.

Jenkinson, D.S. 1988. Determination of microbial biomass carbon and nitrogen in soil. p. 368–386. *In* J.R. Wilson (ed.) Advances in nitrogen cycling in agricultural ecosystems. CAB Int., Wallingford, England.

Johnson, D.L., and D. Watson-Stegner. 1987. Evolution model of pedogenesis. Soil Sci. 143:349–366.

Lavelle, P., E. Blanchart, A. Martin, A.V. Spain, and S. Martin. 1992. Impact of soil fauna on the properties of soils in the humid tropics. p. 157–185. *In* R. Lal and P. Sanchez (ed.) Myths and science of soils of the tropics. SSSA Spec. Publ. 29. ASA and SSSA, Madison, WI.

Linden, D.R., P.F. Hendrix, D.C. Coleman, and P.C.J. van Vliet. 1994. Faunal indicators of soil quality. p. 91–106. *In* J.W. Doran et al. (ed.) Defining soil quality for a sustainable environment. SSSA Spec. Publ. 35. SSSA and ASA, Madison, WI.

Morowitz, H.J. 1970. Entropy for biologists. Academic Press, New York.

Neher, D.A., S.L. Peck, J.O. Rawlings, and C.L. Campbell. 1995. Measures of nematode community structure and sources of variability among and within agricultural fields. Plant Soil 170:167–181.

Niles, R.K., and D.W. Freckman. 1998. Nematode ecology in bioassessment and ecosystem health. p. 65–86. *In* G.A. Pederson et al. (ed.) Plant-nematode interactions. ASA, CSSA, and SSSA, Madison, WI.

Oades, J.M. 1984. Soil organic matter and structural stability: Mechanisms and implications for management. Plant Soil 76:319–337.

Odum, E.P. 1989. Ecology and our endangered life-support systems. Sinauer Assoc., Sunderland, MA.

Parton, W.J., R.L. Sanford, P.A. Sanchez, and J.W.B. Stewart. 1989. Modeling soil organic matter dynamics in tropical soils. p. 153–171. *In* D.C. Coleman et al. (ed.) Dynamics of soil organic matter in tropical ecosystems. Univ. of Hawaii, Honolulu.

Parton, W.J., D.S. Schimel, C.V. Cole, and D.S. Ojima. 1987. Analysis of factors controlling soil organic matter levels in Great Plains grasslands. Soil Sci. Soc. Am. J. 51:1173–1179.

Paul, E.A., and F.E. Clark. 1989. Soil microbiology and biochemistry. Academic Press, San Diego.

Phillips, R.E., and Phillips, S.H. 1984. No-tillage agriculture: Principles and practices. Van Nostrand Reinhold, New York.

Rapport, D.J. 1995. Ecosystem health: Exploring the territory. Ecosyst. Heal. 1:5–13.

Ruess, L., and W. Funke. 1995. Nematode fauna of a spruce stand associated with forest decline. Acta Zool. Fenn. 196:348–351.

Schlesinger, W.H. 1991. Biogeochemistry: An analysis of global change. Academic Press, San Diego.

Sparling, G.P., and A.W. West. 1988. A direct extraction method to estimate soil microbial C: Calibration in situ using microbial respiration and ^{14}C labeled cells. Soil Biol. Biochem. 20:337–343.

Swank, W.T., and D.A. Crossley, Jr. 1988. Forest hydrology and ecology at Coweeta. Springer-Verlag, New York.

Tian, G., L. Brussaard, and B.T. Kang. 1995. An index for assessing the quality of plant residues and evaluating their effects on soil and crop in the (sub-) humid tropics. Appl. Soil Ecol. 2:25–32.

Tian, G., B.T. Kang, and L. Brussaard. 1993. Mulching effect of plant residues with chemically contrasting compositions on maize growth and nutrient accumulation. Plant Soil 153:179–187.

Torsvik, V., J. Goksoyr, F.L. Daae, R. Sorheim, M. J. Salte, and K. Salte. 1994. Use of DNA analysis to determine the diversity of microbial communities. p. 39–48. *In* K. Ritz et al. (ed.) Beyond the biomass—compositional and functional analysis of soil microbial communities. Wiley-Sayce, Chichester, England.

Turco, R.F., A.C. Kennedy, and M.D. Jawson. 1994. Microbial indicators of soil quality. p. 73–90. *In* J.W. Doran et al. (ed.) Defining soil quality for a sustainable environment. SSSA Spec. Publ. 35. SSSA and ASA, Madison, WI.

Vance, E.D., P.C. Brookes, and D.S. Jenkinson. 1987. An extraction method for measuring microbial biomass C. Soil Biol. Biochem. 19:703–707.

Vitousek, P.M., P.R. Ehrlich, A.H. Ehrlich, and P.A. Matson. 1986. Human appropriation of the products of photosynthesis. BioScience 36:368–373.

Wasilewska, L. 1994. The effect of age of meadows on succession and diversity in soil nematode communities. Pedobiology 38:1–11.

Yeates, G.W. 1994. Modification and quantification of the nematode maturity index. Pedobiol. 38:97–101.

Yeates, G.W., T. Bongers, R.G.M. de Goede, D.W. Freckman, and S.S. Georgieva. 1993. Feeding habits in soil nematode families and genera—an outline for soil ecologists. J. Nematol. 25:315–331.

Yeates, G.W., and D.C. Coleman. 1982. Role of nematodes in decomposition. p. 55–80. *In* D.W. Freckman (ed.) Nematodes in soil ecosystems. Univ. of Texas Press, Austin.

2 Molecular Structure–Reactivity–Toxicity Relationships

Paul G. Mezey

Mathematical Chemistry Research Unit
University of Saskatchewan
Saskatoon, Canada

2–1 INTRODUCTION

For most chemists, biochemists, biologists, and toxicologists, the concept of molecular structure means a three-dimensional bonding pattern, often represented by the familiar *ball-and- stick* molecular models of chemistry. Such structural models are very useful tools to convey chemical information, and many molecular properties, such as reactivity, various biochemical effects, including toxicity, are often analyzed using these models; however, these models, by their simplicity, cannot capture the more detailed, and far more complex aspects of the actual molecular features. Molecular bodies are fuzzy, negatively charged electron density clouds that surround various arrangements of a set of small, positively charged nuclei, and there are no sticks, neither elastic springs that correspond to formal *bonds* between the atomic nuclei. Molecules have many properties that cannot be described by classical mechanical theories, and in many respects, molecules do not behave at all like the macroscopic *ball-and-stick* models.

There is nothing else in a molecule but atomic nuclei and an electronic charge density cloud. The shape of the electronic charge cloud fully reflects the arrangement of the nuclei: at the locations where the nuclei are found, the density of the electronic charge cloud shows a local maximum, that is, a local accumulation of the electronic charge cloud is found at the nuclear centers. By analyzing the electronic charge density clouds, the location of the nuclei can be determined, that is fully sufficient for the construction of a ball-and-stick structural model; however, the electronic charge density clouds contain much more information; in particular, the shape of the peripheral, valence regions of these clouds reveal the likely sites of various chemical reactions. In fact, a more faithful and complete description of molecules can be obtained if one replaces the conventional ball-and-stick structural models with a detailed shape description of the three-dimensional electronic charge cloud. In systematic analyses and comparisons of molecular reactivities, such as those required in molecular level toxicology studies, the conventional, detailed molecular shape analysis is expected to serve as a more versatile tool than the somewhat simplistic, ball-and-stick structural models. Hence, in most of the studies serving as the basis of this review, the conventional structural analysis has been replaced by molecular shape analysis (Arteca & Mezey, 1990; Maggiora et al., 1990).

Whereas shape is certainly a fundamental molecular property, involved in nearly all physical and chemical properties of molecules, molecular shape also is of practical significance in highly applied fields, such as soil science and environmental toxicology. An important aspect of molecular shape, chirality, has been identified early as an indicator of molecular activities in soil, with specific role in crop protection (Ramos Tombo & Bellus, 1991).

Toxicological risk is associated with the molecules released to the environment. This risk is a multiparameter property, that cannot be evaluated by a single number: within different environmental moieties a given molecule acts differently and represents a different type of risk. Nevertheless, the properties of the molecule determine the risk for each different environment. The detailed study of the molecule is a much simpler task than a detailed study of all possible environments where the molecule can represent some toxicological risk. For this reason, it appears prudent to increase the efforts aimed at molecular characterizations, and to explore correlations between molecular properties and the various toxicological risks. If sufficiently detailed information on molecular properties, such as the presence of a specific local shape region, is known and if this information is correlated with toxicological activity, then these correlations can be used as risk predictors. The presence of the same molecular property in a newly synthesized molecule can provide a strong indication of specific toxicological risks. Risks can be assigned to molecules before they are released to the environment, even before any harmful effect is detected experimentally. One family of tools for this purpose are quantitative structure— *toxicological* activity relations and a more advanced approach, quantitative shape—*toxicological* activity relations; in the toxicological context, the acronyms QSTAR and QShTAR are used.

2–2 MOLECULAR THEORY: A FRAMEWORK FOR UNDERSTANDING CHEMICAL EFFECTS

Whereas individual experiments provide information on isolated facts, the theoretical and computational models in chemistry are suitable to integrate these isolated facts into a more comprehensive model. The theoretical models (e.g., Smith et al., 1986; Maruani, 1988; Zimpel & Mezey, 1996) and their computational implementations (e.g., Lipkowitz & Boyd, 1990; Leszczynski, 1996), have a dual role: they are expected to explain the earlier experimental observations, and, perhaps more importantly, they are expected to provide predictions concerning potential experiments not yet carried out. In particular, regarding toxic activities, one of the ultimate goals of theoretical models is the reliable prediction of molecular properties. Various actions of molecules depend on the properties of the molecules. Among the molecular properties, toxicologists are interested in the risks associated with releasing a new compound, for example, a new pesticide, to the environment.

One important component of the computer revolution in the last part of the 20th century is the dramatic transfer of methodologies between very diverse fields of science. Theoretical results, which in earlier times served only as guidelines or an intellectual framework for some of the actual advances in applied fields, now have become more directly applicable for the experimentalists. By converting theories into computer programs and applying them to complement experimental studies carried out using various instruments, these theories and computer programs have, in fact, become a new family of instruments, which can be tailor-made to suit the required scientific application. This development has beneficial roles for both theory and experiment: theories can be put to a nearly limitless set of new tests, and experiments in applied fields have gained a whole array of new, computational techniques for both validation and interpretation.

Many of the fundamental aspects of molecular similarity have been identified and various methods have been proposed for quantifying molecular similarity (Johnson & Maggiora, 1990; Carbó, 1995; Dean, 1995; Sen, 1995; Walker et al., 1995a,b). In particular, computational modeling of molecules, an important practical approach to similarity analysis, now has the potential to contribute in a novel way to toxicology. The path between fundamental molecular theory, called quantum chemistry, and applied toxicology, is surprisingly short. From first principles of quantum chemistry, using conventional *ab initio* computer programs, or alternative methods providing *ab initio* quality results, the electronic density can be computed, practically, for any molecule. By analyzing the shape of the calculated electronic charge density, global, and local shape features can be identified and compared, using rigorous shape-similarity measures.

Correlations between the known toxicity and calculated shape features of a sequence of molecules can then be used to extrapolate and predict toxicity properties of new molecules based on their calculated shape.

The first step in this process is the computation of molecular electronic charge densities. All such calculations are based on the solution of the fundamental equation of quantum chemistry, the molecular Schrödinger equation

$$H\psi = E\psi \tag{1}$$

where H, the molecular Hamilton operator, is a complex set of instructions on what to do with the molecular wavefunction ψ, in order to convert it into a multiple $E\psi$ of itself, where E is the energy of the molecule. The above form of the molecular Schrödinger equation is deceptively simple, the solution of this equation using sufficiently accurate approximate methods often involves several millions of integrations and other complex numerical calculations. The development of suitable computer programs for these calculations have taken several years by a large number of computational chemists, and these computer programs are among the major achievements of modern chemistry.

The molecular wavefunction ψ is usually described in terms of linear combinations (LC) of atomic orbitals (AO), and the actual determination of the LCAO wavefunction ψ involves the computation of the relative contributions of various atomic orbitals to the molecular wavefunction ψ.

In this chapter, we denote the *i*th atomic orbital by $\varphi_i(\mathbf{r})$ ($i = 1,2,...,n$), where *n* is the total number of atomic orbitals, and **r** denotes the three-dimensional position vector variable. The relative weights of these contributions determine the so-called density matrix, **P**. If the molecule is described by a wavefunction ψ that involves *n* atomic orbitals, then the dimension of the density matrix **P** is $n \times n$. Most molecular properties can be calculated from the density matrix, and the density matrix itself can be thought of as an alternative representation of the information present in the molecular wavefunction ψ. In particular, the electronic density $\rho(\mathbf{r})$ of the molecule at any location **r** can be computed as

$$\rho(\mathbf{r}) = \sum_{i=1}^{n} \sum_{j=1}^{n} P_{ij}\, \varphi_i(\mathbf{r})\, \varphi_j(\mathbf{r}) \qquad [2]$$

This electron density $\rho(\mathbf{r})$ corresponds to the fuzzy *body* of the electronic charge cloud, providing a representation for the shape of the molecule.

For small molecules, these calculations can be accomplished, for example, using the conventional Gaussian family of *ab initio* computer programs developed by Pople and coworkers (Frisch et al., 1990).

The standard formulation of the *ab initio* method for the generation of molecular wavefunctions followed by an electron density calculation implies that as the size of the molecules increases, the computation of twice as large molecule requires 2^4 , that is, 16 times more computer time. This fourth power dependence all but excludes the study of proteins and other large molecules that would require decades or centuries of computer time on the fastest computers, clearly, not a practical alternative. For large molecules, the conventional *ab initio* computational methods are not applicable.

A family of new techniques, however, based on the additive, fuzzy density fragmentation (AFDF) principle, circumvents this problem. The simplest version of the AFDF approaches, the Mulliken-Mezey approach, is the basis of the MEDLA method developed by Walker and Mezey and applied for a variety of large molecules (Walker & Mezey, 1993a,b).

The general AFDF scheme can be given in terms of membership functions of nuclei within various molecular fragments. The family of all nuclei of the molecule is divided into *m* mutually exclusive subfamilies

$$f_1, f_2, \ldots, f_k, \ldots f_m \qquad [3]$$

These subfamilies of atomic nuclei serve as *anchor points* for *m* fuzzy electron density fragments,

$$F_1, F_2, \ldots, F_k, \ldots F_m \qquad [4]$$

These fuzzy fragments are described by a set of fragment density functions denoted by

$$\rho^1(\mathbf{r}), \rho^2(\mathbf{r}), \ldots, \rho^k(\mathbf{r}), \ldots \rho^m(\mathbf{r}) \qquad [5]$$

We shall use the notation $m_k(i)$ for the membership function of the atomic orbital $\varphi_i(\mathbf{r})$ in the set of orbitals centered on a nucleus of nuclear subfamily f_k of electron density fragment F_k. The formal definition of this membership function is given as

$$m_k(i) = \begin{cases} 1 & \text{if the atomic orbital } \varphi_i(\mathbf{r}) \text{ is centered on any one of the nuclei of subfamily } f_k, \\ 0 & \text{otherwise} \end{cases} \qquad [6]$$

With the help of these membership functions $m_k(i)$, the Mulliken-Mezey version of the $n \times n$ AFDF density matrix $\mathbf{P}^k$ for the kth electron density fragment F_k is defined in terms of its elements P^k_{ij} as

$$P^k_{ij} = 0.5\ [m_k(i) + m_k(j)]\ P_{ij} \qquad [7]$$

The author has proposed a generalized fragmentation scheme that also can be introduced using the membership function formalism of Eq. [6], by taking

$$P^k_{ij} = [m_k(i)\ w_{ij} + m_k(j)\ w_{ji}]\ P_{ij} \qquad [8]$$

In this generalized scheme, the weighting factors w_{ij} and w_{ji} are subject to the following constraint

$$w_{ij} + w_{ji} = 1 \qquad [9]$$

The original, Mulliken-Mezey AFDF scheme corresponds to the choice of

$$w_{ij} = w_{ji} = 0.5 \qquad [10]$$

The Molecular Electron Density Lego Assembler (MEDLA) method (Walker & Mezey, 1993a,b) is based on the Mulliken-Mezey AFDF scheme. When introduced, the MEDLA method was the first technique suitable for the generation of *ab initio* quality electron densities for macromolecules such as proteins and various other natural products such as taxol (Walker & Mezey, 1993a,b, 1994a,b, 1995a,b).

If reliable and detailed computer representations of electronic densities of both small and large molecules are available, then a computer-analysis of many molecular properties, such as local and global shape properties, molecular similarity and complementarity becomes possible, and various correlations with experimental observations can be studied. The computed electronic densities of molecules provide a framework for modeling and understanding molecular behavior.

2–3 MOLECULAR SHAPE AND SHAPE CHANGES: THE SOURCE OF ALL MOLECULAR PROPERTIES

All static molecular properties are determined by the shape of the electronic charge distribution, whereas all dynamic molecular properties depend on the shape changes of the electronic density. Hence, shape analysis methods suitable for the study of both static and dynamic aspects of molecular shape are useful tools for the interpretation and prediction of molecular reactivity.

The fundamental principles of molecular shape analysis techniques, as described in earlier references for general molecules (Mezey, 1986, 1987a,b, 1991, 1993, 1994a), and discussed in later works from the drug design perspective (Mezey, 1989, 1992, 1994b), are applicable to the development of the new computer programs. By analogy with drug design problems, an enhanced predictive power is expected if multiple shape features are considered simultaneously: a given biochemical effect is often triggered by the presence of two or several molecular shape features at distant locations within the molecule. For the detection and predictive use of such multiple shape features a family of new computer programs has been developed.

For reliable molecular modeling, the representation of fuzzy molecular bodies and approximate molecular surfaces must convey the essential shape information. Although molecular electron distributions are fuzzy, quantum mechanical entities and in a rigorous sense they possess neither a formal body nor a surface, certainly not in the same sense as macroscopic objects; nevertheless, formal molecular surfaces and formal molecular bodies can be used for shape analysis.

The arrangement of nuclei within the molecule is often referred to as the nuclear configuration, and is denoted by K. The nuclei have a major influence on the distribution of the electronic density, consequently, the representations of molecular surfaces and molecular bodies also are dependent on the nuclear configuration K. For each choice of an electronic density threshold value a, and nuclear configuration K, the electronic density contour surface $G(K,a)$ of the molecule is defined as the collection of all points of the space where the density happens to be equal to this threshold value a.

In more precise terms, a *molecular isodensity contour surface*, MIDCO $G(K,a)$ of nuclear configuration K and density threshold a is defined as

$$G(K,a) = \{\mathbf{r}: \rho(K,\mathbf{r}) = a\} \qquad [11]$$

Here the standard mathematical notations are used, where { } means a collection of elements specified within the curly brackets, and the symbol : reads *such that*. The electronic density $\rho(K,\mathbf{r})$ is dependent on both the nuclear configuration K and the three dimensional position $\mathbf{r}$ where the density is equal to the threshold value a. For any fixed nuclear arrangement K of the molecule, the molecular electronic density $\rho(K,\mathbf{r})$ is a continuous function of the position variable $\mathbf{r}$, consequently, the MIDCO $G(K,a)$ defined by Eq. [11] is a collection of points $\mathbf{r}$ that form a continuous surface.

The electronic density decreases rapidly with the distance from the nearest nucleus of the molecule. Strictly speaking, the electronic charge density function $\rho(K,\mathbf{r})$ becomes zero only at infinite distance from the nuclei, however, most of the charge cloud is found within a close neighborhood of the family of nuclei of the molecule. In fact, the electronic density becomes negligible at a 10-Å distance from the nearest nucleus of an isolated molecule.

A formal *molecular body* is defined for the given nuclear configuration K and each threshold value a as the collection $F(K,a)$ of all those points $\mathbf{r}$ of the three-dimensional space where the electronic density is greater than the threshold a,

$$F(K,a) = \{\mathbf{r}: \rho(K,\mathbf{r}) > a\} \qquad [12]$$

Note that electrons are negatively charged, however, the density of electrons is considered positive or zero; a negative charge means a positive value for the density function $\rho(K,\mathbf{r})$.

Molecular shapes have both fuzzy and dynamic features, that impose special requirements for the computational methods used for shape analysis in reliable approaches to molecular modeling, in drug design, and in other applications. One family of computational methods suitable for the simultaneous description of both fuzzy and dynamic features is based on topology. A brief review of the basic concepts of the topological shape analysis methods of molecules is given below.

Most topological shape analysis methods involve two main steps:

1. In the first step, the fuzzy aspects of molecular electronic density clouds are described by a whole family of geometrical models, since the fuzziness of the electronic charge cloud cannot be described by a single geometrical body of precise boundaries. Such geometrical models include individual molecular contour surfaces and the molecular bodies enclosed by them. No single geometrical object is selected to represent the molecular shape, instead, the objects within the family are classified based on some physical or geometrical criterion. Often, various ranges of local curvature values along the boundaries of the individual objects are used for classification.
2. In the second step, topological criteria are used for characterization. Common topological properties identified in families of contour surfaces are used for a topological description. For example, common topological properties are the patterns of distributions of intervals of possible curvature values and the range of electron density contour values.

This process involving steps 1 and 2 is usually referred to as *Geometrical Classification and Topological Characterization* (Mezey, 1993).

A related principle used in topological approaches to molecular similarity is the principle of *Geometrical Similarity as Topological Equivalence* (Mezey, 1993), or in short, the GSTE principle.

Using geometrical conditions, for example, ranges of curvatures along a molecular contour surface, a family of geometrical objects is generated. This process leads to a geometrical classification of the surface points of the contour surface into domains. The various interrelations among these domains are topological properties, and these properties are used for a topological characterization in the following step. One should notice that the topological characterization does not change for most small geometrical variations of the actual curvatures. If the geometrical patterns are similar enough, then the topological characterizations based on them remain invariant. That is, geometrical similarity is reflected in topological equivalence; this observation is the basis of the topological approach to similarity analysis. Topology has the role of extracting the essential shape information. This information is topologically stable for small enough geometry variations; a topological change occurs only if the geometrical changes are significant.

The typical components of molecular shape analysis methods based on the GSTE principle are as follows: (i) the family of all the possible geometrical patterns and mutual arrangements of local shape domains of molecular contours or molecular bodies are classified by a combination of geometrical and topological criteria, and (ii) the resulting equivalence classes are characterized using topological tools.

The whole range of possible electron density threshold values is an open-closed interval (0, a_{max}], where the maximum density value a_{max} is the electron density at a nucleus of the largest nuclear charge in the given molecule. In the strict sense, the zero density value is attained only at infinite distance from the nuclei, hence, the interval is open at the low density limit, as indicated by the left-hand side round parenthesis, whereas the interval is closed at the high limit, at the actual density value a_{max}, as indicated by the right-hand side square-bracket symbol.

For every molecule of nonpathological electronic density, there are only a finite number of MIDCOs *G(K,a)* with topologically different curvature patterns. Molecular isodensity contour surfaces with equivalent curvature patterns belong to the same equivalence class. For a topological shape analysis, it is sufficient to take one MIDCO *G(a)* from each of the topological equivalence classes. The essential shape features of the entire molecular electron density can be represented by a finite number of individual MIDCO surfaces.

Formal molecular bodies and various parts of such a molecular body can be represented by electronic density domain *DD(K,a)* of various density thresholds *a*. A density domain (Mezey, 1993) is a formal body enclosed by a MIDCO surface *G(K,a)*, where the boundary, the *G(K,a)* surface itself, is included in the density domain. The level set *F(K,a)* of the same density threshold *a* is regarded as the interior of the density domain *DD(K,a)*. The formal definition of a density domain *DD(K,a)* of boundary *G(K,a)* is given as

$$DD(K,a) = \{\mathbf{r}:\rho(K,\mathbf{r}) \geq a\} \qquad [13]$$

The density domains *DD(K,a)* depend on the electronic state, the density threshold *a*, as well as the nuclear configuration *K* of the molecule; unless stated otherwise, the electronic ground state is considered for each molecule.

The density domains are used for a natural representation of chemical bonding in molecules that is more descriptive than the conventional bonding patterns represented by lines of formal single, double and triple bonds, and the stereochemical *skeletal* diagrams. Following the techniques introduced in density domain analysis (DDA), chemical bonding is described by the interfacing and mutual interpenetration of local, fuzzy charge density clouds (Mezey, 1993).

Within a chemically important range [a_{min}, a_{max}] of density thresholds, one can determine an associated family of topologically representative density domains $[DD(K,a_{min}), DD(K,a_{max})]$ using quantum chemical computational techniques. Within this density range, the topological changes of density domains as a function of the threshold a can be identified. The list of these topological changes provides a detailed description of the bonding pattern of the molecule. The density domain approach also has been used for a detailed quantum chemical representation of formal functional groups of chemistry (Mezey, 1993, 1994a). Functional groups have some limited *autonomy* within molecules, as they are identifiable within a large number of otherwise rather different molecules. This local, limited autonomy is manifested in the existence of a separate density domain representing the functional group within a given molecule. Consequently, the patterns of density domains provide information on functional groups, and the associated reactivities of molecules. The reader may find further details of this approach in the literature (Mezey, 1995a,b, 1996a,b,c).

The changes of density domains of many molecules as a function of the density threshold a follow a common pattern. At high-density thresholds, no density domain contains more than a single nucleus, and only separate, local nuclear neighborhoods appear as individual density domains. At somewhat lower thresholds, the separate pieces of density domains join, that represents a topologically significant change. At very low electron density threshold values, there is only a single density domain, that is usually a simply connected formal body.

A more detailed analysis of the pattern of density domains involves characterization based on curvature properties. Even the more detailed patterns follow some general trends common for most molecules.

If a high density thresholds value a, similar to the value a_{max}, is chosen, then this value usually falls within a subrange where each density domain is an individual nuclear neighborhood. This subrange of density threshold values is called the atomic range. The atomic subrange itself is usually subdivided into two further subranges: the *strictly atomic range* and the *prebonding range*. Within the higher threshold, strictly atomic range, all the atomic density domains are convex sets. Within the lower threshold prebonding range, one finds at least one density domain that is not convex; such a density domain has a local region similar to a water droplet right after breaking off the faucet. This local region is directed towards a nearby density domain; these two density domains form a common domain at a somewhat lower density threshold.

At such lower density thresholds, some nuclear neighborhoods join and form density domains containing two or more nuclei. As long as not all nuclei of the molecule are contained within a single density domain, this subrange is called the functional group range. Within this subrange various functional groups appear as separate, individual density domains. The functional group subrange describes the pattern of interconnection of density domains that represents the bonding pattern of the molecule. The functional group subrange also is called the bonding range for density domains.

If the density threshold is further lowered, then one finds all of the nuclei of the molecule within a single, common density domain $DD(K,a)$. This subrange is called the molecular density range, where the molecular pattern of bonding is fully established. Usually, the molecular density range can be subdivided into further subranges. At density thresholds near the higher limit of the molecular range, the density domain often exhibits at least one local *neck* region, that is, a region where the density domain is not locally convex. Within this subrange the density domain is usually *skinny*; accordingly, this range is called the *skinny molecular range*. If the density threshold is further lowered, then, eventually, no such neck region appears, however, in most cases, the density domain $DD(K,a)$ still has at least one local nonconvex region. Such a shape feature often appears, and the corresponding subrange is called the *corpulent molecular range*. If the density threshold is low enough, then the density domain becomes convex for all molecules. This subrange corresponding to a convex molecular body is called the *quasi-spherical molecular range*. Within the atomic and the functional group ranges, the electronic density has disjoint, local domains; these two ranges together constitute the localized range. By contrast, the molecular density range also is called the global density range.

The Shape Group Methods are topological shape characterization techniques specifically designed for molecules. A detailed review of the shape groups used for molecular shape characterization has been published recently (Mezey, 1993), here only a brief summary will be given.

The shape groups are algebraic groups, describing the shapes of fuzzy objects, such as electron density charge clouds of molecules. These groups are not determined by the point symmetry groups of nuclear arrangements. Nevertheless, the presence of symmetry may influence the shape groups.

In mathematical terms, the shape groups are the homology groups of truncated contour surfaces, where the truncation is determined by local shape properties. In most of the chemical applications, the local shape properties refer to the locally convex, concave, or saddle-type regions of MIDCOs $G(K,a)$; however, more general geometrical or physical conditions also have been used (Mezey, 1993). The geometrical or physical conditions used to define local shape domains on a contour surface $G(K,a)$ are denoted by μ; the corresponding local shape domains are denoted by the symbol D_μ. If the property selected is local convexity, then the local curvature properties of various regions along a MIDCO surface $G(K,a)$ can be compared with a plane. If the plane is moved along the MIDCO as a tangent plane, then the local curvature properties of the entire MIDCO can be compared with zero curvature, that is, to the curvature of the plane. Each point $\mathbf{r}$ of the MIDCO $G(K,a)$ is tested for the local relation between the tangent plane and the density domain enclosed by the surface. At each point $\mathbf{r}$, the MIDCO $G(K,a)$ is regarded locally convex, locally concave, or locally saddle-type, respectively, depending on whether the tangent plane falls on the outside, on the inside, or whether it cuts into the given density domain within any small neighborhood of the surface point $\mathbf{r}$. If this characterization is carried out for all points $\mathbf{r}$ of the MIDCO, then one obtains a subdivision of the molecular contour surface $G(K,a)$ into locally convex, locally concave, and locally saddle-type shape domains, denoted by the symbols D_2, D_0, and D_1, respectively.

More detailed shape description is obtained if the tangent plane is replaced by some other objects, for example, tangent spheres of various radii r, or a series of oriented tangent ellipsoids T, with axes aligned with some reference directions. Oriented tangent ellipsoids can be translated but not rotated as they are brought into tangential contact with the MIDCO surface $G(K,a)$, providing a direction-dependent shape characterization. A directional shape analysis is of importance in studying molecules placed in external magnetic or electric fields, reacting within the confines of catalytic cavities of zeolites, or molecules fitting within enzyme cavities, passing through various membranes, aligned at the interfaces between inorganic and organic materials in soil, or at the interfaces of lipid and aqueous systems. The reader may find details of directional shape analysis in a recent review (Mezey, 1993).

The tangent object T, such as a tangent sphere, may fall locally on the outside, on the inside, or it may cut into the given density domain $DD(K,a)$ within any small neighborhood of the surface point **r** where the tangential contact occurs. These differences lead to another subdivision of the molecular contour surface $G(K,a)$, and to a new family of local shape domains D_2, D_0, and D_1, respectively, relative to the given tangent object T. For tangent spheres as tangent objects, orientation has no role, and it is sufficient to use the curvature $b = 1/r$ of the sphere for specification. For tangent spheres, the local shape domains D_2, D_0, and D_1, represent the local relative convexity domains of the MIDCO, relative to the reference curvature b. The case $b = 0$, a formal sphere of infinite radius, corresponds to the tangent plane.

The local shape domains D_2, D_0, and D_1 determined using any specific reference curvature b generate a shape domain partitioning of the MIDCO surface $G(K,a)$. If all D_μ domains of a specified type μ, for example, all locally convex domains D_2 relative to b, are excised from the MIDCO surface $G(K,a)$, then a topologically different object is obtained. This object, a truncated contour surface denoted by $G(K,a,\mu)$, represents some of the essential shape information of the original MIDCO surface $G(K,a)$. The advantage of the truncated surface is the fact that this shape information is now accessible by simple topological techniques. A topological analysis of the truncated surface $G(K,a,\mu)$ is used as a tool for the shape analysis of the original MIDCO surface $G(K,a)$. One may repeat this procedure for the whole range of chemically relevant reference curvature values b, and a detailed shape analysis of the MIDCO $G(K,a)$ can be obtained.

It is clear that for a given nuclear arrangement K and density threshold a, and for the whole range of chemically relevant reference curvature values b, there are only a finite number of topologically different truncated MIDCOs $G(K,a,\mu)$. By selecting a single truncated surface from each family of topologically equivalent surfaces, the MIDCO $G(K,a)$ can be characterized by the topological invariants of these selected truncated surfaces. The topological invariants provide a numerical shape characterization, that can be used as a partial shape code. Useful topological invariants are defined in terms of the homology groups of truncated surfaces. Homology groups of algebraic topology describe important features of the topological structure of domains on bodies and surfaces. The Betti numbers are important topological invariants, defined as the ranks of these homology groups.

The Shape Groups of the original MIDCO $G(K,a)$ of density threshold a are the homology groups $H^p_\mu(a,b)$ of the various truncated surfaces $G(K,a,\mu)$. For a given MIDCO $G(K,a)$ of density threshold a, for each reference curvature b, and for each shape domain truncation pattern μ, there are three shape groups,

$$H^0_\mu(a,b),\ H^1_\mu(a,b),\ \text{and}\ H^2_\mu(a,b), \qquad [14]$$

of formal dimensions zero, one, and two, respectively, as indicated by index p. These three shape groups describe collectively the essential shape information of the MIDCO $G(K,a)$. For each shape domain truncation type μ, and for each (a,b) pair of parameters, there are three Betti numbers,

$$b^0_\mu(a,b),\ b^1_\mu(a,b),\ \text{and}\ b^2_\mu(a,b) \qquad [15]$$

The family of Betti numbers generates a detailed, numerical description of the essential shape properties of the electron density of the molecule.

One may introduce subdivisions of the three-dimensional space using a range of parameters $p_1, p_2, \ldots, p_t$, describing the local gradient and second derivative properties of the charge distribution at each point **r**. This leads to the extension of shape groups to a direct characterization of the shape of the entire, three-dimensional electronic charge distribution. As in the case MIDCOs, a general range of these parameters is denoted by μ. New, truncated objects are obtained by excising all three-dimensional domains characterized by a specified range μ of these parameters. By definition, the homology groups of the truncated three-dimensional charge distribution are the shape groups

$$H^0_\mu(p_1, p_2, ..., p_t),\ H^1_\mu(p_1, p_2, ..., p_t),$$
$$H^2_\mu(p_1, p_2, ..., p_t),\ H^3_\mu(p_1, p_2, ..., p_t) \qquad [16]$$

of the entire electronic charge density of the molecule. The corresponding Betti numbers

$$b^0_\mu(p_1, p_2, ..., p_t),\ b^1_\mu(p_1, p_2, ..., p_t),$$
$$b^2_\mu(p_1, p_2, ..., p_t),\ b^3_\mu(p_1, p_2, ..., p_t) \qquad [17]$$

are the topological invariants used in this direct, three-dimensional shape characterization. The disadvantage of this alternative is related to visualization: pictorial representations of three-dimensional fuzzy objects are often confusing, when compared with the easily visualizable applications of the shape group method to individual MIDCO surfaces.

As an example for the application of the GSTE principle, the shape group methods combine the advantages of geometry and topology. Whereas the local shape domains and the truncated MIDCOs $G(K,a,\mu)$ are defined in terms of geometrical classification of points of the surfaces, using local curvature properties, the truncated surfaces $G(K,a,\mu)$ are characterized topologically, using the shape groups and the corresponding Betti numbers.

2–4 GLOBAL SHAPE AND LOCAL SHAPE

An important advantage of the topological shape analysis techniques is the numerical representation of shape information. The shape group method has been used to generate numerical shape codes and numerical measures of molecular similarity. These numerical codes and measures can be easily processed, stored and compared by computers, and provide a reproducible and unbiased method for shape comparisons. The conventional, subjective elements of visual shape comparisons are avoided if the shape group method is used.

The shape group analysis of a molecule generates a finite family of Betti numbers $b^p_\mu(a,b)$ for all the shape groups $H^p_\mu(a,b)$ that occur for a given molecule. These Betti numbers $b^p_\mu(a,b)$ are used to generate numerical shape codes for molecular electron density distributions.

The simplest of the numerical shape codes are the grid versions of the (a,b)-parameter maps of Betti numbers calculated for a series of MIDCOs of the molecule. In these maps, a and b refer to the density threshold and the reference curvature values, respectively. By convention, a positive b value indicates that the tangent sphere of radius $r = 1/b$, used as the reference object, is placed on the *exterior side* of the molecular surface, whereas a negative b value indicates that the tangent sphere is placed on the *interior side* of the MIDCO. For a given nuclear arrangement K, the shape groups of a molecule, hence the corresponding Betti numbers, depend on two parameters, the electronic density threshold a and the reference curvature b. The chemically relevant ranges of both of these parameters are finite, and these ranges define a two-dimensional map, called the (a,b)-map. A detailed shape characterization of the electronic density of the molecule is given by the shape group distribution along this (a,b)-map, where the shape groups are represented by the corresponding Betti numbers.

Such (a,b)-maps can be defined in terms of the shape groups of the complete molecule, or one may consider separate shape groups for each disjoint piece of the various density domains at high density thresholds.

Individual (a,b)-maps are generated for each of the three types of Betti numbers, $b^0_\mu(a,b)$, $b^1_\mu(a,b)$, and $b^2_\mu(a,b)$. In order to obtain the (a,b)-maps, the Betti numbers that belong to a pair of values of parameters a and b are assigned to the given locations of the $(a$,b) parameter maps. The one-dimensional Betti numbers of type $b^1_\mu(a,b)$ describe the chemically most relevant shape information.

The grid versions of (a,b)-maps of Betti numbers are generated in a discretized form. A grid of a and b values within a suitable interval $[a_{min}, a_{max}]$ of density thresholds and an appropriate interval $[b_{min}, b_{max}]$ of reference curvature values is generated, and the corresponding $b^1_\mu(a,b)$ Betti numbers are computed. Usually, the range of parameters a and b covers several orders of magnitude, consequently, it is advantageous to use logarithmic scales. For negative curvature parameters b, the $\log|b|$ values are taken. In some earlier applications of this technique to drug design problems (Walker et al., 1995a,b), a 41×21 grid was used, within the ranges of (0.001, 0.1 a.u.; a.u. = atomic unit) for the density thresholds a, and [-1.0, 1.0] for the curvature parameter b. Using either the direct or the logarithmic versions of the (a,b)-maps, the values of the Betti numbers at the grid points (a,b), [or at the points $(\log(a), \log|b|)$], form a matrix, denoted by $\mathbf{M}^{(a,b)}$. This matrix can be regarded as a *numerical shape code* for the molecular electronic charge distribution.

The (a,b)-maps of Betti numbers can be used for the computation of numerical similarity measures. If the matrices $\mathbf{M}^{(a,b)}$ of the grid versions of (a,b)-maps of Betti numbers $b^p_\mu(a,b)$ have been determined for a family of molecules, then a numerical comparison of these shape codes provides numerical similarity measures between molecules. If the matrix forms $\mathbf{M}^{(a,b),A}$ and $\mathbf{M}^{(a,b),B}$ of the shape codes of two molecules, *A* and *B* (of some fixed nuclear configurations) are available, then a numerical shape similarity measure is defined as

$$s(A,B) = m[\mathbf{M}^{(a,b),A}, \mathbf{M}^{(a,b),B}] / t \qquad [18]$$

Here $m[\mathbf{M}^{(a,b),A}, \mathbf{M}^{(a,b),B}]$ is the number of matches between corresponding elements in the two matrices $\mathbf{M}^{(a,b),A}$ and $\mathbf{M}^{(a,b),B}$, and t is the total number of elements in either matrix. If n_a and n_b are the number of grid divisions for parameters *a* and *b*, respectively, then

$$t = n_a n_b, \qquad [19]$$

In earlier applications (Mezey, 1993; Walker et al., 1995a,b), the 41 × 21 grid of the range [0.001, 0.1 a.u.] for *a*, and the range [−1.0, 1.0] for *b* implied that the 41 × 21 matrix $\mathbf{M}^{(a,b)}$ was stored as an integer vector **C** of 861 components.

The corresponding shape similarity measure *s*(*A*,*B*) can be expressed as

$$s(A,B) = \sum_{i=1}^{861} \delta_{j(i),k(i)} / 861, \qquad [20]$$

where $\delta_{j,k}$ is the Kroenecker delta, and the index assignment is given as

$$j(i) = C_i(A), \qquad [21]$$

and

$$k(i) = C_i(B). \qquad [22]$$

The matrices $\mathbf{M}^{(a,b)}$ of (*a*,*b*)-maps serve as descriptors of global shape properties of molecules, where the *s*(*A*,*B*) similarity measures specified in Eq. [20] provide global comparisons for entire molecules.

Local molecular properties are often more significant in determining chemical behavior than the global properties. In particular, shape comparisons of local regions of molecules often indicate more important chemical trends than the evaluation of the global similarities of molecules. If the molecule is large, such as a protein or other macromolecules, then many chemical or biochemical properties are dependent on local shape.

For large molecules, the AFDF principle is the basis of several simple approaches for the study of local shape similarity of molecules. The local electron density of any molecular fragment can be generated using the MEDLA approach for the construction of additive, fuzzy electron density fragments. These fragment electron densities *F* are well defined within any LCAO-based quantum chemical electron density computational framework.

The principles of shape analysis of local molecular moieties is based on a simple analogy with the shape analysis of complete molecules. Molecular isodensity contour (MIDCO) surfaces are replaced by fragment isodensity contour (FIDCO) surfaces.

We shall consider local moiety *A* as the actual fragment for shape analysis, whereas *B* denotes the rest of the molecule. Part B itself may be composed from several fragments, denoted by $B_1, B_2, \ldots, B_{m-1}$.

Two approaches have been proposed for the analysis of the local shapes of local molecular moieties within a molecule. In the simpler approach, a FIDCO for a Fragment A in a Molecule AB is defined as:

$$G_{A\backslash B}(a) = \{\mathbf{r} : \rho_A(\mathbf{r}) = a,\ \rho_A(\mathbf{r}) \geq \rho_{Bk}(\mathbf{r}),\ k = 1,\ldots m-1\}. \qquad [23]$$

According to this approach, a FIDCO $G_{A\backslash B}(a)$ of Fragment A in Molecule AB is defined by locating all those points where the electron density contribution of Fragment A is dominant within the complete Molecule AB.

The advantage of this approach is computational: the actual shape analysis can be carried out on the isolated FIDCO $G_A(a)$. On the fragment isodensity contour $G_A(a)$, one additional domain type is introduced, that represents a departure from the conventional MIDCO analysis of complete molecules. These additional domains represent the connection of Fragment A to the rest B of the molecule, taken within the complete Molecule AB; in formal topological language,

$$D_{-1}(G_{A\backslash B}(a)) = \{\mathbf{r} : \mathbf{r} \in G_A(a),\ k \in \{1,\ldots m-1\}: \rho_A(\mathbf{r}) < \rho_{Bk}(\mathbf{r})\} \qquad [24]$$

Whereas the above notation used for the new domain $D_{-1}(G_{A\backslash B}(a))$ refers to the FIDCO $G_{A\backslash B}(a)$, the actual domain $D_{-1}(G_{A\backslash B}(a))$ exists only on the original $G_A(a)$ contour. This domain $D_{-1}(G_{A\backslash B}(a))$ is a formal *cover* over the hole(s) of the FIDCO $G_{A\backslash B}(a)$ in Molecule AB. Although this new type of domain $D_{-1}(G_{A\backslash B}(a))$ itself is not present on the FIDCO surface $G_{A\backslash B}(a)$, the boundary $DD_{-1}(G_{A\backslash B}(a))$ of domain $D_{-1}(G_{A\backslash B}(a))$ actually exists on $G_{A\backslash B}(a)$.

An even simpler choice for the representation of local Fragment A in Molecule AB is obtained if the Composite B of all Fragments $B_1, B_2, \ldots, B_{m-1}$ is used; in mathematical notations

$$G_{A\backslash SB}(a) = \{\mathbf{r} : \rho_A(\mathbf{r}) = a,\ \rho_A(\mathbf{r}) \geq \rho_B(\mathbf{r})\}. \qquad [25]$$

Here the composite electron density $\rho_B(\mathbf{r})$ for the *rest* of the molecule is defined as

$$\rho_B(\mathbf{r}) = \rho_{B1}(\mathbf{r}) + \rho_{B2}(\mathbf{r}) + \ldots + \rho_{Bm-1}(\mathbf{r}). \qquad [26]$$

The connections between Fragment A and the rest B of the Molecule AB define local domains on both A and B; for local Fragment A one obtains

$$D_{-1}(G_{A\backslash SB}(a)) = \{\mathbf{r} : \mathbf{r} \in G_A(a),\ \rho_A(\mathbf{r}) \leq \rho_B(\mathbf{r})\}. \qquad [27]$$

The boundaries $DD_{-1}(G_{A\backslash SB}(a))$ of these domains are

$$\delta D_{-1}(G_{A\backslash SB}(a)) = \{\mathbf{r} : \mathbf{r} \in G_{A\backslash SB}(a),\ \rho_A(\mathbf{r}) = \rho_B(\mathbf{r})\}. \quad [28]$$

If the interactions among various molecular fragments in a molecule AB are fully taken into account, then a local shape analysis can no longer be carried out on an isolated FIDCO $G_A(a)$. For a detailed description of the interactions, a new contour calculation is needed, determining the fully interactive FIDCO $G_{A(B)}(a)$ in molecule AB. This fully interactive FIDCO $G_{A(B)}(a)$ is defined as follows:

$$G_{A(B)}(a) = \{\mathbf{r} : \rho_A(\mathbf{r}) + \rho_B(\mathbf{r}) = a,\ \rho_A(\mathbf{r}) \geq \rho_B(\mathbf{r})\}. \quad [29]$$

For $G_{A(B)}(a)$, there is no surface defined where the formal *cover* domains of the hole(s) of FIDCO $G_{A(B)}(a)$ would appear. Nevertheless, for uniformity in the notation, the symbol $\delta D_{-1}(G_{A(B)}(a))$ is used for the boundaries of the *holes* on $G_{A(B)}(a)$:

$$\delta D_{-1}(G_{A(B)}(a)) = \{\mathbf{r} : \mathbf{r} \in G_{A(B)}(a),\ \rho_A(\mathbf{r}) = \rho_B(\mathbf{r})\}. \quad [30]$$

The fully interactive FIDCO for a fragment A in a molecule AB describes the local shapes of interacting functional groups within a molecule. In order to determine the fully interactive FIDCO surfaces $G_{A(B)}(a)$, and for the corresponding local shape analysis, an additional contour calculation is needed. This approach requires more computation than the study of the noninteractive FIDCO surfaces $G_{A\backslash B}(a)$.

For both the simple and the fully interactive FIDCO surfaces, the shape group method of electron density shape analysis is applicable with minor modification. The additional domain boundaries $\delta D_{-1}(G_{A\backslash B}(a))$ and $\delta D_{-1}(G_{A(B)}(a))$ can be treated by introducing one additional index, chosen as -1. This new index is treated as if it were an ordinary relative curvature index. Using the one-dimensional homology groups based on truncations for each of the possible index combinations, all the shape groups of FIDCO surfaces can be computed.

In the subsequent steps, the (a,b)-parameter maps and shape codes are generated following the method described for complete molecules (Mezey, 1993). Consequently, the family of MIDCOs, the shape groups, as well as the shape codes $\mathbf{M}^{(a,b),F}$ for such local electron density Fragments F are meaningful. It is natural then to extend the application of similarity measure $s(A,B)$, originally developed for complete molecules, to various local Fragments F and F′, and define the analogous local similarity measure

$$s(F,F') = m[\mathbf{M}^{(a,b),F}, \mathbf{M}^{(a,b),F'}] / t \quad [31]$$

The applications of detailed shape analysis techniques for local molecular regions are new tools for studying correlations between molecular shape and activity.

2–5 QUANTITATIVE SHAPE–ACTIVITY RELATIONS

Most of the recent research efforts aimed at a quantitative interpretation of chemical, biochemical, toxicological properties of molecules in terms of various structural features, were based on bond patterns. Within the conventional framework of QSAR (Quantitative Structure—Activity Relations) analysis, the term *structure* is usually interpreted in terms of molecular graphs of formal bonding patterns or three-dimensional structural diagrams of the arrangements of formal bonds. Since the actual shape features of molecular electron density clouds contain more information, one may expect improved performance if QShAR (Quantitative Shape–Activity Relations) analysis is used instead of the conventional QSAR approach. Data bases containing easily retrievable, comparable, and interpretable three-dimensional shape data on molecular bodies appear to have an important future role in molecular engineering and drug design, and toxicological risk assessment.

The extension of the molecular shape analysis methods to macromolecules, such as proteins, including receptors for toxicants, and their use in toxicological Quantitative Risk Assessment, QRA, have become enhanced by a new MEDLA method developed for the rapid computation of electron densities of large molecular bodies. The detailed shapes of electron density clouds, representing the actual, real molecular bodies, can now be computed efficiently, for molecules of virtually any size (Walker & Mezey, 1993a,b). This method allows one to study in great detail the local and global shape correlations with toxicological activity, and to provide tools for predictive toxicology and toxicological risk assessment (Mezey et al., 1996, 1997).

2–6 PREDICTIVE TOXICITY ANALYSIS BASED ON MOLECULAR SHAPE

Predictive toxicology within quantitative risk assessment can be based on correlations between computed shape codes of molecular shape features and the available toxicological data. By analogy with drug design problems (Mezey, 1993), an enhanced predictive power is expected if multiple shape features of molecules and several toxic effects are considered simultaneously. A given biochemical effect is often triggered by the presence of two or several molecular shape features at distant locations within the molecule. For the detection and predictive use of such multiple shape features, a new computer program was developed. The new method of molecular shape–toxicological activity correlations can be verified using families of well-researched toxicants. Correlations with results obtained using other, bioindicator methodologies can be obtained and assessed. The new methodology is applied to families of various polycyclic aromatic hydrocarbons (PAH) molecules and their derivatives, as well as to other molecular families (Mezey et al., 1996, 1997).

2–7 APPLICATIONS AND EXAMPLES

The toxicological variants, QSTAR (Quantitative Structure–*Toxicological* Activity Relations) and QShTAR (Quantitative Shape–*Toxicological* Activity

Relations) of conventional QSAR and QShAR are rather novel approaches that have required the development of new methodologies. Many currently used approaches of QSAR and QShAR, as applied in contemporary drug design and molecular engineering, can be adapted to toxicological use in a simple and direct manner (Mezey et al., 1996, 1997). One should note, however, that the needs of toxicology are somewhat different from the traditional needs of drug design. In drug design the task is to find new molecules with highly enhanced main effect and highly suppressed side effects. This is a difficult task, as it involves nearly contradictory requirements. By contrast, in toxicology the presence of side effects is of no special concern: if the main toxicological effect is already lethal at some dose, it is rather immaterial if a tenfold increase of the dose would trigger an unpleasant side effect. Consequently, for toxicological risk assessment the task of molecular level predictions is simpler, since specificity is of lesser importance. This suggests that QSTAR and QShTAR can be expected to make an important impact in toxicological risk assessment.

The lesser need for specificity can be exploited in designing new computer programs for detecting and evaluating similarities within molecular families, and correlating shape and structural features with known toxicological information. In particular, a new computer program was developed for the evaluation of local shape codes for chemical functional groups and for generating detailed correlations between local molecular shape features and the available toxicological data.

The new methodology can be applied to families of pesticides and herbicides, various PAH molecules and their derivatives (Mezey et al., 1996, 1997), as well as to other molecular families. In particular, shape–toxicity correlation studies in PAH molecules and oxidation products of substituted PAH molecules are of current importance. Most members of the family of PAH molecules and their various derivatives have a wide range of toxic effects on bacteria, plants, animals, and humans. An important component of environmental damage caused by oil spills is due to the toxic effects of various oxidation products, photo-oxidized, substituted PAH molecules and their metabolites. In oil spills, many of the smaller PAH molecules, e.g., methyl substituted naphtalenes, are the major toxicants for bacteria and plants. Pyrene and its derivatives, benzo[a]pyrene, various substituted and fused anthracene and phenanthrene derivatives are important toxicants where molecular shape of various isomeric forms strongly influences toxicity. Various oxidized products, such as epoxide forms as well as ortho quionones are implicated as the actual toxicants. Clearly, in the family of PAH molecules, shape and toxicological risk correlations are of major importance and these molecules are in the focus of our study within the QSAR– QShAR project (Mezey et al., 1996, 1997).

The special importance of PAHs in the environment have justified their choice as the first molecular family for a toxicological QShAR study using high quality electron density fragments. A shape fragment database of PAHs has been developed (Mezey et al., 1996) that is suitable for the construction of MEDLA electron densities for virtually all PAHs.

In Fig. 2–1 and 2–2, several electron density contours of an important PAH molecule, benzphenanthrene, as well as the matrix form of the associated (*a*,*b*) shape map (see above) are shown, respectively, as calculated at the 6-31G** *ab initio* - MEDLA level. No single contour surface is sufficient by itself to describe the shape of the fuzzy electron density cloud of the molecule, and the whole range of electron density contours is needed for a complete shape analysis. For illustration purposes only two contour surfaces of benzphenanthrene are shown in Fig. 2–1: those at the density threshold of 0.1 a.u. in the top row, and those at the density threshold of 0.01 a.u. in the bottom row, two views for each. The actual shape analysis has involved the entire range from 0.001 a.u. to 0.1 a.u. of infinitely many density thresholds, not only those two shown in Fig. 2–1. The corresponding shape map in Fig. 2–2 represents the distribution of sets of Betti numbers (topological shape descriptors, see above and literature cited there) within the (a,b) parameter plane, where log a is the logarithm of the density threshold a taken from the range from 0.001 a.u. to 0.1 a.u., and log $|b|$ is the logarithm of the absolute value of local reference curvature parameter b. Since b may take both

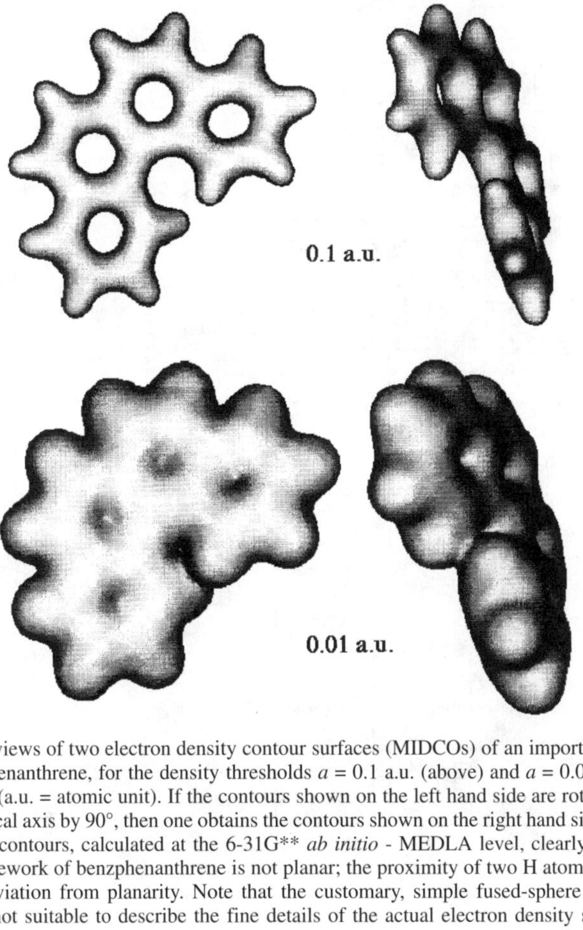

Fig. 2–1. Two views of two electron density contour surfaces (MIDCOs) of an important PAH molecule, benzphenanthrene, for the density thresholds a = 0.1 a.u. (above) and a = 0.01 a.u. (below), respectively (a.u. = atomic unit). If the contours shown on the left hand side are rotated clockwise along a vertical axis by 90°, then one obtains the contours shown on the right hand side. These electron density contours, calculated at the 6-31G** *ab initio* - MEDLA level, clearly show that the nuclear framework of benzphenanthrene is not planar; the proximity of two H atoms causes a considerable deviation from planarity. Note that the customary, simple fused-sphere or dot-surface models are not suitable to describe the fine details of the actual electron density shown in these images.

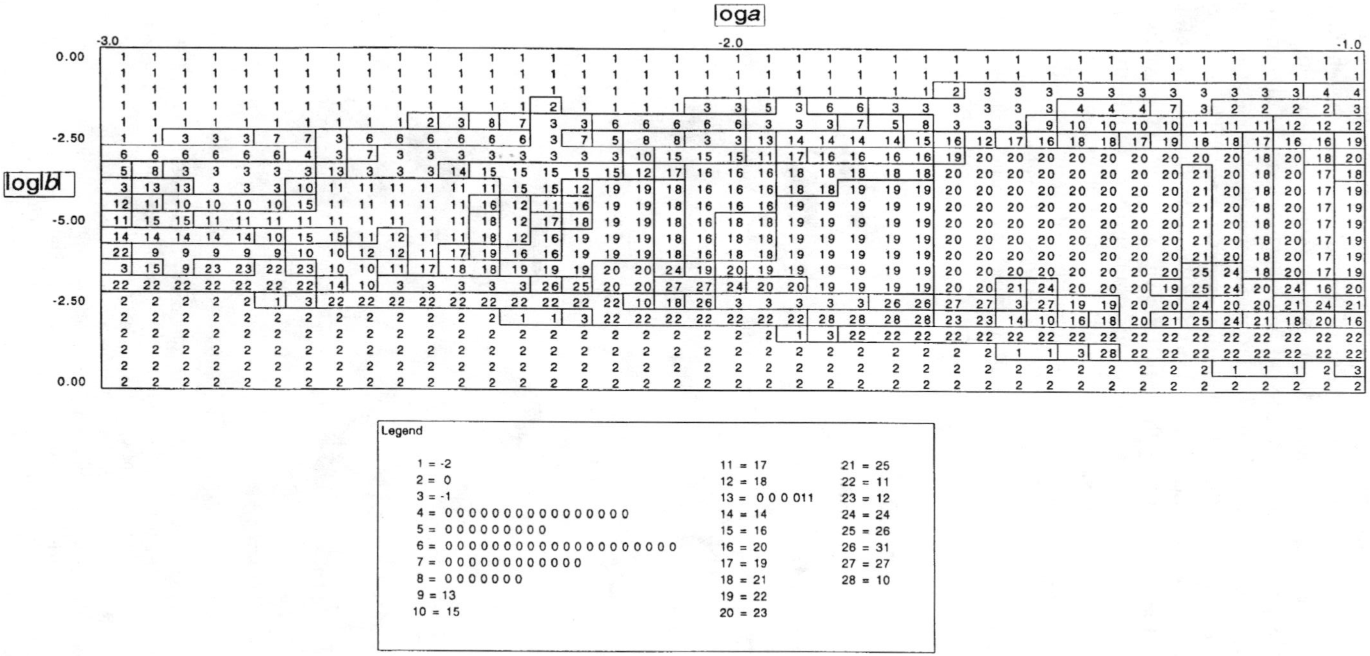

Fig. 2–2. The matrix form of the (*a*,*b*) shape map of the PAH benzphenanthrene shown in Fig. 2–1. The lists of Betti numbers of the shape information are presented in the legend table, and only the simplified, single integer codes (on the left-hand sides of the formal equations in the legend table) are specified in the (*a*,*b*) shape map shown above. Within the QShTAR approach, such numerical shape maps, based on detailed electron density shape analysis, are compared and correlated with toxicological activity.

positive and negative values, the absolute value is used in the argument of the logarithm. The sets of Betti numbers can be rather numerous for each location within the (a,b) map that makes it awkward to indicate these sets directly within the map. Instead, at each point of the (a,b) map a single numerical code is used for these sets; this numerical code is specified in the legend table. For example, the number 13 specified at some locations near the left-hand side edge of the (a,b) map is the numerical code that actually means the set 0, 0, 0, 0, 1, 1 of six Betti numbers, as can be deduced using the legend table. These six Betti numbers indicate the specific shape features within the given density and curvature range of the (a,b) map. These or other shape features correlate with chemical and biochemical properties, that serves as the basis of QShAR. The task of QShAR is to identify those local or global shape features that provide strong correlations with a given type of activity, a task that can be carried out by an automated, computer analysis of the shape maps and known activity data. Although PAHs are relatively simple molecules, they already exhibit intricate shape features throughout the range of density values. Comparisons of shape maps and toxicological activity data have resulted in excellent correlations within families of PAHs (Mezey et al., 1996, 1997).

The molecular shape analysis methods, forming the core of the computer-based shape-activity analysis for toxicological risk assessment, also have been extended to macromolecules, such as proteins, including receptors for toxicants, using a new method for rapid computation of detailed electron density clouds (the actual, real molecular bodies) for molecules of virtually any size. This advance in computational methodology allows one to study in great detail the actual shape of virtually any molecule, find local and global shape correlations with molecular properties, such as toxicological activity (Walker & Mezey, 1993a,b).

2–8 SUMMARY AND CONCLUSIONS

The topological techniques developed for the analysis of the similarities of various molecular shapes can be adapted to the study of the role of molecular similarity in toxicology. In toxicological risk assessment, one of the goals of shape similarity analysis is the establishment of Quantitative Shape-Activity Relations, QShAR, which are expected to become predictive tools. Computer-based multiple toxic effect analysis, combined with multiple shape similarity measures of potential toxicants, is an important component of toxicological QShAR. With the new approaches developed for accurate modeling of the shape of electron distributions of large biomolecules, a new family of biochemical and toxicological problems has become accessible to detailed shape-similarity analysis using reproducible, computer-based, algorithmic techniques.

Molecular shape analysis, based on details of electron densities, is an area of study representing a bridge between basic research and applied research. The methodologies of molecular electron density analysis, using fundamental, quantum chemical approaches, have developed to a level where they can be used to address important, practical problems encountered in applied fields, including toxicology and the diverse disciplines involved in soil science. One may extrapolate the future prospects of the QShAR method as one of the predictive tools of scientists involved in research on ecosystem health from earlier applications in computer-aided drug research. In combination with other approaches, the QShAR method is expected to provide new information that may help to address some of the complex problems faced by soil and environmental scientists.

REFERENCES

Arteca, G.A., and P.G. Mezey. 1990. A method for the characterization of foldings in protein ribbon models. J. Molec. Graphics 8:66–80.

Dean, P.M. (ed.). 1995. Molecular similarity in drug design. Chapman & Hall-Blackie Publ., Glasgow, Scotland.

Carbó, R. (ed.). 1995. Molecular similarity and reactivity: From quantum chemical to phenomenological approaches. Kluwer Acad. Publ., Dordrecht, the Netherlands.

Frisch, M.J., M. Head-Gordon, G.W. Trucks, J.B. Foresman, H.B. Schlegel, K. Raghavachari, M.A. Robb, J.S. Binkley, C. González, D.J. DeFries, D.J. Fox, R.A. Whiteside, R. Seeger, C.F. Melius, J. Baker, R. Martin, L.R. Kahn, J.J.P. Stewart, S. Topiol, and J.A. Pople. 1990. Program Gaussian 90. Gaussian, Pittsburgh, PA.

Lipkowitz, K.B., and D.B. Boyd (ed.). 1990. Reviews in computational chemistry. Vol. 1. VCH Publ., New York.

Johnson, M.A., and G.M. Maggiora (ed.). 1990. Concepts and applications of molecular similarity. John Wiley & Sons, New York.

Leszczynski, J. (ed.). 1996. Computational chemistry: Reviews and current trends. World Sci. Publ., Singapore.

Maggiora, G.M., P.G. Mezey, B. Mao, and K.C. Chou. 1990. A new chiral feature in a-helical domains of proteins. Biopolymers 30:211–215.

Maruani, J. (ed.). 1988. Molecules in physics, chemistry and biology. Reidel, Dordrecht, the Netherlands.

Mezey, P.G. 1986. Group theory of electrostatic potentials: A tool for quantum chemical drug design. Int. J. Quantum Chem. Quant. Biol. Symp. 12:113–122.

Mezey, P.G. 1987a. The shape of molecular charge distributions: Group theory without symmetry. J. Comput. Chem. 8:462–469.

Mezey, P.G. 1987b. Group theory of shapes of asymmetric biomolecules. Int. J. Quantum Chem. Quant. Biol. Symp. 14:127–132.

Mezey, P.G. 1989. The role of shape analysis in drug design. IEEE Eng. Med. Bio. Soc. Ann. Int. Conf. 11:1905–1906.

Mezey, P.G. 1991. The degree of similarity of three-dimensional bodies; applications to molecular shapes. J. Math. Chem. 7:39–49.

Mezey, P.G. 1992. Shape similarity measures for molecular bodies: A 3D topological approach to QShAR. J. Chem. Inf. Comp. Sci. 32:650–656.

Mezey, P.G. 1993. Shape in chemistry: An introduction to molecular shape and topology. VCH Publ., New York.

Mezey, P.G. 1994a. Quantum chemical shape: New density domain relations for the topology of molecular bodies, functional groups, and chemical bonding. Can. J. Chem. 72:928–935.

Mezey, P.G. 1994b. Iterated similarity sequences and shape ID numbers for molecules. J. Chem. Inf. Comp. Sci. 34:244–247.

Mezey, P.G. 1995a. Shape analysis of macromolecular electron densities. Struct. Chem. 6:261–270.

Mezey, P.G. 1995b. Macromolecular density matrices and electron densities with adjustable nuclear geometries. J. Math. Chem. 18:141–168.

Mezey, P.G. 1996a. Functional groups in quantum chemistry. p. 163–222. *In* P.-O. Löwdin et al. (ed.) Advances in quantum chemistry. Vol. 28. Acad. Press, New York.

Mezey, P.G. 1996b. Molecular similarity measures of conformational changes and electron density deformations. p. 89–120. *In* R. Carbó (ed.) Advances in molecular similarity. Vol. 1. Jai Press, New York.

Mezey, P.G. 1996c. Local shape analysis of macromolecular electron Densities. p. 109–137. *In* J. Leszczynsky (ed.) Computational chemistry: Reviews and current trends. Vol. 1. World Sci. Publ., Singapore.

Mezey, P.G., Z. Zimpel, P. Warburton, P.D. Walker, D.G. Irvine, D.G., Dixon, and B. Greenberg. 1996. A high-resolution shape-fragment database for toxicological shape analysis of PAHs. J. Chem. Inf. Comp. Sci. 36:602–611.

Mezey, P.G., Z. Zimpel, P. Warburton, P.D. Walker, D.G. Irvine, X.-D. Huang, D.G., Dixon, and B. Greenberg. 1997. Use of QShAR to model the photoinduced toxicity of PAHs: Electron density shape features accurately predict toxicity. Environ. Toxicol. Chem. (in press).

Ramos Tombo, G.M., and D. Bellus. 1991. Chirality and crop protection. Angew. Chem. Int. Ed. Eng. 30:1193–1215.

Sen, K. (ed.). 1995. Molecular similarity. Topics in current chemistry. Vol. 173. Springer-Verlag, Heidelberg.

Smith, Jr., V.H., H.F. Schaefer III, and K. Morokuma (ed.). 1986. Applied quantum chemistry proceedings of the Hawaii 1985 Nobel Laureate symposium on applied quantum chemistry. Reidel Publ. Co., Dordrecht, the Netherlands.

Walker, P.D, G.M. Maggiora, M.A. Johnson, J.D. Petke, and P.G. Mezey. 1995a. Shape group analysis of molecular similarity: Shape similarity of six-membered aromatic ring systems. J. Chem. Inf. Comp. Sci. 35:568–578.

Walker, P.D, G.M. Maggiora, M.A. Johnson, J.D. Petke, and P.G. Mezey. 1995b. Application of the shape group method to conformational processes: Shape and conjugation changes in the conformers of 2-phenyl pyrimidine. J. Comput. Chem. 16:1474–1482.

Walker, P.D., and P.G. Mezey. 1993a. Program MEDLA 93. Math. Chem. Res. Unit, Univ. Saskatchewan, Saskatoon, Canada.

Walker, P.D., and P.G. Mezey. 1993b. Molecular electron density lego approach to molecule building. J. Am. Chem. Soc. 115:12423–12430.

Walker, P.D., and P.G. Mezey. 1994a. *Ab initio* quality electron densities for proteins: A MEDLA approach. J. Am. Chem. Soc. 116:12022-12032.

Walker, P.D., and P.G. Mezey. 1994b. Realistic, detailed images of proteins and tertiary structure elements: *Ab Initio* quality electron density calculations for bovine insulin. Can J. Chem. 72:2531–2536.

Walker, P.D., and P.G. Mezey. 1995a. A new computational microscope for molecules: High-resolution MEDLA images of taxol and HIV-1 protease, using additive electron density fragmentation principles. J. Math. Chem. 17:203–234.

Walker, P.D., and P.G. Mezey. 1995b. Towards similarity measures for macromolecular bodies: MEDLA test calculations for substituted benzene systems. J. Comput. Chem. 16:1238–1249.

Zimpel, Z, and P.G. Mezey. 1996. A topological analysis of molecular shape and structure. Int. J. Quantum Chem. 59:379–390.

3 Metal Ion Speciation and its Significance in Ecosystem Health

Kim Ford Hayes

Department of Civil and Environmental Engineering
University of Michigan
Ann Arbor, Michigan

Samuel Justin Traina

Ohio State University
Columbus, Ohio

3–1 INTRODUCTION

Metal ion speciation is often cited as the single most important factor in determining the relative reactivity of trace elements in soil environments including nutrient availability and uptake by plants as well as metal ion toxicity (McBride, 1994). In this chapter, a summary of the impact of soil solution properties on metal ion speciation and mobility in the context of ecosystem health is presented. The chapter begins with a discussion of the abundance of metals and their potential impact on ecosystem health. This is followed by an overview of the most relevant soil solution conditions and chemical reactions that control metal ion speciation. Finally, methods for assessing metal ion speciation in complex soil environments as well as highlights of some recent advances in speciation determination are presented.

3–2 OCCURRENCE OF METALS IN SOIL ENVIRONMENTS

The earth's crust is dominated (>99%) by the elements Si, Al, Fe, Ca, Na, K, Mg, Ti, and P. The remainder of the elements are present in trace quantities. These *trace elements* are often of concern from the perspective of ecosystem health. Many are essential elements for plants and animals and microorganisms; however, virtually all exhibit significant toxicities when present at excessive concentrations. Trace metals are present as natural constituents in many soil parent materials. Table 3–1 lists the mean trace metal content of major rock types present at the earth's surface (Alloway, 1995). Representative values for trace element concentrations in uncontaminated soils are presented in Table 3–2 (Wolt, 1994). Metal concentrations in excess of these values typically indicate contam-

Table 3–1. Mean trace metal contents of major rock types (mg kg^{-1}) after Alloway (1995).

	Earth's crust	Igneous rocks			Sedimentary rocks		
		Ultramafic	Mafic	Granitic	Limestone	Sandstone	Shale
Ag	0.07	0.06	0.1	0.04	0.12	0.25	0.07
As	1.5	1	1.5	1.5	1	1	13 (1-900)
Au	0.004	0.003	0.003	0.002	0.002	0.003	0.0025
Cd	0.1	0.12	0.13	0.09	0.028	0.05	0.22 (<240)
Co	20	110	35	1	0.1	0.3	19
Cr	100	2980	200	4	11	35	90 (<500)
Cu	50	42	90	13	5.5	30	39 (<300)
Hg	0.05	0.004	0.01	0.08	0.16	0.29	0.18
Mn	950	1040	1500	400	620	460	850
Mo	1.5	0.3	1	2	0.16	0.2	2.6 (<300)
Ni	80	2000	150	0.5	7	9	68 (<300)
Pb	14	14	3	24	5.7	10	23 (<400)
Sb	0.2	0.1	0.2	0.2	0.3	0.005	1.5
Se	0.05	0.13	0.05	0.05	0.03	0.01	0.5 (<675)
Sn	2.2	0.5	1.5	3.5	0.5	0.5	6
Tl	0.6	0.0005	0.08	1.1	0.14	0.36	1.2
U	2.4	0.03	0.43	4.4	2.2	0.45	3.7 (<1250)
V	160	40	250	72	45	20	130 (<2000)
W	1	0.1	0.36	1.5	0.56	1.6	1.9
Zn	75	58	100	52	20	30	120 (<1000)

ination. In some instances natural levels of trace metals and oxyanions can be great enough to be of environmental concern. Soils formed on sediments of marine origin often contain elevated concentrations of As, Cd, and Se. Soils underlain by Cd-rich Monterey shales in coastal regions of California contain some of the highest natural Cd levels in the world (Alloway, 1995). Similarly, leaching of Se from marine sediments in adjacent upland areas has resulted in the formation of seleniferous soils in California's Central Valley region. Selenium

Table 3–2. Representative natural abundances of trace elements in soil and concentrations in soil solution; from Wolt (1994).

		Soil solution concentration	
Element	Whole soil concentration		At 10% moisture content
	mmol kg^{-1}	mmol L^{-1}	mmol kg^{-1}
Cr	1.0	0.01	0.001
B	0.9	5	0.5
Zn	0.8	0.08	0.01
Be	0.7	0.1	0.01
Ni	0.5	0.17	0.02
Cu	0.3	1	0.06
Co	0.1	0.08	0.008
As	0.1	0.01	0.0013
Sn	0.08	0.2	0.02
Pb	0.05	0.005	0.0005
I	0.04	0.08	0.01
Mo	0.02	0.0004	0.00004
Cd	0.001	0.04	0.004
Se	0.003	0.06	0.006
Hg	0.0001	0.0005	0.00005

rich soils also are found throughout much of the north central USA (Neal, 1995). The presence of ultramafic rocks can lead to elevated concentrations of Ni and Cr, while natural sulfide and coal outcrops, or calcareous marine sediments can lead to uncommonly high levels of other trace metals in soils.

Whereas trace metals are clearly natural constituents in all soils, their concentrations have been greatly increased by anthropogenic inputs. Alloway (1995) lists the dominant sources of trace metal inputs as (a) metalliferous mining and smelting, (b) agricultural and horticultural materials, (c) sewage sludges, (d) fossil fuel combustion, (e) metallurgical industries, (f) electronics industries, (g) chemical and other manufacturing, (h) waste disposal, and (i) sports shooting and fishing. These activities have led to elevated soil concentrations of Pb, Cd, Ni, Hg, Zn, Cu, Cr, Co, Ag, As, Se, and other trace elements. Less extensive in global distribution, nuclear power generation and nuclear weapons production and detonation have resulted in numerous incidences of soil contamination by U, Pu, and several fission products. Mass balances for natural and anthropogenic inputs of metals to soils are available in a recent review edited by Alloway (1995).

3–3 METAL IONS AND ECOSYSTEM HEALTH

The impact of toxic metal ions on the environment can be viewed from a variety of perspectives. From an anthropogenic viewpoint, the ultimate concern of metals in the environment is their impact on human health. Well-publicized cases of the catastrophic consequences of human exposure to metals include the outbreak of *Minamata disease*, which affected people living in the Mimamata Bay region of Japan, and the so-called *Itai-Itai disease* outbreak in the Jintsi River basin of Japan (Laws, 1993). The former was linked to Hg poisoning resulting from decades of discharge into the Bay by a local chemical plant while the later was found to have been caused by Cd-bearing wastewater discharge into the river from the an upstream mining company. These types of incidents have by no means been confined to Japan. Large-scale Hg poisoning of humans also has been documented in Iraq and China. In the USA, Pb poisoning of >2000 and the deaths of >100 children in New York City were directly linked to the consumption of Pb-based paint in dust and peeling paint chips. Although the above incidents indicate the potential harm of metals to humans, less well-documented is the impact of metals and metal-based products that have been introduced to soil environments through years of accidental or improper discharge or past disposal practices. Once introduced into soil environments, because of the strong association of metals with soil minerals and the isolation of many contaminated sites from population centers, human exposure is often less direct. As such, the impact on human health is more insidious. In this case, the potential impact of metals on the viability of other living organisms in the ecosystem may be a better indicator of the potential harm to ecosystem health. For example, metals can have a deleterious impact on plant growth or inhibit the growth of microorganisms. Either may disrupt the natural cycling of the elements and result in the poor overall health of the soil ecosystem. In general, metals may impact the life cycle of any of a number of natural organisms which, in turn, may be associated with and used as indicators of the potential impact on human health.

Regardless of the trophic level under consideration, metal ion speciation has been found to be the single most important factor in determining the relative effects on metals on ecosystem health (e.g., recent compendiums and references therein: Hamelink et al., 1994; Allen et al., 1994; Ross, 1994; McBride, 1994; Adriano, 1992; Sposito, 1989). Although the aquo metal ion species are often considered to be the most bioactive and toxic, with the notable exceptions of species that can be methylated like Pb or Hg, because of the plethora of potential ligands and mineral phases in soil systems, species other than the aquo forms often predominate. When metals are complexed, the toxicity will often be lower compared with the more bioactive forms. Hence, the potential impact of metals in soils must be evaluated from a speciation, rather than a total soil metal concentration point of view. The actual metal species that predominates in soil is highly dependent on the soil constituents and solution conditions. Changes in soil pH, redox conditions, water or mineralogical composition, organic C content or microbiological activity can dramatically change the speciation and, hence, reactivity of metal ions and ligands. This, in turn, affects the relative bioavailability and organism uptake mechanisms, and as a result, the potential for significant disruption of normal life cycles of soil organisms.

In this chapter, the impact of speciation on metal ion mobility, bioavailability, and reactivity will be reviewed. In particular, this chapter will summarize the key soil properties that may control metal ion speciation, and by implication the impact on ecosystem health. A review of soil compositional effects on metal ion speciation will be reviewed first, followed by a discussion of state-of-the-art methods for determining metal ion speciation in complex soil matrices. Finally, examples of the impact of metal ion speciation dominated by solution, surface or solid-state forms will presented, illustrating how metal ion speciation is determined, the type of species that predominate, and the expected impact on ecosystem health.

3–4 CONTROLS OF METAL ION SPECIATION

Several key soil solution parameters have a dramatic impact on metal ion speciation, i.e., pH, pe, and total metal ion concentration. These parameters determine in large measure which solution species will be predominant, and whether hydrolysis, precipitation, complexation, or sorption reactions will take place. In this section, an overview of the impact of these important soil solution parameters and reactions will be given. This will be followed by a discussion of the importance of precipitation, complexation, and surface chemical reactions on metal ion speciation in soil systems. For a more in depth comprehensive coverage of this subject matter, elementary textbooks on the topic should be consulted (e.g., Morel & Hering, 1993; Stumm & Morgan, 1995; McBride, 1994; Sposito, 1989).

3–4.1 Solution pH, pe, and Metal Ion Concentration and Speciation

In the absence of high concentrations of complexing ligands, the most important class of reactions in aqueous systems are protonation (or hydrolysis)

Table 3–3. Chemical characteristics of important cations in soil chemistry.

Cation	Ionic radius†	pK_1‡
Li^+	0.82	13.6
Na^+	1.10	14.1
K^+	1.46	14.5
Rb^+	1.57	>14.5
Cs^+	1.78	>14.5
Be^{2+}	0.35	8.6
Mg^{2+}	0.80	11.4
Ca^{2+}	1.08	12.7
Sr^{2+}	1.21	13.2
Ba^{2+}	1.44	13.4
Ag^+	1.23	12.0
Tl^+	1.58	13.2
Co^{2+}	0.73	10.2
Ni^{2+}	0.77	9.9
Cu^{2+}	0.81	8.0
Zn^{2+}	0.83	9.0
Cd^{2+}	1.03	10.1
Hg^{2+}	1.10	3.7
Pb^{2+}	1.26	8.4
Cr^{3+}	0.70	4.0
$Sb(OH)_2^+$	?	1.4
Tl^{3+}	0.97	1.1

† Distances for Be are for tetrahedral coordination with O_2; for Sb, no ionic radius was calculated for the hydroxide species shown; all other distances assume octahedral coordination with six O_2 anions, each with an ionic radius of 1.32 Å (from G. Faure, 1991).

‡ The first hydrolysis constant of the free metal ion except in the case of Sb for which the free metal ion has never been detected; for Sb, it is the subsequent hydrolysis of the univalent species shown in the table to form the neutral species (from C.F. Baes & R.E. Mesmer, 1986).

reactions. As a function of pH, metals may undergo a series of protonation reactions that for a divalent cation can be represented as follows:

$$Me^{2+} + H_2O = Me(OH)^+ + H^+ \quad [1]$$

$$Me^{2+} + 2H_2O = Me(OH)_2^0 + 2H^+ \quad [2]$$

$$Me^{2+} + 3H_2O = Me(OH)_3^- + 3H^+ \quad [3]$$

As pH increases, a shift in the predominant species from the divalent cation to the less negatively charged metal hydroxide species may result depending on the values of the formation reactions shown above. In general, as pH increases, the tendency to form metal hydroxide ions increases. The value of the first hydrolysis reaction constant (K_1) serves as useful guide for evaluating the relative tendency for metal cations to form the series of metal hydroxide species shown above. Metals with larger K_1 (lower pK_1 in Table 3–3) form the species $Me(OH)^+$ at lower pH.

Redox conditions also can change the reactivity of metals in soil solution. In the soil solution, redox conditions are often described in terms of the master variable, pe, defined as the negative logarithm of the electron activity. Unlike pH, pe cannot be measured directly but can be assessed based on an assumption or knowledge of the redox couples that dominate in a given system. In soils, redox

Table 3–4. Estimates of soil solution pe based on presence of given microorganism.†

Reductants	Oxidants	Products	pe^o_w	Organism
e- donor	e- acceptor			
Organics	$\mathbf{O_2}$	CO_2, $\mathbf{H_2O}$	13.8	heterotrophs
Nitrite	$\mathbf{O_2}$	NO_3^-, $\mathbf{H_2O}$	13.8	nitrobacter
Ammonia	$\mathbf{O_2}$	NO_2^-), $\mathbf{H_2O}$	13.8	nitrosomonas
Fe(II)	$\mathbf{O_2}$	am-Fe(OH)$_3$(s), $\mathbf{H_2O}$	13.8	ferrobacillus
Sulfide	$\mathbf{O_2}$	SO_4^{2-}, $\mathbf{H_2O}$	13.8	thiobacillus
Organics	$\mathbf{NO_3^-}$	CO_2(g), $\mathbf{N_2(g)}$	12.7	denitrifiers
Organics	**am-Fe(OH)$_3$(s)**	CO_2(g), **Fe(II)**	1.0	Fe reducers
Organics	$\mathbf{SO_4^{2-}}$	CO_2(g), $\mathbf{HS^-}$	−3.6	desulfurfibrio
Organics	$\mathbf{HCO_3^-}$	CO_2(g), $\mathbf{CH_4(g)}$	−4.1	methanogenesis

† Standard-state pe value estimated at pH = 7 in water assuming bolded redox species control pe; calculated using thermodynamic constants from Morel and Hering (1993).

conditions have been classified as either oxic (pe >+7 at pH 7) suboxic (+2 < pe < +7 at pH 7) or anoxic (pe < +2 at pH 7; Sposito, 1989). In terms of controlling redox couples, oxic conditions usually prevail when O_2 is present while suboxic and anoxic conditions usually exist in its absence. Because many redox processes are sluggish, in absence of proper facilitators, nonequilibrium redox conditions are more likely the rule than the exception.

Table 3–5. Representative oxidation states and trace element species in soil solution.†

Element	Acid soils	Alkaline soils
Ag(I)	Ag^+, $AgCl^o$	Ag^+, org
As(III)	$As(OH)_3$	AsO_3^{3-}
As(V)	$H_2AsO_4^-$	$HAsO_4^{2-}$
B(III)	$B(OH)_3$	$B(OH)_4^-$
Ba(II)	Ba^{2+}	Ba^{2+}
Be(II)	Be^{2+}	$Be(OH)_3^-$, $Be(OH)_4^{2-}$
Cd(II)	Cd^{2+}, $CdSO_4^o$, $CdCl^-$	Cd^{2+}, $CdCl^-$, $CdSO_4^o$, $CdHCO_3^+$
Co(II)	Co^{2+}, $CoSO_4^o$	$Co(OH)_2^o$
Cr(III)	$Cr(OH)^{2+}$	$Cr(OH)_4^-$
Cr(VI)	CrO_4^{2-}	CrO_4^{2-}
Cu(II)	org, Cu^{2+}, $CuCl^-$	$CuCO_3^o$, org, $CuHCO_3^+$
Hg(II)	Hg^{2+}, $Hg(Cl)_2^o$, CH_3Hg^+, org	$Hg(OH)_2^o$, org
Mn(II)	Mn^{2+}, $MnSO_4^o$, org	Mn^{2+}, $MnSO_4^o$, $MnCO_3^o$, $MnHCO_3^+$
Mo(V)	$H_2MoO_4^o$, $HMoO_4^-$	$HMoO_4^-$, MoO_4^{2-},
Ni(II)	Ni^{2+}, $NiSO^o{}_4$, $NiHCO_3^+$, org	$NiCO_3^o$, $NiHCO_3^+$, Ni^{2+}
Pb(II)	Pb^{2+}, org, $PbSO_4^o$, $PbHCO_3^+$	$PbCO_3^o$, $PbHCO_3^+$, org, $Pb(CO^3)_2^{2-}$), $PbOH^+$
Sb(III)	$Sb(OH)_2^+$, $Sb(OH)_3$	$Sb(OH)_4^-$)
Sb(V)	$Sb(OH)_5^o$, $Sb(OH)_6^-$	$Sb(OH)_6^-$
Se(IV)	$HSeO_3^-$	SeO_3^{2-})
Se(VI)	SeO_4^{2-}	SeO_4^{2-}
Si(IV)	$Si(OH)_4^o$	$Si(OH)_4^o$
Tl(I)	Tl^+	Tl^+
Tl(III)	$Tl(OH)_3^o$	$Tl(OH)_4^-$
V(IV)	VO^{2+}	oxidized to V(V) species
V(V)	VO_2^+, polyvandates	$VO_2(OH)_2^-$, $VO_3(OH)^{2-}$
Zn(II)	Zn^{2+}, $ZnSO_4^o$, org	$ZnHCO_3^+$, $ZnCO_3^o$, org, Zn^{2+}, $ZnSO_4^o$

† "Org" indicates trace element may be predominately in an organic form.

Table 3–6. Representative radionuclide oxidation states and species in soil solutions.

Element	Acid soils	Alkaline soils
low level waste		
$^{137}Cs(I)$	Cs^+	Cs^+
$^{90}Sr(II)$	Sr^{2+}	Sr^{2+}
high level waste		
$^{238}U(IV)$	$UO_2(s)$	$UO_2(s)$
$^{238}U(VI)$	UO_2^{2+}, $UO_2CO^o{}_3$	$UO_2(CO_3)_3^{4-}$, $(UO_2)_3(OH)_7^-$
$^{237}Np(IV)$	$NpO_2(s)$	$Np(OH)_5^-$
$^{237}Np(V)$	NpO_2^+	$NpO_2CO_3^-)$, $NpO_2(CO_3)_2^{3-}$
$^{239}Pu(III)$	Pu^{3+}	oxidizes to Pu(IV)
$^{239}Pu(IV)$	PuO_2	$Pu(OH)_5^-$
$^{239}Pu(V)$	PuO_2^+	soluble Pu-complexes (CO_3^{2-}) or organic)
$^{239}Pu(VI)$	PuO_2^{2+}	soluble Pu-complexes (CO_3^{2-}) or organic)

Redox changes and conditions in soil environments are closely associated with microorganisms that have the needed electron transport cycles to facilitate redox conversion. When suitable electron donors and acceptors are present, the microorganisms and pe_w^o values shown in Table 3–4 are likely to prevail. Hence, the presence of certain solution species and organisms can be useful in assessing the likely redox state of a soil environment. In a common scenario, as water saturates a soil, O_2 is removed by microbiological activity and only slowly diffuses back in from the atmosphere. As such, a sequence of events begins upon water flooding. First microorganisms that use O_2 as the terminal electron acceptor thrive, and pe remains high (pe = 13.8). When O_2 is depleted, pe becomes progressively more reducing depending on the availability of other electron acceptors and microorganisms that use them. As the soil dries out and becomes unsaturated with water, oxygenated air reenters and the cycle may begin again.

Depending on the pe, the predominant oxidation state may vary. Generally speaking, for those species that have several possible stable oxidation states in soil systems, as the pe increases, the predominant form of the species will shift from the lower to higher oxidation states. The most common stable oxidation states for potentially toxic trace elements and radionuclides are shown in Tables 3–5 and 3–6; the most redox-active trace element species (excluding Mn and Fe) in terms of possible number of stable oxidation states are given in Table 3–7.

3–4.2 Solubility Control of Speciation

Dissolution–precipitation equilibria can control the solubility of metals and ligands in natural waters and soil solutions through the general reaction

$$M_aL_b(s) = aM^{m+}(aq) + bL^{n-}(aq) \quad [4]$$

where M^{m+} and L^{n-} are the metal and ligand ions of interest and a and b are the stoichiometric reaction coefficients. Some of the possible and more common solids phases for various trace elements in aerobic and anaerobic soils are shown

Table 3–7. Redox active trace elements in the soil solution.

Element	Low pe	High pe
As (III,V)	AsO_3^{3-} ($AsH_3(g)$ or $As(CH_3)_3(g)$)	AsO_4^{3-}
Co (II, III)	Co^{2+}	Co(3+)
Cr (III, VI)	Cr^{3+}	CrO_4^{2-}
Np (IV, V)	$NpO_2(s)$	NpO_2^{+}
Pu (III, IV, V, VI)	Pu^{3+}, $PuO_2(s)$	PuO_2, PuO_2^{+}, PuO^{2+}
Sb (III, V)	$Sb(OH)_2^{+}$	$Sb(OH)_6^{-}$
Se (IV, VI)	SeO_3^{2-}	SeO_4^{2-}
Tl (I, III)	Tl^{+}	Tl^{3+}
U (IV, VI)	$UO_2(s)$	UO_2^{2+})
V (IV, V)	VO^{2+}	VO_2^{+}

in Tables 3–8 and 3–9. Solubility equilibria are described through a dissolution or precipitation constant, the former being given as

$$K_{dis} = (M^{m+})^a(L^{n-})^b /(M_aL_b(s)) \qquad [5]$$

where ( ) denotes activity. If the solid in Eq. [4] and [5] is in its thermodynamic standard state, then its activity has a value of unity and Eq. [5] can be simplified to

$$^{Ideal}K_{dis} = (M^{m+})^a(L^{n-})^b \qquad [6]$$

where $^{Ideal}K_{dis}$ is an ideal standard state dissolution constant. If a solution is in equilibrium with the solid M_aL_b, then the reaction in Eq. [4] will control the aque-

Table 3–8. Possible soil solution solubility controls for trace elements.

Element	Aerobic soils	Anaerobic soils
Ag	AgCl, Ag_2O (high pH)	Ag, Ag_2S
As	$Ca_3(AsO_4)_2$, $Mg_3(AsO_4)_2$, As_2O_5	As, As_2S_3, As_2O_3
B	B_2O_3	B_2O_3
Ba	$BaSO_4$, $BaCO_3$	$BaSO_4$, $BaCO_3$
Be	$Be(OH)_2$, $BeSO_4$, BeO	$Be(OH)_2$
Cd	$Cd(OH)_2$, $CdCO_3$	Cd, CDs
Co	$Co(OH)_2$, $CoCO_3$, $CoSO_4$	$Co(OH)_2$, $CoCO_3$, $CoSO_4$
Cr	$Cr(OH)_3$ (low to neutral pH)	$Cr(OH)_3$
Cu	CuO, $CuCO_3$, $Cu(OH)_2CO_3$	Cu, CuS, Cu_2S
Hg	$HgCl_2$, HgO, $Hg(OH)_2$	Hg(l), HgS
Mn	MnO_2, MnOOH, Mn_3O_4	$Mn(OH)_2$, $MnCO_3$, MnS
Mo	$FeMoO_4$	$FeMoO_4$
Ni	NiO, $NiCO_3$, $Ni(OH)_2$	Ni, NiS
Pb	PbO, $PbCO_3$, $Pb_3(CO_3)(OH)_2$	Pb, PbS
Sb	Sb_4O_6, $Sb(OH)_3$	Sb, Sb_2S_3, Sb_2O_3
Se	very soluble	Se, SeO_2
Si	SiO_2	SiO_2
Tl	$Tl(OH)_3$, Tl_2O_3	Tl_2S, Tl_2O
V	V_2O_5	V_2O_4, V_2O_3, $V(OH)_3$
Zn	ZnO, $Zn(OH)_2$, $ZnCO_3$, $ZnSO_4$	ZnS, Zn

Table 3–9. Possible soil solution solubility controls for radionuclide elements.

Element	Aerobic soils	Anaerobic soils
^{137}Cs	very soluble	very soluble
^{90}Sr	$SrCO_3$, $Sr(OH)_2$	$SrCO_3$, $Sr(OH)_2$
high level waste		
^{238}U	$UO_2(OH)_2$, UO_2CO_3	UO_2
^{237}Np	NpO_2, $Np(OH)_4$, $NpO(OH)_2$	NpO_2, $Np(OH)_4$, $NpO(OH)_2$
^{239}Pu	$Pu_2(CO_3)_3$, $Pu(OH)_4$, PuO_2	$Pu(OH)_4$, PuO_2

ous activities of the ions M^{m+} and L^{n-}. This control is expressed both through the concentration and the oxidation states of the dissolved species.

Often, Eq. [6] is assumed to hold *a priori* and solid-phase controls on ion speciation are evaluated through the calculation of the ion activity product (IAP), which is given by

$$(M^{m+})^a(L^{n-})^b = \text{IAP} \qquad [7]$$

Equality between the IAP and the K_{dis} is taken as evidence for the existence of solid $M_aL_b(s)$. Implicit in this approach are the assumptions that: (i) the aqueous activity of a given ion can be directly measured or calculated from total solution composition data, (ii) sufficient contact time has occurred between the solution and the solid phases for equilibrium to be attained, and (iii) the solid phase of interest is in its thermodynamic standard state (Sposito, 1981). Unfortunately, this method is faced with several pitfalls. The validity of the first assumption is clearly dependent upon the completeness of the analytical data and the knowledge of thermodynamic stability constants for the formation of solution complexes that also may be present. This becomes particularly problematic when one considers the ubiquity of simple dissolved organic ligands and humic and fulvic acids in soil solutions. Fortunately, recent analytical developments suggest that direct determination of free metal and ligand ion concentrations through the use of separation and detection methods such ion-chromatography-ICP–MS and capillary-electrophoresis-ICP–MS. These methods offer great promise in studying systems in which the separation time is faster than that required for the solution to reestablish ionic equilibrium. The second assumption also is often a source of uncertainty. Because reaction times are typically chosen to be experimentally convenient, they may not reflect the true chemical equilibrium. Pierzynski et al. (1990a) found that the time required for dissolved orthophosphate to reach steady-state in soil slurries was sample dependent.

Even as dubious as the first two assumptions may be in some cases, the third assumption may be the most tenuous and difficult to assess in real soils and sediments. Often the solid phase forms of many elements are present in extremely low concentrations precluding their direct determination. Even if they can be detected with direct physical methods, it is still not certain that they control the solubility and solution phase speciation of a given element because it is not known how pure they are and, hence, the validity of assuming unit activity. The inclusion of foreign ions, the presence of crystal defects, a microcrystalline or amorphous structure, or the presence of organic or inorganic coatings can all prevent the third assumption from being valid. McBride and Bouldin (1984) found that solutions reacted with a Cu contaminated soil were undersaturated with respect to known Cu phases; however, the presence of green coatings on dolomitic limestone collected from the soil and infrared analyses of these materials indicated the presence of malachite. This was attributed to surface enhanced precipitation of the Cu carbonate. Pierzynski and coworkers (1990b) using energy dispersive analytical x-ray spectroscopy (EDAX) found that solid-phase forms of P in soil were apparently mixed solids comprised of Ca and Al phosphates.

This research indicates that solubility can indeed control the solution chemistry of ions in soils; however, nonequilibrium is likely the rule rather than the exception. Caution must be taken in predicting short-term solution speciation based on solid-phase composition and even more care must be taken when using solution-phase chemistry to determine solid-phase speciation in soils. Nevertheless, the chemical and biological liability of specific metals and ligands are often controlled by precipitation–dissolution reactions.

Whether or not the natural solid phases will actually control soil solution concentrations compared with other processes such as sorption (see below) is obviously highly dependent on the value of the solubility product constant. Recently, it has been demonstrated that highly insoluble solid phases may be formed by Pb when apatites are present. For example, Ma and coworkers (1993, 1994a,b, 1995) have shown that reaction of Pb contaminated materials (solutions, soils, and sediments) with natural and synthetic apatites results in aqueous Pb solubilities at or below drinking water limits proscribed by U.S. Environmental protection Agency (15 mg L^{-1}). They attributed these results to the dissolution of apatite followed by the precipitation of Pb-phosphates, most notably chloro-, fluoro-, or hydroxypyromorphite ($Pb_{10}(PO_4)_6(X)_2$, where $X = Cl^-$, F^-, or OH^-). This process is described by the general reactions:

$$Ca_{10}(PO_4)_6(X)_2(c) + 14H^+ = 10Ca^{2+} + 6H_2PO_4^- + 2H_2X \quad [8]$$

$$10Pb^{2+} + 6H_2PO_4^- + 2H_2X = Pb_{10}(PO_4)_6(X)_2(c) + 14H^+ \quad [9]$$

Discrete reaction products were detected by powder x-ray diffraction (XRD) and by scanning electron microscopy (SEM) in model aqueous systems when the initial aqueous Pb concentrations were in excess of 5 mg L^{-1}. The discrete Pb-bearing solids had different crystal structures and particle morphologies than the original hydroxyapatites and did not contain Ca (within the detection limits of EDAX), discounting the possibility of Pb sorption through ion substitution (Ma et al., 1993). XRD-detectable pyromorphites also were found after reaction of Pb-saturated cation exchange resins (which supported dissolved Pb concentrations <1 mg L^{-1}) with hydroxyapatite (Ma et al., 1993).

Ma et al. (1994a) also examined the effects of NO_3, Cl, F, SO_4, and CO_3 on the immobilization of aqueous Pb by hydroxyapatite. Pb concentrations were reduced from initial levels of 5 to 100 mg L^{-1} to below 15 μg L^{-1} except at very high CO_3. Hydroxyapatite was transformed to hydroxypyromorphite after reaction with $Pb(NO_3)_2$ in the presence of NO_3, SO_4, and CO_3, to chloropyromorphite

Table 3–10. Particle-size distribution of Pb in contaminated soil (S.J. Traina, 1995, unpublished data).

Size fraction	Total Pb	Pb solids by x-ray diffraction†
mm	mg kg^{-1}	
2000–500	36 855	Cerussite‡
500–250	30 889	Cerussite
250–100	32 793	Cerussite
100–50	34 747	Cerussite
50–20	46 682	Cerussite
20–2	34 198	Cerussite
2–0.2	38 001	Cerussite
>0.2	65 563	Cerussite
Whole soil	37 024	

† Soil from Oakland, CA.
‡ $PbCO_3(c)$.

($Pb_5(PO_4)_3Cl$) after reaction with $PbCl_2$, and to fluoropyromorphite ($Pb_5(PO_4)_3F$) after reaction with PbF_2, respectively. These reaction products were identified by XRD and SEM. Reactions of natural carbonated fluoro- and chloroapatites with dissolved Pb also resulted in the formation of XRD detectable pyromorphites (Ma et al., 1995). Amendment of contaminated soils with these materials again reduced dissolved Pb to concentrations <15 mg L^{-1} (the U.S. EPA drinking water limit).

Many soils and sediments also contain trace metals such as Zn, Cd, Ni, Cu, Fe(II), and Al. These ions have varying effects on Pb immobilization by apatite (Ma et al., 1994b). At initial solution concentrations of <20 mg L^{-1}, these ions had no discernible effect on the immobilization of dissolved Pb by hydroxyapatite. Additionally, significant quantities of these metals also were removed from solution. At greater initial concentrations (>20 mg L^{-1}), dissolved Cu was the most effective in inhibiting Pb immobilization by hydroxyapatite, followed by Fe(II), Cd, Zn, Al, and Ni. Hydroxypyromorphite was the only mineral detected by XRD besides hydroxyapatite after Pb reaction with hydroxyapatite in the presence of these metals. The amounts of hydroxypyromorphite formed (as determined from the XRD patterns) decreased with an increase in competing metal concentrations. The order of inhibition of hydroxypyromorphite formation was positively correlated with the solubility of the other metal phosphates supporting a mechanism of competitive precipitation. No other metal-phosphate precipitates, however, could be detected with XRD or SEM (Ma et al., 1994b).

In order to assess the potential impact on Pb bioavailability, Laperche and coworkers (1995) examined the effects of apatite additions on the uptake of Pb by Sudan grass, in Pb-contaminated soil. This soil, from an urban garden in an inner city area in Oakland, CA, is thought to have resulted from a large paint spill. The total Pb content in the whole soil and in different grain size classes is given in Table 3–10. The XRD data in Fig. 3–1a are from the >3.3 g cm^{-3} density fraction reacted with hydroxyapatite for 9 d. Diffraction data also are shown for the >3.3 g cm^{-3} density fraction of the same soil not reacted with apatite. It is evident that apatite addition caused a decrease in cerrusite peaks and the appearance of a peak consistent with hydroxypyromorphite. The pyromorphite peaks, however, are asymmetric and coprecipitation of Pb-phosphates with other metals can not be ruled out.

Figure 3–1b shows the effect of apatite additions on the Pb content of shoots from Sudan grass grown in the Oakland soil for 9 d. It is evident that reaction of the contaminated material with apatite resulted in significant decreases in the shoot concentrations of Pb. Similar decreases were observed in the concentrations of Pb in root tissues (Laperche et al., 1995). Apparently, conversion of Pb from a soluble mineral, cerrusite, to a less soluble form, pyromorphite, resulted

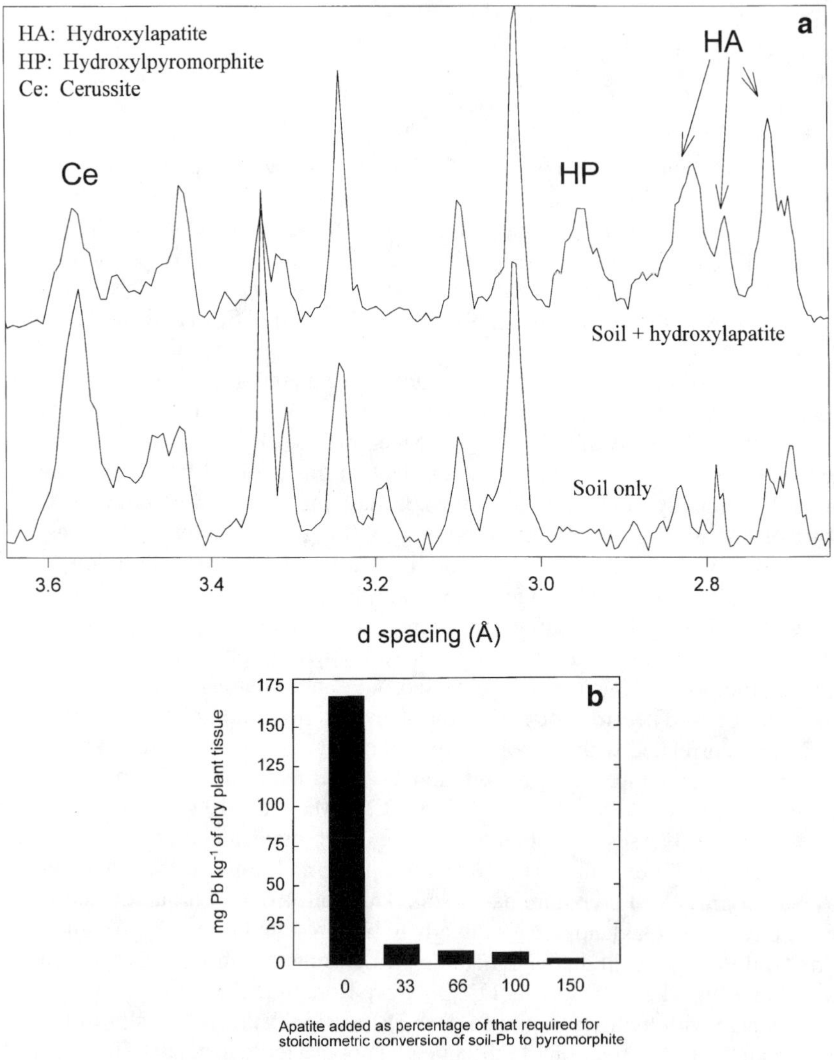

Fig. 3–1. Effect of apatite additions on (A) the solid-phase speciation of Pb in a contaminated soil, as detected by x-ray diffraction (XRD) and (B) Pb uptake (by Sudan grass) from a Pb-contaminated soil. Total Pb concentrations in excess of 30 g kg^{-1} (S.J. Traina, 1995, unpublished data).

in dramatic declines in bioavailability. These data provide strong evidence that solid phase speciation can effect the phytoavailability of contaminants in soils.

Solid phase controls of speciation also have been shown to influence the gastrointestinal availability of Pb in ingested soil materials. For example, using an in vitro assay to simulate the human digestive tract, Ruby et al. (1992) found that the gastrointestinal bioavailability of Pb decreased with decreases in the dissolution rate constants of different Pb solids. This suggests that highly insoluble forms of toxic metals may have a beneficial effect in limiting bioavailability and toxicity to organisms at higher trophic levels.

3–4.3 Complexing Ligands and Speciation

The fate and impact of dissolved metals are strongly influenced by the formation of complex ions with organic and inorganic ligands. Many metals and ligands exist in soil solution as complex ions, and not dissociated ionic species (Tables 3–5 and 3–6); however, the most biologically active species are usually the free, unassociated ions (Table 3–4). Clearly, a knowledge of solution species is important in assessing the environmental impact of a specific element.

Typical inorganic ligands and some selected chemical properties are given in Table 3–11. The relative concentrations of these ions will vary from soil to soil, but the dominant inorganic ligands in most soil solutions are nitrate, sulfate, chlo-

Table 3–11. Chemical characteristics of important anions in soil chemistry (adapted from McBride, 1994).

Oxyanion	Formula	Shared charge†	Electronegativity	pK_{A1}‡
Borate	$B(OH)_4^-$	0.75	2.0	9.2
Antimonate	$Sb(OH)_6^-$	0.83	1.9	2.7
Arsenite	AsO_3^{3-}	1.0	2.0	9.3
Silicate	SiO_4^{4-}	1.0	1.8	9.9
Hydroxyl	OH^-	1.0	2.1	14.0
Phosphate	PO_4^{3-}	1.25	2.1	2.1
Arsenate	AsO_4^{3-}	1.25	2.0	2.2
Selenite	SeO_3^{2-}	1.33	2.4	2.8
Carbonate	CO_3^{2-}	1.33	2.5	3.5§
Molybdate	MoO_4^{2-}	1.5	1.8	3.6
Chromate	CrO_4^{2-}	1.5	1.6	0.2
Sulfate	SO_4^{2-}	1.5	2.5	(−2)
Selenate	SeO_4^{2-}	1.5	2.4	(−2)
Nitrate	NO_3^-	1.67	3.0	(−1)
Perchlorate	ClO_4^-	1.75	3.0	(−10)
Fluoride	F^-	--	4.0	3.2
Chloride	Cl^-	--	3.0	(−7)
Bromide	Br^-	--	2.8	(−9)
Iodide	I^-	--	2.5	(−11)

† Shared charge calculated by taking metal valence and dividing by number of oxygens.

‡ The first acidity constant for the completely protonated or neutral species (values for oxyanions from C.F. Baes and R.E. Mesmer (1986)). Negative p*K* values shown in parentheses are estimates from dissolution in organic acids and are indicative of strong acids that completely dissociate in water.

§ True carbonic acid.

ride, carbonate and bicarbonate, and the hydroxyl ion. Significant inorganic ions present at lower concentrations include orthophosphate, boric acid, silicic acid, arsenate, selenate, molybdate, and fluoride. Dominant organic ligands in soil solutions are best described by their functional groups. These are dominated by carboxyl, carbonyl, amino, imidazole, phenolic OH, alcoholic OH, and sulfhydryl groups. The propensity of these species to form stable solution complexes with metal ions can be estimated from the hard–soft, Lewis acid–Lewis base principle (HSAB). Simply stated, hard Lewis acids will tend to from solution complexes with hard Lewis bases and soft Lewis acids will exhibit a greater tendency to form complexes with soft Lewis bases (Sposito, 1981). The relative softness of a Lewis acid is estimated by the Misono softness parameter, Y, (Sposito, 1989) as

$$Y = 10 \frac{I_z\, r}{(Z)^{1/2} I_{z+1}} \qquad [10]$$

where I_z is the ionization energy for a cation of valence Z and radius r. Hard bases in Table 3–11 include hydroxyl, sulfate, nitrate, orthophosphate, carbonate, and fluoride. It also is of interest to note that water is the hardest Lewis base commonly found in aqueous environments such as the soil solution. Hard organic bases include the carboxyl and organic hydroxyls. Borderline or intermediate species are comprised of amines, imidazoles, bromide, and chloride, and soft bases include sulfhydryls and iodide (Sposito, 1981).

Most of the trace metals in soils are borderline to soft Lewis acids, so they generally show a tendency to form aqueous complexes with borderline to soft Lewis bases. Thus, $CdCl^-$ can be a significant solution complex in many soils. The formation of chloro-complexes can result in decreased retardation of trace metals by soil media. Doner (1978) found greater leaching of Ni, Cu, and Cd through soil columns bathed in a NaCl mobile phase, relative to the amounts of trace metal migration observed in ClO_4^- solutions. This was attributed to reduced affinity of the univalent metal-chloride complexes for ion adsorption sites.

Anthropogenic introductions of complexing ligands can have a profound effect on the fate and transport of cationic contaminants in soils and sediments. Synthetic organic chelates are used extensively in the formulation of detergents and shampoos, as food preservatives, and as decontamination reagents. These enter the environment through disposal of municipal and industrial wastes. Xue (1995) has shown that complexation by EDTA resulting form municipal sewage disposal, controls the solubility and speciation of several metals in aquatic environments.

EDTA has been used extensively in the production of nuclear energy and weapons. Inadequate disposal practices have resulted in EDTA-facilitated transport of radioactive solutes in some subsurface sediments at sites managed by the U.S. Department of Energy (Szecsody et al., 1994). One example of particular interest is the facilitated transport of ^{60}Co. In the absence of complexing ligands, Co(II) is a highly sorbing solute. Oxidation of Co(II) to Co(III) by Mn oxides has been reported by a number of investigators (McKenzie, 1967; Murray, 1975; Traina & Doner, 1985). This reaction dramatically reduces the mobility of Co in soils and sediments through the formation of Co(III) surface rinds or the incorporation of Co(III) into the Mn oxide lattice. Little attention has be paid to redox reactions between Co(II) and Fe(III). The oxidation of Co(II) by Fe(III) is thermodynamically spontaneous but kinetically limiting, and thus is not a dominant processes in natural environments; however, the situation changes dramatically in the presence of EDTA. The formation of stable Co-EDTA adducts greatly reduces the adsorbtivity of Co to soil minerals, facilitating its transport. Additionally, the oxidation of Co(II)-EDTA to Co(III)-EDTA, by Fe oxides such as goethite, is rapid (Xue, 1995). The Co(III)-EDTA complex has a log K in excess of 40, thus this anionic species is stable and likely to promote the migration of Co. In the event that a significant fraction the Co is present as ^{60}Co, the formation of Co-EDTA complexes can lead to significant transport of radioactive solutes. Clearly, anthropogenic ligands can greatly alter the chemistry and ecotoxicology of toxic metals in soils and natural waters and are of environmental concern.

3–4.4 Sorption Controls of Speciation

The reaction of metal ions with soil mineral surfaces typically causes a significant fraction of the metal to sorb to the solids, often lowering the metal ion aqueous phase concentration well below the solubility limits of the solid phases that may form. A number of processes can occur during sorption including absorption by diffusion into the solid matrix or adsorption at the surface–water interface by surface complexation, surface polymerization, and surface precipitation. The type of sorption process that occurs is highly dependent on the quantity and type of mineral and organic phases present. Solution conditions such as pH, ionic strength, metal ion concentration, and the presence and concentration of other sorbing species or complexing agents also may play a major role in determining the extent and type of operative sorption process.

3–4.4.1 Surface Complexation

In oxic soil systems, the most reactive surfaces are those composed of surface hydroxyl functional groups. These types of surface sites form when minerals are exposed to water, either in the vapor or liquid form. Metal oxide, oxyhydroxide, or hydroxide solids (e.g., Fe, Al, and Mn oxides), and noncrystalline clays (e.g., allophanes) are examples of minerals bearing significant concentrations of surface hydroxyl sites. Many divalent and trivalent trace metals form strong, inner-sphere surface complexes with these type of sites. When they do, they are not readily desorbed and, as such, are effectively removed, immobile, and unavailable. In contrast, those metals that form weaker outer-sphere surface complexes may be more readily desorbed and remain available when conditions that favor desorption in the soil solution result.

Surface complexation reactions of surface hydroxyl sites with metal cations can be represented analogously to the solution protonation reactions shown above (Eq. [1] to [3]) with the surface hydroxyl acting in a similar fashion to the hydroxide ion in solution:

$$\equiv SOH + Me^{2+} = \equiv SOMe^{+} + H^{+} \text{ (inner-sphere)} \quad [11]$$

$$\equiv SOH + Me^{2+} = \equiv SO^{-} - Me^{2+} + H^{+} \text{ (outer-sphere)} \quad [12]$$

In general, sorption of metal cations on metal oxides, hydroxides, and oxyhydroxides (hereafter collectively referred to as metal oxides) increases with pH (Fig. 3–2). Because of the weaker affinity associated with outer-sphere complexation, when this kind of surface association is predominant, increases in the concentration of competing cations tends to reduce the amount of metal sorbed (Fig. 3–2; Hayes & Leckie, 1987). The preference of metal ions for metal oxide surfaces has been correlated with acid-base properties of the surface hydroxyl sites and the metal ions (McBride, 1994). In the case of mineral surfaces, more basic surface sites have a stronger affinity for a given metal ion than more acidic sites. A correlation has been found to exist between the value of the first hydrolysis constant (Table 3–3) of the metal ions and sorption affinity. Typically, metal ions having a lower value of pK_1, sorb to a greater extent for a given metal oxide surface (McBride, 1994).

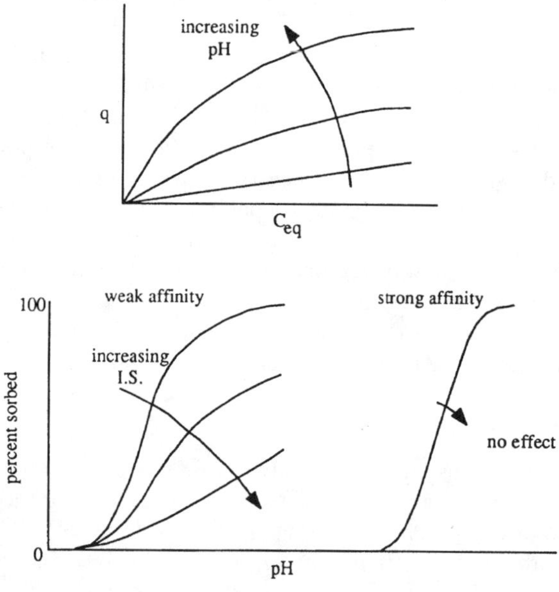

$$\equiv SOH + Me^{2+} = \equiv SOMe^{+} + H^{+}$$
(inner-sphere)

$$\equiv SOH + Me^{2+} = \equiv SO^{-} \cdots Me^{2+} + H^{+}$$
$$\equiv SOH + Na^{+} = \equiv SO^{-} \cdots Na^{+} + H^{+}$$
(outer-sphere)

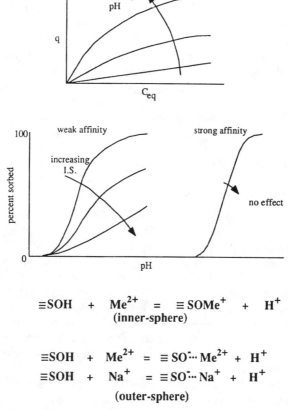

Fig. 3–2. Typical pH dependent sorption behavior for metal cations sorbing to surface hydroxyl groups on soil oxide minerals.

3–4.4.2 Ion Exchange

Another important class of sorption reactions in soil systems are those that take place at so-called fixed-charge sites of aluminosilicate clay minerals. Negative, fixed-charge sites are found on clay minerals that contain a significant amount of isomorphic substitution. For example, Al substitution for silicon in the tetrahedral sheets or Mg for Al in the octahedral sheets leads to a net negative charge delocalized over the oxygens of the siloxane ditrigonal cativities exposed at such sites. Depending on the extent and location of the isomorphic substitution, metal cations may form inner- or outer-sphere complexes with fixed-charge sites. Soils with significant quantities of these types of sites have high cation-exchange capacity (CEC) and low metal cation mobility. In the case of montmorillonitic soils exchanged extensively with monovalent cations, sorption to fixed-charged

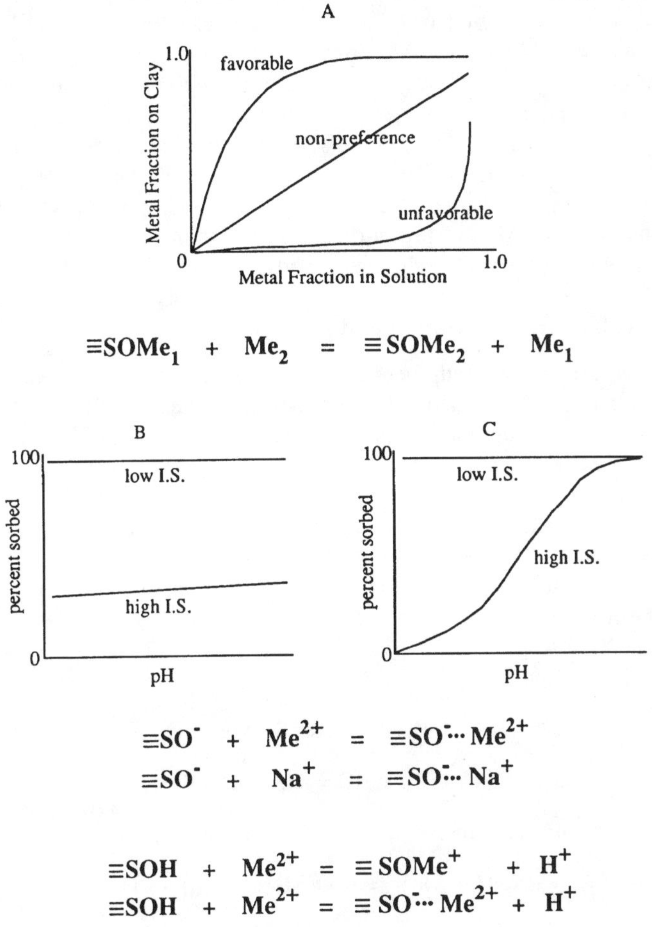

Fig. 3–3. Typical ion-exchange behavior for metal cations sorbing to fixed-charge sites on clay minerals. (A) binary exchange; (B) pH dependence of the effects of ionic strength for metal cations that sorb weakly to surface hydroxyl edge sites on clays; and (C) pH dependence of the effects of ionic strength for metal cations that sorb strongly to surface hydroxyl edge sites on clays.

sites by many multivalent metals is often readily reversible and indicative of outer-sphere complex formation. Structural exchange cations in more highly-charged clays, however, leave metal ions more tightly complexed or *fixed* and much less available.

Unlike sorption reactions with surface hydroxyl sites, ion exchange reactions at fixed-charge sites tend to be much less pH dependent (Fig. 3–3b). Depending on the exchanging cations, the exchange reactions can be favorable, unfavorable or show relatively no preference (Fig. 3–3a). The preference seems to correlate with charge and ionic radii (Table 3–3) of the sorbing cations, with more highly-charged cations and those with larger ionic radii having a higher affinity for this type of reaction site. This is thought to be due to the ease with which a metal cation can dehydrate (McBride, 1994). Because cations with smaller ionic radii tend to be more strongly hydrated, they are less able to dehydrate upon sorption, and as a result, are less tightly held to fixed-charge sites.

Although the presence of fixed-charge sites usually results in pH independent sorption, when high concentrations of competing ions are present, the trace metal cations may be effectively excluded from fixed-charged sites. Under these circumstances, metal cations may sorb more favorably to surface hydroxyl sites exposed at the edges of clay minerals leading to the pH-dependent sorption observed on metal oxide surfaces (Fig. 3–3c; Zachara et al., 1992; Papelis & Hayes, 1996). This type of sorption behavior may be important when high concentrations of Ca or Mg are present, especially for metal ions that form inner-sphere complexes with surface hydroxyl sites.

3–4.4.3 Surface Polymerization and Precipitation

As sorption density (number of sites per unit area) on metal oxides surfaces increases, changes in the reaction stoichoimetry and the controlling sorption reaction may result. For example, as surface coverage becomes quite high (>5%), the sorption process may change from simple sorption of monomeric complexes to the formation of surface multinuclear polymers or precipitates (Katz & Hayes, 1995). This can result in significantly more metal ion sorption over a narrow pH range, and potentially a less reversible sorbed complex. Surface complexation reaction stoichiometries and species consistent with this type of sorption process include the following:

Surface Polymer Reactions

$$\equiv SOH + 2Me^{2+} + 2H_2O = \equiv SOMe_2(OH)_2^{+} + 3H^{+} \qquad [13]$$

$$\equiv SOH + 4Me^{2+} + 5H_2O = \equiv SOMe_4(OH)_5^{2+} + 6H^{+} \qquad [14]$$

Surface Precipitate Reaction

$$\equiv SOH + Me^{2+} = \equiv SO(Me(OH)_{2(s)}) + 2H^{+} \qquad [15]$$

Each of the above reactions has a reaction stoichiometry of greater than one proton released per metal ion sorbed. This type of reaction leads to an enhancement in metal sorption at any given pH compared with the one or two proton release reactions that are often used to describe monomer complex sorption of metal cations to surface hydroxyl sites. Multinuclear reaction processes of this type have only been observed for metal ions that form inner-sphere surface complexes at lower coverage. While not completely understood at this time, it appears that the close proximity and template match of reactive surface sites with metal ions also is a prerequisite to the formation of surface polymers and precipitates. No evidence has been found for multinuclear complex formation of metal ions sorbed to fixed-charge sites of clays. Apparently, the site density and juxtaposition of sites is too sparse for hydroxyl bridging to take place among sorbed mononuclear species.

3–4.4.4 Impact of Organic Matter

When soluble organic compounds are present in soil systems, metal ion sorption may be enhanced, unaffected, or diminished. The relative impact has been found to depend on the type of mineral surfaces, metal ions, and organic functional groups and the relative distribution. Because of the multiplicity of reactions that may occur, it is often difficult to predict *a priori*, what the controlling reaction process will be. This has not, however, precluded the ability to systematically identify the potential impact of organic matter on metal ion sorption in relatively simply systems (Sposito, 1983). In cases where surface hydroxyl sites are involved in metal ion sorption reactions (Fig. 3–4), the impact can be easily assessed. For example, when metal ions (Me^{2+}) form cationic complexes (e.g., MeL^+) with organic functional groups (L^-) that can then subsequently sorb to metal oxide sites, the extent of sorption can be reduced. This results in a shift in sorption edge to higher pH (Fig. 3–4a). It also is possible for chelating agents with multiple negatively charged surface functional groups (e.g., L^{3-}) to form negative charged metal ions complexes (MeL^-). These may then subsequently

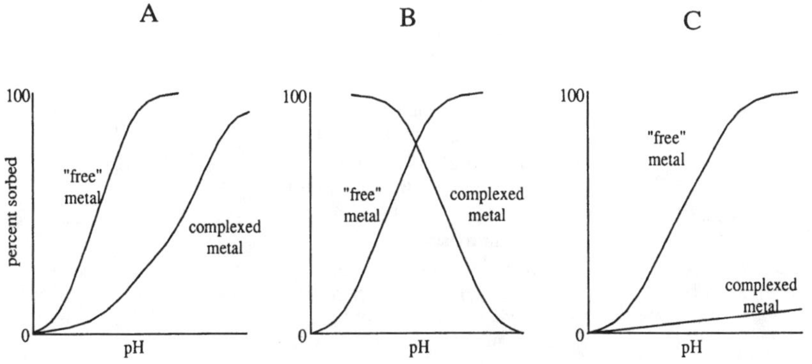

$SOH + Me^{2+} = SOMe^+ + H^+$ $\quad$ $SOH + Me^{2+} = SOMe^+ + H^+$ $\quad$ $SOH + Me^{2+} = SOMe^+ + H^+$

$SOH + Me^{2+} + L^- = SOMeL + H^+$ $\quad$ $SOH + Me^{2+} + L^{3-} = SLMe + OH^+$ $\quad$ $Me^{2+} + L^{2-} = MeL$

(nonsorbing complex)

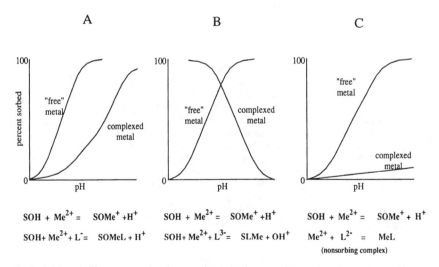

Fig. 3–4. Ligand effects on metal cation sorption behavior to surface hydroxyl groups on soil oxide minerals.

sorb to surface hydroxyl sites in a fashion typical of anions (Fig. 3–4b), with sorption decreasing with increasing pH rather than increasing as do cations. Alternatively, soluble complexes may form that have little or no affinity for surface hydroxyl sites (Fig. 3–4c). Similarly, sorption may enhanced, unaffected or diminished when the fixed-charge surface sites are involved. Because fixed-charge sites are located in the interlayer of expanding 2:1 layer clays, larger organic-complexed metal ions may not be able to penetrate to these sites due to stearic constraints. Sorption behavior, in this case, can be more difficult to anticipate or predict under these circumstances.

Solid forms of organic humic material may also extensively sorb metal ions. When soil humus contains large quantities of carboxylic, phenolic, amines, carbonyls, and sulfhydryl substituents, significant quantities of trace metals may be retained through inner- and outer-sphere complexation (Senesi, 1992). As with sorption to metal oxides and clay minerals, the relatively mobility of humus-bound metals depends on the strength and type of the complex formed.

Table 3–12. Impact of soil properties on trace element mobility.

Soil property	Basis for impact	Impact on mobility
Dissolved inorganic ligands	Increasing trace metal solubility	Increase
Dissolved organic ligands	Increasing trace metal solubility	Increase
Low pH	Decreasing sorption for cations to oxides of Fe, Al, and Mn	Increase
	Increasing sorption for anions to oxides of Fe, Al, and Mn	Decrease
	Increasing precipitation of oxyanions	Decrease
High pH	Increasing precipitation of cations as metal carbonates and hydroxides;	Decrease
	Increasing sorption for cations to oxides of Fe, Al, and Mn	Increase
	Increasing complexing of some trace cations by dissolved ligands	Increase
	Increasing sorption of cations by solid humus material	Decrease
	Decreasing sorption of anions	Increase
High clay content	Increasing ion-exchange for trace cations (at all pH)	Decrease
High humus content	Increasing complexation for most trace cations	Decrease
Presence of Fe, Al, or Mn oxides or oxide coatings	Increasing sorption of trace metal cations with increasing pH	Decrease
	Increasing sorption of trace anions with decreasing pH	Decrease
Redox	Decreasing solubility at low pe as metal sulfides when S present	Decrease
	Increasing solubility of divalent metal oxides, hydroxides compared with trivalent	Increase
	Decreasing solution complexation with lower oxidation state	Increase
	Decreasing sorption with lower oxidation state	Increase

3–5 RELATIVE MOBILITY, AVAILABILITY, AND TOXICITY OF METAL IONS

As the previous section has indicated, metal ion speciation depends strongly on the composition of the soil system. Soil conditions that promote precipitation or sorption tend to reduce metal ion mobility while those that promote dissolution or desorption enhance mobility (Table 3–12). Metals which tend to be the most mobile and bioavailable are those that form weak outer-sphere associations with solid soil phases (mineral or organic) or complex with ligands in solution and are not sorbed; metals that form inner-sphere complexes with soil solids are much less likely to desorb and are, hence, not expected to be generally mobile or bioavailable (Table 3–12). Because metals which form strong inner-sphere surface complexes with surface hydroxyl sites are often also the ones that form strong soluble complexes with organic functional groups, it is possible for mobility and bioavailability to be higher than expected based on sorption affinity, especially when a high concentration of soluble organic matter is present.

While relative mobility and bioavailability are directly related in soil systems, a less direct relationship exists between mobility and toxicity. For example, the most toxic form of a given trace element (Table 3–13) may not be very mobile in soil systems (Table 3–14). This is particularly true for toxic cations like Ag^+, Cu^{2+}, Pb^{2+}, or Cr^{3+} that form strong inner-sphere complexes with soil minerals. While the tendency to form strong complexes is what makes metal cations highly toxic (Table 3–15), allowing them to disrupt normal metabolic functions, it also may cause them to be less available in soil systems where reactions with minerals are prevalent. In contrast, anionic species that are relatively more mobile

Table 3–13. Speciation and toxicity of important trace elements.†

Element	Most toxic form	Phytotoxicity	Mammalian toxicity
Ag	Ag^+	H	H
As	AsO_4^{3-}, $As(CH_3)_3(g)$	M	L
B	$B(OH)_3$	M	L
Ba	Ba^{2+}	L	H
Be	Be^{2+}	MH	H
Cd	Cd^{2+}	MH	H
Co	Co^{2+}	MH	M
Cr	CrO_4^{2-} (Cr^{3+})	MH	H (M)
Cu	Cu^{2+}	MH	M
Hg	Hg^{2+}, $(CH_3)_2Hg$ (g)	H	H
Mo	MoO_4^{2-}	M	M
Ni	Ni^{2+}	MH	M
Pb	Pb^{2+}	M	H
Sb	$Sb(OH)_6^-$?	M	H
Se	SeO_4^{2-}	MH	H
Tl	$Tl(OH)_3^0$?	MH	H
V	$VO_2(OH)_2^-$?	H	H
Zn	Zn^{2+}	LM	LM

† Letters stand for relative toxicity: low (L), medium (M), and high (H). Adapted from McBride (1994).

Table 3–14. Relative mobility of selected trace elements.†

Element	Basis for relative mobility	Mobility
Ag(I)	Cation sorbs strongly to metal oxides, clays, humus;	L
	forms insoluble sulfides	M
As(III)	Oxyanion sorbs more weakly than arsenate	
	to metal oxides and only at higher pH	M
As(V)	Oxyanion sorbs strongly to metal oxides;	L
	forms relatively insoluble precipitates with Fe	
B(III)	Borate anion is not a strong sorber or complexer	H
Ba(II)	Cation is insoluble as sulfate or carbonate;	L
	can be fixed by exchange in clays	L
Be(II)	Cation sorbs strongly on metal oxides, clays;	L
	forms insoluble hydroxides	
Cd(II)	Cation sorbs moderately to metal oxides, clays;	M
	forms insoluble carbonate and sulfide precipitates;	L
Co(II)	Cation sorbs strongly to metal oxides and clays;	L
	forms insoluble sulfides at high pH	L
Cr(III)	Cation sorbs strongly to metal oxides, clays;	L
	forms insoluble metal oxide precipitates	L
Cr(VI)	Oxyanion sorbs moderately to metal oxides at low pH;	M
	weaker sorption at high pH	H
Cu(II)	Cation sorbs strongly to humus, metal oxides, clays;	L
	forms insoluble metal oxides, sulfides;	L
	forms soluble complexes at high pH	M
Cs(I)	Cation sorbs weakly to metal oxides, clays;	H
	fixation by vermiculites and micacious clays	L
Hg(II)	Cation sorbs moderately to oxides, clays at high pH;	L
	relatively high hydroxide solubility; forms volatile organics	M
Mn(II)	Highly dependent on soil conditions of pH, pe	L-H
Mo(V)	Oxyanion sorbs moderately to metal oxides at low pH;	M
	soluble and weak sorber at other pH values	H
Ni(II)	Cation similar to Cu(II) above	L
Np(IV)	Forms very insoluble oxides at low to neutral pH;	L
	forms soluble hydroxide anion at high pH	M
Np(V)	Cation sorbs strongly to metal oxides, clays;	L
	forms soluble carbonate complexes	M
Pu(III)	Cation sorbs strongly to metal oxides, clays;	L
	oxidizes to insoluble Pu(IV) oxides	L
Pu(IV)	Cation sorbs strongly to metal oxides, clays;	L
	forms very insoluble oxides;	L
	forms soluble hydroxide and carbonate complexes	M
Pu(V)	Reduces to form insoluble Pu(IV) oxides;	L
	cation sorbs strongly to metal oxides, clays;	L
	forms soluble organic complexes	M
Pu(VI)	Reduces to form insoluble Pu(IV) oxides;	L
	forms soluble organic complexes	M
Pb(II)	Similar to Cu(II) above	L
Sb(III)	Similar to As(III)	M
Sb(V)	Similar to As(V)	M
Se(IV)	Oxyanion sorbs weakly to metal oxides	H
Se(VI)	Oxyanion sorbs strongly to metal oxides;	L
	forms insoluble precipitates with Fe	L
Sr(II)	Sorbs moderately to metal oxides and clays;	M
	forms insoluble carbonates	L
Tl(I)	Cation sorbs weakly to metal oxides and clays;	H
	fixation by clays?; forms insoluble sulfides	L
Tl(III)	Forms insoluble metal oxides at all pH	L

(continued on next page)

Table 3–14. Continued.

Element	Basis for relative mobility	Mobility
U(IV)	Forms insoluble oxides at all pH	L
U(VI)	Cation sorbs strongly to metal oxides, clays;	L
	forms soluble carbonate, fluoride complexes	M
V(IV)	Cation sorbs strongly to metal oxides and clays;	L
	cation complexes strongly with humus;	L
V(V)	Oxyanion does not sorb strongly at neutral or high pH;	H
	can sorb moderately at lower pH; reduced to V(IV) at low pH	M
Zn(II)	Cation sorbs strongly to metal oxides, clays; forms insoluble sulfides;	L
	at low pH, sorbs weakly; at high pH forms soluble complexes	H

† Letters stand for relative mobility: L, low; M, medium; H, and high.

in soils may be of more concern. For example, when Se is in the form of selenate, SeO_4^{2-}, it tends to be both mobile and toxic (Tables 3–13 and 3–14). Although these type of generalizations may apply under normal soil solution conditions, changes locally or temporally that enhance metal solubility can cause normally nonbioavailable elements to become more so and make them a greater threat to ecosystem health. Such changes may include presence of soluble complexing agents or pH extremes. Water solubility, however, is not necessarily the only important component to consider in assessing bioavailability. For example, when metal species can be microbially converted to organic-soluble forms, such as methylated forms of Hg and As, their health threat may be increased dramatically due to the enhanced uptake of these forms directly into the cell membranes of organisms.

Even when potentially toxic metals are bioavailable, toxicity may not result, particularly for organisms that have adapted the ability to accumulate metals without deleterious effects on growth. Both plants and microorganisms have been found that can minimize the impact of the potentially harmful effects of metals (Ross, 1994). Since the visual warning signs that an unhealthy plant would give are absent in plants that have acquired this ability, the potential for bioaccumulation and harm to higher trophic levels is great. Furthermore, the levels of metal available in the soil system and amount of metal taken up by an organism may be unrelated. For example, plants have been categorized as: (i) accumulators, which take up metals effectively regardless of whether low or high concentrations are present, (ii) indicators, which tend to take up metal quantities in amounts proportional to the amount available, and (iii) excluders, which tend to completely exclude metals until a critical solution concentration is reached at which point unrestricted uptake results. Because of these qualities, assessing the

Table 3–15. Representative metal toxicity sequences.†

Organisms	Toxicity sequence‡
Algae	Hg>Cu>Cd>Cr>Zn>Co
Flowering plants	Hg>Pb>Cu>Cd>Cr>Ni>Zn
Fungi	Ag>Hg>Cu>Cd>Cr>Ni>Pb>Co>Zn
Phytoplankton	Hg>Cu>Cd>Zn>Pb

† Adapted from Sposito (1989), Nieboer and Richardson (1980), and Eichenberger (1986).

‡ All metals are assumed in oxidation state of (II) except Ag(I) and Cr(III); highest toxicity based on the smallest concentration required to produce a toxic effect.

metal burden in the plant tissue at a given time may not be indicative of the level of contamination or potential harm of metals to other species in an ecosystem.

3–6 DETERMINATION OF METAL ION SPECIATION

As has been pointed out above, in order to assess the potential threat of metals to ecosystem health, metal speciation must be determined. In complex soil systems, this has not been an easy task. A common approach for assessing the potential availability of metals in soil has been to conduct a series of sequential extractions using phase-specific or increasingly stronger extractants with each step (Ross, 1994). In this manner, either the specificity or relative strength of the extraction conditions have been used as an indicator of speciation and bioavailability. The interpretation of sequential extraction data, however, is often fraught with difficulty. For example, during the course of sequential extractions, the pH, pe, mineral solubility, and organic matter solubility may be changed, resulting in speciation information being lost in the attempt to gain it. In addition, depending on extraction times, extractant and solid and their concentrations, and other experimental protocols that are followed, results may differ from one study to the next. These issues suggest that more direct means are needed for determining the chemical form of a metal ion in soils, a method in which speciation is not altered in the attempt to assess it. Only by observing metal speciation directly and over time is it possible to discern the specific chemical forms that are mobile, bioavailable, transported across cell membranes, and toxic. This section details the current spectroscopic methods that are available to evaluate speciation directly for metals in solution, solids, and sorbed at mineral-water interfaces.

3–6.1 Solution Speciation

Assessment of the nature of solution species can occur at a number of different levels including (i) the identification of the element, (ii) the physical state, (iii) the oxidation state, (iv) the empirical formula, and (v) the detailed molecular structure. The methods listed in Table 3–16 can provide some or all this information. The analytical sensitivities of these methods range from 10^{-10} to 10^{-1} mmol $^{-1}$ (Brown, 1995). Not listed, but particularly promising, coupled separation-spectroscopic techniques such as ion-chromatography-plasma emission spectroscopy (IC-PES) and capillary electrophoresis-plasma emission-mass spectroscopy (CE-PE-MS). These methods depend on the separation of individual, chemically distinct ionic species for each element of interest. These are then each detected with plasma-emission spectroscopy or plasma-emission mass spectroscopy. A recent study on the feasibility of CE-PE-MS found good agreement between this method, ion selective electrodes, and published stability constants (Olesik, 1995). Clearly, separation-spectroscopic methods will find greater use as their application in studies of aqueous ion speciation becomes better known and more routine. The invasive nature of the methods, however, makes it likely that they will only be useful if the time scale of the separation kinetics is short relative to the time required for solution equilibria to be reestablished. The most robust methods listed in Table 3–16 are ESR and XAS. These methods are capa-

Table 3–16. General methods for solution speciation.

Method	Lower limit	Application†
	M	
Chemoluminescence	$>10^{-10}$	1,3
Polarography	$>10^{-9}$	1,3,4
Radiochemical trace analysis	10^{-8} to 10^{-1}	1,2,3
Electron spin resonance (ESR)	10^{-5} to 10^{-12}	1,3,4,5
Laser induced fluorescence	10^{-5} to 10^{-9}	1, 3,4
Electronic spectroscopy (UV-VIS-NIR)	10^{-3} to 10^{-6}	1,3,4
Nuclear magnetic resonance (NMR)	10^{-1} to 10^{-4}	3,4,5
Vibrational spectrsocopy (Raman, FTIR)	10^{-1} to 10^{-3}	3,4,5
X-ray absorption spectroscopy (XANES, EXAFS)	10^{-1} to 10^{-4}	1,3,4,5

† 1, elemental identification; 2, physical state; 3, oxidation state; 4, empirical formula; and 5, molecular structure.

ble of providing detailed information on element identification, oxidation, empirical formula, and structure information in situ without the need for separation. At present, ESR is more sensitive than XAS techniques; however, it is limited to a small number of elements. XAS can be used on virtually any element in the periodic table, but its sensitivity is presently limited by the configuration of currently existing first and second generation light sources. With the advent of new third generation synchrotrons with enhanced brightness and flux, and anticipated improvements in x-ray optics and detectors, the sensitivity of XAS is expected to increase by 1 to 3 log units.

The utility of XAS-based methods was recently illustrated in a particularly elegant study of Se speciation in soils from the Kesterson Reservoir region of California. This is a natural wetland area in central California that was contaminated with Se-bearing agricultural drainage waters. Figure 3–5 contains a conceptual model of the chemical speciation of Se at the Kesterson Reservoir (T. Tokunaga, 1997, unpublished data). Most of the Se entered the wetland as SeO_4^{2-}, later to be reduced to lower valence species. These redox transformations are of great interest because the SeO_3^{2-} ion is much more strongly sorbed to oxide surfaces than SeO_4^{2-}. Conversion to Se^{o} produces an insoluble species while the selenide species can undergo vapor-phase transport (Neal, 1995). Tokunaga and coworkers (1997, unpublished data) have made in situ measurements with XAS of the distribution and oxidation state of Se in experimental soil columns. In these studies, uncontaminated soil from the Kesterson region was placed in small plastic columns. These were then ponded with solutions containing 240 gm^{-3} of Se with 98% as SeO_4^{2-} and 2% as SeO_3^{2-} (Pikering et al., 1995). The soil microcosms were placed in a XAS microprobe and the *k*-alpha fluorescence of Se was recorded. Rastering of the x-ray beam down the column provided a measure of the Se concentration with depth in the column reactor. Se residing in the water column overlying the soil remained as selenate Se(VI) regardless of time (Fig. 3–6). In contrast, at depth of 1.0 mm into the sediment, partial conversion of selenate to elemental Se(0) was evident (Fig. 3–7) and at a depth of 5.5 mm, conversion to Se (0) was nearly complete (Fig. 3–8) after 50 h. For data shown (Fig. 3–6 to 3–8), the speciation of Se was ascertained by comparing model compound spectra with the data collected. For the pounded water, a model spectrum of selenate

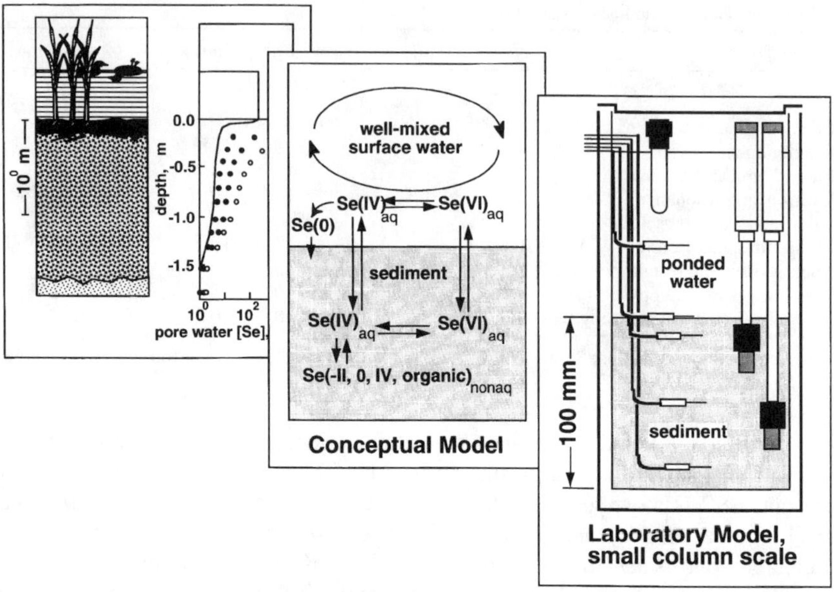

Fig. 3–5. Schematic representation of natural and laboratory setup of Se transport and speciation in a wetlands.

fit the ponded water sample data well. For the sediment samples, a mixture of selenate and elemental selenium model spectra was necessary to fit the sample data.

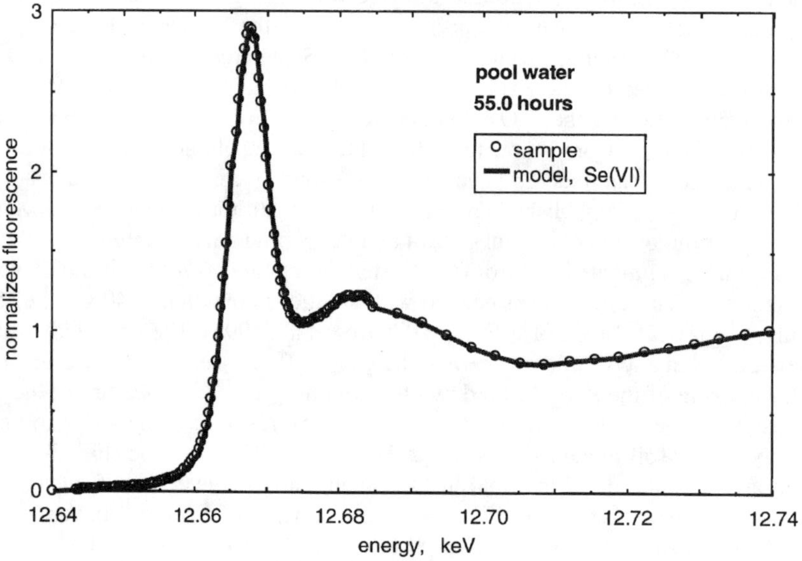

Fig. 3–6. Normalized fluorescence spectrum for pool water over sediment sample after 55 h compared with the model spectrum of Se(VI) as selenate.

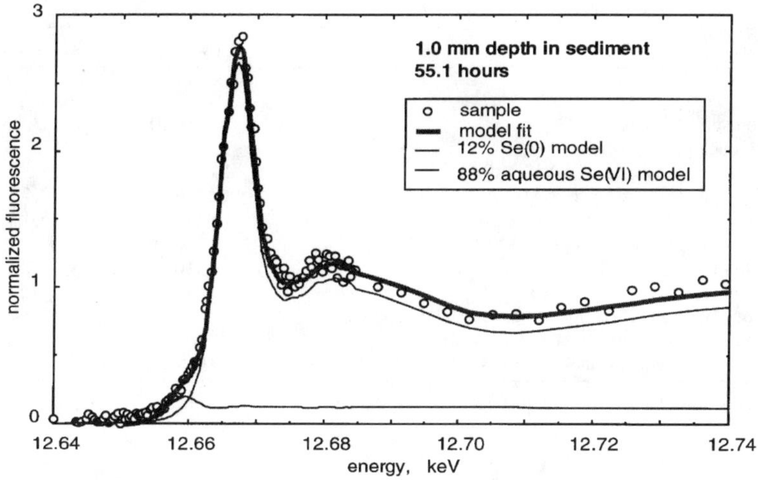

Fig. 3–7. Normalized fluorescence spectrum for Se sediment sample at a depth of 1.0 mm from sediment-pool water interface after 55.1 h compared with the model spectra of a mixture of selenate [Se(VI)] and elemental selenium [Se(0)].

3–6.2 Solid Phase Metal Ion Speciation

Identifying the specific solid responsible for the aqueous solubility and the chemical and biological lability of a given metal can be an arduous task. Identifying the presence of a specific solid in a soil or sedimentary matrix does not completely meet this need since it is possible that small quantities of a par-

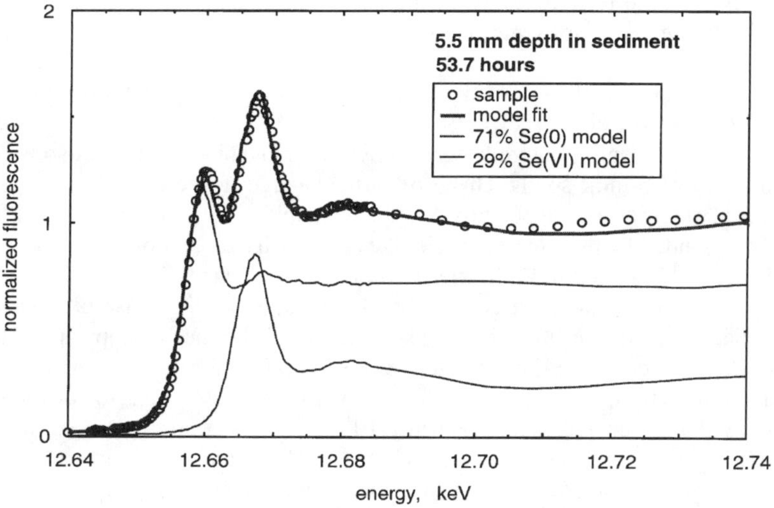

Fig. 3–8. Normalized fluorescence spectrum for Se sediment sample at a depth of 5.5 mm from sediment–pool water interface after 53.7 h compared with the model spectra of a mixture of selenate [Se(VI)] and elemental selenium [Se(0)].

Table 3–17. General methods for solid speciation.

Method	Lower limit	Application†
	M	
Electron spin resonance (ESR)	10^{-5} to 10^{-12}	1,3,4,5
Laser induced fluorescence	10^{-5} to 10^{-9}	1, 3,4
Electronic spectroscopy (UV-VIS-NIR)	10^{-3} to 10^{-6}	1,3,4
Nuclear magnetic resonance (NMR)	10^{-1} to 10^{-4}	3,4,5
Vibrational spectrsocopy (Raman, FTIR)	10^{-1} to 10^{-3}	3,4,5
X-ray absorption spectroscopy (XANES, EXAFS)	10^{-1} to 10^{-4}	1,3,4,5
X-ray fluorescence (XRF)	$<10^{-3}$	1,3
X-ray fluorescence (microprobe)	$<10^{-8}$	1,3
X-ray and neutron diffraction (single crystal and powder)	10^{0}	5
Electron diffraction	--	1
Energy dispersive x-ray analysis (EDX)	<1000 mg kg^{-1}	1

† 1, elemental identification; 2, physical state; 3, oxidation state; 4, empirical formula; 5, molecular structure.

ticular solid, present as a surface coating or an inclusion with another mineral may be responsible for the aqueous-phase chemistry, yet this solid may be present in quantities insufficient to detect by direct physical methods such as XAS, x-ray diffraction, electron microscopy or vibrational spectroscopy. Nevertheless, it is apparent that specific solid phase forms of a given element can control its chemical and biological reactivity and thus its impact on ecosystem health.

Table 3–17 lists a number of methods suitable for the identification of solid-bearing metal phases in soils. The advantages and limitations of these methods are discussed by Brown (1990). Many of these methods can provide quite detailed information in pure systems, but their applications to soils can be problematic. The greatest limitation results from the typically low sensitivity of the majority of these techniques. With the exception of extreme cases, most toxic metals are present in soils and sediments at trace concentrations, rendering most of the methods listed in Table 3–17 unsuitable for applications to real systems. Nevertheless, many of these methods have been used to study metal speciation in soils. Cotter-Howells et al. (1994) used EXAFS and analytical electron microscopy to identify naturally formed pyromorphites in separates of soils contaminated by Pb-mine spoils. This is of particular significance in light of the geochemical stability and apparently low bioavailability of this form of Pb. This study was aided by the use of particle size and density separations to obtain mineral fractions enriched in Pb (Cotter-Howells et al., 1994).

Additional gains in sensitivity can be obtained with the use of advanced analytical microscopies. Synchrotron x-ray fluorescence spectroscopy and micro-XANES have been used to determine the solid-phase forms of U in soil from Fernald, OH (Bertsch et al., 1994). The former method is similar to conventional x-ray fluorescence, but it is typically 10 to 100 times more sensitive. Micro-XANES facilitated in situ distinctions between U(IV) and U(VI), in soils contaminated by U milling activities. U(VI) forms an number of stable, soluble solution species and is more susceptible to solution phase transport. Fortunately, the micro-XANES studies have indicated that much of the soil at this site is present as the less soluble U(IV).

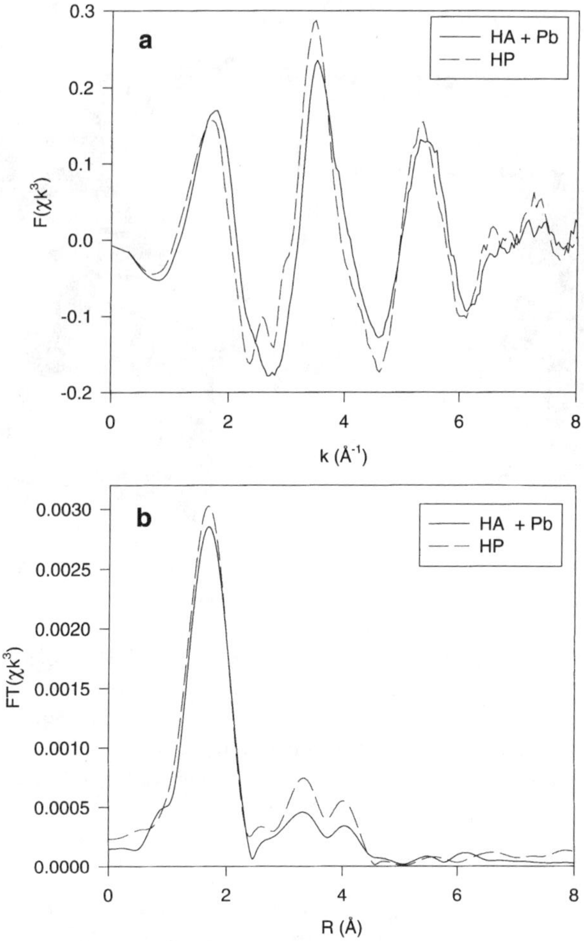

Fig. 3–9. Extended x-ray absorption fine structure (EXAFS) radial structure functions of pyromorphite ($Pb_{10}(PO_4)_6(OH)_2$) and Pb-reacted apatite (S.J. Traina, 1995, unpublished data).

Ruby et al. (1994) used conventional analytical electron microscopy to study the solid-phase speciation of Pb in a contaminated soil. They identified naturally formed pyromophites and other Pb-bearing solid-phases with EDAX.

In most cases, XAS can detect the solid phase species of a given element at concentrations below the threshold detection limits of XRD. Figure 3–9 shows radial structure functions (RSF) of Pb-EXAFS spectra collected from a pure synthetic pyromorphite, and a hydroxyapatite reacted with a solution of $PbNO_3$. The reaction product is not detectable by XRD but comparisons of the RSFs suggest that the Pb-reacted apatites do indeed contain Pb in a coordination environment similar to that of pyromorphite.

The local coordination chemistry of a given element can strongly influence its chemical lability. Myneni (1995) examined the solid-phase and surface chem-

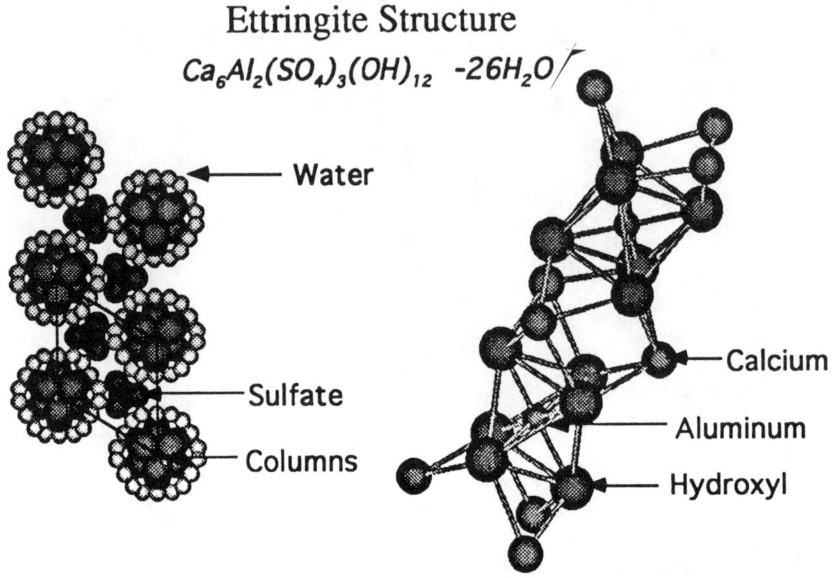

Fig. 3–10. The molecular structure and coordination environment of oxyanions in ettringite.

istry of AsO_4^{3-} in the mineral ettringite ($Ca_6Al_2(SO_4)_3(OH)_{12}$-$26H_2O$). This solid forms in sulfate-rich, alkaline environments such as flue-gas desulfurization wastes from coal-burning power plants. The structure of ettringite is comprised of Al-hydroxide octahedral linked to square antiprisms of Ca-hydroxide polyhedra. These columnar units are surrounded by water molecules, and the columns

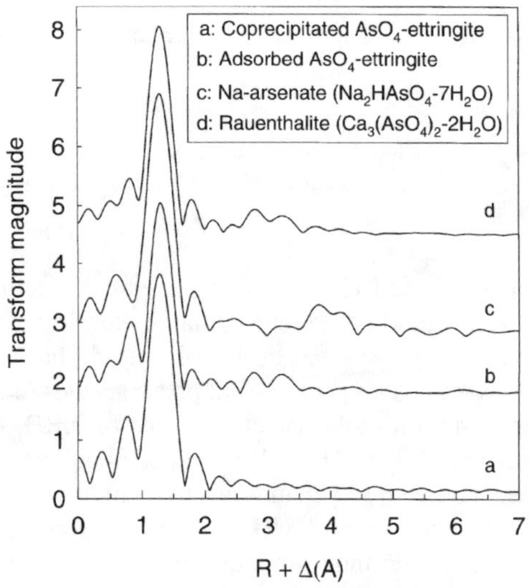

Fig. 3–11. Extended x-ray absorption fine structure(EXAFS) radial structure functions for arsenate adsorbed on or coprecipitated in ettringite, and in two model solid phase arsenates (S.J. Traina, 1995, unpublished data).

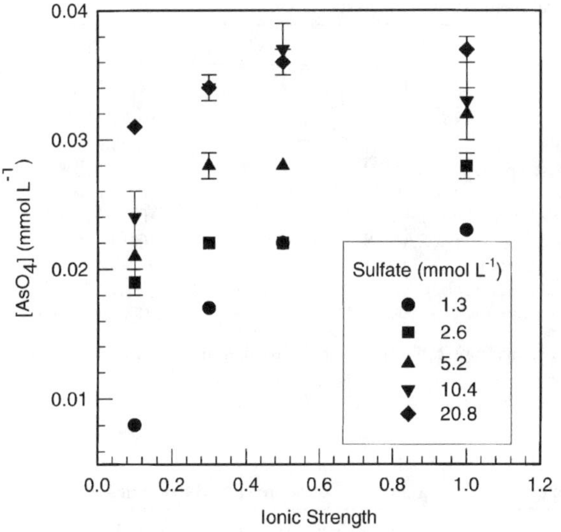

Fig. 3–12. The ionic strength and sulfate-dependent desorption of arsenate from coprecipitated ettringite (S.J. Traina, 1995, unpublished data).

are held together by sulfate tetrahedra, in outer-sphere complexes (Fig. 3–10). Adsorption of arsenate to this solid results in the formation of an innersphere complexes with Ca. This results in the presence of two back-scattering peaks in the RSF of As EXAFS and is similar to the RSF of As in the model Ca-arsenate, rauenthalite ($Ca_3(AsO_4)_2$-$2H_2O$; Fig. 3–11). Arsenate in this coordination can not be extracted from ettringite when it is reacted with concentrated sulfate solutions. In contrast, if arsenate is incorporated into the solid during coprecipitation it resides in outer-sphere complexes, as evidenced by the EXAFS spectra. Thus, AsO_4^{3-} that is coprecipitated into ettringite can be displaced by sulfate ions, and the amount of AsO_4^{3-} released is proportional both to the total sulfate concentration and the solution ionic strength (Fig. 3–12).

3–6.3 Speciation at the Mineral–Water Interface

A variety of spectroscopic techniques are available to study the structure of metal ion surface complexes in situ at mineral-water interfaces. These techniques include Raman and Fourier transform infrared spectroscopy (FTIR), magnetic resonance spectroscopies, including electron spin or paramagnetic resonance (ESR or EPR), electron-nuclear double resonance (ENDOR), electron spin-echo envelope modulation (ESEEM) and nuclear magnetic resonance (NMR), Mossbauer spectroscopy (MOSS), and x-ray absorption spectroscopy (XAS). A review that details the advantages and limitations of each of these methods can be found in Brown (1990). Table 3–18 summarizes the type of information on the chemical form of an element that can be obtained from the various methods and the concentration ranges of application. Based on the in situ spectroscopic studies of ion sorption at mineral-water interfaces that have been conducted to date,

Table 3–18. General methods for surface speciation.

Method	Lower limit	Application†
	M	
Electron spin resonance (ESR)	10^{-5} to 10^{-12}	1,3,4,5
Laser induced fluorescence	10^{-5} to 10^{-9}	1, 3,4
Electronic spectroscopy (UV-VIS-NIR)	10^{-3} to 10^{-6}	1,3,4
Nuclear magnetic resonance (NMR)	10^{-1} to 10^{-4}	3,4,5
Vibrational spectrsocopy (Raman, FTIR)	10^{-1} to 10^{-3}	3,4,5
X-ray absorption spectroscopy (XANES, EXAFS)	10^{-1} to 10^{-4}	1,3,4,5
X-ray fluorescence (XRF)	$<10^{-3}$	1,3
X-ray fluorescence (microprobe)	$<10^{-8}$	1,3
Energy dispersive x-ray analysis (EDX)	<1000 mg kg^{-1}	1

† 1, elemental identification; 2, physical state; 3, oxidation state; 4, empirical formula; 5, molecular structure.

Table 3–19. Extended x-ray absorption fine structure (EXAFS) studies of metals on minerals.

System	Type of complex	Reference
Se(IV)/α-FeOOH	Inner-sphere	Hayes et al., 1987
Se(VI)/α-FeOOH	Outer-sphere	Hayes et al., 1987
Pb(II)/γ-Al_2O_3	Inner-sphere, monodentate multinuclear	Chisholm-Brause et al., 1990a
Co(II)/γ-Al_2O_3 and TiO_2 (rutile)	Inner-sphere multinuclear	Chisholm-Brause et al., 1990b
Pb(II)/α-FeOOH	Inner-sphere multinuclear	Roe et al., 1990
Cr(III)/α-FeOOH and hydrous ferric oxide	Inner-sphere bidentate, multinuclear solid solution	Manceau & Charlet, 1992
Ur(VI)/silica and montmorillonite	Inner-sphere bidentate mono/multinuclear	Dent et al., 1992
Ur(VI)/vermiculite	Outer-sphere	Hudson et al., 1994
Co(II)/$CaCO_3$(calcite)	Solid solution	Xu, 1993
Np(V)/α-FeOOH	Inner-sphere mononuclear	Combes et al., 1992
Cd(II)/aluminas	Inner-sphere surface precipitation	Papelis et al., 1995
As(VI)/α-FeOOH and assorted Fe oxides	Inner-sphere mono-,bidentate	Waychunas et al., 1993b
Cr(III)/silica	Inner-sphere monodentate, multinuclear	Fendorf et al., 1994
Co(II)/montmorillonite	Inner-sphere, multinuclear outer-sphere in interlayer	Papelis & Hayes, 1996
Sr(II)/montmorillonite and hectorite	Outer-sphere monodentate	Papelis et al., 1994
Co(II)/kaolinite	Inner-sphere, bidentate mono-, multinuclear	O'Day et al., 1994a,b
Zn(II)/ferrihydrate	Inner-sphere, bidentate surface precipitation	Wachunas et al., 1993a
As(V)/ettringite	Inner-sphere, bidentate outer-sphere, monodentate	Myneni et al., 1994
Cd/ferric oxides	Inner-sphere, monodentate	Spadini ct al., 1994

a clearer picture of the types of reaction processes and surface complexes that may result during sorption is emerging (Hayes & Katz, 1996). Because the most definitive in situ determination of surface structure and composition of sorbed species have come from the results of x-ray absorption spectroscopy, and due to the fact that new third generation synchrotron sources that will provide enhanced sensitivity and spatial resolution are nearly complete, emphasis on these studies is given here.

XAS provides unique information about the oxidation state and the chemical environment surrounding a central metal atom. A metals oxidation state can be gleaned through the analysis of the x-ray absorption near-edge structure (XANES); coordination number and bonding distances of neighbor atoms in the vicinity of a central metal atom come from the detailed analysis of the x-ray absorption fine structure (XAFS). To date, XAS has been used to determine the oxidation state, location and coordination environment for a number of metal ions sorbed to various oxide minerals and clays (Table 3–19).

Because XAFS data analysis allows the coordination number and bonding distances for different coordination shells surrounding a sorbing metal ion out to about 6Å to be determined, it is possible to distinguish among inner-sphere or outer-sphere surface complexes, surface polymers and precipitates, and solid solution formation. For example, for outer-sphere coordination, the sorbing cation remains completely hydrated, no surface metal atoms enter into the near coordination environment, and the XAS spectrum looks identical to the aqueous phase metal ion spectrum. When inner-sphere complexes form, however, water is displaced from the inner coordination shell of the sorbing cation, bringing surface metal atoms from an oxide sorbent to within about 3 or 4 Å of the sorbing cation. The presence of these surface metal ions leads to the appearance of spectral features in the XAFS spectrum known as beat patterns. Likewise, it is possible to distinguish mononuclear surface species from multinuclear species based on the presence or absence of additional metal atoms in the near coordination environment surrounding the sorbing cations. When surface polymers or precipitates form, bridging hydroxyls connect adjacent sorbing metal cations reducing the distance between them. The appearance of spectral beat patterns, indicative of additional metal atoms within 6Å of a sorbing metal ion, distinguishes multinuclear species from mononuclear ones. Hence, it is possible to distinguish among the variety of near coordination environments for metal ions sorbed at mineral-water interfaces by XAS.

To illustrate the utility of XAS in assessing the impact of sorbed structure on trace metal ion mobility in soil environments, two examples will be given here. In the first, the effects of solution pH and competing ions on the sorption of Co(II) to montmorillonite is described. This examples illustrates how changes in the controlling sorption process can dramatically change the mobility of Co(II) as a function of changing solution conditions. In the second example, Cs(I)/Ca(II) binary exchange to an illite and the effects of exchange extent of Cs(I), a potentially highly mobile radionuclide at waste repositories, on its mobility is described.

Metal cations are thought to sorb to smectite clays either at permanent-charge sites on the interlayer basal planes or at surface hydroxyl sites at the clay

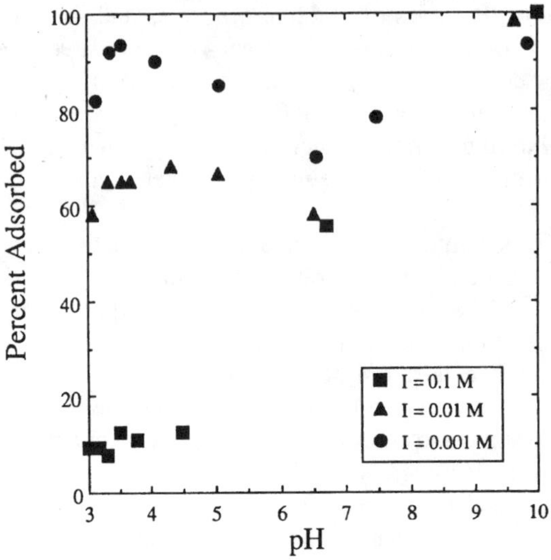

Fig. 3–13. Fractional uptake of 1×10^{-4} *M* Co(II) by 0.5 g L^{-1} montmorillonite as a function of pH and Na ion concentration (Papelis & Hayes, 1996).

edge where the crystal structure is interrupted. Although most of the surface area and CEC in smectites is found in the interlayer region, recent studies suggest that the less-abundant sites on the external surfaces may be even more important, depending on solution conditions and the type of metal cation (Zachara et al., 1992). This multiple site nature of smectites is well established, based on macroscopic sorption experiments, but in the absence of spectroscopic information, it has not been possible to fully delineate the impact of each site type on metal ion mobility. In a series of experiments in which pH, background electrolyte and the relative concentration of surface sites to metal ion were varied, using XAS, Papelis and Hayes (1996) identified the surface species and conditions that may increase divalent cation mobility in high CEC soils.

As shown in Fig. 3– 13, in the 1.0×10^{-4} *M* Co(II) system, at higher Na ion concentration (0.1 *M*), Co(II) sorption is substantially depressed at lower pH values. At low Na ion concentrations (0.001 or 0.01 *M*), Co(II) sorption is largely pH independent. These data suggest that Co(II) is selectively desorbed from permanent-charge sites at low pH and high Na concentration, but remains sorbed at higher pH regardless of the Na concentration. The pH-dependence of Co(II) sorption at 0.1 *M* Na ion concentration is characteristic of nearly exclusive sorption of Co on surface hydroxyl sites.

XAS data have been used to confirm the basis for these changes (Fig. 3–14). At the lowest Na concentration (0.001 *M*) and pH 7, only one peak is evident in the Fourier transforms, indicating a Co atom coordinated by six oxygens in an outer-sphere complex with permanent-charge sites. At pH 10, two additional peaks in the Fourier transforms at approximately 2.75 and 5.75 Å (uncorrected for phase shift) appear, corresponding to Co–Co shells and the formation of polynuclear species or a surface precipitate. At higher Na concentrations (0.1

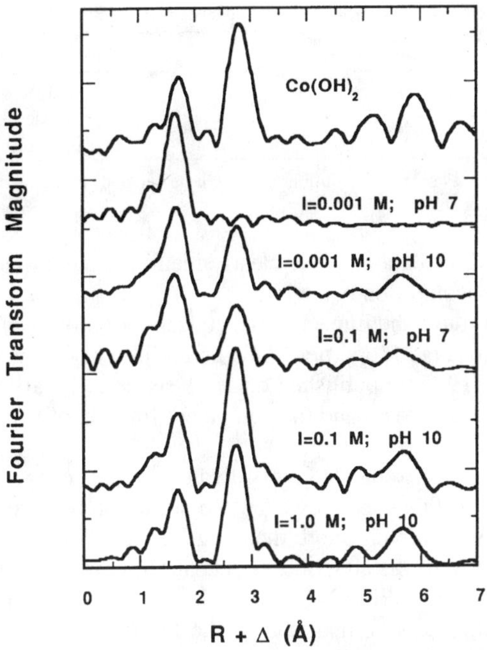

Fig. 3–14. Extended x-ray absorption fine structure (EXAFS) radial structure functions of the model compound $Co(OH)_2$ and Co(II) sorbed to montmorillonite as a function of pH and ionic strength (Papelis & Hayes, 1996).

M), Na can exclude Co(II) from permanent-charge sites at all pH values. Therefore, at high Na concentration, it was expected that Co(II) would bond predominantly to external surface hydroxyl sites. The appearance of the two peaks near 2.75 and 5.75 Å in the Fourier transforms of the 0.1 *M* Na, pH 7 sample (Fig. 3–14) confirms that multinuclear complexes with surface hydroxyl groups are forming for this sample. At still higher pH (10) and Na concentration (0.1 or 1.0 *M*), as more Co(II) sorbs, the amplitude of the second and fourth shell peaks (Co–Co) increases (Fig. 3–14), compared with the sample at pH 7. The increased amplitudes are consistent with growing Co polymers, or a more ordered surface precipitate on the surface hydroxyl sites as surface coverage increases. The lower amplitude of the second and fourth coordination shells for pH 10 and Na at 0.001 compared with 0.1 *M*, suggests a lower surface hydroxyl coverage at the lower Na concentration. Because changes in the Na concentration are not expected to affect the amount or coordination environment of Co sorbed exclusively to surface hydroxyl sites (Katz & Hayes, 1995), the explanation for the lower CN at the lower Na concentration is that some of the Co is also sorbed, as an outer-sphere complex, in the interlayer region.

In terms of the effects of solution conditions on metal ion mobility, these results have significant ramifications. For example, trace metal cations, which are initially associated with clays as exchangeable outer-sphere complexes, can be converted into a much less mobile inner-sphere surface complexes or polymers at edge sites on these clays by increasing pH or concentration of the major exchang-

Table 3–20. Binary exchange of Cs(I)/Ca(II) in illite system (P.M. Bertsch, 1995, unpublished data).

Cs equation solution fraction†	Cs equation exchanged fraction
1.0 (100%)	1.0
0.5 (50%)	0.8
0.1 (10%)	0.53

† Numbers in parentheses represent solution equivalent percentages based on a total solution equivalent concentration of 0.5 *N*.

ing cations. The XAS data analysis clearly identifies the basis for the observed pH and Na ion concentration behavior.

In a related study, the impact of the concentration of exchanging cation Ca on the relative exchange properties of Cs on an illite has been investigated with XAS by Bertsch (1995, unpublished data). A series of binary Cs–Ca exchange experiments were performed and the equivalent fraction of sorbed Cs calculated (Table 3–20).

Typically, binary exchange favors divalent cations over a monovalent ones (McBride, 1994). In this case, however, the strong preference for Cs is clearly indicated, with Cs occupying more than one-half the exchange sites even when making up only 10% of the equivalent concentration in solution. Analysis of the XAS Fourier transform (Fig. 3–15) provides a structural explanation for this behavior. For example, a comparison of the Fourier transform peak near 2 Å among the three solution conditions shows an increase in amplitude and a narrowing of the width as the fraction of Cs exchanged decreases. This indicates both a closer association with tetrahedral oxygens and a more definitive structure in the siloxane ditrigonal cavity of illite for the lower compared with the higher Cs-exchanged samples. Apparently, a distribution of Cs exchange sites exist with the more mobile and readily-exchanged Cs coming off the sites which bond Cs relatively weakly, leaving behind the more tightly bound Cs that fits snuggly into the ditrigonal cavities. A more complete analysis of the XAS data is required to completely verify this interpretation. Nonetheless, this illustrates the importance of using both equilibrium and XAS data simultaneously, showing that metal ion mobility depends strongly on solution conditions and the presence of competing cations. In this case, the combined data shows that Cs becomes less and less eas-

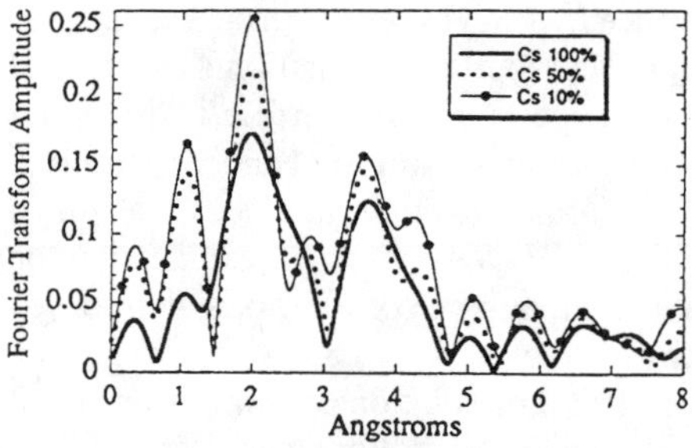

Fig. 3–15. Extended x-ray absorption fine structure (EXAFS) radial structure functions of the Cs exchanged with a Silver Hill Illite as a function of the Cs percentage of the total equivalence in the binary Cs–Ca system. (Total equivalence in the system was 0.5 N: Cs^{+} plus Ca^{2+}; P.M. Bertsch, 1995, unpublished data).

ily removed from illite as the Ca concentration increases due to the presence of a fraction of surface sites that form inner-sphere surface complexes. While the high selectivity of illites for Cs have been know for some time, XAS makes it possible to distinguish the solution conditions in which inner- or outer-sphere surface complexes predominate.

By using sorption and XAS data from model soil systems and studies such as these to create a library of data, it may be possible in the future to identify soils that have similar metal XAS characteristics by going through a computer fingerprinting exercise much like is done with more established spectroscopic methods. Likewise, XAS model studies may be used to define the soil conditions under which metal ion species are likely to be the most mobile, bioavailable, or toxic. By creating a library of XAS data for various soil solutions and system conditions, it may be possible some day to better assess the relative impact of a given metal ion on ecosystem health based on system properties. At the present, a good use of XAS would be to refine speciation extraction procedures, by identifying which species are actually extracted and determining whether the unextracted and extracted fractions are truly unaffected by the sequential treatments. With the current limitation on access to synchroton resources and the relatively high cost, this may be the best way in which to initially maximize XAS applications for assessing the impact of metal ion speciation on ecosystem properties and health.

3–7 MODELING METAL ION SPECIATION IN SOILS

A wide array of computational methods are available for modeling of metal ion speciation in soils. These models are cable of describing chemical species structure and distribution in solution, at the solid-solution interface and in the solid phase. A comprehensive discussion of computational modeling of metal ions speciation is outside the scope of this chapter. The reader is directed to recent comprehensive reviews by Loeppert et al. (1995), Dzombak and Morel (1990) and Melchior and Basset (1990). Computational methods can be attractive for their ease of use, their low cost, and their ability to generate a large number of postulated species; however, their value is limited by the applicability of the specific chemical constants, reactions and surface species invoked in their algorithms and data bases. It is incumbent on the user to judge the suitability of each model to the problem in question.

3–8 SUMMARY AND CONCLUSIONS

This chapter and the work cited herein provides a compelling case for the effects of trace metal speciation on ecotoxicology and environmental health. The fate, mobility, and impact of a given toxic metal is virtually always controlled more by its chemical form than by its total concentration. This paradigm has long been know in the area of soil fertility an plant nutrition. It has only recently come to the attention of the environmental health community.

The task then becomes the development of accurate and sensitive methods for the determination of trace metal speciation in soils and sediments. The examples cited in this chapter illustrate some of the best case situations where modern instrumental techniques have allowed a rigorous determination of discrete chemical forms. Unfortunately these are indeed best case examples that have typically benefited from excess concentrations of the metals of interest or from the construction of clean model systems. The typical situation in most contaminated soils is likely to be more problematic. The presence of two or more interfering metals, microcrystalline, and amorphous solids and commonly low contaminant concentrations (<0.1% by weight) will often render the use of direct physical methods impractical. Nevertheless, chemical speciation or lability data is still needed for rational evaluation of the severity of particular contamination events. Thus, there is still a role for indirect, chemical extraction methods; however, development and refinement of these techniques should follow the example provided by soil testing programs, namely the calibration of chemical extractions with empirical bioassays. If this data is coupled to careful studies using direct physical methods on model systems, then it is likely that our understanding of the role of speciation on trace metal toxicology and impact will have both practical and heuristic value.

REFERENCES

Adriano, D.C. (ed.), 1992. Biogeochemistry of trace metals. Lewis Publ., Ann Arbor, MI.

Allen, H.E., C.P. Huang, G.W. Bailey, and A.R. Bowers, 1994. Metal speciation and contamination of soil. Lewis Publ., Ann Arbor, MI.

Alloway, B.J., 1995. Heavy metals in soils. Blackie Academic & Professional, New York.

Baes, C.F., and R.E. Mesmer, 1986. The hydrolysis of cations. Krieger Publ. Company, Malabar, FL.

Bertsch, P.M., D.B. Hunter, S.R. Sutton, S. Bajt, and M.L. Rivers, 1994. In situ chemical speciation of uranium in soils and sediments by micro x-ray absorption spectroscopy. Environ. Sci. Technol. 28:980–984.

Brown, G.E., Jr., 1990. Spectroscopic studies of chemisorption reaction mechanisms at oxide–water interfaces. p. 309–363. *In* M.F. Hochella and A.F. White (ed.) Mineral water interface geochemistry. Mineralogical Soc. of Am., Washington, DC.

Brown. G.E. Jr. (ed.). 1995. Molecular environmental science: Speciation, reactivity, and mobility of environmental contaminants. An assessment of research opportunities and the need for synchrotron radiation facilities. *In* U.S. Department of Energy Workshop Report. 5–8 July 1995. USDOE, Airlie Center, VA.

Chisholm-Brause, C.J., P.A. O'Day, G.E. Brown, Jr., and G.A. Parks, 1990b. Evidence for multinuclear metal ion complexes at solid/solution interfacs from x-ray absorption spectroscopy. Nature (London) 348:528–530.

Chisholm-Brause, C.J., A.L. Roe, K.F. Hayes, G.E. Brown, Jr., G.A. Parks, and J.O. Leckie. 1990a. XANES and EXAFS study of Pb(II) adsorbed on oxides surfaces. Geochim. Cosmochim. Acta. 54:1897–1909.

Combes, J.M., C.J. Chisholm-Brause, G.E. Brown, Jr., G.A. Parks, S.D. Conradson, P.G. Eller, I.R. Triay, D.E. Hobart, and A. Meijer. 1992. EXAFS spectroscopic study of neptunium (V) sorption at the a-FeOOH/water interface. Environ. Sci. Technol. 26:376–342.

Cotter-Howells, J.D., P.E. Champness, J.M. Charnock, and R.A.D. Pattrickk, 1994. Identification of pyromorphite in mine-waste contaminated soils by ATEM and EXAFS. Europ. J. Soil Sci. 132:335–342.

Dent, A.J., J.D.F Ramsay, and S.W. Swanton, 1992. An EXAFS study of uranyl ion in solution and sorbed onto silica and montmorillonite clays colloids. J. Colloid Interface Sci. 150:45–60.

Doner, H. 1978. Chloride as a factor in the mobilities of Ni(II), Cu(II), and Cd(II) in soil. Soil Sci. Soc. Am. J. 42:882–885.

Dzombak, D.A., and F.M.M. Morel. 1990. Surface complexation modeling of hydrous ferric oxide. Wiley Interscience, New York.

Eichenberger, E., 1986. The interrelation between essentiality and toxicity of metals in aquatic ecosystems. Metal Ions Biol. Syst. 20:67–100.

Faure, G. 1991. Principles and applications of inorganic geochemistry. Macmillian Publ. Company, New York.

Fendorf, S.E., G.M. Gamble, M.G Stapleton, M.J. Kelley, and D.L Sparks, 1994. Mechanisms of chromium(III) sorption on silica: I. Cr(III) surface structure derived by extended x-ray absorption fine structure (EXAFS) spectroscopy. Environ. Sci. Technol. 28:284–289.

Hamelink, J.L., P.F. Landrum, H.L. Bergman, and W.H. Benson, 1994. Bioavailability: Physical, chemical, and biological interactions. Lewis Publ., Ann Arbor, MI.

Hayes, K.F., and L.E. Katz. 1996. Application of x-ray absorption spectroscopy for surface complexation modeling of metal ion sorption. p. 147–223. *In* P.V. Brady (ed.) The physics and chemistry of mineral surfaces. CRC Press, Boca Rotan, FL.

Hayes, K.F., and J.O. Leckie, 1987. Modeling ionic strength effects on cation adsorption at hydrous oxide/solution interfaces. J. Colloid Interface Sci. 115: 564–572.

Hayes, K.F., L.A. Roe, G.E. Brown, Jr., K.O. Hodgson, J.O. Leckie, and G.A. Parks, 1987. In situ x-ray absorption spectroscopy study of surface complexes: Selenium oxyanions on a-FeOOH. Science (Washington, DC) 238:783–786.

Hudson, E.A., L.J. Terminello, B.E. Viani, T. Reich, J.J. Bucher, D.K. Shuh, and N.M. Edelstein. 1994. X-ray absorption studies of uranium sorption on mineral substrates. p. 177–176. *In* K.C. Cantwell and L. Dunn (ed.) 1994 Activity Report. Stanford Synchrotron Radiation Lab., Stanford, CA.

Katz, L.E., and K.F. Hayes, 1995. Surface complexation modeling: II. strategy for modeling polymer and precipitation reactions at high surface coverage. J. Colloid Interface Sci. 170:491–501.

Laperche, V., S.J. Traina, P. Gaddam, and T.J. Logan. 1996. Chemical speciation of Pb in a contaminated soil: Effects of apatite amendment. Environ. Sci. Technol. 30:3321–3326.

Laws, E.A., 1993. Aquatic pollution: An introductory text, second edition. Wiley Interscience, New York.

Loeppert, H.R., A.P. Schwab, and S. Golberg (ed.). 1995. Chemical equilibrium and reaction models. SSSA Spec. Publ. 42. SSSA, Madison, WI.

Ma, Q.Y., T.J. Logan, and S.J. Traina. 1994a. Effects of NO_3^-, Cl^-, F^-, SO_4^{2-}, and CO_3^{2-} on Pb^{2+} immobilization by hydroxyapatite. Environ. Sci. Technol. 28:408–419.

Ma, Q.Y., T.J. Logan, and S.J. Traina. 1995. Lead immobilization from aqueous solutions and contaminated soils using phosphate rocks. Environ Sci. Technol. 29:1118–1126.

Ma, Q.Y., S.J. Traina, and T.J. Logan, 1993. In situ Pb immobilization by apatite. Environ. Sci. Technol. 27:1803–1810.

Ma, Q.Y., S.J. Traina, and T.J. Logan, 1994b. Effects of aqueous Al, Cd, Cu, Fe(II), Ni, and Zn on Pb immobilization by hydroxyapatite. Environ. Sci. Technol. 28:1219–1228.

Manceau, A.L., and L. Charlet. 1992. X-ray absorption spectroscopic study of the sorption of Cr(III) at the oxide-water interface: I. Molecular mechanisms of Cr(III) oxidation on Mn oxides. J. Colloid Interface Sci. 148:425–442.

McBride, M.B. 1994. Environmental chemistry of soils. Oxford Univ. Press, New York.

McBride, M.B., and D.R. Bouldin, 1984. Long-term reactions of copper(II) in a contaminated calcareous soil. Soil Sci. Soc. Am. J. 48:56–59.

McKenzie, R.M. 1967. The sorption of cobalt by manganese oxide minerals in soils. Aust. J. Soil Res. 5:235–246.

Melchior, D.C., and R.L. Bassett. 1990. Chemical modeling of aqueous systems II. p. 1–14. *In* D.C. Melchior and R.L. Bassett (ed.) 196th Natl. Meeting of the Am. Chemical Soc., Los Angeles. 25–30 Sept. 1990. Am. Chem. Soc., Washington, DC.

Morel, F.M.M., and J.G. Hering, 1993. Principals of aquatic chemistry. Wiley Interscience, New York.

Murray, J.W., 1975. The interaction of cobalt with hydrous manganese dioxide. Geochim. Cosmochem. Acta. 43:781–787.

Myneni, S.C.B. 1995. Oxyanion–mineral surface interactions in alkaline environments: AsO_4 and CrO_4 sorption and desorption in ettringite. Ph.D. diss. Ohio State Univ., Columbus.

Myneni, S.C.B., S.J. Traina, G.A. Wachunas, and T.J. Logan. 1994. Sorption and coprecipitation of arsenate and chromate. p. 180–183. *In* K.C. Cantwell and L. Dunn (ed.) 1994 Activity Report. Stanford Synchrotron Radiation Lab., Stanford, CA.

Neal, R.H., 1995. Selenium. p. 260–283. *In* B.J. Alloway (ed.) Heavy metals in soils. Blackie Academic & Professional, New York.

Nieboer, E., and D.H.S. Richardson, 1980. The replacement of the nondescript term "heavy metal" by a biologically and chemically significant classification of metal ions. Environ. Pollut. (Ser. B). 1:2–26.

O'Day, P.A., G.E. Brown, Jr., and G.A. Parks, 1994b. X-ray absorption spectroscopy of cobalt(II) multinuclear surface complexes and surface precipitates on kaolinite. J. Colloid Interface Sci. 165:269–289.

O'Day, P.A., G.A. Parks, and G.E. Brown, Jr., 1994a. Molecular structure and binding sites of cobalt(II) surface complexes on kaolinte from x-ray absorption spectroscopy. Clays Clay Miner. 42:337–355.

Olesik, J. 1995. Capillary electrophoresis-ICP spectrometry for rapid elemental speciation. Anal. Chem. 67:1–12.

Papelis, C., G.E. Brown, Jr., G.A. Parks, and J.O. Leckie, 1995. X-ray absorption spectroscopic studies of cadmium and selenite adsorption on aluminum oxides. Langmuir 11:2041–2048.

Papelis, C., C.C. Chen, and K.F. Hayes, 1994. XAS study of metal ion partitioning at the clay–water interface. p. 192–196. *In* K.C. Cantwell and L. Dunn (ed.) 1994 Activity Report. Stanford Synchrotron Radiation Lab., Stanford, CA.

Papelis, C., and K.F. Hayes, 1996. Distinguishing between interlayer and external sorption sites of clay minerals using x-ray absorption spectroscopy. Colloid Surf. A 107:89–96.

Pierzynski, G.M., T.J. Logan, S.J. Traina, and J.M. Bigham, 1990a. Phosphorus chemistry and mineralogy in excessively fertilized soils: Quantitative analysis of phosphorous rich particles. Soil Sci. Soc. Am. J. 54:1576–1583.

Pierzynski, G.M., T.J. Logan, S.J. Traina, and J.M. Bigham, 1990b. Phosphorus chemistry and mineralogy in excessively fertilized soils: Descriptions of phosphorous rich particles. Soil Sci. Soc. Am. J. 54:1583–1589.

Pikering, I.J. , G.E. Brown Jr., and T.K. Tokunaga, 1995. Quantitative speciation of selenium in soils using x-ray absorption spectroscopy. Environ. Sci. Technol. 29:2456–2458.

Roe, L.A., K.F. Hayes, C. Chisholm, G.E. Brown, Jr., K.O. Hodgson, G.A. Parks, and J.O. Leckie. 1990. In situ x-ray absorption study of lead ion surface complexes at the goethite–water interface. Langmuir 7:367–373.

Ross, S.M. (ed.). 1994. Toxic metals in soil–plant systems. Wiley Interscience, New York.

Ruby, M.V., A. Davis, J. Houiston Kempton, J.W. Drexler, and P.D. Bergstrom, 1992. Lead bioavailability: Dissolution kinetics under simulated gastric conditions. Environ. Sci. Technol. 26:1242–1248.

Ruby, M.V., A. Davis, and A. Nicholson. 1994. *In situ* formation of lead phosphates in soils as a method to immobilize lead. Environ. Sci. Technol. 28:646–654.

Senesi, N. 1992. Metal-humic substance complexes in the environment. Molecular and mechanistic aspects by multiple spectroscopic approaches. p. 429–496. *In* D.C. Adriano (ed.) Biogeochemistry of trace metals. Lewis Publ., Ann Arbor, MI.

Spadini, L., L. Manceau, P.W. Schindler, and L. Charlet, 1994. Structure and stability of Cd^{2+} surface complexes on ferric oxides: 1. Results from EXAFS spectroscopy. J. Colloid Interface Sci. 168:73–86.

Sposito, G. 1981. The thermodynamics of soil solutions. Oxford Univ. Press, New York.

Sposito, G. 1983. The chemical form of trace metals in soils. p. 123–170. *In* I. Thorton (ed.) Applied environmental geochemistry. Academic Press, Geol. Ser., London.

Sposito, G. 1989. The chemistry of soils. Oxford Univ. Press, New York.

Stumm, W., and J.J. Morgan, 1995. Aquatic chemistry: Chemical equilibria and rates in natural waters. 3rd ed. Wiley Interscience, New York.

Szecsody, J.E., J.M. Zachara, and P.L. Bruckhart, 1994. Adsorption–dissolution reactions affecting the distribution and stability of Co^{II}EDTA in iron oxide-coated sand. Environ. Sci. Technol. 28:1706–1718.

Traina, S.J., and H.E. Doner, 1985. Heavy metal induced releases of Mn(II) from a hydrous manganese dioxide. Soil Soc. Am. J. 49:317–321.

Waychunas, G.A., C.C. Fuller, and J.A. Davis. 1993a. Sorption of aqueous Zn(II) on ferrihydrite: Observation of the onset of precipitation. p. 78–80. *In* K.C. Cantwell and L. Dunn (ed.) 1994 Activity Report. Stanford Synchrotron Radiation Lab., Stanford, CA.

Waychunas, G.A., B.A. Rea, C.C. Fuller, and J.A. Davis 1993b. Surface chemistry of ferrihydrate: I. EXAFS studies of the geometry of coprecipitated and adsorbed arsenate. Geochem. Cosmo. Acta. 57:2251–2269.

Wolt, J. 1994. Soil solution chemistry. Application to environmental science and agriculture. Wiley Interscience, New York.

Xu, N. 1993. Spectroscopic and solution chemistry studies of cobalt (II) sorption mechanisms at the calcite/water interface. Ph.D. diss. Stanford Univ., Stanford, CA.

Xue, Y. 1995. Co-solvent effects on metal-ligand interactions in the environment. Ph.D. diss. Ohio State Univ., Columbus.

Zachara, J.M., S.C. Smith, C.T. Resch, and C.E. Cowan, 1992. Cadmium sorption to soil separates containing layer silicates and iron and aluminum oxides. Soil Sci. Soc. Am. J. 56:1074–1084.

4 Dynamics and Transformations of Radionuclides in Soils and Ecosystem Health

R. J. Fellows, C. C. Ainsworth, C. J. Driver, and D. A. Cataldo

Pacific Northwest Laboratory
Richland, Washington

4–1. INTRODUCTION

Soils are an essential element of terrestrial and aquatic ecosystems both as an abiotic component (matrix) and a source of nutrients for the ecosystem's biotic components (microbes, plants, and animals). Soil composition, either naturally or anthropogenically derived, can and does directly affect the status of biotic populations.

Radionuclides are well recognized as potentially toxic to living organisms both from a chemical (heavy metal toxicity) and/or radiological dose (ionizing radiation) aspect; however, the mere presence of a radionuclide in soil may not adversely affect either a single organism, or a community of organisms if that radionuclide is in a biologically, chemically, or physically unavailable chemical form or if concentration, distance, or shielding limit the potential for absorbed dose. It is therefore important in any review of soils and ecosystem health to evaluate the potential results (chemical and biological) of radionuclides as soil constituents and/or contaminants.

4–1.1 Radionuclide Sources

Radioactive elements have been a component of the earth's geosphere since the formation of the planet. While the raw material for the formation of these elements was provided by stellar fusion reactions, or by cataclysmic supernovae explosions, the ultimate accumulation and dispersion of this material was a function of the same chemical and physical processes that promoted the distribution of all the other elements within the earth's crust (Cowart & Burnet, 1994, and the references therein). These naturally occurring radioactive materials have been, and continue to remain, components of the ecosystems in which they reside.

Radionuclides are isotopes that possess the property of spontaneous emission of radiation (particles and/or electromagnetic waves) through the decay of their unstable nuclei. The most biologically important of these emissions include

 Soil Chemistry and Ecosystem Health. Special Publication no. 52.

positively (alpha, α), negatively (beta, β^-, β^+), and noncharged (neutrons) particles, as well as electromagnetic radiation in the form of gamma (γ) and x-rays. These emissions are referred to as ionizing radiation. They contain sufficient energy to cause the ejection of electrons (or protons in the case on neutrons) from atoms they interact with thus producing positively and negatively charged ion pairs that are capable of disrupting important chemical bonds within living cells.

Many of the >2000 recognized radioisotopes are short lived and found only within the confined spaces of reactors and accelerators; however, events both accidental and deliberate over the last 50 yr have intermittently permitted the release of some of the longer lived man-made radionuclides and/or their progeny to the external environment. These anthropogenic sources, and some of the more renowned isotopes, include fallout or waste disposal from weapons testing and production (^{239}Pu, ^{134,137}Cs, ^{125,129}I, ^{90}Sr, ^{3}H), industrial and medical wastes (^{137}Cs, ^{235,238}U, ^{239}Pu, ^{89,90}Sr, ^{60}Co, ^{99}Tc), mining operations (^{235,238}U), and nuclear reactor accidents (^{239,240,241}Pu, ^{134,137}Cs, ^{89,90}Sr, ^{125,129}I; Talmage & Meyers-Schöne, 1995).

There are a number of pathways for the transport–distribution of these radionuclide releases within the environment as shown in Fig. 4–1. While the material may be released directly to terrestrial, aquatic, or atmospheric environs most will ultimately interact with a soil or sediment. Here chemical, physical, and biological interactions will render the radionuclide mobile, or immobile within the environment.

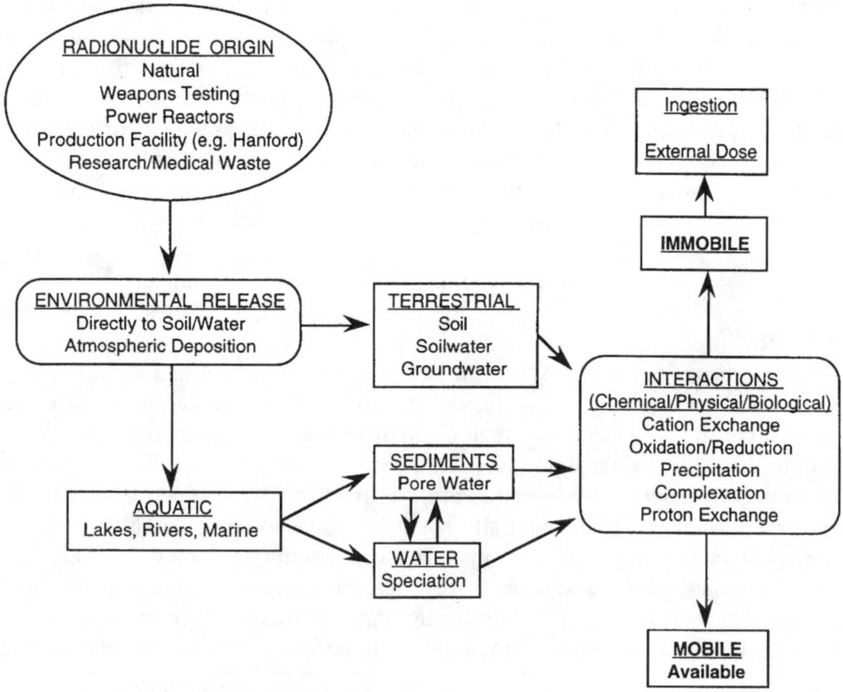

Fig. 4–1. Potential transport routes for radionuclides released to the environment.

Mobile radioisotopes that have chemical similarities to (e.g., Cs/K, Sr/Ca, Tc/S, As/P, Cr/P, and V/P), or are isotopes of, essential elements (e.g., ^{3}H, ^{60}Co, ^{35}Cl, Mn, and ^{65}Zn) may be available in the soil or groundwater for uptake, transport, and accumulation by living organisms (Cataldo & Wildung, 1978). Movement of immobile radionuclides is generally restricted to physical dispersal such as wind driven resuspension–deposition onto surfaces of organisms but also may include mass transport as a result of burrowing organisms (Bishop, 1989). Both routes may still prove deleterious to biota through direct radiological dosage (external or internal following ingestion–inhalation) or chemical toxicity.

4–1.2 Focus of Present Discussion

As mentioned above there are >2000 radionuclides and a myriad of geochemical environments available. To address even a small portion of them in all possible environmental pathways would be beyond the limit of this presentation. We have therefore restricted our discussion to terrestrial pathways and sought to focus on a selected group of radionuclides whose chemical interactions encompass the soil partitioning mechanisms mentioned above, are important from a public perception, may have nonradiological toxicity considerations, and may have enhanced potential to become incorporated into food chains. These selections include U ($^{235,238}U$), Co (^{60}Co), Cs ($^{134,137}Cs$), and Sr ($^{89,90}Sr$). For a more detailed and extensive treatment of post-Chornobyl radioecology the reader is referred to Warner and Harrison (1993).

4–2 SOIL CHEMISTRY OF RADIONUCLIDES

The mechanisms (or chemical processes) that govern the dynamics and transformations of radionuclides in soils are essentially the same as those that govern nonradioactive elements and isotopes. Hence, soil mobility–availability of any isotope is variable and depends on the geochemical environment in question. Radionuclides do, however, have several important characteristics. First, the rate these materials emit radiation cannot be modified by any physical, chemical, or biological process. The process(s) governing mobility–immobility result from the physical transfer of radionuclides as chemical elements from a dissolved form to a solid form or vice-versa. During this transition, the nature of their radioactivity remains unchanged except for any radioactive decay that occurs during this process. Second, like their stable counterparts, radionuclides react chemically. For example, if a process removes the stable form of Co, it will generally also remove the radioactive form of Co. This property is important because relevant information regarding the possible behavior of a radionuclide can be inferred from its stable counterpart (or chemical analogs). Third, the concentrations of radioactive material present in natural systems are exceedingly low and their measurement is only possible through analytical techniques with extremely low radiation detection limits. The detection limits for their nonradioactive counterparts is often several orders of magnitude higher; however it also must be stated that high concentrations of their stable counterparts may significantly affect the speciation of the low concentration radionuclide.

While total concentrations of radionuclides in soils indicate the extent of contamination, they give little insight into the forms in which the metals are present or the potential for their mobility, or bioavailability in the environment.

Radionuclides can conceivably occur in soils as soluble-free, inorganic-soluble-complexed, organic-soluble, complexed, adsorbed, precipitated, coprecipitated, or solid structural species. Sposito (1989) calculated that a typical soil solution will easily contain 100 to 200 different soluble species and 50 to 100 surface-associated or precipitated species. Consequently, an understanding of the processes affecting chemical speciation is necessary to understand how and under what conditions radionuclides are retarded or mobilized in soil environments. Those geochemical processes that contribute to the formation of these species–solids include complexation–precipitation, redox reactions, sorption, and ion exchange.

4–2.1 Complexation and Precipitation

A complex is said to be formed whenever a molecular unit, such as an ion, acts as a central group to attract and form a close association with other atoms or molecules. Thc aqucous species $Th(OH)_4^0$, $UO_2(OH)^+$, and HCO_3^- are complexes, with Th^{4+}, U^{6+}, and CO_3^{2-}, respectively, acting as the central group. The associated ions, OH^- or H^+, in these complexes are termed *ligands*. If two or more functional groups of a single ligand are coordinated to a metal cation in a complex, the complex is termed a *chelate*. If the ligands in a complex are water molecules (as, for example, in $Ca(H_2O)_6^{2+}$), the unit is called a *solvation complex* or, more frequently, a *free species*.

As an example, consider the formation of a neutral sulfate complex with a bivalent metal cation (M^{2+}) as the central group

$$M^{2+}{}_{(aq)} + SO_4^{2-}{}_{(aq)} = MSO^0_{4(aq)} \qquad [1]$$

where the metal M can be Co, Sr, Ca, Mn, or Cu. The conditional equilibrium constant, ${}^cK$, corresponding to the above reaction is

$${}^cK = [MSO^0_{4(aq)}]/[M^{2+}{}_{(aq)}][SO_4^{2-}{}_{(aq)}] \qquad [2]$$

where [ ] represents the concentration of the species. The conditional equilibrium constant can describe the distribution of a given constituent among its possible chemical forms if complex formation and dissociation reactions are at equilibrium. The conditional stability constant is affected by a number of factors, including the ionic strength of the aqueous phase and the presence of competing reactions, such as hydrolysis or complexation with competing reactants.

Complexing anions present in substantial amounts in the environment include HCO_3^-/CO_3^-, Cl^-, SO_4^{2-}, PO_4^{3-}, and organic materials. Their relative propensity to form complexes with many metals are: $CO_3^- > SO_4^{2-} > PO_4^{3-} > Cl^-$ (Stumm & Morgan, 1981). Uranium, as the UO_2^{2+} species, forms extremely strong complexes with CO_3^{2-} and SO_4^{2-} ions (Grenth, 1992). The presence of these anions, even at trace levels, has dramatic effects on the chemical behavior of U

(Kim, 1986). A large number of dissolved, small-chain organic materials are present in the soil solution and their complexation properties with metals–radionuclides are not well understood. Because of the presence of organic materials in most natural systems and their tendency to form strong complexes, complexation of many radioactive metals by organic materials is likely an important chemical reaction (Kim, 1986). The chelate anion, ethylenediaminetetraacetic acid (EDTA), forms strong complexes with many cations, much stronger than carbonate and organic materials (Kim, 1986).

The precipitation reaction of dissolved species is a special case of the complexation reaction (Eq. [1] and [2]) in which the complex formed by two or more aqueous species is a solid. As an example, consider the formation of a sulfide precipitate with a bivalent metal cation as the central group

$$M^{2+}{}_{(aq)} + HS^{-}{}_{(aq)} = MS_{(s)} + H^{+}{}_{(aq)} \qquad [3]$$

where M can be a radionuclide metal. The solubility product constant, K_{sp}, corresponding to Eq. [3] is

$$K_{sp} = [MS_{(s)}][H^{+}{}_{(aq)}]/[M^{2+}{}_{(aq)}][HS^{-}{}_{(aq)}] \qquad [4]$$

It must be emphasized that the aqueous species of Eq. [4] are the activities of the free species, not the total concentration. Hence, complexation, as illustrated in Eq. [1], has a direct impact on the propensity of a given species to precipitate. Precipitation of most radionuclides is not likely to be a dominant reaction in most soils because the radionuclide activities, even in the absence of complexation, are not likely to be high enough to cause the quotient of Eq. [4] to exceed the K_{sp}. Precipitation is more likely to occur if the radionuclide has a stable counterpart in the soil that exists at an activity that would cause the K_{sp} to be exceeded. Under such conditions, the radioactive isotope will precipitate with, and in direct proportion (on a mass basis) to, the stable isotope because the two are chemically identical; however, it is very likely that coprecipitation of radionuclides in soils could occur.

Coprecipitation is the simultaneous precipitation of a chemical element with other elements (Sposito, 1989). The three broad types of coprecipitation are inclusion, surface precipitation, and solid solution formation. Inclusion is the occasional substitution of a foreign ion for the matrix ion in the solid. Surface precipitation is the formation of a three-dimensional solid phase at the surface of an existing solid. Solid solution formation is the formation of a solid phase containing more than one central atom where substitution for one atom–molecule is generally greater than that which occurs via inclusion.

4–2.2 Redox Chemistry

An oxidation–reduction (redox) reaction involves the complete transfer of electrons from one species to another. The chemical species that loses electrons in this charge-transfer process is called oxidized, while the one receiving electrons is called reduced. For example, in the reaction involving Fe species

$$FeOOH_{(s)} + 3H^+_{(aq)} + e^-(aq) = Fe^{2+}_{(aq)} + 2H_2O_{(aq)} \quad [5]$$

the solid phase, goethite, is the oxidized species, and Fe^{2+}(aq) is the reduced species. Equation [5] is a reduction half-reaction, in which an electron in aqueous solution, denoted $e^-_{(aq)}$, serves as one of the reactants. This species, like the proton in aqueous solution, is understood in a formal sense to participate in charge-transfer processes. Radionuclides capable of undergoing redox reactions include Co, I, Ir, Np, Pu, Tc, and U.

Redox chemistry has a direct effect on radionuclide chemistry in soils and sediments. The oxidation state and, therefore, the potential mobility–immobility of Co, I, Ir, Np, Pu, Tc, and U are affected. For example, the reduction of Pu

$$^{239}Pu^{4+} + e^- = {}^{239}Pu^{3+} \text{ , pE} = 1.7 \quad [6]$$

makes ^{239}Pu appreciably less reactive to complexation (that is, $^{239}Pu^{3+}$ stability constants are much less than those of $^{239}Pu^{4+}$) and sorption–partitioning reactions (Kim, 1986). The reduction of U(VI) as the $UO^{2+}{}_2$ ion to U(IV) has the opposite effect [that is, U(IV) forms stronger complexes and sorbs more strongly to surfaces than U(VI)]. In addition, the U(IV) species precipitates much more readily than the U(VI) species. Therefore, changes in redox may increase or decrease the tendency for radionuclides to undergo retardation reactions in soils, depending on the aqueous-phase chemical composition and the radionuclide in question.

4–2.3 Sorption

Sorption, as discussed here, is a generic term, devoid of a mechanism used to describe the partitioning of aqueous-phase constituents to a solid phase; that is, the sorbed material may be adsorbed onto the surface of a solid, absorbed into the structure of a solid, precipitated as a three-dimensional molecular structure on the surface of the solid, or partitioned into an organic structure. Sorption is often quantified through, the term distribution coefficient, K_d, and is defined as

$$K_d = q_i/C_i \quad [7]$$

where q_i is the amount of constituent sorbed to the surface and C_i is the concentration of constituent i in the equilibrium aqueous phase that is in contact with the solid phase (Sposito, 1989).

4–2.4 Adsorption

Adsorption is described as the net accumulation of matter at the interface between a solid phase and an aqueous-solution phase. Adsorption differs from precipitation because it does not include the development of a three-dimensional molecular structure. Adsorption on clay particle surfaces can take place via three mechanisms: (i) the inner-sphere complex that resides in the Stern layer, (ii) the outer-sphere surface complex that has a solvation shell (at least one water molecule between it and the surface), and (iii) if a solvated ion does not form a complex with a charged surface functional group but instead neutralizes surface charge only in a delocalized sense, it is said to be adsorbed in the diffuse-ion swarm. The diffuse-ion swarm and the outer-sphere surface complex mechanisms of adsorption involve electrostatic bonding almost exclusively; whereas inner-sphere complex mechanisms are likely to involve ionic, as well as covalent, bonding.

As a rule of thumb, the relative affinity of an absorbent for a free-metal cation will increase with the tendency of a cation to form inner-sphere surface complexes. The higher the valence or the ionic potential (the ratio of the valence to the ionic radius) of a cation, the greater the tendency the cation has of forming an inner-sphere complex (Sposito, 1989). Based on these considerations and laboratory observations, the relative-adsorption affinity of metals has been described as follows (Sposito, 1989):

$$Cs^{+} > Rb^{+} > K^{+} > Na^{+} > Li^{+} \quad [8]$$

$$Ba^{2+} > Sr^{2+} > Ca^{2+} > Mg^{2+} \quad [9]$$

$$Hg^{2+} > Cd^{2+} > Zn^{2+} \quad [10]$$

$$Fe^{3+} > Fe^{2+} > Fe^{+} \quad [11]$$

With respect to transition metal cations, however, ionic potential is not adequate as a single predictor of adsorption affinity because electron configuration plays a very important role in the complexes of these cations. Their relative affinities tend to follow the Irving-Williams order:

$$Cu^{2+} > Ni^{2+} > Co^{2+} > Fe^{2+} > Mn^{2+} \quad [12]$$

The molecular basis for this ordering is discussed in Huheey (1983).

4–2.5 pH

The effect of pH on metal cation adsorption is principally the result of changes in the variable-charge component of the surface charge and hydrolysis of metal cations. As the pH increases, the surface charge decreases toward negative values, and the electrostatic attraction of an adsorbent for a metal cation is enhanced. The presence of complex-forming ligands in solution greatly complicates the prediction of the relative ion affinity of a metal cation. For example, radioactive $^{60}Co^{2+}$ usually adsorbs very strongly to inorganic surfaces, but if EDTA is present to form nonadsorbing-soluble complexes with the $^{60}Co^{2+}$, its electrostatic attraction to a clay surface will be diminished appreciably (Means et al., 1978). Decreases in Am, Pu, and U sorption also have been observed when these radionuclides form complexes with naturally occurring humic substances (Means et al., 1978; Kim, 1986).

The mechanisms by which nonpolymeric anions adsorb are surface complexation and diffuse-ion swarm association. Outer-sphere surface complexation of anions involves coordination with a protonated hydroxyl or amino group or to a surface metal cation (such as water-bridging mechanisms). Inner-sphere surface complexation of anions, such as borate, phosphate, and carboxylate involves coordination with created or native Lewis acid sites. Almost always, the mechanism of this coordination is hydroxyl-ligand exchange (Sposito, 1989). In general, ligand exchange is favored by pH < zero point of charge.

4–2.6 Ion Exchange

In its most general meaning, an ion-exchange reaction involves the replacement of one ionic species in or on a solid phase by another ionic species taken from an aqueous solution in contact with the solid. In the cation-exchange reaction

$$CaX_{(s)} + Sr^{2+}{}_{(aq)} = SrX_{(s)} + Ca^{2+}{}_{(aq)} \quad [13]$$

Sr^{2+} replaces Ca^{2+} from an exchange site, X. The exchange coefficient (K_{ex}) for this exchange reaction is defined by

$$K_{ex} = [SrX_{(s)}][Ca^{2+}{}_{(aq)}] \,/\, [CaX_{(s)}][Sr^{2+}{}_{(aq)}] \quad [14]$$

Numerous ion-exchange models are described by Sposito (1989) and Stumm and Morgan (1981).

This brief discussion was meant to identify the various soil physicochemical interactions to be considered while assessing the potential for a radionuclide's mobility, or bioavailability in the environment. Table 4–1 attempts to condense these considerations into a thumbnail sketch of some common radionuclides; these sketches are by no means complete.

4–3 RADIONUCLIDE TRANSPORT AND POTENTIAL ROUTES INTO FOOD CHAINS

Those chemical and physical processes that may render a radionuclide mobile within soil or groundwater environs, or soil surface events such as resuspension–deposition function to provide a source term to the biotic components of terrestrial and aquatic ecosystems (Cataldo & Wildung, 1978, 1983; Wildung et al., 1979). A schematic of some (but most certainly not all) of the potential interactions that may subsequently occur is presented in Fig. 4–2.

Transfer to the biotic components of terrestrial or aquatic ecosystems may be accomplished by mineral uptake and partitioning by plants through root (soil solution) or shoot (cuticular penetration) absorption and subsequent translocation to herbivore accessible tissues, or through direct ingestion of particulates.

4–3.1 Plant Uptake and Partitioning

Plants can represent >90% of the terrestrial biomass in many ecosystems (Odum, 1971) and form the basis of most food chains. They are therefore critical

Table 4–1. List of isotopes†, their oxidation states, soil–water mechanisms that impact transport in soils and sediments‡, important complexes and complexing ligands§, and an estimate of their soil mobility¶.

Isotope(s) half-lives	Common oxidation states	Important mechanisms in soil–water systems	Important ligands in aqueous speciation	Soil mobility	References
				K_d,mL g^{-1}	
$^{235,238}U$ (7.1 × 10^8, 4.51 × 10^9 yr)	U(IV)	Hydrolysis, precipitation (insoluble; present in reducing environments)	$U(OH)_n^{4-n}$; humic, fulvic acids and other organic ligands	Low (1000, est.)	Grenthe, 1992 Parks & Pohl, 1988
	U(VI)	Hydrolysis, cation exchange, surface complexation, precipitation	UO_2^{2+}, $UO_2(OH)^+$, polynuclear hydrolysis species [i.e., $(UO_2)_3(OH)_5^+$]; multiple CO_3^{2-} and PO_4^{3-} species; potential SO_4^{2-}, F^-, and Cl^- species	Medium-high (0.08–79.3)	Grenthe, 1992 Idiz et al., 1986 Shanbhag & Choppin, 1981 Zachara & McKinley, 1993
$^{57,60}Co$ (270 d, 5.26 yr)	Co(II)	Hydrolysis, cation exchange, surface complexation, precipitation	Co^{2+} and its hydrolysis species, CO_3^{2-}, SO_4^{2-}, Cl^-, humic and fulvic acids, and other organic ligands	Low (1200–12500)	Trischen et al., 1981 Morel, 1983 Ainsworth et al., 1994 Zachara et al., 1994, 1991
	Co(III)	Unstable in uncomplexed form	Forms very stable organic complexes	High	Trischen et al., 1981 Jardine et al., 1993
$^{134,137}Cs$ (2.05, 30.23 yr)	Cs(I)	Cation exchange	Cs^+ forms weak complexes with SO_4^{2-}, Cl^-, $NO^-{}_3$, and organic ligands	Low (540–3180)	Bruggenwert & Kamphorst, 1979 Gast, 1969 Gee et al., 1983
$^{89,90}Sr$ (52 d, 28.1 yr)	Sr(II)	Cation exchange, precipitation	Sr^{2+} forms weak complexes with SO_4^{2-}, Cl^-, NO_3^-, CO_3^{2-}, and organic ligands	Medium-High (5–173)	Zachara et al., 1991 Gee et al., 1983
^{239}Pu (2.44 × 10^4 yr)	Pu(III)	Hydrolysis, precipitation (insoluble; only present in extremely reducing environments)	Low (1000 est.)		Cleveland, 1979 Rai & Serne, 1977
	Pu(IV)	Hydrolysis, precipitation (insoluble; stable across a wide redox range)	OH^-, CO_3^{2-} and organic Chelates (EDTA)	Low (1000 est.)	Gee et al., 1983 Cleveland, 1979 Rai & Serne, 1977

(continued on next page)

Table 4–1. Continued.

Isotope(s) half-lives	Common oxidation states	Important mechanisms in soil–water systems	Important ligands in aqueous speciation	Soil mobility	References
				K_d,mL g^{-1}	
	Pu(V)	Cation exchange (stable in highly oxidizing environments)	PuO_2^+, forms only weak inorganic and organic complexes	Medium¶ (80–>1980)	Cleveland, 1979 Rai & Serne, 1977
99Tc (2.12 × 10^5 yr)	Tc(IV)	Precipitation (insoluble; stable only in highly reduced environments)	Soluble organic matter (?) low molecular weight organic acids	Low (1000 est.)	Kaplan et al., 1995 Gu & Shultz, 1991
	Tc(VII)	Anion exchange	TcO_4^-, does not form strong complexes	High (0–1.3)	Kaplan, et al., 1995 Gu & Shultz, 1991
^3H (12.26 yr)		Proton exchange			Sweet & Murphy, 1984

† The radionuclides listed are the most often associated with environmental concerns at U.S. Department of Energy sites such as Hanford, Oakridge, Los Alamos, Rocky Flats, and others. This table provides an overview of selected radionuclide species interactions in soils, soil water components, and potential mobilities. Readers are directed to the listed references for more detailed discussion.

‡ The soil–water mechanisms considered here are precipitation (and coprecipitation), surface complexation (to oxides, carbonates, organic matter, and clay edge sites), cation exchange, anion exchange, and proton exchange (in the case of ^3H). No attempt is made to distinguish between particular sorbing surfaces, precipitates–coprecipitates, or other mechanisms such as solid phase incorporation, micropore diffusion, and solid phase diffusion.

§ The formation of aqueous complexes can greatly influence the total aqueous concentration of a given element and its mobility. We have attempted to give the reader a sense for the aqueous ions typically associated with soil environments that could impact the total aqueous concentration of a given element. It is beyond the scope of this discussion to detail the association constants of these complexes; readers are directed to the listed references for more detailed discussion.

¶ The soil mobility classification is purely empirical and is based on K_d values reported in Kaplan et al. (1995). All the K_d values were determined or estimated for the Hanford Sediment in neutral-to-high pH, low salt (ionic strength <0.01 *M*), low organic, oxic solutions except U(IV), Pu(III,IV), and Tc(IV), which were estimated under anoxic conditions. The K_d measurement is dependent on the mineralogic components of the sorbent, pH, Eh, ionic strength, ionic composition of the aqueous phase, and often the equilibration time. Changes in any of these factors can greatly influence the magnitude of the K_d; however, in general, the mobility classification will not be affected in a relative sense. For instance, at higher salt concentrations (≥0.01 *M*) the listed K_d values tend to decrease: Cs 64 - 1360, Pu(V) 10->98, U(VI) 0 - 4, Co(II) 222 - 4760, Tc(VII) 0 - 0.01, Sr 0.3 -42. The Pu(V) values listed at both ionic strengths may be overestimations due to reduction to Pu(IV); therefore the lower value only was used to classify its soil mobility.

in the transport of radionuclides from the soil to biota. Surface vegetation derives its nutrient and trace element requirements (except for C and O_2) from the available ion pools in soils associated with the roots (Marschner, 1995). The rooting depth and volume from which plants absorb these ions varies with plant species and soil structure–soil moisture characteristics. This generally extends downward from 0.2 to 2 m below the soil surface and may have a diameter of two to four or more times that of the projected canopy.

Root absorption is highly dependent on the plant species, soil conditions, and the chemical form of the inorganic species or organically complexed form. In general however, it falls into several distinct classes based on the soil/plant concentration ratios (CR values, or mg isotope g^{-1} dry weight of plant tissue divided by the mg isotope g^{-1} dry weight soil). Ions such as U, Pu, Am, Cs, Cr, and Ba can be grouped as low availability (CR values <0.1), while Pb, As, Co, T, Hg, and Sb are classed as moderate availability (CR value 0.1 to 2; Table 4–2) and Sr, Tc, Mn, Zn, and Cl are classed as high availability (CR value >2). Soil contaminant concentrations tend to be less important for the low availability ions than for the high availability ions. In many instances, the CR values for high availability ions is directly proportional to soil concentration.

Plant root absorption of radionuclides and nonnutrient contaminants is generally related to their chemical and physical behavior as analogs of essential elements (Cataldo et al., 1978a, 1983; Cataldo & Wildung 1983). Root absorption involves membrane transport of ionic species in most cases. The classical analog pairs of interest include $^3H/H$ (as $^3HHO/HHO$), Cs/K, Sr/Ca, Tc/S, As/P, Cr/P, and V/P. Many elements (stable or radioactive species) are accumulated because they are nutritionally essential; these include Co, Cl, I, Mn, and Zn. Other nonnutritional elements such as U, Am, Np, Pu, Pb, Sb, Ba, Cd, Ni, and Hg also are absorbed by processes in place for nutrient species, with rates of uptake being dependent on the hydrated radius and associated charges (Garland et al., 1983, 1987).

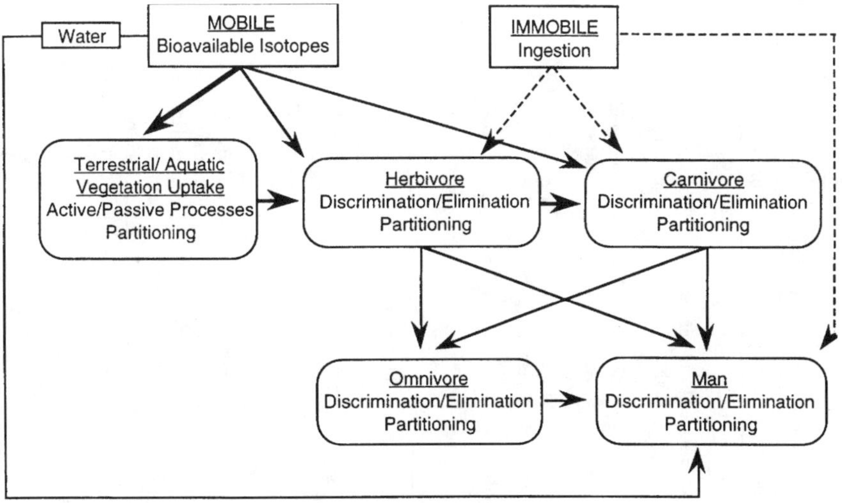

Fig. 4–2. Potential transport routes for radionuclides within food chains.

Table 4–2. Selected terrestrial plant concentration ratios (Bq g^{-1} dry weight tissue/Bq g^{-1} dry weight soil).

Metal/isotope	Plant species	Soil type	Environment	Concentration ratio	Reference
Uranium					
^{238}U	Mixed species	U tailings	Semiarid	0.8	Ibrahim & Whicker, 1988
U	Radish	Neutral sand	Growth chamber	0.047	Sheppard & Evenden, 1992
U	Radish	Acid sand	Growth chamber	0.237	Sheppard & Evenden, 1992
U	Radish	Limed sand	Growth chamber	0.094	Sheppard & Evenden, 1992
U	Bean	Limed sand	Field lysim.	0.066	Sheppard & Evenden, 1992
U	Trees	Corse	Field	0.024	Sheppard & Evenden, 1988
U	Shrubs	Corse	Field	0.009	Sheppard & Evenden, 1988
U	Vegetation	Unnamed	LANL	0.02–0.034	Hanson & Miera, 1976
Cobalt					
Co	Soybean	Ritzville (Hanford)	Growth chamber	0.159	Cataldo & Wildung, 1978
Cesium					
^{137}Cs	Tumbleweed	Burbank sandy loam	Growth chamber	0.025–0.053	Routson & Cataldo, 1978
^{137}Cs	Bean	Unnamed (sand)	LANL	0.15	White et al.,1981
^{137}Cs	Squash	Unnamed (sand)	LANL	0.25	White et al., 1981
^{137}Cs	Barley	Ephrata silt loam	Growth chamber	$3.0–6.0 \times 10^{-3}$	Cline, 1981
^{137}Cs	Trees	Unnamed (WA)	Hanford	0.02–0.06	Landeen & Mitchell, 1986
^{137}Cs	Crested wheatgrass	Various	INEL	0.619–0.81	Arthur, 1982
^{137}Cs	Trees	Various	Northern Europe	0.027–0.28	Livens et al., 1991
^{137}Cs	Bilberry	Various	Northern Europe	0.25–1.62	Livens et al., 1991
^{137}Cs	Tumbleweed	Various	INEL	5.4–35.1	Arthur, 1982
^{134}Cs	Cheatgrass	Rupert (Hanford)	Growth chamber	0.064	Cataldo et al., 1978b
^{134}Cs	Cheatgrass	Ritzville (Hanford)	Growth chamber	0.031	Cataldo et al., 1978b
^{134}Cs	Cheatgrass	Lickskillet (Hanford)	Growth chamber	0.015	Cataldo et al., 1978b
^{134}Cs	Tumbleweed	Rupert (Hanford)	Growth chamber	0.014	Cataldo et al., 1978b
^{134}Cs	Tumbleweed	Ritzville (Hanford)	Growth chamber	0.066	Cataldo et al., 1978b
^{134}Cs	Tumbleweed	Lickskillet (Hanford)	Growth chamber	0.0078	Cataldo et al., 1978b
Cs	*Atriplex canescens*	Sandy (Mojave)	Greenhouse	0.116–0.48	Wallace et al., 1971

Strontium					
^{90}Sr	Tumbleweed	Burbank sandy loam	Growth chamber	9.6–19.0	Routson & Cataldo, 1978
^{90}Sr	Barley	Ephrata silt loam	Growth chamber	1.63–2.15	Cline, 1981
^{90}Sr	Trees	Unnamed (WA)	Hanford	63.8–85.3	Landeen & Mitchell, 1986
^{90}Sr	Crested wheatgrass	Various	INEL	0.85	Arthur, 1982
^{90}Sr	Tumbleweed	Various	INEL	4.5	Arthur, 1982
^{85}Sr	Barley	Ephrata silt loam	Growth chamber	0.62–2.60	Cline, 1981
^{85}Sr	Cheatgrass	Rupert (Hanford)	Growth chamber	12	Cataldo et al., 1978b
^{85}Sr	Cheatgrass	Burbank sandy loam	Growth chamber	12	Cataldo et al., 1978b
^{85}Sr	Cheatgrass	Lickskillet (Hanford)	Growth chamber	3.5	Cataldo et al., 1978b
^{85}Sr	Tumbleweed	Rupert (Hanford)	Growth chamber	16	Cataldo et al., 1978b
^{85}Sr	Tumbleweed	Burbank sandy loam	Growth chamber	8.7	Cataldo et al., 1978b
^{85}Sr	Tumbleweed	Lickskillet (Hanford)	Growth chamber	4.4	Cataldo et al., 1978b
Plutonium					
^{239}Pu	Barley	Ephrata silt loam	Growth chamber	1.3×10^{-4}	Wilson & Cline, 1966
^{238}Pu	Bean	Unnamed (sandy)	LANL	7.2×10^{-2}	White et al., 1981
^{238}Pu	Squash	Unnamed (sandy)	LANL	0.12	White et al., 1981
^{238}Pu	Crested Wheatgrass	Various	INEL	0.013–0.042	Arthur, 1982
^{238}Pu	Tumbleweed	Various	INEL	0.083–0.66	Arthur, 1982
^{239}Pu	Bean	Unnamed (sandy)	LANL	5.8×10^{-2}	White et al., 1981
^{239}Pu	Squash	Unnamed (sandy)	LANL	9.7×10^{-2}	White et al., 1981
^{239}Pu	Vegetation	Unnamed	Field (NV)	0.1	Garten, et al., 1987
^{239}Pu	Landino Clover	Unnamed	Greenhouse	1.4×10^{-4}	Romney et al., 1970
^{239}Pu	Crested Wheatgrass	Various	INEL	0.022–0.125	Arthur, 1982
^{239}Pu	Tumbleweed	Various	INEL	0.33–1.6	Arthur, 1982
^{239}Pu	Trees	Unnamed	Hanford	0.002–0.005	Landeen & Mitchell, 1986
^{239}Pu	Alfalfa	Unnamed	Growth chamber	$1.1–8.26 \times 10^{-4}$	Romney et al., 1985
^{239}Pu	Wheat	Unnamed	Growth chamber	$1.1–8.26 \times 10^{-4}$	Romney et al., 1985
^{239}Pu	Bushbean (Leaves)	Unnamed	Growth chamber	$9.9–72.4 \times 10^{-4}$	Romney et al., 1985
^{239}Pu	Carrot (Lvs)	Unnamed	Growth chamber	$0.81–4.75 \times 10^{-3}$	Romney et al., 1985
^{239}Pu	Cheatgrass	Burbank sandy loam	Growth chamber	0.017×10^{-3}	Price, 1972
^{239}Pu	Tumbleweed	Burbank Sandy loam	Growth chamber	0.046×10^{-3}	Price, 1972

(continued on next page)

Table 4–2. Continued.

Metal/isotope	Plant species	Soil type	Environment	Concentration ratio	Reference
Technetium					
^{99}Tc	Tumbleweed	Burbank sandy loam	Growth chamber	213–232	Routson & Cataldo, 1977
^{99}Tc	Tumbleweed	Ritzville (Hanford)	Growth chamber	314–390	Routson & Cataldo, 1977
^{99}Tc	Tumbleweed	Lickskillet (Hanford)	Growth chamber	76–127	Routson & Cataldo, 1977
^{99}Tc	Cheatgrass	Burbank sandy loam	Growth chamber	104–112	Routson & Cataldo, 1977
^{99}Tc	Cheatgrass	Ritzville (Hanford)	Growth chamber	158–192	Routson & Cataldo, 1977
^{99}Tc	Cheatgrass	Lickskillet (Hanford)	Growth chamber	54–114	Routson & Cataldo, 1977
^{99}Tc	Cheatgrass	Rupert (Hanford)	Growth chamber	91–185	Cataldo, 1979
^{99}Tc	Tumbleweed	Rupert (Hanford)	Growth chamber	82–148	Cataldo, 1979
^{99}Tc	Fescue	ORNL	Field	37	Garten et al., 1984
Tritium					
^{3}H	Tomatoes	Unnamed	Growth chamber	0.67–1.41	Spencer, 1984
	Vegetation	Unnamed	Field	0.003	Murphy, 1990

For example, for Pu deposited to soils from nuclear or nonnuclear detonation of devices, the principle form is either high fired oxides–particles or larger fragments of metal material (Olafson et al., 1957; Hanson, 1975). Conversely, Pu from processing operations, simulated in many pot studies, starts out as a nitrate or other acid soluble salt (Garland et al., 1987). When this latter form is added to soil it generally forms monomeric or polymeric Pu hydroxides depending on conditions and concentrations (Wildung et al., 1979; Cataldo & Wildung, 1983). The soil solubility and thus plant availability of these two diverse chemical forms can account for a factor of 100 difference in soil/plant CR values [$Pu(OH)_4 > PuO_2$] (Garland et al., 1987). A similar situation exists for nearly all di-and polyvalent cations and many anions. A comparison of the relative availability of 15 elements endogenous in single soil (Ritzville, coarse-silty, mixed, mesic calciorthidic Haploxeroll) to recently amended radioelements demonstrated the competing effects of endogenous versus amended elements for plant uptake (Cataldo & Wildung, 1978). In some soil types, however, this amended bioavailability also decreases rapidly with time. Grebenshchikova et al. (1991) reported a 50% reduction in Cs content of grain crops grown in the Gomal area of Belarus the first year following the Chornobyl accident and an additional 50% within the next 3 yr.

Plant accumulation is further modified by soil amendments (e.g., lime, fertilizer, chelators) and moisture availability. Russian scientists modified the pH of some Ukranian soils to three different levels, 4.5 to 5.5, 5.5 to 6.5, and 6.5 to 7.5 following the Chorobyl accident in an attempt to moderate the uptake of radioactive Cs by crop plants. They later reported a concomitant reduction of Cs uptake in several crop species with the increased soil pH at a ratio of 4:2:1 (Prister, 1991).

4–3.2 Subsequent Bioaccumulation

Reported bioaccumulation ratios (the amount of radionuclide accumulated in the organism's tissues divided by the concentration of the radionuclide in the soil, sediment, or water) and food chain transfer coefficients (the tissue concentration of radionuclide in a higher trophic species divided by the tissue concentration of radionuclide in its food base species) vary widely within taxa (see Table 4–3). Ranges of several orders of magnitude have been observed (Swanson, 1985, Reichle et al., 1970a,b). These variations arise from factors such as the difference in radionuclide bioavailability in the various test systems, or sites from which the data was gathered. While the chemistry in these systems was the major determinant of bioavailability, other factors include: the variety of trophic and spatial niches within taxa; species differences in radionuclide absorption, distribution, and retention; food source; behavior; and organism habitat (Takeda & Iwakura, 1992; Swanson, 1985; Taylor et al., 1981, 1983; Osburn, 1968; Krumholz, 1964).

An example of feeding behavior and habitat on bioconcentration factors is the concentration of radionuclides in aquatic birds. Highest concentration factors were seen in herbivorous ducks (*Anas* sp.) and larvae-feeding shorebirds that probe for food in soil or sediment (direct uptake of contaminated soils). Omnivorous dabbling ducks had intermediate levels in their tissues while fish-

Table 4–3. Bioconcentration and food chain transfer of radionuclides in aquatic and terrestrial ecosystems.

Isotope(s) (effective half-life†)	Bioconcentration: Aquatic‡ Algae	Aquatic‡ Invert	Aquatic‡ Fish	Terrestrial Plant§	Terrestrial Invert	Terrestrial Mammal	Food chain entry¶	Food chain transfer¶	Reference
$^{235,238}U$ (15 d)	1800	100–1000 (0.1–0.3)	2–50 (0.02–0.05)	$1–10^{-5}$	--	$0.5–10^{-5}$	A: Adsorption to plants and small animals T: Foliar deposition; water uptake	A: One order of magnitude decline in concentration with each tropic level T: forage to herbivore transfer is minimal	Mahon, 1982 Stegner & Kobal, 1982 Anderson et al., 1963 Reichle et al., 1970a,b Coughtrey et al.,1983–1985 Canadian Standards Association, 1987 Myers, 1989 Poston et al., 1984 Sheppard & Evenden, 1985 Ng et al., 1982
$^{57,60}Co$ (10 d)	16 000	30–325	20–1000 (0.005)	0.16	0.5	0.3	A,T: Readily accumulated in plant tissue	A: Trophic transfer is low T: Generally one-half the plant concentration after two trophic exchanges	Cataldo & Wildung, 1978 Reichle et al., 1970a,b Coughtrey et al.,1983–1985 Voshell et al., 1985 Baudin & Nucho, 1992 Stewart et al., 1992 Reichle & Crossley, 1969
$^{134,137}Cs$ (70 d)	500–4000 (0.9–1.8)	60–11 000 (0.07–0.009)	125–9500 (0.01–0.01)	$10^{-4}–35$	0.1–0.5	0.3–7.0	A: Physical absorption to algae and zooplankton in water compartment; detritivore from sediment T; plant uptake, but only from sandy soil and low cation-exchange capacity	A: No biomagnification; concentration decrease with successive trophic levels; increase in areas with sandy sediment T: Decrease to one- half plant concentration after two trophic exchanges; biotic accumlation at higher trophic levels but only in food chains tied to sandy soil	Reichle et al., 1970a,b Coughtrey et al., 1983–1985 Voshell et al., 1985 Dunford et al., 1985 Cushing & Watson, 1974 Rickard & Sweany, 1977

$^{89,90}Sr$ (15 y)	10–3000	1	1–150	0.85–20	0.1	0.5–4.5	A: Entry at all trophic levels T: Readily accumulates in plants	A,T: Does not increase with trophic level (metabolic control of uptake by Ca); stored in non-conumed tissues (bone, shell)	Routson & Cataldo, 1978 Arthur, 1982 Reichle et al., 1970a,b Coughtrey et al., 1983–1985 Horsic et al., 1982 Carraca et al., 1990
^{239}Pu (197 y)	280–5 × 10^6	320–5 × 10^6	10–350	10^{-4}–0.1	0.0034	0.006–0.17	A: Concentrates in algae T: Only minimally incorporated (fossorial animals) T: Not accumulated along terrestrial food webs	A: Although accumualtion occurrs in algae, transfer decreases by a factor of 10 at each subsequent trophic level	Garten et al., 1987 Noshkin et al., 1973 Hanson, 1975 Eyman & Trablaka, 1980 Poston & Klopher, 1985 NRCC, 1982 Trabalka & Eyman, 1976 Olafson et al., 1957 Coughtrey et al., 1983–1985)
^{99}Tc (20 d)	--	--	15	37–390	--	0.015	A: Little is known about aquatic entry into food webs; direct uptake from water at all trophic levels likely T: Rapid uptake by plants	A: Concentration in fish; may be direct from water T: Although assimilation of ingested Tc can be high, retention is low; no food chain magnification	Routson & Cataldo, 1978 Cataldo, 1979 Garten et al., 1984 Myers, 1989 Poston & Klopher, 1985 Sullivan et al., 1979 Thomas et al., 1984 Coughtrey et al., 1983–1985
^{3}H (12 d)	--	--	--	0.003–1.4	--	0.2–1	A: Direct uptake of treated water by all trophic levels T: Plants and/or water uptake	A,T: Metabolic turn-over rate rapid; enrichment in food chains not observed	Spencer, 1984 Murphy, 1990 Takeda & Iwakura, 1992

† Bioconcentration factors for the aquatic biota are water-based (i.e., the concentration of radionuclide in the organism's tissues–water concentration). Values in parentheses are sediment-based concentration ratios (i.e., concentration of radionuclide in the organism's tissues–concentration in the sediment).

‡ Plant bioconcentration ratios with wide ranges often include plants for which surface contamination by radionuclides was not separated from radionuclide transfer from soil to root and incorporation into edible aboveground tissue.

§ Effective half-life is the time required for a radionuclide in a biological system to be reduced to half of its initial value by both radioactive decay and biolgical processes (e.g., elimination).

¶ A, aquatic ecosystem; T, terrestrial ecosystem.

eating birds had the lowest radionuclide burden (Osburn, 1968; Krumholz, 1964; Willard, 1960; Silker, 1958).

Although radionuclides may concentrate to high levels in low trophic positions (i.e., algae, plants, and invertebrates), the movement of radionuclides from lower to higher trophic levels appears to diminish with each trophic level in both terrestrial and aquatic systems. In a broad sense this can be observed with the diminution of bioconcentration ratios from plant to higher vertebrate categories in Table 4–3. The cause of radionuclide loss at successive trophic levels are varied. For example, trophic transfer of ^{239}Pu and ^{235,238}U in aquatic systems is limited despite significant adsorption or absorption in algae because of low absorption–assimilation by successive trophic organisms whereas in terrestrial systems, very little of these two radionuclides are translocated to edible portions of the vegetation base of food webs (see Table 4–3 for references). Assimilation of ^{137}Cs in organisms is high, but rapid metabolic turn-over and low retention greatly reduce tissue concentrations that may be available to predator species and reducing transfer to higher trophic levels. In contrast, ^{90}Sr, a chemical analog of the essential element Ca, and to a lesser extent, ^{239}Pu, are stored in tissues seldom consumed by other organisms (e.g., exoskeleton, shells, and bone).

In terrestrial systems, a route less influenced by radionuclide speciation in soil and water is the consumption of radionuclides that may be directly deposited to foliar or skin surfaces from fall-out or resuspended particles. This route of entry along with plant uptake has been important in the transfer of radionuclides to humans via consumption of cow's milk (Belli et al., 1989) and hunted herbivores such as mule deer (*Odocoileus hemionus*) and caribou (*Rangifer tarandus*) following atmospheric releases of nuclear material (Hove et al., 1990; Howard et al., 1991; Voors & Van Weers, 1991).

4–4 RADIONUCLIDE TOXICITY TO BIOTA

In addition to knowledge of the relative chemical availability of a particular radioisotope within the soil, porewater, and groundwater as well as the possible routes of exposure, uptake and bioaccumulation, actual impacts to the ecosystem will depend on the sensitivity of a receptor species (the subsequent deleterious effects the radioisotope may produce). These negative effects may be both from a chemical or radio toxicity perspective. Relative chemical toxicity is often based on the concentration and duration–frequency of exposure to the organism(s) that promotes deleterious effects. For the assessment of possible radiological impacts to living organisms, however, there are four important points including, the amount of the material present (expressed in Becquerels, Bq), the persistence of the radioisotope (physical half-life), the energy that the emissions may contain (expressed in MeV), and the resulting absorbed dose (Grays, Gy), or dose equivalent (Sievert, Sv).

4–4.1 Mechanisms

When ionizing radiation enters a biological system from external or internal exposure, it induces the formation of free radicals and other reactive molecules (Fridovich, 1976; Upton, 1982; Halliwell & Gutteridge, 1985). These chemical products may then react with biologically important molecules within the cell. The altered macromolecules may adversely affect cell function and viability (unless restored by cellular repair mechanisms). Resulting cytopathological effects are numerous and varied but are indistinguishable from those induced by other types of cellular injury. The outcome of the ionizing damage encompasses growth, development and reproductive abnormalities, genetic changes, life-span shortening, cancer, and death.

Biota can receive damaging exposures to ionizing emissions without coming into direct contact with the radionuclide. It is often this remotely induced damage that inspires much of the public's concern for environmental contamination by radioactive materials; however, only those isotopes that emit high energy gamma rays and/or beta particles pose a external radiation hazard (Table 4–4). Low energy alpha particles are the least penetrating of the ionizing emissions (e.g., unable to penetrate clothing or paper) and seldom cause damage to organisms from external exposure. Beta particles, although capable of greater penetration than alpha particles, cannot penetrate skin and therefore pose a limited hazard from external exposure to soft-bodied animals within close proximity of the radionuclides. Only gamma emitters (Co and Cs) are capable of traveling long distances and penetrating materials (sediment//soil, water, wood, skin) that otherwise form natural barriers to alpha and beta particles (Krivolutzkii & Pokarzheviskii, 1992) in the environment.

Significant damage, however, can result from ingested or absorbed radionuclide emissions, particularly the highly ionizing alpha particles. Because the radionuclide continues to radiate the organism until excreted in some way (feces, urine, milk, fetus, or egg), or until its radioactivity decays (Cerveny & Cockerham, 1986), the distribution and retention kinetics of the radionuclide within the organism greatly affect the toxic response.

These kinetics are dependent on the physical and chemical characteristics of the radionuclide, rate of metabolism, age of the organism, and mineral–element content of the diet or surrounding medium. Strontium-90, as a metabolic analog of Ca, is readily absorbed by the very young (McClellan, 1964). Absorbed ^{90}Sr is deposited primarily in Ca rich tissues such as bone and has a long biological half-life (15 yr in mammals) as well as a long physical half-life (29 yr) with associated damage to marrow cells and high incidence of neoplasia in bone and bone-related tissues (McClellan & Jones, 1969). In contrast, ^{137}Cs, which behaves as an analog of K, also is rapidly absorbed but is distributed throughout the entire organism (whole-body irradiation), particularly to metabolically active tissues (Boecker, 1972). Excretion is rapid and though ^{137}Cs has a longer physical half-life (30 yr), its biological half-life is short (4.4–70 d in wild and domestic mammals; Winsor & O'Farrell, 1970) with the resultant damage a consequence of irradiation of active, proliferating cells of the hematopoietic system and gastrointestinal tract (Boecker, 1972).

Although human populations measure deleterious impact of radiation exposure by the incidence of genetic damage and/or cancer, these molecular and cellular changes cannot be used to characterize population and community health in the face of radiation exposure. This is because most organisms are too short-lived

Table 4–4. Chemical and radiological toxicity of radionuclides in aquatic and terrestrial ecosystems.

				Chemical toxicity†						
				Aquatic‡			Terrestrial‡			
Isotope(s) (radiation)§	Chemical analog	Exposure external	Hazard internal	Algae	Invertebrate	Fish	Plant	Invertebrate	Mammal–Bird	References
$^{235,238}U$ (alpha)	None	No	Yes	**H**	**H**	**M**	**H**	**L**	**VH**	Poston et al., 1984 Gus'kova et al., 1966 Gross & Heller, 1946 Bringman & kuhn, 1959 Murthy et al., 1984 Sheppard & Evenden, 1992 Sheppard et al., 1983 Hyne et al., 1991 Tazwell & Henderson, 1960 Parkhurst et al., 1984 Davies, 1980 Bywater et al., 1991 ATSDR, 1990
^{60}Co (beta gamma)	Co	Yes	Yes	RT	**H**	RT	**H**	RT	**M**	Baudoin & Scopa, 1974 Berry, 1978 NIOSH, 1987
$^{134,137}Cs$ (beta gamma)	K	Yes	Yes	RT	**L**	RT	RT	RT	**M**	Baudoin & Scoppa, 1974 Cochran et al., 1950

$^{89,90}Sr$ (beta)	Ca¶	Yes	Yes	U	**L**	RT	RT	RT	**L-M**	Baudoin & Scoppa, 1974 Venugopal & Luckey, 1978
^{239}Pu (alpha)	None	No	Yes	RT	RT	RT	RT	RT	RT	Myers, 1989 Whicker et al., 1973 Fritsch et al., 1987 King, 1964 Woodwell, 1970 Underbrink & Sparrow, 1974 Oakberg & Clark, 1964 Kushniruk, 1964 Abraham, 1972
^{99}Tc (beta)	S	Yes	Yes	**L**	U	U	**VH**	RT	RT	Landa et al., 1977 Cataldo et al., 1987 Masson et al., 1989
^{3}H (weak beta)	H	No	Yes	RT	RT	RT	RT	RT	RT	Osborne, 1972 Stannard, 1973

† Chemical toxicity classes: algae (mg L^{-1}) and plants (μg g^{-1} of soil), low (L) = >100, moderate (M) = 1–100, high (H) = 1–0.1, very high (VH) = <0.1 invertebrates 96h LC50 (mg L^{-1} or mg kg^{-1}), low (L) = >1.0, moderate (M) = 0.01–1.0, high (H) = 0.001–0.01, very high (VH) = <0.001; fish (96 h LC50 (mg L^{-1}), low (L) = >100, moderate (M) = 1–100, high (H) = 0.1–1.0, very high (VH) = <0.1; birds and mammals LD50 (mg kg^{-1}), low (L) = >1000, moderate (M) = 201–1000, high (H) = 41–200, very high (VH) = <40. Radiation toxicity (RT) = radiological toxicity greater than chemical toxicity. See references for ^{239}Pu for discription of ionizing radiation toxicity. U, unknown toxicity. Chemical toxicity given in bold letter.

‡ Varies with water/or soil chemistry.

§ Relative internal radiation toxicity = alpha >> beta >> gamma.

¶ High radiation toxicity risk in vertebrates because of long effective half-life of Sr in bone.

in nature for cancer to be expressed and damaged or debilitated organisms are rapidly removed from populations. Therefore, measurable population–community attributes such as reproductive success and species diversity are commonly used to assess the impact of environmental radionuclides on ecosystem health.

4–4.2 Relative Toxicity Responses

Organism sensitivity to ionizing radiation varies greatly within taxa (Table 4–4). In general, larger herbivorous mammals have proven to be the most vulnerable to ionizing radiation followed by (in descending order of sensitivity) the smaller mammals, birds, herbivorous insects, filter-feeding aquatic invertebrates, higher plants, and lower plants. Simpler life forms including unicellular plants and animals, bacteria, and viruses are more resistant to the lethal effects of radiation (Krumholz et al., 1957; Whicker & Schultz, 1982; Bond et al., 1965; Gleiser, 1953; Rust et al., 1954). Other biological factors that modify the toxicologic response of organisms to ionizing radiation include age, sex, O_2 tension, and nutritional and metabolic status of the organism. Organisms in the rapid growth stage of their life cycle are more radiosensitive than mature organisms.

Plants are relatively resistant to ionizing radiation. The most resistant species were the ones commonly found in disturbed places, i.e., generalists capable of surviving a wide range of conditions. Mosses and lichens survived exposures >1000 Roentgen (R)/d (1 R is 2.58×10^{-4} Coulomb kg^{-1} in air from x-rays or gamma rays) while no higher plants survived >200 R/d. Oak (*Quercus* sp.) trees survived up to 40 R/d, whereas pine (*Pinus* sp.) trees were killed by 16 R/d (Woodwell, 1970). At Chornobyl the pine forests exhibited death at exposures calculated to have ranged from 80 to 100 Gy (8000 to 10 000 rad)and abnormal growth patterns at levels as low as 3 to 4 Gy (300 to 400 rad; Eisler, 1995). The sensitivity of various plant species appears to be related to the cross-sectional area of the nucleus in relation to cell size; the larger the nucleus and chromosome volume, the more sensitive the plant (Underbrink & Sparrow, 1974).

Although viability and reproduction are reduced in insects exposed to ionizing radiation, the level of exposure required to induce these effects is quite high. Sterilization of the screw worm fly (*Callitroga*) occurred at 500 R, whereas the fruit fly (*Drosophila*) required an exposure of 16 000 R to induce sterilization, and the powder post beetle (*Lyctus*) required 32 000 R. The LD50 (dosage required to kill 50% of a test population) for adult fruit flies was about 10^5 R. The LD50 for fly eggs was about 190 R (Packard, 1936). Reduction of egg viability was observed in the European corn borer after exposure to 2500 R (Walker & Brindley, 1963). In long-term field experiments with high levels of ^{239}Pu in chernozem soils [65 Tbq m^{-2}, plowed to a depth of 25 to 30 cm and sown in wheat (*Triticum aestivum* L.)], the radionuclide was shown to decrease the population density of earthworms (*Lumbricidae* sp.) and insect larvae by 50% during a period of 3 yr (Krivolutskii et al., 1992). Microarthropod populations were decreased by a factor of 7.5. Plutonium was particularly radiotoxic to those micorarthropod species that have a fast rate of development such as gamasid and throglyphoid mites. The density of small acariform (*Acarina* sp.) mites decreased by a factor of 18 in the Pu contaminated plots. Finally, after 18 yr, macrofaunal populations were comparable to those in control plots when the major portion of Pu-241 had been transformed to Am-241 (Krivolutskii et al., 1992). Following the Chornobyl accident Krivolutzkii and Pokarzhevskii (1992) reported that soil mesofaunae [mites, juvenile specimens of macrofaunal groups, and springtails (*Collembola* sp.)] located 3 km from the reactor exhibited a precipitous decline in population (>90%) within the first 3 mo after receiving dosages higher than 89 Gy. Those most sensitive appeared to be the eggs and juveniles (Krivolutzkii & Pokarzhevskii, 1992); however, within 3 yr population levels had returned to control levels (most likely through migration activity) even though gamma radiation levels remained at greater than 1 mR h^{-1} (Krivolutzkii & Pokarzhevskii, 1992).

Lethal effects are observed in most mammals at acute radiation doses in excess of 2 Gy (Myers, 1989). Mortality results from failure of the hematopoietic system. Radiation injuries most significant to animal populations are those affecting life span and reproduction. In laboratory rodents, survival time was shortened by 10% when the radiation dose was more than half the LD50/30 dose (dosage required to kill 50% of the population in 30 d) for that species (Bacq & Alexander, 1961). Reproductive cells are the most radiosensitive cells of the mammalian system, and fertility in laboratory animals has been reduced by ionizing radiation. Acute exposures of a few hundred R rendered male mice temporarily sterile. Smaller doses (1 Gy or 100 rad) resulted in permanent sterility in females (Oakenberg & Clark, 1964). The difference in reproductive sensitivity to radiation is due to the presence of spermatogonia in males and the lack of a comparable regenerative stem cell in the ovary (i.e., oocytes cannot be replaced once they are destroyed). Male organisms as diverse as the grasshopper, fruit fly, silkworm, and guinea pig have shown very similar responses to ionizing radiation because of the similarity of spermatogenesis (French, 1965). In addition to causing sterility, acute exposure can decrease the number of young produced by the irradiated parent. An acute dose of 30 R in young female mice significantly reduced the number of offspring produced. Exposures of about 400 R in males decreased production of young (Rugh, 1964). The reduction in young is attributable to dominant lethal mutations. Somatic effects on the female parent also contribute to loss of young (Touchberry & Verley, 1964).

As a group, birds appear to be at greater risk of beta-gamma radiation exposure than other wild animals. About 33% of birds collected from a contaminated area had radiation counts above the background level, whereas only 7% of the mammals collected, and 5% of the reptiles collected had higher-than-background counts. The higher rate of contamination was attributed to the grit-use behavior of birds (Bellamy et al., 1949). The LD50/30s for wild bird species exposed to ionizing radiation range from 4.9 to 25 Gy (average 7.9). External exposure to gamma radiation of up to 600 R did not result in the mortality of embryos or nestlings of passerine species (Zach & Mayoh, 1984, 1986a,b) while chronic irradiation with 9.60 Gy >20 d reduced hatchability in domestic chicken (*Gallus domesticus*) and black-headed gull (*Larus ridibundus*) eggs (Phillips & Coggle, 1988). In pheasants (*Phaisanus colchincus*), single exposures to the ovaries of 500 to 2025 R and cumulative exposures from 500 to 5316 R did not affect egg production, plumage coloration, or ovarian tissue structure (Greb & Morgan,

1961). Chronic exposure to 1.0 Gy resulted in far more severe growth depression of nestling passerines than single doses of 3.2 Gy (Zach & Mayoh, 1982, 1986a; Guthrie & Dugle, 1983).

Most of these toxicity studies have dealt with a single species or type of organism and most have relied on laboratory settings. The transfer of radionuclides from soil to an ecosystem will obviously affect more than one species. When an ecosystem is viewed as an assemblage of several diverse species, it is the interaction between these biota that ultimately determines the ecosystem's "health" and thus it is essential to consider a larger scope.

4–5 POPULATION–COMMUNITY IMPACTS

Historically a number of studies using large gamma sources (^{137}Cs and ^{60}Co) for short-term and chronic exposures of ecosystems have provided insight into the response of population assemblages to large source irradiation. In these, plant communities generally changed significantly while animals moved out of the area in response to the habitat changes (Woodwell, 1967). A reduction in species diversity was observed in this terrestrial system at dose rate as low as 1 Gy d^{-1}. More recently a very similar response was observed following the Chornobyl accident. Here also stands of pine trees died, large animals moved out of the area, and the abundance and species diversity of the pine-forest-litter fauna [mites, springtails, beetles (*Coleptera* sp.), earthworms, spiders (*Araneae* sp.)] decreased. Soil organisms were less affected because of shielding from the soil (Krivolutzkii & Pokarzhevskii, 1992). Contamination of terrestrial ecosystems at several other nuclear facilities, however, have not resulted in measurable population or community level effects (Dunaway & Kaye, 1961; Childs & Cosgrove, 1976; Evenson et al., 1978; Millard et al., 1990; Zach & Mayo, 1984). In these cases the threshold level for adverse effects on natural mammalian populations appeared to be about 10 mGy d^{-1} (Howard et al., 1991). The International Atomic Energy Agency recognizes dose rates of ≤mGy d^{-1} as protective of terrestrial ecosystems (IAEA,1992).

4–6 SELECTED RADIONUCLIDES

Radionuclides such as ^{238}U, ^{60}Co, ^{137}Cs, and ^{90}Sr vary widely in their decay rates, sources, and chemical behaviors in soils and sediment environments. Uranium-238 is of natural origin and its native occurrence is largely dependent on the regional mineralogy. The other radionuclides are primarily man-made and their occurrence in the environment will be dependent almost entirely on manufacture and release. The objective of this section is to present a more detailed discussion of the soil chemistry governing the mobility of four radionuclides (^{238}U, ^{60}Co, ^{137}Cs, and ^{90}Sr), their primary routes of transport within differing trophic levels, their relative toxicity to individual organisms, and observed effects within biological communities and populations.

4–6.1 Uranium

Uranium can exist in the +3, +4, +5, and the +6 oxidation state; however, only the +4 and +6 oxidation states are typically observed in the environment. Naturally occurring U typically contains 99.283% ^{238}U, 0.711% ^{235}U, and 0.0054% ^{234}U by weight. The half-lives of these U isotopes are 4.51 × 10^9 yr, 7.1 × 10^8 yr, and 2.47 × 10^5 yr, respectively. Uranium is of great importance as a nuclear fuel for reactors, as starting material in the production of Pu (^{239}Pu), and as depleted U for inertial guidance devices, shielding materials, and military ordinance. Geologically, U occurs most often as U(IV) minerals, such as pitchblende, uraninite, carnotite, and autunite and as U(VI) uranophane. It also is found in phosphate rock, lignite, and monazite sands at levels that can be commercially recovered. In the presence of lignite and other sedimentary carbonaceous substances, U enrichment is believed to be the result of UO_2^{2+} transport, adsorption, or complexation by humic materials, and reduction of dissolved U(VI) to U(IV), followed by the formation of uraninite.

4–6.1.1 General Soil Chemistry

The aqueous U(VI) uranyl cation (UO_2^{2+}) is the most stable ion in oxidizing solutions; the U(III) species easily oxidizes to U(IV) under most environmental conditions, while the U(V) aqueous species (UO_2^+) readily disproportionates to U(IV) and UO_2^{2+}. In aqueous systems, the U(IV) species will not be present to any great degree as a result of precipitation; probably uraninite (UO_2) or some higher O/U solids ratio with an O/U ratio between 2.3 and 2.7 (Maynard, 1983 as reported by Bruno et al., 1991). The average U concentrations in natural waters under reducing conditions are between 3 and 30 ppb (Bruno et al., 1991); this is consistent with equilibrium concentrations supported by $UO_2(s)$ (Bruno et al., 1988). In the absence of any complexing agents, U(IV) is expected to hydrolyze to form mononuclear hydroxo complexes, such as $U(OH)_n^{4-n}$ (Langmuir, 1978). It has been suggested that complexation of (IV) actinides form stable complexes with natural organic humic and fulvic acids (log K of 12-16; Allard & Persson as reported by Birch and Bachofen, 1990); hence U(IV) could form stable organic complexes, increasing the aqueous concentration of U(IV).

Aqueous uranyl tends to form strong complexes with inorganic O-containing ligands such as hydroxide, carbonate, and phosphate. Aqueous UO_2^{2+} hydrolyses to form a number of aqueous hydroxo complexes including UO_2OH^+, $(UO_2)_2(OH)_2^{2+}$, $(UO_2)_3(OH)_5^+$, and $UO_2(OH)_3^-$. In aqueous systems equilibrated with air or higher pCO_2 waters at near neutral to high pH, the carbonate complexes [$UO_2CO_3^0$, $UO_2(CO_3)_2^{2-}$, $UO_2(CO_3)_3^{4-}$] will dominate, but at lower pH the hydrolysis species will dominate as $CO_{2(g)}$ solubility in water decreases. Phosphate-UO_2^{2+} complexes ($UO_2HPO_4^0$,$UO_2PO_4^-$) could be important in aqueous systems with a pH between 6 and 9 when the total concentration ratio $(PO_4^{3-})_T/(CO_3^{2-})_T$ is greater than 10^{-1} (Sandino & Bruno, 1992; Langmuir, 1978). Complexes of SO_4^{2-}, F, and possibly Cl^-, are potentially important U(VI) species where concentrations of these ions are high; however, their stability is considerably less than the carbanato and phosphato complexes (Grenthe, 1992).

Complexation of UO_2^{2+} by dissolved fulvic acid has been suggested to facilitate U transport to and in groundwaters (Bonotto, 1989). The stability of a U(VI)-humic acid aqueous complex has been determined to be at least 10^3 greater than that of a Ca^{2+}-humic complex (Shanbhag, 1979 as reported by Idiz et al., 1986). Other than this information, little is known definitively about aqueous U(VI)-humic acid (or fulvic acid) complexation; however, the carbonate complexes discussed before are very competitive ligands above pH 6. Shanbhag and Choppin (1981) concluded that the humic acid complexes with uranyl would not be of importance in sea water because of the significant presence of bicarbonate and carbonate at pH 8. Likewise, solutions of Na-bicarbonate-carbonate at pH 7 to 10 proved to be very effective at leaching U(VI) out of Holocene peat (Zielinski & Meier, 1988), suggesting a U(VI)-carbanato complex competes well with humic substances. In a study of purified Pettit peat bog humic acid, Idiz et al. (1986) found the adsorption of UO_2^{2+} increased as pH increased until about pH 5 and then decreased as a result of the formation of the carbonato complexes.

Because of the presence of organic and inorganic ligands, it is not believed that precipitation or coprecipitation of a solid phase will control the solubility of U(VI) in most natural systems; however, if reduction of U(VI) to U(IV) occurs, precipitation of UO_2 is conceivable. Reduction has been suggested to be the result of (i) reduction of U(VI) by organic matter itself, (ii) indirect reduction by hydrogen sulfide, and (iii) microbial reduction of U(VI) (Andreyev & Chumachenko, 1964; Disnar & Trichet, 1983). Regardless of mechanism, the product of reduction has been observed to be uraninite (UO_2).

Early studies by Szalay (1964) demonstrated that peat-derived humic acid could concentrate U by a factor of about 10 000 times over the associated waters. In studies of UO_2^{2+} movement through peat columns, measured K_d values were observed to range from 700 to 18 600 L kg^{-1}. Competition for exchange or complexation sites associated with particulate organic matter by Ca^{2+} or other macroelement species (such as Mg, Na, or K) should not be significant, if the U(VI)-humic acid complex determined for dissolved materials is indicative of the particulate-U(VI) stability. As with dissolved organic matter complexation, however, competition from carbonate may diminish the effectiveness of particulate organic matter to sequester U from solution.

Formation of complexes between U(VI) and organic ligands, such as humic and fulvic particulates, has been studied most frequently because of interest in ore-forming environments. At ambient temperatures (~25°C), U is adsorbed to humic substances through rapid ion exchange and complexation processes with carboxylic and other acidic functional groups without reduction to the U(IV) species (Idiz et al., 1986; Boggs et al., 1985; Shanbhag & Choppin, 1981; Nash et al., 1981; Borovec et al., 1979; Szalay, 1964). It has been suggested that U(VI) adsorbed (or fixed) to organic matter may undergo reduction to U(IV) followed by precipitation of UO_2 (Andreyev & Chumachenko, 1964; Disnar & Trichet, 1983). In studies with lignite, however, the uranyl species formed a stable complex with the lignite without subsequent reduction; reduction occurred only at elevated temperatures (Nakashima et al., 1984); however, organic matter does have the capacity to act as a reductant, most notably because of the presence of quinone, sulfone, and reduced metal porrphryn (tetrapyroles) moeities (Macalady et al., 1986).

Uranium sorption to Fe oxides and smectite clay has been shown to be extensive in the absence of CO_3^{2-} (Kent et al., 1988; Hsi & Langmuir, 1985; Ames et al., 1982; McKinley et al., 1996), but in the presence of CO_3^{2-} and organic complexants sorption was shown to be substantially reduced or severely inhibited (Bond et al., 1990; Hsi & Langmuir, 1985; Kent et al., 1988; Ames et al., 1982). The importance of U(VI) sorption to inorganic solid phases, such as clays and Fe oxides will be dependent on the quantity of these materials, the pH, the presence of dissolved and particulate organic matter, and dissolved carbonate.

4–6.1.2 Uranium Transport and Bioconcentration

Transport of U from soil to biota has been well documented (Dreesen et al., 1982; Moffett & Tellier, 1977; Mahon, 1982). It has been assumed in these studies that the nature of the soil determined the amount of bioavailable U. For example, soil conditions that favored decreased sorption or formation of soluble complexes with U enhanced uptake. Swiss chard (*Beta vulgaris* L.) grown in sandy soils contained U at concentrations 80 times higher than chard grown in peat (Sheppard et al., 1983); however, in a more recent study of the effect of 11 different soil types on bioavailability indices for uranium (Sheppard & Evenden, 1992), no correlation between plant or invertebrate uptake and soil parameters was observed. The soils were treated with up to 10 000 mg U kg^{-1} soil and varied with regard to texture, clay, organic content, pH, background U content, and cation-exchange capacity. Uranium concentrations in plants and earthworms were not linearly related to U concentrations in the soil. Thus, a single value for use as a conservative concentration ratio for a soil type could not be determined, and the implication is that other reported concentration ratios for U in plants should not be applied to soil concentrations outside those for which the concentration ratio was determined.

Uranium appears to be restricted to the root system of plants and may be precipitated on the outer root membrane rather than accumulated in the interior of the root (Sheppard, 1985). Little U enters the root sap system and virtually none was reported to be translocated from the soil to the aboveground plant tissue (Sheppard, 1985; Van Netten & Morley, 1983). Several concentration ratios have been reported for shoots, leaves, fruits and seeds, but all of these ratios were <1. Ng et al. (1982) reported U concentration ratios for edible portions of food crops (wet plant/dry soil) from 1.7×10^{-7} to 2.0×10^{-2}. The range of concentration ratios for the edible portions of pasture plants was 1.6×10^{-6} to 8.5×10^{-1} (Ng et al., 1982).

Mahon (1982) studied the trophic transfer of U in several wildlife food chains. For the terrestrial food chains studied, there was a drop in body burden of U by one order of magnitude for each trophic level (Mahon, 1982). The transfer coefficient from vegetation [grouseberry forb (*Vaccinium scoparium*), lichens (*Bryoria freemontia* and *Alectoria sarmentosa*), and grass (*Calamagrostis rubescens*)] to deer (*Odocoileus hemionus*) was 0.7. Bioconcentration ratios for chipmunks (*Eutamius amoenus*) and herbivorous mice feeding on fireweed seed heads (*Epilobium angustifolium*) and grouseberry were 0.5 and 0.26, respectively. The top predator in this food chain was an avian predator, the raven (*Corvus corvus*), which was found to have <5 ppb U in its tissues. Similar U food chain transfers were seen in domestic animal foragers. Transfer coefficients from soil surface layer (0 to 30 cm) to forage grass ranged from 2.67×10^{-5} to 2.98×10^{-4}. Forage grass to sheep transfer coefficients were 2.5×10^{-5} to 2.4×10^{-4} in meat.

4–6.1.3 Uranium Toxicity

Some of the radionuclides also have a chemical toxicity unrelated to radioactivity. Of the radionuclides that display chemical toxicity (Table 4–4), only U is highly toxic to a broad range of organisms. Chlorosis, early leaf abscission, and reduction in root growth were the toxic symptoms induced by U in plants. These symptoms were probably caused by reduced root absorption capacity and dysfunction of xylem and phloem tissue from U precipitation (Cannon, 1960). In soil-grown plants, overt effects on growth and survival were not seen below 1000 mg U kg^{-1} of soil (Sheppard et al., 1992). Earthworm survival also was decreased at the same concentrations or higher (Sheppard & Evenden, 1992).

In terrestrial organisms, mammals are particularly sensitive to U. Kidney and bone tissues are the main targets of both the radiation and chemical toxicity. Of these two, kidney tissue, particularly the terminal segment of the renal proximal tubule (Avasthi et al., 1980; Haley, 1982), is the most sensitive and is considered to be the key target organ for hazard assessment (Diamond, 1989). The critical target organ for chronic exposure to U is the skeletal system (Adams & Spoor, 1974; Guglielmotti et al., 1984); however, skeletal burden–exposure relationships have not been determined for laboratory or wild mammal species. Although the sensitivity of many organisms to U chemical toxicity has been established, most studies have used chemical forms of U nitrate that are highly absorbable species not found outside the laboratory setting. Under environmental conditions very little U is available for uptake or in a form that will be absorbed from the gastrointestinal track of vertebrate animals.

4–6.2 Cobalt

Cobalt occurs naturally in the environment and its crustal abundance is about 25 mg kg^{-1}. Cobalt is an essential nutrient and has a number of industrial applications ranging from catalysts to use in alloys. Radioactive isotopes of cobalt (^{57}Co, ^{58}Co, ^{60}Co) result mainly from neutron activation and have been used for both medical and industrial uses. For medical diagnostic uses, radiopharmaceuticals are made using ^{57}Co and ^{58}Co, namely ^{57}Co-bleomycin, ^{57}Co-vitamin B_{12}, and ^{58}Co-vitamin B_{12} (National Council on Radiation Protection and Measurements, 1989). Teletherapy units using thousands of GBq of ^{60}Co are employed for external beam radiation therapy. The radioactive source unit used in radiation therapy units is made up of thousands of ^{60}Co pellets where a single pellet may contain up to 2.8 GBq of ^{60}Co activity (Mettler & Ricks, 1990). In industry, ^{60}Co sources (370–3700 GBq) are used as radiography units for evaluating the integrity of metallic objects, such as pipe welds and airplane propellers

(Mettler & Ricks, 1990). Cobalt-60 sources also are used for sterilization of medical supplies, chemicals, or other objects and these stationary units may contain up to 185 PBq of ^{60}Co.

4–6.2.1 Soil Chemistry of Cobalt

In the uncomplexed form, only the cobaltous ion, Co(II), is stable in aqueous and soil environments; however, the cobaltic ion, Co(III), does form very stable complexes with certain ligands (Trischan et al., 1981).

$$[Co(H_2O)_6]^{3+} + e^- = [Co(H_2O)_6]^{2+} \; E^0 = 1.84 \text{ V} \qquad [15]$$

$$[Co(NH_3)_6]^{3+} + e^- = [Co(NH_3)_6]^{2+} \; E^0 = 0.1 \text{ V} \qquad [16]$$

As can be seen from Eq. [15] and [16], complexation with N-containing organic ligands or NH^3 could conceivably make the Co(III) ion a viable oxidation state in the soil environment. One such species that is known to be present is the cyanocobalamin, or vitamin B_{12}, which is found in soil, water, sewage sludge, manure, and dried estuarine mud (Trischan et al., 1981; Beck & Brink, 1976). Inorganic ligand-Co(II) ion pairs that could be of importance in soils are $CoOH^+$, $Co(OH)_2^0$, $Co(OH)_3^-$, $CoCl^+$, $CoSO_4^0$, and $CoCO_3^0$. All of these species could act to decrease the free divalent metal Co^{2+} species (Trischan et al., 1981; Morel, 1983). Most inorganic species of Co(III) would probably not be stable in soils.

Both NTA and EDTA form strong complexes with Co(II) and Co(III) ions; Co(II)NTA and Co(II)EDTA stability constants are, in general, between 0.2 to 3 log units lower than other divalent transition metals, such as Pb, Ni, Hg, and Cu. The log stability constant for Co(II)EDTA is approximately 16.2, however, the log stability constant for Co(III)EDTA is 36.0. As the stability constant suggests, the Co(III) complex is extremely strong (Trischan et al., 1981). The conversion of Co(II) to Co(III) in Eq. [16], and other complexes such as Co(III)EDTA, may be accomplished easily in aerated systems (Trischan et al., 1981) or in the presence of an oxidant like MnO_2 (Crowther et al., 1983; Murray & Dillard, 1979). The stability of Co(III)EDTA [and presumably other Co(III) organic complexes] in environments where one would expect Co(II) is evidenced by the persistence of Co(III)EDTA in the soil and subsoil environment (Jardine et al., 1993). While NTA and EDTA are not common in natural systems, they are often used in manufacturing and decommissioning processes and have been codisposed with radioactive Co at several U.S. Department of Energy sites (Zachara & Riley, 1992).

The majority of dissolved organic ligands in soils are of a complex nature that is poorly defined in terms of the ligand structure, metal-ligand stability, and other chemical-physical parameters; however, the interaction of metals with dissolved organic fractions (such as fulvic acids) in the natural environment has been studied and could serve as an estimate of potential Co-DOC complexation. Conditional stability constants have been determined for a variety of dissolved humic and fulvic material (Stevenson & Fitch, 1986). Cobalt and other metals tend to be complexed by the weakly acidic, phenolic, carboxyl, and enol groups of fulvic acid (Ephraim et al., 1989). Importantly, the stabilities of the complexes at pH 8 follow the approximate order of the Irving-Williams stability series (Matuora et al., 1978). Complexation with fulvic acid-type ligands could play an important role in of Co isotope mobility in soils.

Precipitation of pure Co solid phases is not likely to be a dominant process in most soils because the activity of the free metal species (Co^{2+}) is not likely to exceed the K_{sp} of any solid Co phase. Morel et al. (1975) using a chemical equilibrium model suggested that the two most likely Co solid phases, CoS and $CoCO_3$, would be relatively soluble even when considering only inorganic speciation. In the presence of dissolved organic ligands, the free metal species would be expected to decrease further. The potential for Co to be coprecipitated with other solid phases (such as Fe oxide, $CaCO_3$, and other carbonates) is much more feasible (Ainsworth et al., 1994).

Cobalt sorption in soils and sediments may occur as a result of interactions between the dissolved Co and (i) inorganic particulates (Fe and Al oxide, carbonates), (ii) particulate organic matter, and (iii) clay minerals. The magnitude of these interactions will vary, dependent on pH, aqueous speciation, ionic strength, and ionic composition. Where present, calcium carbonate sorption of Co is thought to be important in both soils and groundwater environments and the sorption process has been described as an exchange between the Co free metal species and the surface Ca (Zachara et al., 1991).

Oxides and hydroxides of Al and Fe have long been recognized as important sorbents for a number of trace metal contaminants whose adsorption to, and desorption from, the surface of oxides influence sorbate behavior in the environment. Adsorption of trace metals, such as Co, to the oxide and hydroxide surface–water interface has been well characterized (Dzombak & Morel, 1990; Girvin et al., 1993; Ainsworth et al., 1994). Cobalt adsorption to these oxides is cation-like (that is, sorption increases with increasing pH). The adsorption edge (fractional sorption as a function of pH) is very steep, typically going from 0 to 100% sorption over a pH range of about two units. Sorption of Co is minimally affected by changes in ionic strength; however, Co sorption to Fe and Al oxides, in the presence of a strong complexing ligand, such as EDTA, reverses the Co sorption edge to where the CoEDTA behavior is more ligand-like; that is, sorption increases as pH decreases (Girvin et al., 1993). Additionally, the Co(II)–oxide interaction is thought to be an inner-sphere complex, but the CoEDTA–oxide interaction is characterized as an outer-sphere complex. The latter can be greatly affected by ionic strength changes. The presence of Co(III) as a complex can have significant consequences on Co adsorption to oxides. In studies of Co(III)EDTA to Al_2O_3 surface, this species behavior was observed to be more ligand-like and sorption was reduced by about 70% compared with Co(II)EDTA (Girvin et al., 1993). It is assumed that other complexing ligands, including organic ligands associated with soils, could affect Co(II) sorption in a similar manner as that observed with EDTA.

Cobalt like Mn, is poorly sorbed to particulate humic materials. Schnitzer and Kerndorff (1980) investigated the sorption of a number of trace metals on particulate humic acids and found that sorption followed the order Hg >Fe > Pb >Cu = Al >Ni > Cr = Zn =Cd = Co = Mn at pH 3.7. While the order shifted slightly with increasing pH, Co and Mn were always the least sorbed metals. Data using spectroscopic techniques have suggested the interaction between Mn and humic materials is an outer-sphere complex, the Mn retains its waters of hydration $[Mn(H_2O)_6^{2+}$, and the complex is electrostatic in nature (Gamble et al., 1977; McBride, 1978). This may explain the weak sorption of Mn and, by association Co, to humic materials. Because of the quantity of particulate organic material present in soils, however, even weak interactions between Co and particulate organic material could influence Co sorption.

4–6.2.2 Cobalt Transport and Bioaccumulation

There is no evidence that Co has any direct role in the metabolism of higher plants (Marschner, 1995). It is, however, critical to N_2 fixation in leguminous species both for *Rhizobium* infection and N_2 fixation enzyme synthesis (Dilworth et al., 1979). In legumes Co is primarily retained in the roots (up to 25 times that of the shoot; Robson et al., 1979) but may achieve concentrations of up to 730 ng g^{-1} in the seed (Robson & Mead, 1980).

Cobalt progressively decreases in concentration through terrestrial invertebrate food chains, generally averaging one-half the concentration of plants after two trophic exchanges (Reichle & Crossley, 1969). The general terrestrial concentration factors for Co are listed in Table 4–3. Voshell et al. (1985) determined the uptake of Co at the various trophic levels of an aquatic insect community and reported that where the algae and plankton concentrated Co from water and had transfer coefficients of 11 800 and 20 600, respectively, herbivorous insects (coleopteran adults) feeding on the filamentous algae had an algae-to-insect transfer coefficient of 0.1. The transfer coefficient of saprovores (mayflies and chronomid midges) was 0.04. Saprovore-to-carnivore (damsel flies and dragon flies) transfer was 0.01. The transfer coefficient for predators (*Notonectidae*) that consumed only the body fluids of their prey was 0.13.

4–6.2.3 Cobalt Toxicity

As described above ^{60}Co is a strong gamma emitter and thus poses significant radiotoxic properties. As a heavy metal there also are associated toxicities although the literature in terrestrial ecosystem effects is limited. Cobalt exposure is detrimental to seedlings of terrestrial plants (Berry, 1978) and is highly toxic to zooplanktonic species (Baudouin & Scoppa, 1974) in aquatic systems (Table 4–4). The ecological significance of the sensitivity of these taxa to Co lies in their prominent position in aquatic and terrestrial food webs. Ingestion of excessive amounts of Co results in polycythemia (excess formation of red blood cells) in most mammals, however, Co is an essential element and is found in vitamin B12 (0.0434 μg of Co μg^{-1} vitamin B_{12}).

4–6.3 Cesium

Cesium exists in the environment in the 1^+ oxidation state. Stable Cs is ubiquitous in the environment with a crustal abundance of approximately 3.2 mg kg^{-1}; in soils, Cs concentrations range between 0.3 and 25 mg kg^{-1} (Lindsay,

1979). Industrial uses for Cs include use as a catalyst for hydrogenation of certain organic compounds, photoelectric cells, and as O_2 scrubbers. A number of unstable isotopes of Cs exist, but in the context of the present topic only ^{134}Cs and ^{137}Cs are important; half-lives for these two isotopes are 2.05 and 30.23 yr, respectively. The production of these isotopes is through neutron activation and fission (United Nations Scientific Committee on the Effects of Atomic Radiation, 1982). Medical use of ^{137}Cs in the USA is limited to sealed teletherapy sources (Ault, 1989). Sources of ^{134}Cs and ^{137}Cs in the natural environment may be from research, decontamination, and production of sealed sources; however, the greatest sources (and most wide spread) of these radionuclides are typically the result of fallout from atmospheric nuclear testing and nuclear accidents.

The dominant aqueous species in soil and aquatic systems is thought to be the free Cs^+ species. The Cs^+ ion forms only extremely weak aqueous complexes with SO_4^{2-}, Cl^-, and NO_3^- and therefore formation of inorganic complexes is not believed to a major influence on Cs speciation. Cesium is poorly complexed by organic substances, as shown by the relative order of ion complexation to organic substances: Ce > Fe > Mn > Co ≥ Ru ≥ Sr > Cs > I (Bovard et al., 1970). Further, complexation of Cs by chelates (such as EDTA) is believed to be poor due to generally low stability of Cs-chelates and the presence of other elements (such as Ca) at significantly higher concentrations than Cs. Therefore, aqueous speciation and complexation is not thought to greatly influence Cs behavior in the natural environment.

Neither precipitation nor coprecipitation are expected to affect Cs aqueous concentrations or its mobility in soils. Sorption of Cs to organic colloids should follow a relationship similar to that of dissolved organic humic materials discussed above and should therefore not be an important sink for Cs in most soils. It has been frequently demonstrated that Cs becomes associated with the clay mineral fraction of soils. The association of Cs with clay minerals is characterized by high selectivity.

Sorption of Rb, K, NH_4^+, and Cs to clay minerals has been well-studied. The clay interlattice fixation of these cations has long been recognized as the result of their ability to dehydrate and, in this state, fit into the hexagonal holes (radius 1.40 Å) in the exterior O_2 plane of the tetrahedral layer (Bruggenwert & Kamphorst, 1979). The energy gain associated with fixation, in the form of electrostatic attraction, must be balanced against the energy loss due to the dehydration process; this fact, in conjunction with cation radii, explains why cations with high energies of hydration like Ca, Mg, Na, and Li are not fixed and why Rb, K, NH_4^+, and Cs with low energies of hydration are fixed. It has been suggested, however, that fixation of ^{137}Cs in soils, in carrier-free quantities, is an essentially different process from those that predominate in the fixation of macro quantities of Cs, Rb, K, and NH_4^+ (Schultz et al., 1960).

The extent to which exchange predominates would appear to be dependent on (i) the amount and type of clay minerals found in soils, (ii) the quantity of macrocations that can effectively compete with Cs for exchange sites (K, NH_4^+), and (iii) the total mass of Cs. In a number of studies of Cs contaminated environments, however, the presence of only small quantities of clay minerals have been related to the high Cs selectivity and its minimal mobility. In a study of the distribution of radionuclides in streambed gravels, Cerling and Spalding (1982) found that ^{137}Cs was held very selectively by illite in shale fragments within the sediments. Estimated ^{137}Cs K_d's varied widely but the average was 8 460 mL g^{-1} (±5,130 mL g^{-1}) for a series of streambed gravels of differing lithologies and water chemistry. The ^{137}Cs K_d for a contaminated fluvial-sand aquifer material was estimated at 250 mL g^{-1} (±180 mL g^{-1}); the high K_d value was suggested to be related to the presence of the layer silicate mineral vermiculite (Jackson et al., 1980). Essentially no movement of ^{137}Cs was observed during an 18-yr period in a sandy loam soil (6% clay) and arid environment; under the same conditions, however, ^{90}Sr (see below) was distributed through the entire soil profile (Cline & Rickard, 1972; Cline, 1981). Similarly, very low migration rates for global fallout from weapons testing [0.1 to 0.4 cm yr^{-1} for the A horizons of a forest (podzolic Parabrown earth soil) and a grassland soil (Parabrown earth pseudogley) in Germany] have been observed in higher rain fall areas (Schimmack et al., 1989).

4–6.3.1 Cesium Transport and Bioaccumulation

Cesium uptake from soil by a single crop is <0.1% of the soil's content (Menzel, 1963). Prairie grasses concentrate Cs by factors of 0.02 to 5.0, depending on soil conditions and grass species (Schuller et al., 1993). On the basis of Menzel's classification of concentration factors of elements in plants, Cs is considered slightly excluded (Menzel, 1963). Concentration factors for emergent seed plants range from 50 to 600. On the other hand, Voight et al. (1991) reported root transfer factors for ^{137}Cs of 0.002 for grains, 0.002 for potatoes (*Solanum tubersium* L.), 0.0047 for lettuce (*Lactuca sativa* L.), and 0.003 for bush beans (*Phaseolus* sp.). Garland et al. (1983) found concentration factors of 3×10^{-4} for tumble mustard and 0.5 for cottonwood (*Populus* sp.) and willow (*Salix* sp.) leaves. Much of the transfer of ^{137}Cs from plants to herbivores following Chornobyl was through the ingestion of fallout derived surface contamination (Eisler, 1995).

Assimilation of ^{137}Cs from detritus is low (53–65%) because it is incorporated into poorly digested tissue structures (Reichle et al., 1970a). Cesium in herbaceous foliage is more readily available (73–94%), especially in sap-sucking animals such as aphids (about 100%). Cesium progressively decreases in concentration through invertebrate food chains, generally averaging one-half the concentration of plants after two trophic exchanges (Reichle & Crossley, 1969); however, Cs concentrations increase at the higher trophic levels in mammals in part from its chemical similarity to K. A several-fold increase of ^{137}Cs has been reported in plant–mule deer (*Dama* sp.)–cougar food chains (Pendleton et al., 1964) and in lichen–reindeer food chains in Sweden following the Chornobyl accident (Rissanen et al., 1990 as cited in Eisler, 1995). Flesh-eating predators also show high assimilation efficiencies for Cs (i.e., 79–94%). Transfer of Cs from forage to milk in ruminants is about 0.25% (Voors & VanWeers, 1991). Fielitz (1991) reports a feed-to-meat transfer of 0.045 in fallow deer. In the lichen–caribou–wolf chain, ^{137}Cs increased two-fold at each successive link in the food chain (Hanson et al., 1967). The general terrestrial food-chain concentration factors for Cs are listed in Table 4–3.

4–6.3.2 Cesium Toxicity

The toxic response of mammals to Cs resembles Rb and K toxicity. Liver injury, neuroendocrine and neuromuscular disturbance leading to irritability, and convulsions are clinical signs of Cs toxicity (Venugopal & Luckey, 1978). The oral LD50 of Cs (without regard to its radioactive toxicity) is 84.6 mg kg^{-1} body weight as Cs hydroxide. The parenteral LD50s of Cs nitrate, Cs carbonate, and Cs halide compounds range from 716 to 1330 mg kg^{-1} (Venugopal & Luckey, 1978). The greater toxicity of Cs hydroxide is probably due to its caustic action (Cochran et al., 1950). The lowest oral LD50 of noncaustic forms of Cs was 710 mg kg^{-1} in mice (Lewis & Tatken, 1980). Irradiation of female cotton rats in enclosed areas of a natural habitat showed that LD50/15 was 1130 R and that survival time and dose were directly related. At 500 R, a 91% survival rate was observed, whereas only 25% survived 1200 R (Pelton & Provost, 1969). Following the Chornobyl accident Skogland and Espelien (1990) reported a 25% decline in reindeer calf survival in a herd in Norway heavily contaminated with ^{137}Cs. During the same time period a positive correlation between erythrocyte mutagenicity and ^{137}Cs body burdens was reported in Swedish bank voles (*Clethrionomys glareolus* Schreb.) exposed to doses ranging from 4.2 to 39.4 μGy d^{-1} (Cristaldi et al., 1991).

Birds environmentally exposed to ^{137}Cs during breeding season received total dose equivalent rates to the whole body of 9.8×10^{-7} Sv h^{-1} or 2.8 mSv for the whole period of 120 d (breeding season). No reproductive or population effects were observed in even the most contaminated individuals and species (Lowe, 1991). The number of eggs and chicks produced by American coot (*Fulica americana*) colonizing a cooling pond that received low levels of ^{137}Cs were similar to number produced on uncontaminated ponds (Rickard et al., 1981). The coots consumed aquatic plants containing about 407 Bq of Cs g^{-1} dry weight and, inadvertently, sediments containing about 1036 Bq of Cs g^{-1} dry weight (Rickard et al., 1981).

4–6.4 Strontium

Strontium exists in the environment in the Sr(II) oxidation state. While the crustal abundance of Sr is approximately 150 mg kg^{-1}, the total concentration range in soils is between 50 and 1000 mg kg^{-1} (Lindsay, 1979). Strontium is usually present in the surface environment as a carbonate or a sulfate mineral, and can isomorphically substitute for Ca in carbonates and sulfates. The most important of the 12 nonstable isotopes of Sr are ^{85}Sr, ^{89}Sr, and ^{90}Sr. As a result of nuclear weapons testing, ^{90}Sr is distributed widely in the geosphere. The average activity in surface soils in the USA is approximately 3.7 GBq per square mile. The chemistry of Sr and Ca are closely related. As a Ca analog, Sr tends to accumulate in bone (UNSCEAR, 1982).

The dominate aqueous Sr species in natural waters across a broad pH range (2 to 9) is the free divalent Sr^{2+} species. The solubility of Sr^{2+} is not greatly affected by the presence of most inorganic anions because Sr^{2+} forms only weak aqueous complexes with CO_3^{2-}, SO_4^{2-}, Cl^-, and NO_3^-; however, large concentrations of

SO_4^{2-} or elevated alkalinity could play an important role in Sr mobility. Since Sr and Ca form humic acid complexes of similar stability (Stevenson & Fitch, 1986), Sr should not effectively compete with Ca for humic acid ligands because Ca is typically present at much greater concentrations. Therefore, natural organic ligand-Sr species are not thought to greatly affect the solubility of Sr.

While Sr aqueous concentrations in some soil environments may be expected to be controlled by inorganic solid phases, it is unlikely that these phases would be a pure Sr precipitate. More likely, Sr would form coprecipitates with $CaCO_3$, $CaSO_4 \cdot 2H_2O$, or $BaSO_4$ (Ainsworth & Rai, 1987; Felmy et al., 1993).

Strontium partitioning to organic soils [as measured by K_d (L kg^{-1})] has been reported to be 150, compared with 90 for Ca (Sheppard & Thibault, 1990); however, in nonorganic soils and sediments, cation exchange and specific adsorption would appear to be the dominate mechanisms. Kipp et al. (1986) modeled ^{90}Sr transport in glacial outwash sediments and found that an ion exchange model described the data well. Studies of radionuclide sorption in a fluvial-sand aquifer (discussed in regards to Cs) found the ^{90}Sr K_d to be about 10 mL g^{-1} ± 5 mL g^{-1} (almost two orders of magnitude less than that observed for Cs; Jackson et al., 1980). Importantly, the greatest radioactivity was associated with those fractions with the greatest abundance of micaceous minerals; suggesting that exchange was the dominant mechanism responsible for Sr attenuation. Similarly, studies of streambed gravels concluded that ^{90}Sr was associated with clay minerals through an ion exchange mechanism (Cerling & Spalding, 1982). In contrast to these studies, Jackson and Inch (1989) investigated ^{90}Sr adsorption in a sand aquifer at the Chalk River Nuclear Laboratories. They concluded that specific adsorption to Al, Fe, and Mn oxides rather than ion exchange dominated the attenuation process. It is interesting to note that the aqueous concentration of Ca was shown to be about 10 to 100 times greater than Sr in the Chalk river sand, yet data by Cowan et al. (1991) estimated that the Fe oxide surface complexation constants for Sr and Ca are very similar. This would suggest that specific adsorption would not be a factor in Sr attenuation. The ^{90}Sr studies of Cline (1981) and Cline and Rickard (1972) during a 26-yr period showed that Sr became distributed throughout the soil profile with time (Fig. 4–3); Cs, on the other hand, has remained in essentially the same location (top 15 cm) since its placement in 1962. Even though the mobility of these two radionuclides have been suggested to be controlled by the same mechanism (ion exchange) one can see from the above data and discussion that they behave in very differently.

4–6.4.1 Strontium Transport and Bioaccumulation

The uptake of ^{90}Sr from sediment or soil to plants and from plants to animals is affected by the presence of Ca in the systems. The observed ratio described by Comar et al. (1956) relates the amount of ^{90}Sr and Ca in a sample to the amount of the radionuclide and competing element in the precursor. This empirically determined relationship has proven to be consistent and has been successfully applied to modeling the passage of ^{90}Sr through food webs (Comar, 1965). Most of the reported observed ratios have been determined for food chains leading to human consumers. Strontium uptake by plants is therefore greatest from soils of low Ca content and in many cases increased organic matter content (Paasikallio et al., 1994). Plant crops assimilate from 0.2 to 3% of the strontium in the soil (Comar, 1965; Menzel, 1963). It also should be noted that the availability to plants of soil ^{90}Sr decreases only slightly with time (Cline, 1981; Paaskallio et al., 1994).

Strontium is greatly reduced, relative to plant levels, in whole-body concentration in insects and other invertebrates. A concentration factor of about 0.1 has been observed for second-order consumers and predators (excluding species with calcified exoskeletons; Reichle et al., 1970a,b). Calcium sink invertebrates (e.g., millipedes, isopods, snails) concentrate Sr by factors >150 (Reichle & Crossley, 1969). Assimilation of ^{90}Sr from detritus is low (77%) because it is incorporated into poorly digestible tissue structures (Reichle et al., 1970a). Some general terrestrial concentration factors for strontium are listed in Table 4–3.

An experimentally determined concentration factor [Sr concentration in the bird (g)/Sr concentration in water (mL)] of 1500 (bone) for coot (*Fulica* sp.) was reported by Hanson and Kronberg (1956). The transfer coefficient for Sr from the diet into milk of cows, goats, and pigs as related to the Ca content of the diet is 0.1 (0.08–0.16; Comar 1965). The observed ratio valuefor body/diet of 0.2 to 0.5 has been reported for cattle, goats, sheep, and pigs. Observed ratio values for egg yolk, egg shell, and femur in chickens was 0.6 (Comar, 1965).

4–6.4.2 Stontium Toxicity

The Sr ion has a low order of chemical toxicity (Venugopal & Luckey, 1978) in comparison to U. Moderate to large doses are required to cause nausea, diarrhea, electrocardiographic changes, and death due to respiratory paralysis

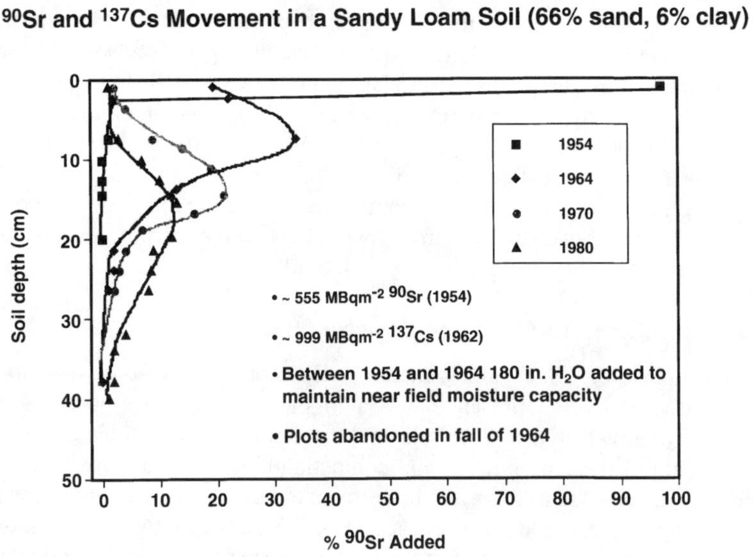

Fig. 4–3. Strontium-90 movement in a sandy loam soil (66% sand, 6% clay; Cline & Richard, 1972; Cline, 1981).

(Venugopal & Luckey, 1978). The oral LOAEL (Lowest Observable Adverse Effects Level) of Sr dichloride hexahydrate in rats is 405 mg kg^{-1}.; however, the radioisotope, ^{90}Sr, is highly dangerous, and the radiation hazard is well established. Because Sr is a metabolic analog of Ca, ^{90}Sr is readily absorbed from the lung, gastrointestinal tract, or bloodstream (dermal exposure). The Sr that is retained in the body, in large part, is deposited in the bone. Therefore, exposure to ^{90}Sr via any exposure route results in a high incidence of neoplasia on bone and related tissues (Harley, 1991). Bone sarcoma generation does not fit a linear dose-response relationship over a wide dose range. Low levels of exposure are better fit by sigmoid dose-response relationships (Mays & Lloyd, 1972).

4–7 SUMMARY AND CONCLUSIONS

As shown in Fig. 4–4 there are a number of factors that will affect the potential for environmental risk for any radionuclide. These include the physical and chemical nature of the element, its potential for bioconcentration, and the potential for partitioning to sensitive tissues within organisms. Public and scientific perception of the potential for risk from a given radionuclide in a given ecosystem must be a combination of all of the accompanying factors.

The chemical behavior of radionuclides can vary widely in soil and sediment environments. Equally important, for a given radionuclide the physicochemical properties of the solids and aqueous phase can greatly influence a radionuclide's behavior. Radionuclides can conceivably occur in soils as soluble-free, inorganic-soluble-complexed, organic-soluble, complexed, adsorbed, precipitated, coprecipitated, or solid structural species. Sposito (1989) calculated that a typical soil solution will easily contain 100 to 200 different soluble species and 50 to 100 surface-associated or precipitated species. Consequently, an understanding of the processes affecting chemical speciation is necessary to understand how and under what conditions radionuclides are retarded or mobilized in soil environments. While it is clear that an assessment of a radionuclide's soil chemistry and potential shifts in speciation will yield a considerable understanding of

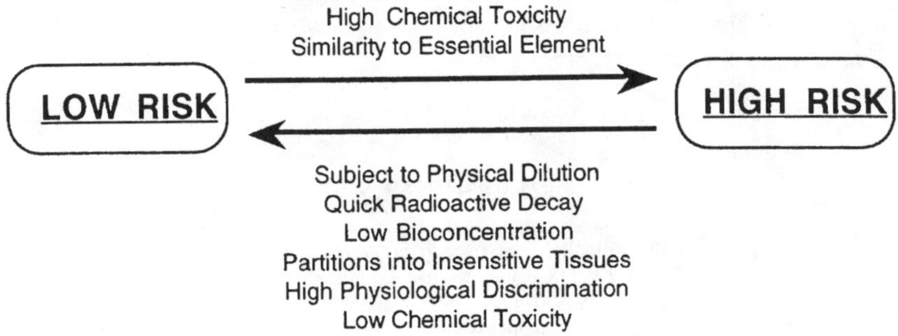

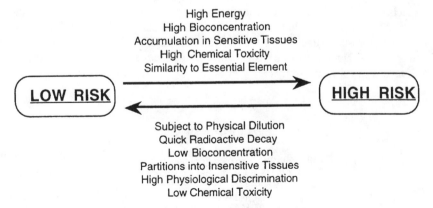

Fig. 4–4. Physical, chemical, and biotic factors affecting ecosystem risk.

its behavior in the natural environment, it does not directly translate to bioavailability or its impact on an ecosystem's health. The soil chemical factors have to be linked to food chain considerations and other ecological parameters that directly tie to an analysis of ecosystem health.

In general, the movement of radionuclides from lower to higher trophic levels diminishes with each trophic level in both aquatic and terrestrial systems (Table 4–3). In some cases, transfer is limited because of low absorption–assimilation by successive trophic organisms (Pu, U); for other radionuclides (Tc, H) assimilation may be high but rapid metabolic turnover and low retention greatly reduce tissue concentrations available to predator species. Still others are chemical analogs of essential elements whose concentrations are maintained under strict metabolic control in tissues (Cs) or are stored in tissues seldom consumed by other organisms (Sr storage in exoskeleton, shells, and bone). Therefore, the organisms that receive the greatest ingestion exposures are those in lower trophic positions or are in higher trophic levels but within simple, short food chains. The majority of lower trophic organisms are those with relatively high resistance to radiation toxicity that may contribute to the lack of ecosystem damage observed at most contaminated sites. Harmful levels of radionuclides have been transferred to humans (concentrations that may increase cancer incidence) largely via food chains with less than three trophic transfers such as hunter consumption of caribou or moose, or ingestion of milk from cows that have consumed plants with both incorporated and surface radionuclide contamination from nuclear accident or testing fallout. Fish also may concentrate radionuclides directly from the water.

Food source, behavior, and habitat influence the accumulation of radionuclides in animals. For example, several studies have shown that birds that are herbivorous or larvae-feeding and probe for food in soil or sediment have much higher concentration factors than fish-eating or omnivorous birds (Osburn, 1968; Krumholz, 1964; Willard, 1960; Silker, 1958).

Ecosystem damage from the impacts of radionuclides introduced into animal food webs results from either accumulated radioactivity and/or chemical toxicity. While in most cases, it is the radiation dose that is of concern, there are instances where chemical toxicity can be demonstrated (i.e., U and Co). The impacts of radionuclides on environmental health depend to a large degree on the routes through which radionuclides enter the food chain, the amount of analog species in the soil–water or successive trophic levels, the chemical forms of the contaminants, any chemical transformations that occur as a result of bioaccumulation, and species sensitivity.

REFERENCES

Abraham, R.L. 1972. Mortality of mallards exposed to gamma radiation. Radiat. Res. 49(2):322–327.

Adams, N., and N.L. Spoor. 1974. Kidney and bone retention functions in the human metabolism of uranium. Phys. Med. Biol. 19:460–471.

Ainsworth, C.C., J.L. Pilon, P.L. Gassman, and W.G. Van Der Sluys. 1994. Cobalt, cadmium, and lead sorption to hydrous iron oxide: Residence time effect. Soil Sci. Soc. Am. J. 58:1615–1623.

Ainsworth, C.C., and D. Rai. 1987. Selected chemical characterization of fossil fuel wastes. EPRI EA-5321. Electric Power Res. Inst., Palo Alto, CA.

Ames, L.L., J.E. McGarrah, B.A. Walker, and P.F. Salter. 1982. Sorption of uranium and cesium by Hanford basalts and associated secondary smectites. Chem. Geol. 35:205–225.

Anderson, J.B., E.C. Tsivoglou, and S.D. Shearer. 1963. Effects of uranium mill wastes on biological fauna of the Animas river (Colorado–New Mexico). p. 373. *In* V. Schultz and A.W. Klement (ed.) Radioecology. Van Nostand Reinhold, New York.

Andreyev, P.F., and A.P. Chumachenko. 1964. Reduction of uranium by natural organic substances. Geochem. Int. 1:3–7.

Arthur, W.J., III. 1982. Radionuclide concentrations in vegetation at a solid radioactive waste-disposal area in southeastern Idaho. J. Environ. Qual. 11(3):394–399.

ATSDR. 1990. Toxicological profile for uranium. TP-90-29, U.S. Department of Health and Human Services, Public Health Services, Agency for Toxic Substances and Disease Registry, Atlanta, GA.

Ault, M.A. 1989. Gamma emitting isotopes of medical origin detected in sanitary waste samples. Rad. Protection Manage. 6:48–56.

Avasthi, P.S., A.P. Evans, and D. Hay. 1980. Glomerular endothelial cells in uranyl nitrate-induced acute renal failure in rats. J. Clin. Incest. 65:121–127.

Bacq, Z.M., and P. Alexander. 1961. Fundamentals of radiobiology. Pergamon Press, New York.

Baudin, J.P., and R. Nucho. 1992. ^{60}Co accumulation from sediment and planktonic algae by midge larvae (*Chironomus luridus*). Environ. Pollut. 76:133–140.

Baudouin, M.F., and P. Scoppa. 1974. Toxicity of heavy metals for fresh water zooplankton: Influence of some environmental factors. Bol. Zool. 41(4):457.

Beck, R.A., and J.J. Brink. 1976. Sensitive chemical method for routine assay of colalamins in activated sewage sludge. Environ. Sci. Technol. 10:173–175.

Berry, W.L. 1978. Comparative toxicity of VO^{-3}, CrO^{-4}_2 Mn^{2+}, CO^{2+}, Ni^{2+}, Cu^{2+}, Zn^{2+}, and Cd^{2+} to lettuce seedlings. p. 582–589. *In* D.C. Adriano and I.L. Brisbin, Jr. (ed.) Environmental chemistry and cycling processes. CONF-760429. U.S. Department of Energy, Washington, DC.

Bellamy, A.W., J.L. Leitch, K.H. Larson, and D.B. Dunn. 1949. The 1948 radiological and biological survey of areas in New Mexico affected by the first atomic bomb detonation. UCLA-32, Univ. of California, Los Angeles.

Belli, M., A. Drigo, S. Menegon, A. Menin, P. Nazzi, U. Sansone, M. Toppano. 1989. Transfer of Chornobyl fall-out caesium radioisotopes in the cow food chain. Sci. Total Environ. 85:169–177.

Birch, L.D., and R. Bachofen. 1990. Effects of microorganisms on the environmental mobility of radionuclides. p. 438–516. *In* J-M. Bollag and G. Stotzky (ed.) Soil biochemistry. Vol. 6. Marcel Dekker, New York.

Bishop, G.P. 1989. Review of biosphere information: Biotic transport of radionuclides as a result of mass movement of soil by burrowing animals. NSS/R194. U.K. Nirex, Harwell.

Boecker, B.B. 1972. Toxicity of $^{137}CsCl$ in the beagle: Metabolism and dosimetry. Radiat. Res. 50:556–573.

Boggs, S., Jr., D. Livermore, and M.G. Seitz. 1985. Humic substances in natural waters and their complexation with trace metals and radionuclides: A review. ANL-84-78, Argonne Natl. Lab., Argonne, IL.

Bond, K.A., J.E. Cross, and F.T. Ewart. 1990. Thermodynamic modeling of the effect of organic complexants on sorption behavior. Radiochim. Acta 52/53:433–437.

Bond, V.P., T.M. Fliedner, and J.O. Archambeau. 1965. Mammalian radiation lethality: A disturbance in cellular kinetics. Academic Press, NewYork.

Bonotto, D.M. 1989. The behavior of dissolved uranium in groundwaters of the Morro Do Ferro thorium deposit, Brazil. J. Hydrol. 107:155–171.

Borovec, Z., B. Kribek, and V. Tolar. 1979. Sorption of uranyl by humic acids. Chem. Geol. 27:39–46.

Bovard, P., A. Grauby, and A. Sass. 1970. Chelating effects of organic matter and its influence on the migration of fission products. p. 471–495. *In* Proc. of Symp. Isotopes and Radiation in Soil Organic-Matter Studies. Vienna, Austria. 15–19 July 1968. CONF-680725, STI/PUB-190, NSA. IAEA, Vienna.

Bringman, G., and R. Kuhn. 1959. Water toxicity studies with protozoa as test organisms. Gesund. Ing. 80:239–242.

Bruno, J., I. Casas, and I. Puigdomenech. 1988. The kinetics of dissolution of $UO_{2(s)}$ under reducing conditions. Radiochim. Acta 44/45:11–16.

Bruno, J., I. Casas, and I. Puigdomenech. 1991. The kinetics of dissolution of UO_2 under reducing conditions and the influence of an oxidized surface layer (UO_{2+x}): Application of a continuous flow-through reactor. Geochim. Cosmochim. Acta. 55:647–659.

Bruggenwert, M.G.M., and A. Kamphorst. 1979. Survey of experimental information on cation exchange in soil systems. *In* G.H. Bolt (ed.) Soil chemistry B. Physico-chemical models. Elsevier Scientific Publ., New York.

Bywater, J.F., R. Banaczkowski, and M. Bailey. 1991. Sensitivity to uranium of six species of tropical freshwater fishes and four species of *Cladocerans* from Northern Australia. Environ. Toxicol. Chem. 10:1449–1458.

Canadian Standards Association. 1987. Guidelines for calculating derived release limits for radioactive material in airborne and liquid effluents for normal operation of nuclear facilities. Natl. Standard of Canada. CAM/CSA-N288.1-M8. Canadian Standards Association, Rexdale, Toronto, Canada.

Cannon, H.L. 1960. The development of botanical methods of prospecting for uranium on the Colorado plateau. U.S. Geol. Surv. Bull. 1085A:1–50.

Carraca, S., A. Ferreira, and J. Coimbra. 1990. Sr transfer factors between different levels in the trophic chain in two dams of Douro River (Portugal). Wat. Res. 24(12):1497–1508.

Cataldo, D.A. 1979. Behavior of technetium and iodine in a Hanford sand and associated subsoil: Influence of soil aging on uptake by cheatgrass and tumbleweed. PNL-2740, Pacific Northwest Lab., Richland, WA.

Cataldo, D.A., T.R. Garland, and R.E. Wildung. 1978a. Nickel in plants: I. Uptake kinetics using intact seedlings. Plant Physiol. 62:563-565.

Cataldo, D.A., R.C. Routson, D. Paine, and T.R. Garland. 1978b. Relationships between properties of Hanford area soils and the availability of ^{134}Cs and ^{85}Sr for uptake by cheatgrass and tumbleweed. PNL-2496, Pacific Northwest Laboratories, Richland, WA.

Cataldo, D.A., and R.E. Wildung. 1978. Soil and plant factors influencing the accumulation of heavy metals by plants. Environ. Health Persp. 27:149–159.

Cataldo, D.A., and R.E. Wildung. 1983. The role of soil and plant metabolic processes in controlling trace element behavior and bioavailability to animals. Sci. Total Environ. 28:159–168.

Cataldo, D.A., R.E. Wildung, and T.R. Garland. 1983. Root absorption and transport behavior of technetium in soybean. Plant Physiol. 73:849-852.

Cataldo, D.A., R.E. Wildung, and T.R. Garland. 1987. Speciation of trace inorganic contaminants in plants and availability to animals: An overview. J. Environ. Qual. 16:289-295.

Cerling, T.E., and B.P. Spalding. 1982. Distribution and relationship of radionuclides to streambed gravels in a small watershed. Environ. Geol. 4:99–116.

Cerveny, T.J., and L.G. Cockerham. 1986. Medical management of internal radionuclide contamination. Med. Bull. U.S. Army, Europe 43:24–27.

Childs, H.E., and G.E. Cosgrove. 1976. A study of pathological conditions in wild rodents in radioactive areas. Am. Midl. Nat. 76:309.

Cleveland, J.E. 1979. Critical review of plutonium equilibria of environmental concern. *In* E.A. Jenne (ed.). Chemical modeling in aqueous systems. ACS Symp. Ser. 93. Am. Chem. Soc., Washington, DC.

Cline, J.F. 1981. Aging effects of the availability of strontium and cesium to plants. Health Phys. 41:293–296.

Cline, J.F., and W.H. Rickard. 1972. Radioactive strontium and cesium in cultivated and abandoned field plots. Health Phys. 23:317–324.

Cochran, K.W., J. Doulls, M. Mazur, and K.P. Du Bois. 1950. Acute toxicity of zirconium, columbium, strontium, lanthium, cesium, tantalium, and yttrium. Arch. Ind. Hyg. Occup. Med. 1:637–650.

Comar, C.L. 1965. Movement of fallout radionuclides through the biosphere and man. Annu. Rev. Nucl. Sci. 15:175–206.

Comar, C.L., R.H. Wasserman, and M.M. Nold. 1956. Strontium–calcium discrimination factors in the rat. Proc. Soc. Exp. Biol. Med. 92:859.

Coughtrey, P.J., M.C. Thorne, D.C. Jackson, C.H. Jones, and P. Kane. 1983–1985. Radionuclide distribution and transport in terrestrial and aquatic ecosystems. Vol. 1 to 6. A.A. Balkema, Rotterdam, the Netherlands.

Cowan, C.E., J.M. Zachara, and C.T. Resch. 1991. Cadmium adsorption on iron oxides in the presence of alkaline-earth elements. Environ. Sci. Technol. 25:437–446.

Cowart, J.B., and W.C. Burnett. 1994. The distribution of uranium and thorium decay-series radionuclides in the environment: A review. J. Environ. Qual. 23:651–662.

Cristaldi, M., L.A. Ieradi, D. Mascanzoni, and T. Mattei. 1991. Environmental impact of the Chornobyl accident: Mutagenises in bank voles from Sweden. Int. J. Rad. Biol. 59:31–40.

Crowther, D.L., J.G. Dillard, and J.W. Murray. 1983. The mechanism of Co(II) oxidation on synthetic birnessite. Geochim. Cosmochim. Acta 47:1399–1403.

Cushing, C.E., and D.G. Watson. 1974. Aquatic studies of Gable Mountain pond. BNWL-1884. Pacific Northwest Lab., Richland, WA.

Davies, P.H. 1980. Acute toxicity to brook trout (*Salvelinus fontinalis*) and rainbow trout (*Salmo gairdneri*) in soft water. Water Pollut. Stud. Proj. F-33-R. Federal aid in fish and wildlife restoration. Job Progress Rep. F-33-R-15. Colorado Div. of Wildlife, Fort Collins, CO. Cited in guidelines for surface water quality. Vol. 1. Inorganic Chemical Substances. Inland Waters Directorate, Ottawa, Canada.

Diamond, G.L. 1989. Biological consequences of exposure to soluble forms of natural uranium. Radiat. Prot. Dosimet. 26(1/4):23–33.

Dilworth, M.J., A.D. Robson, and D.L. Chatel. 1979. Cobalt and nitrogen fixation in *Lupinus angustfolius* L: II. Nodule formation and functions. New Phytol. 83:63–79.

Disnar, J.R., and J. Trichet. 1983. Pyrolyse de complexes organo-metalliques formes entre un materiau organique actuel d'origine algaire et divers cations metalliques divalents ($UO^{2+}{}_2$, Cu^{2+}, Pb^{2+}, Ni^{2+}, Mn^{2+}, Zn^{2+}, et Co^{2+}). Chem. Geol. 40:203–223.

Dreesen, D.R., M.J. Williams, M.L. Marple, E.S. Gladney, and D.R. Perrin. 1982. Mobility and bioavailability of uranium mill tailings contaminants. Environ. Sci. Technol. 16:702–709.

Dunaway, P.B., and S.V. Kaye. 1963. Effects of ionizing radiation on mammal populations on the White Oak lake bed. p. 333–338. *In* V. Schultz and A.W. Klement (ed.) Proc. of the Natl. Symp. on Radioecology, 1st, Fort Collins, CO. 10–15 Sept. 1961. Rheinhald Publ. Co., New York and AIBS, Washington, DC.

Dunford, W.E., O.E. Acres, and R.W. Pollock. 1985. Concentration of ^{137}Cs in water and fish from the Winnipeg River, Canada. AECL-8098, IAEA-SR-85/13. Atomic Energy of Canada Limited, Whiteshell Nuclear Res. Establishment, Pinawa, Manitoba, Canada.

Dzombak, D.A., and F.M.M. Morel. 1990. Surface complexation modeling: Hydrous ferric oxide. John Wiley & Sons, New York.

Eisler, R. 1995. Ecological and toxicological aspects of the partial meltdown of the Chornobyl nuclear power plant reactor. p. 549–564. *In* D.J. Hoffman et al. (ed.) Handbook of ecotoxicology. Lewis Publ., Boca Raton.

Ephrain, J.H., J.A. Marinsky, and S.J. Cramer. 1989. Complex-forming properties of natural organic acids: Fulvic acid complexes with cobalt, zinc and europium. Talanta 36:437–443.

Evenson, L.M., D.P. Olson, D.K. Halford, and O.D. Markham. 1978. Systemic effects of radiation exposure on rodents inhabiting liquid and solid radioactive waste disposal areas. p. 117–143. *In* O.D. Markham (ed.). Ecological studies on the Idaho Engineering Laboratory Site, Idaho Falls. IDO-12087. U.S. Dep. of Energy, Washington, DC.

Eyman, L.D., and J.R. Trablaka. 1980. Patterns of transuranic uptake by aquatic organisms: Consequences and implications. p. 612–624. *In* W. Hanson (ed.) Transuranic elements in the environment. DOE/TIC-22800. NTIS, Springfield, VA.

Felmy, A.R., D. Rai, and D.A. Moore. 1993. The solubility of $(Ba,Sr)SO_4$ precipitates: Thermodynamic equilibrium and reaction path analysis. Geochim Cosmochim. Acta 57:4345–4363.

Fielitz, U. 1991. Schriftenreihe Reaktosicherheit und Strahlenshutz. Bundesministerium fuer Umwelt, Naturshcutz und Reaktorsicherheit. BMU-1991-294, Inst. Fuer Wildbiologie und Jagdkunde, Bonn, Germany.

French, N. R. 1975. Chronic low-level gamma irradiation of a desert ecosystem for five years. p. 1151–1165. *In* Actes du Symposium International de Radioecologie, Aix-en-Providence, France. Sept. 1969. Vol. 2. Centre d'Etudes Nucleaires de Cadarche, France.

Fridovich, I. 1976. Oxygen radicals, hydrogen peroxide, and oxygen toxicity. p. 239–277. *In* Free radicals in biology. Vol. I. Academic Press, New York.

Fritsch, P., M. Beauvallet, and K. Moutairou. 1987. Acute lesions induced by alpha-irradiation of intestine after plutonium gavage of neonatal rats. Int. J. Radiat. Stud. Phys. Chem. Med. 52:1–6.

Gamble, D.S., M. Schnitzer, and D.S. Skinner. 1977. Mn(II)-fulvic acid complexing equilibrium measurements by electron spin resonance spectrometry. Can. J. Soil Sci. 57:47–53.

Garland, T.R., D.A. Cataldo, K.M. McFadden, R.G. Schreckhise, and R.E. Wildung. 1983. Comparative behavior of 99-Tc, 129-I, 127-I and 137-Cs in the environment adjacent to a fuels reprocessing facility. Health Phys. 44:658–662.

Garland, T.R., D.A. Cataldo, K.M. McFadden, and R.E. Wildung. 1987. Factors affecting absorption, transport, and form of plutonium in plants. p. 83–95. *In* J.E. Pinder III et al. (ed.)

Environmental research on actinide elements. CONF-841142 (DE86006713) U.S. DOE OHER Symp. Ser. 59.

Garten, C.T., Jr., E.A. Bondietti, J.R. Trabalka, R.L. Walker, and T.G. Scott. 1987. Field studies on the behavior of actinide elements in east Tennessee. p. 109–119. *In* J.E. Pinder III et al. (ed.) Environmental research on actinide elements. CONF-841142 (DE86006713) U.S. DOE OHER Symp. Ser. 59.

Garten, C.T., Jr., F.O. Hoffman, and E.A. Bondietti. 1984. Field and greenhouse experiments on the fate of technetium in plants and soil. Health Phys. 46:647–656.

Gast, R.G. 1969. Standard free energies of exchange for alkali metal cations on Wyoming bentonite. Soil Sci Soc. Am. Proc. 33:37–41.

Gee, G.W., D. Rai, and R.J. Serne. 1983. Mobility of radionuclides in soils. p. 203–226. *In* D.W. Nelson (ed.) Chemical mobility and reactivity in soil systems. SSSA. Spec. Publ. 25. SSSA, Madison, WI.

Girvin, D.C., P.L. Gassman, and H. Bolton, Jr. 1993. Adsorption of aqueous cobalt ethylenediaminetetraacetate by Al_2O_3. Soil Sci. Soc. Am. J. 57:47–57.

Gleiser, C.A. 1953. The determination of the lethal dose 50/30 of total body x-radiation for dogs. Am. J. Vet. Res. 14:284–286.

Greb, R.J., and W. Morgan. 1961. Treatment of pheasant ovaries with x-rays and gamma rays. Proc. S.D. Acad. Sci. 40:112.

Grebenshchikova, N.V., S.K. Timofeev, S.F. Firsakora, A.A. Novik, and G.I. Palekshanova. 1991. Transfer patterns of radionuclides into crops in the condition of radioactive contamination after the Chornobyl accident. p. 465–472. *In* Comparative Assessment of the Environmental Impact of Radionuclides Released During Three Major Accidents: Kyshtym, Windscale and Chornobyl. CEC Luxembourg, EUR 13574. Commission of the European Countries, Directorate General XI, Environment, Civil Protection, and Nuclear Safety, Luxembourg, Belgium.

Grenthe, I. 1992. Chemical thermodynamics of uranium. North-Holland, New York.

Gross, W.G, and U.G Heller. 1946. Chromium in animal nutrition. J. Ind. Hyg. Toxicol. 28:52.

Gu, B.R., and K. Shultz. 1991. Anion retention in soil: Possible application to reduce migration of buried technetium and iodine. A review. NUREG/CR-5464. U.S. Nuclear Regulatory Commission, Washington, DC.

Guglielmotti, M.B., A.M. Ubios, B.M. de Rey, and R.L Cabrini. 1984. Effects of acute intoxication with uranyl nitrate on bone formation. Experimentia 40:474–476.

Gus'kova, V.N., L.N. Gurfein, and A.I. Tikhonova. 1966. Uranium action in a reservoir. Gidrobiol. Zh. 2(6):53–57. Cited in guidelines for surface water quality. Vol. 1. Inorganic Chemical Substances. Inland Waters Directorate, Ottawa, Canada.

Guthrie, J.E., and J.R. Dugel. 1983. Gamma-ray irradiation of a boreal forest ecosystem: The field irradiator-gamma (FIG) facility and research programs. Can. Field Nat. 97:120–128.

Haley, D.P. 1982. Morphologic changes in uranyl nitrate-induced acute renal failure in saline-and water-drinking rats. Lab. Invest. 46:196–208.

Halliwell, B., and J.M.C. Gutteridge. 1985. Free radicals in biology and medicine. Clarendon Press, Oxford, England.

Hanson, W.C. 1975. Ecological considerations of the behavior of plutonium in the environment. Health Phys. 28:529–537.

Hanson, W.C., and H.A. Kronberg. 1956. Radioactivity in terrestrial animals near an atomic energy site. p. 385. *In* Proc. Int. Conf. Peaceful Uses of Atomic Energy, Geneva. 8–20 Aug. 1955. United Nations, New York.

Hanson, W.C., and F.R. Miera, Jr. 1976. Long-term ecological effects of exposure to uranium. LA-6269. Los Alamos Scientific Lab., Los Alamos, NM.

Hanson, W.C., D.G. Watson, and R.W. Perkins. 1967. Concentration and retention of fallout radionuclides in Alaskan arctic ecosystems. p. 233–245. *In* B. Aberg and F. P. Hungate (ed.) Radioecological concentration processes. Pergamon Press, Oxford, England.

Harley, N.H. 1991. Toxic effects of radiation and radioactive materials. p. 723–752. *In* M.O. Amdur et al. (ed.) Casarett and Doull's toxicology: The basic science of poisons. 4th ed. McGraw-Hill, New York.

Horsic, E., Z. Milsevic, R. Kljajic, and A. Bauman. 1982. Concentration factors and absorbed doses of Sr-90 and Cs-137 in the Sava River fishes. p. 110–113. *In* Proc. of the Int. Symp. on Radiological Protection. Advances in theory and practice. Soc. for Radiol. Protec., Inverness, Scotland. Radiological Protection, Berkeley, England.

Howard, B.J., N.A. Beresford, and K. Hove. 1991. Transfer of radiocesium to ruminants in natural and semi-natural ecosystems and appropriate countermeasures. Health Phys. 61:715.

Hove, K., O. Pedersen, T. Garmo, H. Hansen, and H. Staaland. 1990. Fungi: A major source of radiocesium contamination of grazing ruminants in Norway. Health Phys. 59:189.

Hsi, C-K.D., and D. Langmuir. 1985. Adsorption of uranyl onto ferric oxyhydroxides: Application of the surface complexation site-binding model. Geochim. Cosmochim. Acta 49:1931–1941.

Huheey, J.E. 1983. Inorganic chemistry. 3rd ed. Harper & Row, New York.

Hyne, R.V., G.D. Rippon, and G. Ellender. 1991. pH dependent uranium toxicity to freshwater hydra. Sci. Total Environ. 125:159–174.

International Atomic Energy Agency. 1992. Effects of ionizing radiation on plants and animals at levels implied by current radiation protection standards. Int. Atomic Energy Agency, Vienna.

Ibrahim, S.A., and F.W. Whicker. 1988. Comparative uptake of U and Th by native plants at a U production site. Health Phys. 54(4):413–419.

Idiz, E.F., D. Carlisle, and I.R. Kaplan. 1986. Interaction between organic matter and trace metals in a uranium rich bog, Kern County, California. Appl. Geochem. 52:573–581.

Jackson, R.E., and K.J. Inch. 1989. The in-situ adsorption of ^{90}Sr in a sand aquifer at the Chalk River Nuclear Laboratories. J. Contam. Hydro. 4:27–50.

Jackson, R.E., K.J. Inch, R.J. Patterson, K. Lyon, T. Spoel, W.F. Merritt, and B.A. Risto. 1980. Adsorption of radionuclides in a fluvial-sand aquifer: Measurement of the distribution coefficients $K^{Sr}{}_{d}$ and $K^{cs}{}_{d}$ and identification of mineral adsorbents. p. 311–328. *In* R.A. Baker (ed.) Contaminants and sediments. Vol. 1. Science Publ., Ann Arbor, MI.

Jardine, P.M., G.K. Jacobs, and J.D. O'Dell. 1993. Unsaturated transport processes in undisturbed heterogeneous porous media: Co-contaminants. Soil Sci. Soc. Am. J. 57:954–962.

Kaplan, D.I., R.J. Serne, and M.G. Piepho. 1995. Geochemical factors affecting radionuclide transport through near and far fields at a low-level waste disposal site. PNL-10379. Pacific Northwest Lab., Richland WA.

Kent, D.B., V.S. Tripathi, N.B. Ball, J.O. Leckie, and M.D. Siegel. 1988. Surface -complexation modeling of radionuclide adsorption in subsurface environments. Final report. NUREG/CR-4807. U.S. Nuclear Regulatory Commission, Washington, DC.

Kim, J.J. 1986. Chemical behavior of transuranic elements in aquatic systems. p. 413–455. *In*. A.J. Freeman and C. Keller (ed.). Handbook on the physics and chemistry of the actinides. Elsevier Science Publ., Amsterdam.

King, S.F. 1964. Uptake and transfer of cesium-137 by chlamydomonas, daphnia, and bluegill fingerlings. Ecology 45:852–193.

Kipp, K.L., Jr., K.G. Stollenwerk, and D.B. Grove. 1986. Groundwater transport of strontium 90 in a glacial outwash environment. Water Resour. Res. 22:519–530.

Krivolutskii, D.A., T.L. Kozhevnikova, V.Z. Martjushov, and G.I. Antonenko. 1992. Effects of transuranic (^{239}Pu, ^{239}Np, ^{241}Am) elements on soil fauna. Biol. Fert. Soils 13:79–84.

Krivolutzkii, D.A., and A. Pokarzhevskii. 1992. Effects of radioactive fallout on soil animal populations in the 30 km zone of the Chornobyl atomic power station. Sci. Total Environ. 112:69–77.

Krumholz, L.A. 1964. A summary of findings of the ecological survey of White Oak Creek, Roane County Tennessee Valley Authority. Vol III. ORO-587. USAEC, Washington, DC.

Krumholz, L.A., E.D. Goldberg, and H. Boroughs. 1957. Ecological factors involved in the uptake, accumulation, and loss of radionuclides by aquatic organisms. p. 135–157. *In* Effects of atomic radiation on oceanography and fisheries. Publ. 551. Natl. Acad. Sci., Natl. Res. Council, Washington, DC.

Kushniruk, V.A. 1964. The radiosensitivity of birds. Biological effects of radiation: 1. Laboratory of radiobiological problems. JPRS 23:169.

Landa, E.R., L.J. Hart Thorvig, and R.G. Gast. 1977. Uptake and distribution of technetium-99 in higher plants. p. 390–401. *In* Biological implications of metals in the environment. Proc. of the 15th Annual Hanford Life Sciences Symp., Richland, WA. 29 Sept.–1 Oct. 1975. CONF-750959. U.S. Dep. of Energy, Washington, DC.

Landeen, D.S., and R.M. Mitchell. 1986. Radionuclide uptake by trees at a radwaste pond in Washington state. Health Phys. 50(6):769–774.

Langmuir, D. 1978. Uranium solution-mineral equilibria at low temperatures with application to sedimentary ore deposits. Geochim. Cosmochim. Acta 42:547–564.

Lewis, R.J., and R.L. Tatken (ed.). 1979–1980. Registry of toxic effects of chemical substances. U.S. Public Health Service, NIOSH, Cincinnati, OH.

Lindsay, W.L. 1979. Chemical equilibria in soils. John Wiley & Sons, New York.

Livens, F.R., A.D. Horrill, and D. L. Singleton. 1991. Distribution of radiocesium in the soil–plant systems of upland areas of Europe. Health Physics 60:539–545.

Lowe, V.P.W. 1991. Radionuclides and the birds at Ravenglass. Environ. Pollut. 70:1–26.

Macalady, D.L., P.G. Tratnyek, and T.J. Grundl. 1986. Abiotic reduction reactions of anthropogenic organic chemicals in anaerobic systems: A critical review. J. Contam. Hydrol. 1:1–28.

Mahon, D.C. 1982. Uptake and translocation of naturally-occurring radionuclides of the uranium series. Bull. Environ. Contam. Toxicol. 29:697–703.

Marschner, H. 1995. Mineral nutrition of higher plants. 2nd ed. Academic Press, New York.

Masson, M., F. Patti, C. Colle, P. Roucoux, A. Grauby, and A. Saas. 1989. Synopsis of French experimental and *in situ* research on the terrestrial and marine behavior of Tc. Health Phys. 57:269–279.

Matoura, R.F.C., A. Dickson, and J.P. Riley. 1978. The complexation of metals with humic materials in natural waters. Estuarine Coastal Mar. Sci. 6:387–408.

Mays, C.W., and R.D. Lloyd. 1972. Bone sarcoma incidence vs. alpha particle dose. p. 79–96. *In* B.J. Stover, and W.S.S. Jee (ed.) The radiobiology of plutonium. J.W. Press, Salt Lake, UT.

McBride, M.B. 1978. Transition metal bonding in humic acid. An ESR study. Soil Sci. 126:200–209.

McClellan, R.O. 1964. Calcium–strontium discrimination in miniature pigs as related to age. Nature (London) 202:104–106.

McClellan, R.O., and R.K. Jones. 1969. ^{90}Sr-induced neoplasia: A selective review. p. 78–92. *In* C.W. Mays (ed.) Delayed effects of bone-seeking radionuclides. Univ. of Utah Press, Salt Lake City.

McKinley, J.P., J.M. Zachara, S.C. Smith, and G.D. Turner. 1995. The influence of uranyl hydrolysis and multiple site binding reactions on adsorption of U(VI) to montmorillonite. Clays Clay Miner. 43:586–598.

Means, L.M., D.A. Crerar, and J.O. Duguid. 1978. Migration of radioactive wastes: Radionuclide mobilization by complexing agents. Science (Washington, DC) 200:1477–1479.

Menzel, R.G. 1963. Factors influencing the biological availability of radionuclides for plants. Federation Proc. 22:1398–1401.

Mettler, F.A., and R.C. Ricks. 1990. Historical aspects of radiation accidents. p. 18–26. *In* F.A. Mettler et al. (ed.) Medical management of radiation accidents. CRC Press, Boca Raton, FL.

Millard, J.B., F.W. Whicker, and O.D. Markham. 1990. Radionuclide uptake and growth of barn swallows nesting by radioactive leaching ponds. Health Phys. 58:429.

Moffett, D., and M. Tellier. 1977. Uptake of radioisotopes by vegetation growing on uranium tailings. Can. J. Soil Sci. 57:417–424.

Morel, F.M.M. 1983. Principles of aquatic chemistry. Wiley-Interscience, New York.

Morel, F.M.M., J.C. Westall, C.R.O. Melia, and J.J. Morgan. 1975. Fate of trace metals in Los Angeles County wastewater discharge. Environ. Sci. Technol. 9:756–764.

Murphy, C.E., Jr. 1990. The transport, dispersion, and cycling of tritium in the environment. Rep. WSRC-RP-90-462. Westinghouse Savannah River Co., Savannah River Site, Aiken, SC.

Murray, J.W., and J.G. Dillard. 1979. The oxidation of cobalt(II) adsorbed on manganese dioxide. Geochim. Cosmochim. Acta 43:781–787.

Murthy, T.C.S., P. Weinberger, and M.P. Measures. 1984. Uranium effects on the growth of soybean (*Glycine max* (L.) Merr.). Bull. Environ. Contam. Toxicol. 32:580–586.

Myers, D.K. 1989. The general principles and consequences of environmental radiation exposure in relation to Canada's nuclear fuel waste management concept. AECL 9917. Atomic Energy of Canada Limited, Chalk River Nuclear Lab., Chalk River, Ontario, Canada.

Nakashima, S., J.R. Disnar, A. Perruchot, and J. Trichet. 1984. Experimental study of mechanisms of fixation and reduction of uranium by sedimentary organic matter under diagenrtic or hydrothermal conditions. Geochim. Cosmochim. Acta 48:2321–2329.

Nash, K., S. Fried, A.M. Freidman, and J.C. Sullivan. 1981. Redox behavior, complexing, and adsorption of hexavalent actinides by humic acid and selected clays. Environ. Sci. Technol. 15:834–837.

National Council on Radiation Protection and Measurements. 1989. Exposure of the U.S. population from diagnostic medical radiation. NCRPM Rep. 100. NCRP, Bethesda, MD.

National Institute for Occupational Safety and Health. 1987. Registry of toxic effects of chemical substances. U.S. Dep. of Health and Human Services, Public Health Serv., Washington, DC.

National Research Council of Canada. 1982. Data sheets on selected toxic elements. 19252. Natl. Res. Council, Ottawa, Canada.

Ng, Y.C., C.S. Colsher, and S.E Thomspon. 1982. Soil to plant concentration factors for radiological assessments. NUREG/CR-2975, UCID-19463. Lawrence Livermore Lab., Livermore, CA.

Noshkin, V.E., V.T. Bowen, K.M. Wong, and J.C. Burke. 1973. Plutonium in North Atlantic Ocean organisms. Ecological relationships. p. 681–688. *In* Radionuclides in ecosystems. Vol. 2. USAEC Rep. CONF-710501. NTIS, Springfield, Virginia.

Oakberg, E.F., and E. Clark. 1964. Effects of ionizing radiation on the reproductive system. p. 11. *In* W.D. Carlson, and F.X. Gassner (ed.) Int. Symp. on the Effects of Ionizing Radiation on the Reproductive System. Pergamon Press, New York.

Odum, E.P. 1971. Fundamentals of ecology. 3rd. ed. W.B. Saunders Co. Philadelphia, PA.

Olafson, J.H., H. Nishita, and K.H. Larson. 1957. The distribution of plutonium in the soils of central and northeastern New Mexico as a result of the atomic bomb test of July 16, 1945. UCLA-406. Univ. of California, Los Angeles.

Osborne, R.V. 1972. Permissible levels of tritium in man and the environment. Radiat. Res. 50:197–211.

Osburn, W.S., Jr. 1968. Forecasting long-range recovery from nuclear attack. p. 107. *In* Proc. from Postattack Recovery from Nuclear War Symp., Fort Monroe, MI. 15–17 May 1967. Natl. Academy of Sciences, Natl. Academy of Engineering, Natl. Res. Council, Washington, DC.

Paasikallio, A., A. Rantavaara, and J. Sippola. 1994. The transfer of cesium-137 and strontium-90 from soil to food crops after the Chornobyl accident. Sci. Total Environ. 155:109–124.

Packard, C. 1936. Biological effectiveness of x-ray wavelength. p. 459–471. *In* B.M. Duggar (ed.). Biological effects of radiation. McGraw-Hill, New York.

Parkhurst, B.R., R.W. Pennak, and W.T. Waller. 1984. An environmental hazard evaluation of uranium in a Rocky Mountain stream. Environ. Toxicol. Chem. 3:113–124.

Parks, G.A., and D.C. Pohl. 1988. Hydrothermal solubility of uraninite. Geochim Cosmochim. Acta 52:863–875.

Pendleton, R.C., R.D. Lloyd, C.W. Mays, and B.N. Church. 1964. Trophic level effect on the accumulation of caesium 137 in cougars feeding on mule deer. Nature (London) 204:708–709.

Pelton, M.R., and E.E. Provost. 1969. Effects of radiation on survival of wild cotton rats (*Sigmodon hispidus*) in enclosed areas of natural habitat. p. 39–45. *In* D.J. Nelson and F.C. Evans (ed.) Symp. on radioecology, Ann Arbor, MI. 15–17 May 1967. CONF-670503, CFSTI. Natl. Bureau of Standards, Springfield, VA.

Phillips, L.J., and J.E. Coggle. 1988. The radiosensitivity of embryos of domestic chickens and black-headed gulls. Int. J. Radiat. Biol. 53:309–317.

Poston, T.M., R.W. Hanf, Jr., and M.A. Simmons. 1984. Toxicity of uranium to *Daphnia magna*. Water Air Soil Pollut. 22:289–298.

Poston, T.M., and D.C. Klopher. 1985. A literature review of the concentration factors of selected radionuclides in freshwater and marine fish. PNL-5484. Pacific Northwest Lab., Richland, WA.

Price, K.R. 1972. Uptake of Np-237, Pu-239, Am-241 and Cm-244 from soil by tumbleweed and cheatgrass. BNWL-1688. Battelle, Pacific Northwest Lab., Richland, WA.

Prister, B.S. 1991. Agricultural aspects of the radiation situation in the areas contaminated by the South Ural and Chornobyl accidents. p. 449–464. *In* Comparative assessment of the environmental impact of radionuclides released during three major accidents: Kyshtym, Windscale and Chornobyl. Commission of the European Communities, London.

Rai, D., and R.J. Serne. 1977. Plutonium activities in soil solutions and the stability and formation of selected plutonium minerals. J. Environ. Qual. 6:89–95.

Reichle, D.E., and D.A. Crossley, Jr. 1969. Trophic level concentrations of cesium-137, sodium, and potassium in forest arthropods. p. 678–686. *In* D.J. Nelson and F.C. Evans (ed.) Proc. of the Ecological Society of America's Natl. Symp. on Radioecology, 2nd, Ann Arbor, MI. 15–17 May 1967. CONF-670503, USAEC, Washington, DC.

Reichle, D.E., P.B. Dunaway, and D.J. Nelson. 1970a. Turnover and concentration of radionuclides in food chains. Nuclear Safety 11:43–55.

Reichle, D.E., D.J. Nelson, and P.B. Dunaway. 1970b. Biological concentration and turnover of radionuclides in food chains. ORNL-TM-2492. Oak Ridge Natl. Lab., Oak Ridge, TN.

Rickard, W.H., R.E. Fitzner, and C.E. Cushing. 1981. Biological colonization of an industrial pond: Status after two decades. Environ. Conserv. 8:241–247.

Rickard, W.H., and H.A. Sweany. 1977. Radionuclides in Canada goose eggs. p. 623–627. *In* Biological implications of metals in the environment. Proc. of the Annual Hanford Life Sciences Symp., 15th, Richland, WA. CONF-750929, U.S. Dep. of Energy, Washington, DC.

Robson, A.D., M.J. Dilworth, and D.L. Chatel. 1979. Cobalt and nitrogen fixation in *Lupinus angustifolius* L: I. Growth nitrogen concentrations and cobalt distribution. New Phytol. 83:53–62.

Robson, A.D., and G.R. Mead. 1980. Seed cobalt in *Lupinus angustifolium*. Aust. J. Exp. Agric. 27:657–660.

Romney, E.M., A. Wallace, J.E. Kinnear, and R.A. Wood. 1985. Plant root uptake of Pu and Am. p. 185–199. *In* W.A. Howard et al. (ed.) The radioecology of transuranics and other radionuclides in desert ecosystems. NVO-224 (DE86001243).U.S. DOE, Nevada Operations Office, Las Vegas.

Romney, E.M., H.M. Mork, and K.H. Larson. 1970. Persistence of plutonium in soil, plants, and small mammals. Health Phys. 19:487–491.

Routson, R.C., and D.A. Cataldo. 1977. Tumbleweed and cheatgrass uptake of ^{99}Tc from five Hanford project soils. BNWL-2183, Battelle Pacific Northwest Lab., Richland, WA.

Routson, R.C., and D.A. Cataldo. 1978. A growth chamber study of the effect of soil concentration and plant age on the uptake of Sr and Cs by tumbleweed. Commun. Soil Sci. Plant Anal. 9(3):215–229.

Rugh, R. 1964. Effects of ionizing radiation on the reproductive system. p. 25. *In* W.D. Carlson, and F.X. Gassner (ed.) Pergamon Press, New York.

Rust, J.H., B.F. Trum, and U.S.G. Kuhn, III. 1954. Physiological aberrations following total body irradiation of domestic animals with large doses of gamma rays. Vet. Med. 49:318.

Sandino, A., and J. Bruno. 1992. The solubility of $(UO_2)_3(PO_4)_2 \cdot 4H_2O$ and the formation of U(VI) phosphate complexes: Their influence in uranium speciation in natural waters. Geochim. Cosmochim, Acta 56:4135–4157.

Schimmack, W., K. Bunzl, and L. Zelles. 1989. Initial rates of migration of radionuclides from the Chornobyl fallout in undisturbed soils. Geoderma 44:211–218.

Schnitzer, M., and H. Kerndorff. 1980. Sorption of metals on humic acid. Geochim. Cosmochim. Acta 44:1701–1715.

Schultz, R.K., R. Overstreet, and I. Barshad. 1960. On the soil chemistry of cesium-137. Soil Sci 89:16–22.

Schuller, P., C Lovengreen, and H. Handl. 1993. ^{137}Cs concentration in soil, prairie plants, and milk from sites in southern Chile. Health Phys. 64:157–161.

Shanbhag, P.M., and G.R. Choppin. 1981. Binding of uranyl by humic acid. J. Inorg. Nucl. Chem. 43:3369–3373.

Sheppard, M.I. 1985. The plant concentration ratio concept as applied to natural U. Health Phys. 48:494–500.

Sheppard, M.I., and D.H. Thibault. 1990. Default soil solid/liquid partition coefficients, K_ds, for four major soil types: A compendium. Health Phys. 59:471–482.

Sheppard, M.I., T.T. Vandergraaf, D.H. Thibault, and J.A. Keith Reid. 1983. Technetium and uranium: Sorption by and plant uptake from peat and sand. Health Phys. 44(6):635–643.

Sheppard, S.C., and W.G. Evenden. 1985. Mobility and uptake by plants of elements placed near a shallow water table interface. J. Environ. Qual. 14:544–560.

Sheppard, S.C., and W.G. Evenden. 1988. Critical compilation and review of plant/soil concentration ratios for uranium, thorium and lead. J. Environ. Radioactivity 8:255–285.

Sheppard, S.C., and W.G. Evenden. 1992. Bioavailability indices for uranium: Effect of concentration in eleven soils. Arch. Environ. Contam. Toxicol. 23:117–124.

Sheppard, S.C., W.G. Evenden, and A.J. Anderson. 1992. Multiple assays of uranium toxicity in soil. Environ. Toxicol. Water Qual. 7:275–294.

Silker, W.B. 1958. Strontium-90 concentrations in the Hanford environs. HW-55117. Hanford Atomic Products Operation, General Electric Co., Richland, WA.

Skogland, T., and I. Espelien. 1990. The biological effects of contamination of wild reindeer in Norway following the Chornobyl accident. Trans. Congr. Int. Union Game Biologist. 19(1):275–279.

Spencer, F.S. 1984. Tritiated water uptake kinetics in tissue-free water and organically- bound fractions of tomato plants. Rep. 84-69-K. Ontario Hydro Res. Div., Ontario Canada.

Sposito, G. 1989. The chemistry of soils. Oxford Univ. Press, New York.

Stannard, J.N. 1973. Toxicology of radionuclides. Annu. Rev. Pharmacol. 13:325–357.

Stegner, P., and I. Kobal. 1982. Uptake and distribution of radium and uranium in the aquatic food chain. p. C-69. *In* Abstracts and program for the Proc. of the Int. Conf. on Heavy Metals in the Environment, Amsterdam. Sept. 1981. CEP Consultants, Edinburgh, England.

Stevenson, F.J., and A. Fitch. 1986. Chemistry of complexation of metal ions with soil solution organics. p. 29–58. *In* P.M. Huang and M. Schnitzer (ed.) Interactions of soil minerals with natural organics and microbes. SSSA Spec. Publ. 17. SSSA, Madison, WI.

Stewart, A.J., G.J. Haynes, and M.I. Martinez. 1992. Fate and biological effects of contaminated vegetation in a Tennessee stream. Environ. Toxicol. Chem. 11:653–664.

Stumm, W., and J.J. Morgan. 1981. Aquatic chemistry: An introduction emphasizing chemical equilibria in natural waters. 2nd ed. Wiley-Interscience, New York.

Sullivan, M.F., T.R. Garland, D.A. Cataldo, and R.G. Schreckhise. 1979. Absorption of plant-incorporated nuclear fuel cycle elements from the gastrointestinal tract. p. 447–457. *In* Biological implications of radionuclides released from nuclear industries. Vol. II. IAEA-SM-237/58. Int. Atomic Energy Agency, Vienna.

Swanson, S.M. 1985. Food chain transfer of U-series radionuclides in Northern Saskatchewan aquatic system. Health Phys. 49:747–770.

Sweet, C.W., and C.E. Murphy, Jr. 1984. Tritium deposition in pine trees and soil from atmospheric releases of molecular tritium. Environ. Sci. Technol. 18:358–361.

Szalay, A. 1964. Cation exchange properties of humic acids and their importance in the geochemical enrichment of $UO^{2+}{}_2$ and other cations. Geochim. Cosmochim. Acta 28:1605–1614.

Takeda, H., T. Iwakura. 1992. Incorporation and distribution of tritium in rats exposed to tritiated rice or tritiated soybean. J. Rad. Res. 33(4):309–318.

Talmage, S.S., and L. Meyers-Schöne. 1995. Nuclear and thermal. p. 469–491. *In* D.J. Hoffman et al. (ed.) Handbook of ecotoxicology. Lewis Publ., Boca Raton, FL.

Taylor, G.N., C.W. Jones, P.A. Gardner, R.D. Lloyd, C.W. Mays , and K.E. Charrier. 1981. Two new rodent models for actinide toxicity studies. Radiat. Res. 86:115–122.

Taylor, G.N., C.W. Mays, R.D. Lloyd, P.A. Gardner, L.R. Talbot, D. VanMoorhem, D. Brammer, T.W. Brammer, G. Ayoroa, and D. Taysum. 1983. Comparative toxicity of Ra^{226}, Pu^{239}, Am^{241}, Cf^{249}, and Cf^{252} in C57BL/Do black and albino mice. Radiat. Res. 95:584–601.

Tazwell, C.M., and C. Henderson. 1960. Toxicity of less common metals in fish. Ind. Wastes 5:52–67.

Thomas, J.M., L.L. Cadwell, D.A. Cataldo, T.R. Garland, and R.E. Wildung. 1984. Concentration of orally administered and chronically fed ^{95m}Tc in Japanese quail eggs. Health Phys. 46(3):657–663.

Touchberry, R.W., and F.A. Verley. 1964. Some effects of x-radiation in successive generations of an inbred and a hybrid population of mice. Genetics 50:1187.

Trabalka, J.R., and L.D. Eyman. 1976. Distribution of plutonium-237 in a littoral freshwater microcosm. Health Phys. 31:390–393.

Trischen, G.M., R.R. Wilkinson, and B.L. Carson. 1981. III Chemistry. p. 53–117. *In* I.C. Smith and B.L. Carson (ed.) Trace metals in the environment. Vol. 6. Cobalt. Ann Arbor Science Publ., Ann Arbor, MI.

Underbrink, A.G., and A.H. Sparrow. 1974. The influence of experimental end-points, dose, dose rate, neutron energy, nitrogen ions, hypoxia, chromosome volume and ploidy on RBE in *Tradscantia* stamen hairs and pollen. p. 185–214. *In* Biological effects of neutron irradiation. Int. Atomic Energy Agency, Vienna, Austria.

United Nations Scientific Committee on the Effects of Atomic Radiation. 1982. Ionizing radiation: Sources and biological effects. UNIPUB E.82.IX.8, 06300P. UNIPUB, New York.

Upton, A.C. 1982. The biological effects of low-level ionizing radiation. Sci. Am. 246:41–49.

Van Netten, C., and D.R. Morley. 1983. Uptake of uranium, molybdenum, copper, and selinium by the radish from uranium rich soils. Arch. Environ. Health 38:172–175.

Venugopal, B., and T.D. Luckey. 1978. Metal toxicity in mammals. p. 21–24. *In* Chemical toxicity of metals and metalloids. Vol. 2. Plenum Press, New York.

Voight, G., G. Prohl, and H. Mueller. 1991. Experiments on the seasonality of cesium translocation in cereals, potatoes, and vegetables. Radiat. Environ. Biophys. 30:295–304.

Voors, P.I., and A.W. Van Weers. 1991. Transfer of Chornobyl radiocaesium (^{134}Cs and ^{137}Cs) from grass silage to milk in dairy cows. J. Environ. Radioactivity 13:125–140.

Voshell, J.R., Jr., J.S. Eldridge, and T.W. Oakes. 1985. Transfer of Cs-137 and Co-60 in a waste retention pond with emphasis on aquatic insects Health Phys. 49:777–789.

Walker, J.R., and T.A. Brindley. 1963. Effect of x-ray exposure on the European corn borer. J. Econ. Entomol. 56:522–525.

Wallace, A., E.M. Romney, and R.A. Wood. 1971. Cycling of stable cesium in a desert ecosystem. p. 183–186. *In* D.J. Nelson (ed.) Radionuclides in ecosystems. USAEC CONF-710501-P1. USAEC, Washington, DC.

Warner, F., and R.M. Harrison. 1993. Radioecology after Chornobyl. Biogeochemical pathways of artificial radionuclides. John Wiley & Sons, West Sussex, England.

Whicker, F.W., C.A. Little, and T.F. Winsor. 1973. Symposium on environmental surveillance around nuclear installations, Warsaw, Poland. 5–9 Nov. 1973. Int. Atomic Energy Agency, Vienna.

Whicker, F.W., and V. Schultz. 1982. Radioecology: Nuclear energy and the environment. Vol. 1 and 2. CRC Press, Boca Raton, FL.

White, G.C., T.E. Hakonson, and A.J. Ahlquist. 1981. Factors affecting radionuclide availability to vegetables grown at Los Alamos. J. Environ. Qual. 10(3):294–299.

Wildung, R.E., T.R. Garland, and D.A. Cataldo. 1979. Environmental processes leading to the presence of organically bound plutonium in plant tissues consumed by animals. p. 319–334. *In* Biological implications of radionuclides released from nuclear industries. Vol. II. IAEA-SM-237/37. Int. Atomic Energy Agency, Vienna, Austria.

Willard, W.K. 1960. Avian uptake of fission products from an area contaminated by low-level atomic wastes. Science (Washington, DC) 132(3420):148.

Wilson, D.O., and J.F. Cline. 1966. Removal of plutonium-239, tungsten-185, and lead-210 from soil. Nature (London) 5026:941–942.

Winsor, T.F., and T.P. O'Farrell. 1970. The retention of ^{137}Cs by Great Basin pocket mice. p. 217–218. *In* Pacific Northwest Lab. Annual Rep. for 1969. USAEC Rep. BNWL-1306. Part 2. Battelle, Pacific Northwest Lab., Richland, WA.

Woodwell, G.M. 1967. Radiation and the patterns of nature. Science (Washington, DC) 156:461.

Woodwell, G.M. 1970. Effects of pollution of the structure and physiology of ecosystems. Science (Washington, DC) 168(3930):429–433.

Zach, R., and K.R. Mayoh. 1982. Breeding biology of tree swallows and house wrens in a gradient of gamma-radiation. Ecology 63:1721–1728.

Zach, R., and K.R. Mayoh. 1984. Gamma radiation effects on nestling tree swallows. Ecology 65:1641–1647.

Zach, R., and K.R. Mayoh. 1986a. Gamma irradiation of tree swallow embryos and subsequent growth and survival. Condor 88:1–10.

Zach, R., and K.R. Mayoh. 1986b. Gamma-radiation effects on nestling house wrens: A field study. Radiat. Res. 105:49–57.

Zachara, J.M., C.E. Cowan, and C.T. Resch. 1991. Sorption of divalent metal ions on calcite. Geochim. Cosmochim. Acta. 55:1549–1562.

Zachara, J.M., and J.P. McKinley. 1993. Influence of hydrolysis on the sorption of metal cations by smectites: Importance of edge coordination reactions. Aquatic Sci. 55:250–261.

Zachara, J.M., C.T. Resch, and S.C. Smith. 1994. Influence of humic substances on Co^{2+} sorption on a subsurface mineral sperate and its mineralogic components. Geochim. Cosmochim. Acta 58:553–562.

Zachara, J.M., and R.G. Riley. 1992. Chemical contaminants on DOE lands and selection of contaminant mixtures for subsurface science research. DOE/ER-0547T. U.S. Dep. of Energy, Office of Energy Res., Subsurface Science Program, Washington, DC.

Zielinski, R.A., and A.L. Mier. 1988. The association of uranium with organic matter in Holocene peat: An experimental leaching study. Appl. Geochem. 3:631–649.

5 Adsorption of Dissolved Organic Ligands Onto (Hydr)Oxide Minerals[1]

Dharni Vasudevan and Alan T. Stone

Johns Hopkins University
Baltimore, Maryland

5–1 INTRODUCTION

A number of organic chemicals used in the pharmaceutical, dyestuff, photographic, agrochemical and metal plating industries possess Lewis base functional groups, and can therefore be classified as ligands. In addition to forming bonds to protons and metal ions in solution, these compounds can form bonds with metals that reside on the surfaces of (hydr)oxides and other naturally-occurring minerals. This process is termed *surface complexation* or simply adsorption.

We begin by discussing a few illustrative examples of synthetic aromatic organic compounds that possess ligand donor groups and their possible routes of entry into the environment. This is followed by a discussion of naturally occurring aromatic organic compounds with ligand donor groups. (Hydr)oxide minerals encountered within natural aquatic environments and engineered systems are then discussed. The environmental significance of the adsorption of organic ligands is presented. This is followed by an overview of the adsorption process. Characteristics of ligand structure, mineral surface structure and aqueous medium composition likely to be important in developing structure–activity relationships for the adsorption process are described. Finally, we propose a methodology that seeks to examine the effects of specific ligand, (hydr)oxide and aqueous medium characteristics on the adsorption process.

5–2 SYNTHETIC AND NATURAL AROMATIC ORGANIC LIGANDS

We have chosen to study aromatic ligands because (i) a number of synthetic and naturally occurring organic ligands are aromatic and (ii) it is easier to develop structure–activity relationships for aromatic compounds. Examples of common ligand donor groups are shown in Fig. 5–1. Typically, ligand donor

[1] Support by the Environmental Engineering Program of the National Science Foundation (Grant BES9317842).

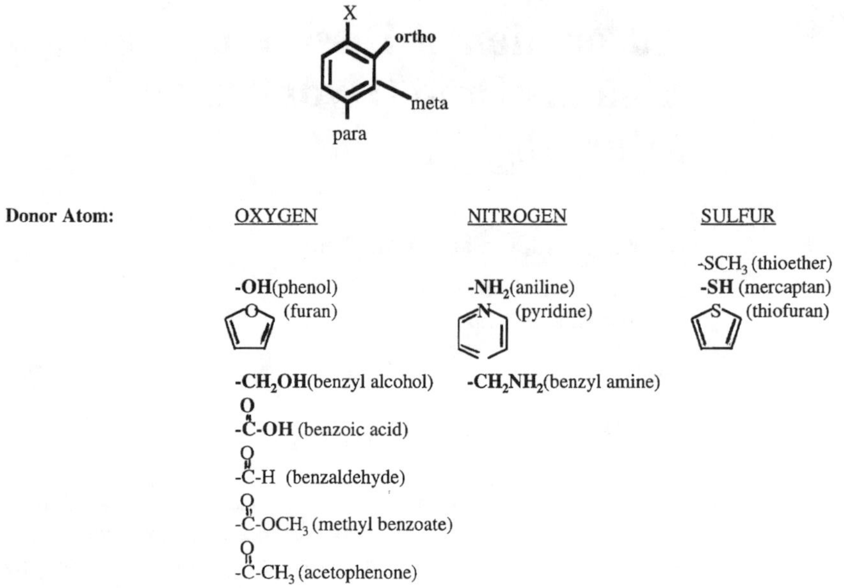

Fig. 5–1. Common ligand donor groups possessing O, N, and S atoms.

groups contain N, O, and S atoms capable of coordinating protons and dissolved and surface bound metal ions. A ligand donor group placed in Position *X* (Fig. 5–1) may have one or more ligand donor groups placed in the ortho, meta, or para positions. The groups shown in bold are capable of protonation–deprotonation at a pH defined by their pK_a(s).

5–2.1 Illustrative Examples of Synthetic Organic Ligands

Synthetic organic ligands are precursors, intermediates, and final products of many industrial processes and have various routes of entry into the natural environment. It is important to understand the fate of these synthetic organic ligands in the natural environment because many of them are toxic and carcinogenic. Lewis base and other compound properties are selected to match the intended use of the compound. These same properties also determine their chemical behavior in the environment. Here, we are concerned primarily with the ability of synthetic organic ligands to form complexes with dissolved and mineral bound metal species that are likely to be found in natural and engineered aquatic environments.

Catechols, 2-aminophenols and 1,2-phenylenediamines (Fig. 5–2) are known to bind strongly to metals and are used as chelating agents in the metal plating industry and as reducing agents in developer solutions in the photographic industry. These compounds can enter the environment at high concentrations as a result of inadvertent spillage into soils and streams and at low concentrations via industrial waste streams.

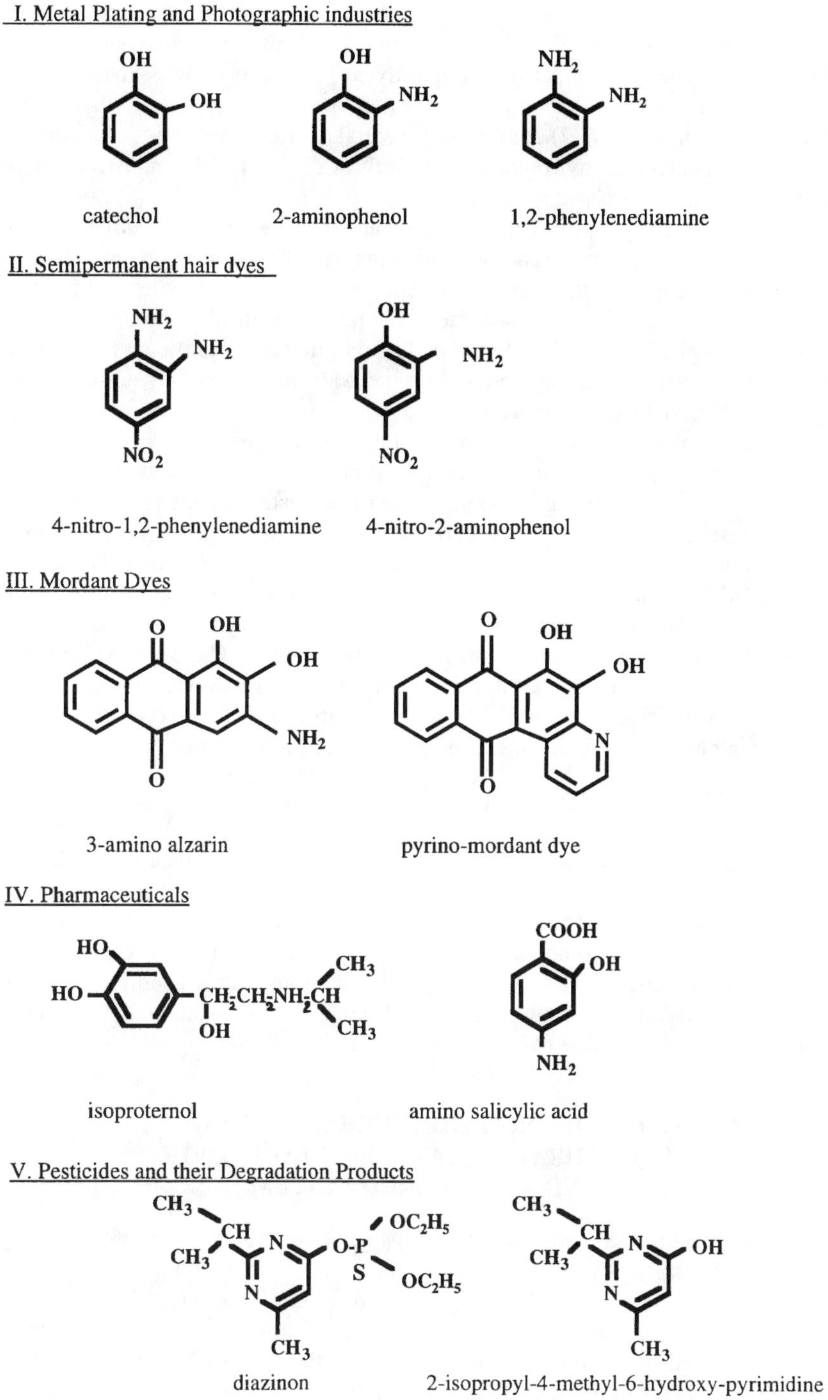

Fig. 5–2. Illustrative examples of organic ligands used commercially.

Dyes often possess N containing ligand donor groups intended for imparting color. For example, 4-nitro-2-aminophenol and 4-nitro-1,2-phenylenediamine (Fig. 5–2) are used in semipermanent hair dyes. The ability of some dyes to complex metal ions in solution is important for their action; mordant dyes (their structures are shown in Fig. 5–2) combine with metals to form insoluble colored complexes. Dyes enter the environment as a result of disposal at the industrial site or disposal into the household waste stream.

Pharmaceuticals possess the chemical structure and appropriate donor groups that allows them to react with enzymes and other biochemicals, thereby altering physiological functions. Isoproterenol and amino-salicylic acid are two examples (Fig. 5–2). Disposal of wastes from pharmaceutical manufacturing may pose an ecological risk. It also should be kept in mind that significant fractions of many pharmaceuticals are excreted from the body in biologically active forms, thereby posing risks to receiving waters downstream.

Unlike the classes of compounds just mentioned, pesticides are intentionally released into the environment. Most pesticides are designed to degrade via hydrolysis of an ester linkage. The structure of this ester linkage is selected such that the pesticide persists long enough to reach the target organism, but not long enough to leave a residue in the field for the following year's planting. The degradation products of a pesticide also may be toxic, and hence, the fate of these products is of concern. Pesticides are typically degraded by hydrolysis, nucleophilic substitution, and other chemical transformation processes. The degradation products may have a greater metal-binding ability than their parent compounds. The hydrolysis of diazinon, for example, yields a substituted hydroxypyridine (Fig. 5–2) and a phosphate diester anion, both capable of binding metals.

5–2.2 Natural Organic Ligands

Natural organic ligands capable of binding metals and adsorbing onto surfaces are abundant in the environment. Structural components of terrestrial plants are comprised of lignin and other aromatic compounds, which are converted by microbial decay into a complex mixture of compounds collectively termed *humic substances*. Although a variety of ligand donor groups have been identified in these mixtures, carboxylic acids and phenols are most abundant, and play a dominant role in metal binding (Morel & Hering, 1993).

5–3 MINERAL (HYDR)OXIDE SURFACES ENCOUNTERED IN NATURAL AQUATIC ENVIRONMENTS AND ENGINEERED SYSTEMS

Soils, sediments, and aquifers consist of source rocks along with primary and secondary minerals generated by weathering. As shown in Fig. 5–3, divalent metal ions and silica are selectively leached as the duration and intensity of weathering is increased. As a result, soils in the intermediate stages of weathering are enriched in aluminosilicates, and soils in the advanced stage of weathering are enriched in Al(III), Fe(III) and Ti(IV) (hydr)oxides (Sposito, 1989; Elmsley, 1989).

Common Source Rocks		
Granites	Sandstones	Shales
Basalts	Limestones	Schists
Gneisses		
Elemental	Si 27.2%	Ti 0 .63%
Crustal	Al 8.3%	Mn 0.11%
Abundance	Fe 6.2%	

Early Weathering Stage	Intermediate Weathering Stage	Advanced Weathering Stage
Gypsum	Quartz	Kaolinite
Carbonates	Dioctahedral Mica/Illite	Al Oxides (Gibbsite)
Olivine/Pyroxene/Amphibole	Vermiculite/Chlorite	Fe Oxides (Goethite, Hematite)
Fe(II)-Bearing Silica	Smectites	Ti Oxides (Anatase, Rutile, Ilmenite)
Feldspars		

Fig. 5–3. Earth surface rocks and minerals encountered by organic pollutants (adapted from Sposito, 1989; Elmsley, 1989).

Because of their high surface area-to-volume ratio, (hydr)oxide minerals are more important adsorptive surfaces in soils than their weight percentage would indicate. (Hydr)oxide minerals are frequently found as coatings on sand-sized particles and as grain cements, yielding high surface areas in contact with interstitial waters (Schwertmann & Taylor, 1981). Most (hydr)oxide minerals do not possess the permanent layer charge and interlayer cavities found in kaolinites, smectites and other complex aluminosilicates. (Hydr)oxide minerals do, however, exhibit a wide range of protonation–deprotonation behaviors, making them useful in laboratory adsorption studies.

In engineered systems such as water treatment plants, the addition of alum ($Al_2(SO_4)_3$), polymeric aluminum chloride and ferric chloride generates hydrous oxide flocs. These flocs serve as coagulants to remove colloidal particles in the water supply. They also can serve as potential sorbents for natural and synthetic organic ligands present in the water supply.

5–4 ENVIRONMENTAL SIGNIFICANCE OF ADSORPTION

As mentioned earlier, the study of organic ligand adsorption improves our understanding of the transport and chemical transformations of these compounds. Studies of this kind also are important because organic ligands affect the environmental chemistry of other important pollutants and affect environmental processes in a number of ways.

5–4.1 Pollutant Transport

Within soils, sediments and aquifers, adsorption retards downfield transport of organic pollutants by interstitial waters. In surface waters, adsorption onto settling particles provides a means of removing pollutants from the water column and transporting them onto the sediments (McCarthy & Zachara, 1989). In order to evaluate the effect of adsorption on mass transport, the abundance and available surface areas of naturally-occurring minerals, along with their affinity towards various pollutant ligands, must be fully characterized.

5–4.2 Pollutant Transformation

Adsorption brings organic pollutants into contact with sparingly-soluble solid reactants. Manganese (III, IV) and iron(III) (hydr)oxide minerals are capable of oxidizing adsorbed phenols (Stone, 1987), anilines (Laha & Luthy, 1990) and mercaptans. Fe(II) within biotite and vermiculite surfaces is believed to participate in the reduction of tetrachloromethane (Kriegman-King & Reinhard, 1992).

Even when the surface does not act as a reactant, adsorption alters organic pollutant reactivity. Complexation by surface-bound metals alters bond lengths and the overall electronic structure of the organic ligands, thereby altering their susceptibility towards chemical reaction. Several phosphorothionate pesticides (Torrents & Stone, 1991, 1994) and alkyl sulfate detergents (Lukenheimer et al., 1993) have been found to hydrolyze more quickly when adsorbed onto (hydr)oxide mineral surfaces.

5–4.3 Dissolution of Mineral Surfaces

When an organic ligand forms a complex with a metal ion bound to a mineral surface, the dissolution of the underlying mineral can take place. Studies conducted have identified both dissolution-promoting and dissolution-inhibiting organic ligands, but the differences between them are not yet fully understood (Rubio & Matijevic, 1979; Stumm & Furrer, 1987).

5–4.4 Effect of Natural Organic Compounds on Pollutant Sorption

As mentioned earlier, natural organic compounds can adsorb onto mineral surfaces via carboxylic and phenolate ligand donor groups. Surface sites occupied by natural organic compounds are no longer available for pollutant adsorption. Pollutants that bind minerals strictly by adsorption have to compete with natural organic compound for available surface sites. This competitive adsorption effect is particularly pronounced for inorganic oxyanions and organic ligand pollutants.

Medium- and high-molecular weight natural organic compounds possess hydrophobic core regions into which hydrophobic organic pollutants can partition (Karickhoff et al., 1979; Means et al., 1982; Schellenberg et al., 1984). When natural organic compounds adsorb onto mineral surfaces, these hydrophobic core regions travel along with them. As a result, hydrophobic organic pollutants are removed from bulk solution via the adsorption of natural organic compounds (Murphy et al., 1990).

Some ligand groups of natural organic compounds become attached to mineral surfaces during the adsorption process, others remain unattached. These unattached groups are capable of forming complexes with metal ion pollutants. In this way, adsorbed natural organic compounds can serve as a bridge, facilitating metal ion sorption (Tipping et al., 1983; Davis & Leckie, 1978).

5–4.5 Organic Removal by Water Treatment Plants

Dissolved natural organic compounds are removed from raw water prior to chlorination to minimize the formation of organochlorine compounds. Coagulation basins are used for this purpose; (hydr)oxide flocs generated by the addition of alum ($Al_2(SO_4)_3$), polymeric aluminum chloride and ferric chloride to the coagulation basins sweep out particulate natural organic matter and adsorb dissolved organic matter.

Pesticides and other synthetic organic compounds are also present in some raw water sources. Coagulation basins employing (hydr)oxide flocs may play an important role in removing hydrophilic and ligand-like organic compounds. This process has not, however, been extensively explored.

5–5 OVERVIEW OF THE ADSORPTION PROCESS

Of the many classes of organic ligands, carboxylic acids have received the greatest attention in adsorption studies. Literature reports that provide a description of carboxylic acid adsorption are listed in Table 5–1. Literature reports that deal with the adsorption of other classes of organic ligands are listed in Table 5–2.

Adsorption experiments have been done under varying pH and ionic strength conditions, and in the presence of a number of different (hydr)oxide surfaces. The experimentally observed adsorption behavior of aromatic ligands is illustrated using the adsorption of benzoic acid onto Al_2O_3 and the adsorption of 2,4-dinitrophenol, and 2-aminophenol onto TiO_2 as examples (organic compound structures are shown in Fig. 5–4). To date, quantitative descriptions of experimentally observed adsorption behavior have used a common approach, which has met with considerable success. The elements of this approach are now summarized.

5–5.1 Experimentally Observed Adsorption Behavior

Figures 5–5A and 5–5B show the adsorption of benzoic acid onto Al_2O_3 and 2,4-dinitrophenol onto TiO_2 as a function of pH. Reaction conditions and literature sources for the experimental data are presented in Table 5–3. The adsorption behavior of the two compounds are similar; adsorption increases as the pH is decreased and reaches a maximum near the pK_a of the ligand donor group (Kummert & Stumm, 1980; Stone et al., 1993). The strong electron withdrawing nature of the nitro substituent on 2,4-dinitrophenol generates a *hard* phenolate anion with a basicity similar to the carboxylate ion, resulting in the comparable adsorption behavior of the two compounds.

Several studies have shown that compounds possessing a single carboxylate group adsorb in the same fashion. There is some speculation, however, as to whether one or two O atoms within the carboxylate group are involved in bonding (Kummert & Stumm, 1980; Biber & Stumm, 1994). For a series of substituted phenols, adsorption increases as the pK_a of the phenolate group is decreased (McBride & Kung, 1991). Adsorption also increases as the ionic strength of the

Table 5–1. Adsorption of organic ligands possessing carboxylic acid groups.

Organic ligand	Hydr)Oxide	Reference
Amino acids	TiO_2	Tentorio & Canova, 1989
Aspartic acid	TiO_2	Giacomelli et al., 1995
Benzoic acid	Al(OH)3 γ-Al_2O_3 α-FeOOH	Parfitt et al., 1977a,b; Kummert & Stumm, 1980; Stumm et al., 1980; Ballion & Jaffrezic-Renault, 1985; Djafer et al., 1991
4-Substituted benzoic acids	α-FeOOH $Fe(OH)_3$ (amorphos)	Kung & McBride, 1989a,b
Citric acid	α-Fe_2O_3 γ-AlOOH	Zhang et al., 1985; Kallay & Matijevic, 1985; Cambier & Sposito, 1991
2,4-Dihydroxybenzoic acid	α-FeOOH	Tejedor-Tejedor et al., 1990, 1992
Dipicolinic acid	α-Fe_2O_3	Pope et al., 1981
EDTA, HEDTA, DPTA, CTPA NTA	β-FeOOH α-Fe_2O_3	Rubio & Matijevic, 1979; Chang et al., 1983; Torres et al., 1990
Glutamic acid	$Fe(OH)_3$ (amorphous) γ-Al_2O_3	Davis & Leckie, 1978; Ballion & Jaffrezic-Renault, 1985
4-Hydroxybenzoic acid	α-FeOOH $Fe(OH)_3$ (amorphous) $FeOOH.3H_2O$ α-Fe_2O_3	Kung & McBride, 1989a,b; Tejedor-Tejedor et al., 1990, 1992
Iminodiacetic acid	α-Fe_2O_3	Torres et al., 1988, 1990
Lactic acid	α-FeOOH	Balistrieri & Murray, 1987
Nicotinic acide	α-Fe_2O_3	Pope et al., 1981
Oxalic acid	$Al(OH)_3$ α-FeOOH α-Fe_2O_3 TiO_2	Parfitt et al., 1977a,b; Zhang et al., 1985; Balistrieri & Murray, 1987; Mesuere & Fish, 1992; Torres et al., 1990; Djafer et al., 1991; Hug & Sulzberger, 1994; Kallay & Matijevic, 1995
Phthalic acid	γ-Al_2O_3 α-FeOOH	Kummert & Stumm, 1980; Stumm et al., 1980; Balistrieri & Murray, 1987; Lovgren, 1991; Tejedor-Tejedor et al., 1990, 1992
Picolinic acid	$Fe(OH)_3$ (amorphous) α-Fe_2O_3	Davis & Leckie, 1978; Pope et al., 1981
Polyacetic amino acid	γ-Al_2O_3	Bowers & Huang, 1985
2,3-Pyrazine dicarboxylic acid and 3,4-Dihydroxybenzoic Acid	$Fe(OH)_3$ (amorphous)	Davis & Leckie, 1978
Salicylic acid	δ-Al_2O_3 γ-Al_2O_3; α-Fe_2O_3; α-FeOOH; γ-FeOOH $Fe(OH)_3$ (amorphous)	Davis & Leckie, 1978; Kummert & Stumm, 1980; Stumm et al., 1980; Balistrieri & Murray, 1987; Machesky et al., 1989; Thomas et al., 1989; Biber & Stumm, 1994

medium is decreased (Stone et al., 1993). These observations indicate that adsorption arises from a combination of near-range interactions (ionic bonding between the phenolate O and the surface bound metal) and long-range electrostatic interactions (accumulation of phenolate anions near positive surfaces).

According to Fig. 5–5C the experimentally observed adsorption behavior of 2-aminophenol differs from that of benzoic acid and 2,4-dinitrophenol.

Table 5–2. Adsorption of organic ligands not possessing carboxylic acid groups.

Organic ligand	(Hydr)oxide	Reference
Aminonapthalene	SiO_2 (amorphous)	Zachara et al., 1990
2-Aminophenol, 4-Methyl-2-aminophenol and 4-Nitro-2-aminophenol	Al_2O_3; α-FeOOH; TiO_2	Stone et al., 1993
8-Aminoquinoline	TiO_2	Ludwig & Schindler, 1995
2,2′-Bipyridyl	TiO_2	Ludwig & Schindler, 1995
Catechol	$Al(OH)_3$; γ-AlOOH $Al(OH)_3$ (amorphous)	McBride & Wesselink, 1988
4-Hydroxybenzaldehyde	$Fe(OH)_3$ (amorphous) α-FeOOH; α-Fe_2O_3	McBride & Kung, 1991
8-Hydroxyquinoline-5-sulfonate	Al_2O_3	Hering & Stumm, 1991
Hydroquinone	$Al(OH)_3$; γ-AlOOH; $Al(OH)_3$ (amorphous); α-FeOOH; α-Fe_2O_3; $Fe(OH)_3$ (amorphous)	McBride & Wesselink, 1988; McBride & Kung, 1991
Phenol	$Al(OH)_2$; γ-AlOOH $Al(OH)_3$ (amorphous); α-FeOOH; α-Fe_2O_3; $Fe(OH)_3$ (amorphous)	McBride & Wesselink, 1988; McBride & Kung, 1991
4-Nitrophenol, 4-Aminophenol, and 4-Methoxyphenol	$Fe(OH)_3$ (amorphous); α-FeOOH; α-Fe_2O_3	McBride & Kung, 1991
2,4-Dinitrophenol and 2,6-Dinitrophenol	Al_2O_3; α-FeOOH; TiO_2	Stone et al., 1993
2-Pyridinemethanol	Al_2O_3; α-FeOOH; TiO_2	Stone et al., 1993
Quinoline	SiO_2 (amorphous)	Zachara et al., 1990

Table 5–3. Reaction conditions and literature sources for the experimental data presented in Fig. 5–5.

Organic ligand	(Hydr)oxide	Ionic Strength	Adapted from
Benzoic acid ($L_T = 1.95 \times 10^{-3}$ *M*)	11.04 g L^{-1} Al_2O_3	0.1 *M*	Kummert & Stumm, 1980
2,4-Dinitrophenol ($L_T = 5.1 \times 10^{-5}$ *M*)	10.0 g L^{-1} TiO_2	0.1 *M*	Stone et al., 1993
2-Aminophenol ($L_T = 5.28 \times 10^{-5}$ *M*)	10.0 g L^{-1} TiO_2	0.1 M	Stone et al., 1993

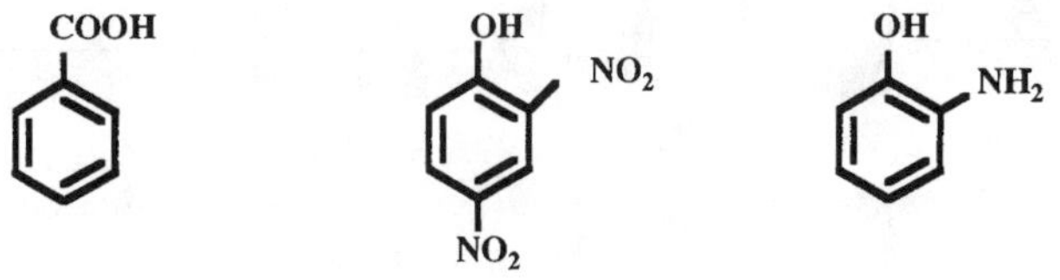

benzoic acid 2,4-dinitrophenol 2-aminophenol

Fig. 5–4. Structures of compounds discussed in Fig. 5–5.

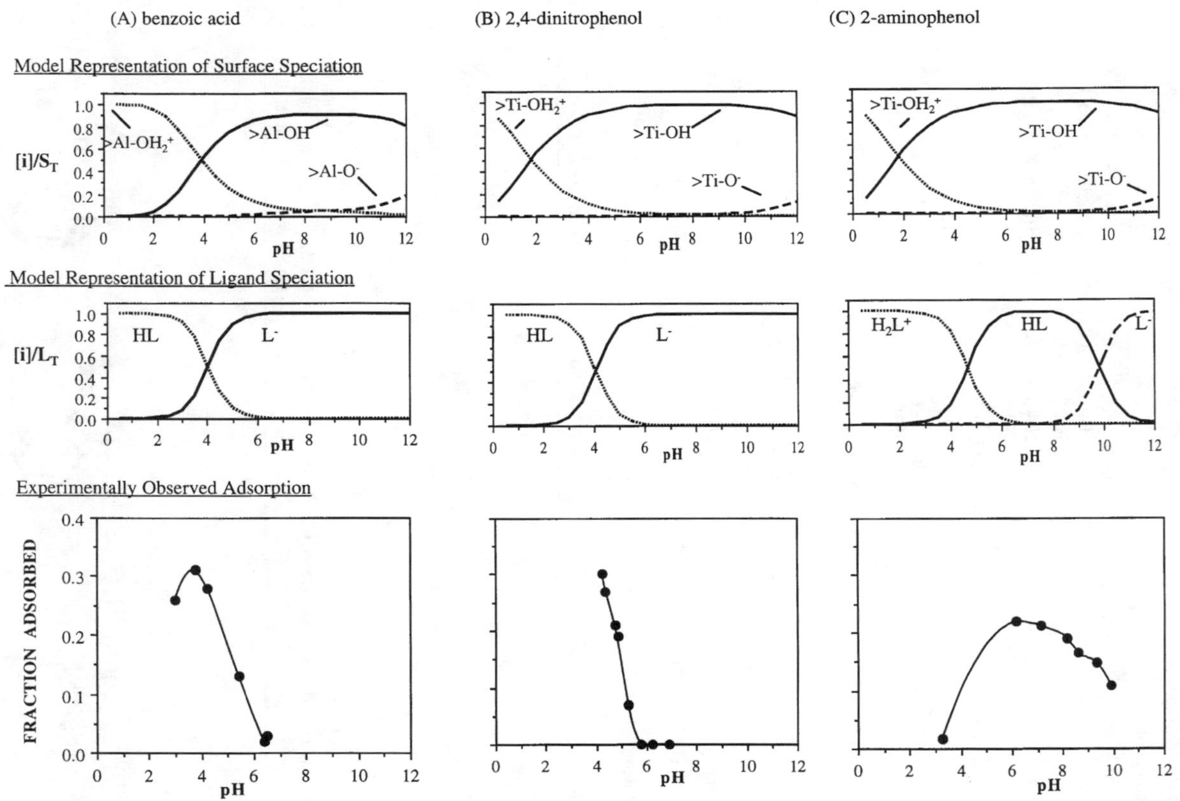

Fig. 5–5. Comparison of the adsorption behavior of (A) benzoic Acid, (B) 2,4-dinitrophenol, and (C) 2-aminophenol onto (hydr)oxides. Reaction conditions and literature sources are listed in Table 5–3. The lines through the data points for experimentally observed adsorption are intended to show the adsorption trend as a function of pH and do not depict model results.

Adsorption increases as the pH is increased, reaches a maximum between the two pK_as that define the acid–base behavior of the compound, and then decreases as the pH becomes alkaline.

The presence of two ligand donor groups yields a protonation behavior for 2-aminophenol that is more complex than that of the other two organic compounds. More importantly, the two ligand donor groups are suitably placed for simultaneous coordination of a surface-bound metal atom, giving rise to a *surface chelate*. In analogy to solution complexes, such chelates are believed to be unusually strong. The favorable chelate effect is believed to arise from the following phenomena: (i) the capacity to simultaneously coordinate two donor groups gives an advantage to metal centers in comparison to protons; (ii) once one ligand donor group has been bound, the frequency of encounter between the metal and the second donor group is increased, favoring complex formation; and (iii) bidentate chelate formation releases two inner-sphere water molecules for every incoming ligand, yielding a favorable reaction entropy.

Long-range electrostatic forces affecting the distribution of ions near charged surfaces are relatively well understood, and can be estimated from the surface charge, solution species charge, and medium ionic strength. Subtracting these effects from experimental adsorption measurements allows us to look at the near-range physical and chemical phenomena governing adsorption in greater detail. The next few sections will describe how this is achieved, using a widely-applied model for (hydr)oxide surfaces and the adsorption process.

5–5.2 Surface Protonation–Deprotonation Reactions

The coordinative requirements of metal ions residing within the interior of (hydr)oxide minerals are fully satisfied by oxo (O^{2-}) or hydroxo (OH^-) lattice ions. At the mineral surface, however, lattice ions cannot complete the coordinative requirements of the surface-bound metal ions. Instead, hydroxide ions, water

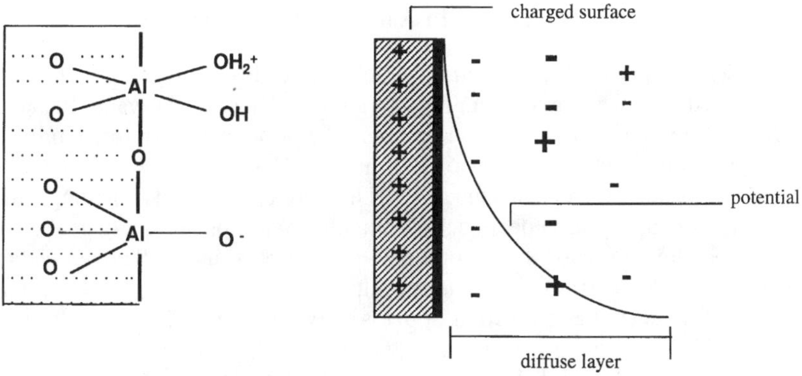

(A) The protonation/deprotonation behavior of the Al_2O_3 surface hydroxyl groups (B) Representation of the Electric Double Layer

Fig. 5–6. Schematic representations of metal (hydr)oxide–water interfaces.

molecules, and other Lewis bases are drawn from solution to fulfill the coordinative requirements.

The degree of protonation of this hydrated layer changes as a function of pH. Although site-to-site variations may exist, it is common to model surface protonation–deprotonation reactions using a single type of site that is capable of existing in three proton levels (Schindler & Stumm, 1987; Fig. 5–6). A model representation of the protonation–deprotonation of the Al_2O_3 surface is shown below:

$$>\text{Al-OH}_2^+ \rightarrow \quad >\text{Al-OH} + H^+ \quad pK^{s}_{a1} \qquad [1]$$

$$>\text{Al-OH} \rightarrow \quad >\text{Al-O}^- + H^+ \quad pK^{s}_{a2} \qquad [2]$$

In the absence of other adsorbing species, the surface charge is determined by the degree of surface protonation. The pH where number of protonated and deprotonated sites are equal to one another is termed the pH_{zpc} (for *zero point of charge*). TiO_2, exhibiting a pH_{zpc} in the vicinity of pH 6.0, is considerably more acidic than Al_2O_3 (pH_{zpc} near 9.0). The speciation of the surface hydroxyl groups of Al_2O_3 and TiO_2 as a function of pH is shown in Fig. 5–5. This representation of the (hydr)oxide surface is based upon results from acid–base titrations in solutions of varying ionic strength. Neutral surface sites are the predominant species throughout a wide range in pH centered at the pH_{zpc}, particularly in low ionic strength solutions.

According to the representation presented in Reactions 1 and 2, only one water molecule is required to meet the coordination requirements of the surface metal ion. This approach is somewhat simplistic, since sites requiring two or more coordinated water molecules (i.e., step, kink, terrace, or edge site) have long been postulated and have recently been observed using atomic microscopy (Hochella, 1990). Since the number of coordinated water molecules and their protonation level is not known for the predominant neutral surface sites, they will be denoted by the symbol >S in this chapter.

5–5.3 Ligand Protonation–Deprotonation

The ability of a compound to serve as a Lewis base is termed *basicity*. As the basicity increases among a group of structurally-related organic ligands, the pK_a is increased, and the domain in which the protonated form predominates in homogeneous solution extends towards higher pHs.

Benzoic acid and 2,4-dinitrophenol are monoprotic acids, yielding a monovalent anion upon deprotonation. 2-Aminophenol protonates under acidic conditions, yielding a monovalent cation and deprotonates under alkaline conditions, yielding a monovalent anion. The speciation of benzoic acid, 2,4-dinitrophenol and 2-aminophenol as a function of pH is shown in Fig. 5–5.

5–5.4 Mass Balance and Speciation

Examining the adsorption behavior of the three compounds in Fig. 5–5 with reference to the model representation of the surface, we see that adsorption is dominant in the pH range where the concentration of neutral surface sites is high and the neutral form of the ligand dominates. The experimentally observed adsorption behavior can be understood in terms of mass balance and speciation of the system constituents. The mass balance equations for benzoic acid adsorption onto Al_2O_3 is an illustrative example:

$$S_T = [\text{>Al-OH}_2\text{+}] + [\text{>Al-OH}] + [\text{>Al-O}^-] + [\text{>Al-L}] \quad [3]$$

$$L_T = [\text{HL}] + [\text{L}^-] + [\text{>Al-L}] \quad [4]$$

S_T is the total number of sites, L_T is the total ligand concentration and >Al-L denotes the surface-ligand complex. In order for a change in the system to yield an increase in adsorption, the last term in both equations must grow at the expense of all preceding terms.

From the Eq. [3] and [4] we observe two competing effects: (i) The deprotonated ligand must compete with protons and hydroxide ions for binding available surface sites and (ii) the surface sites must compete with protons for binding the deprotonated ligand. In the case of benzoic acid, at high pH hydroxide ions outcompete the ligand for binding the surface, while at low pH, competing effects favor binding of the ligand to the surface metal ion.

5–5.5 Ligand Adsorption

We have chosen to classify interactions at the interface into two categories: (i) those arising from long-range electrostatic forces and (ii) all other interactions that we refer to as near range physical and chemical forces.

5–5.5.1 Long Range Electrostatic Forces

Because of long range electrostatic forces, anion activities near a positive (hydr)oxide surface are higher than in bulk solution, and cation activities are lower than in bulk solution. (Near a negative surface, the opposite is true.) The net result is the formation of an electric double layer consisting of a fixed layer of surface charge and a closely associated layer of diffuse charge (Fig. 5–6).

The Boltzmann Distribution Function (Verwey & Overbeek, 1948) accounts for the balance of two opposing phenomena: diffusion of ions in response to a concentration gradient and the diffusion of ions in response to an electrical potential gradient. Accordingly, the activities of ions at the plane of closest approach ($\{i\}_o$) are related to the activities of ions in bulk solution ($\{i\}_{bulk}$) through the following equation:

$$\{i\}_o = \{i\}_b \exp(-z_i F \psi_o / RT) \quad [5]$$

where z_i is the charge on species i, F is the Faraday constant, ψ_o is the electrical potential at the plane of closest approach, R is the gas constant and T is the temperature.

Long range electrostatic forces have the greatest effect on adsorption under the following conditions: (i) the pH is much higher or much lower that the pH_{zpc}, yielding high surface charge; and (ii) the ionic strength is low, creating a greater perturbation in the distribution of ions near the surface.

5–5.5.2 Adsorption Stoichiometry

As with all chemical reactions, the extent of ligand adsorption is governed by reaction stoichiometry and a corresponding equilibrium constant. The typical approach is to postulate several alternative stoichiometries and determine which one yields the best fit to experimental data. The neutral hydrated surface site and the lowest proton level for the ligand provide the starting point for evaluating the effect of pH on adsorption.

As an example, we explore three possible stoichiometries for the adsorption of 2-aminophenol onto TiO_2, yielding adsorption complexes bearing net charges of +1, 0, and −1:

$$>S + 2H^+ + L^- \rightarrow (>S\text{-}L)^+ + xH_2O \quad [6]$$

$$>S + H^+ + L^- \rightarrow (>S\text{-}L)^o + xH_2O \quad [7]$$

$$>S + L^- \rightarrow (>S\text{-}L)^- + xH_2O \quad [8]$$

Model fits developed for each of the three stoichiometries are presented in Fig. 5–7. The solid line corresponding to the stoichiometry of Eq. [7] yields the best fit to experimental data: adsorption is highest between the two pK_as of 2-aminophenol, and drops to negligible levels at very high and low pH. It should be kept in mind that this approach ignores site-to-site variations that may yield dissimilar adsorption stoichiometries. The exercise establishes however, that one

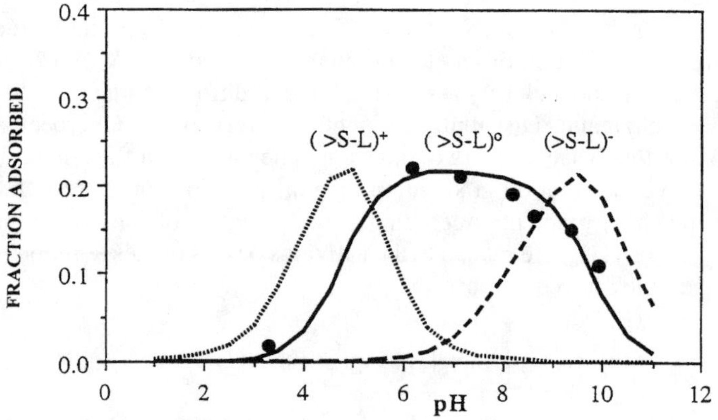

Fig. 5–7. Diffuse layer modeling results for the adsorption of 2-aminophenol. The filled circles represent experimental data. The three lines are the modeling results of the three absorption stoichiometries represented by Reactions 6–8.

proton on average is consumed for every 2-aminophenol molecule adsorbed and that no net change in surface charge takes place.

The shape of each curve is defined by the reaction stoichiometry, while the height of each curve is determined by the numerical value of the equilibrium constant (which is systematically varied until the height most closely matches the experimental data).

5–5.5.3 Equilibrium Constant for Adsorption

In order for the equilibrium constants for adsorption to be independent of surface charge, they should be expressed in terms of ion activities at the plane of closest approach. An example of an *intrinsic* equilibrium constant ($K^{intr.}{}_{ML}$) of this kind is shown below, corresponding to Eq. [7]:

$$K^{intr.}{}_{ML} = \frac{\{>S\text{-}L\}}{\{>S\}\{L^-\}_o\{H^+\}_o} \quad [9]$$

$K^{intr.}{}_{ML}$ is reflective of the near range physical and chemical forces.

Bulk solution concentrations (and activities) are the experimentally measurable quantities in most adsorption experiments, and provide the basis for the mass balance equations and the calculation of equilibrium speciation. Using the Boltzmann Distribution Function, $K^{intr.}{}_{ML}$ can be written in terms of bulk activities

$$K^{intr.}{}_{ML} = \frac{\{>S\text{-}L\}}{\{>S\}\ \{L^-\}_{bulk}\ \exp(F\psi_o/RT)\ \{H^+\}_{bulk}\exp-(F\psi_o/RT)} \quad [10]$$

$$K^{intr.}{}_{ML} = \frac{\{>S\text{-}L\}}{\{>S\}\ \{L^-\}_{bulk}\ \{H^+\}_{bulk}} \quad [11]$$

It should be noted that the exponential terms do not always cancel out in this way. A comparable treatment of Eq. [6], for example, yields the term $\exp(-z_i F\psi_o/RT)$ in the denominator.

5–5.5.4 Computation

Several computer programs are available for calculating the extent of adsorption as a function of adsorption stoichiometry, $K^{intr.}{}_{ML}$, ligand concentration, oxide loading, surface site density, and ionic medium composition (Westall et al., 1976; Papelis et al., 1988). Intrinsic adsorption constants for surface protonation–deprotonation (Eq. [1] and [2]) are obtained by acid–base titration (James & Parks, 1982). Surface site density is inferred from B.E.T N_2 adsorption measurements or from Langmuir Adsorption Isotherms. Computer programs dealing with adsorption typically offer several alternative representations of the (hydr)oxide–water interface, such as the Constant Capacitance Model, the Diffuse Layer Model, and the Triple Layer Model (Westall & Hohl, 1980). The approach used in Fig. 5–7 is based upon the Diffuse Layer Model, which can account for organic ligand adsorption with the fewest fitting parameters.

The quantitative treatment of adsorption just described has been used extensively by a number of researchers. In order to improve our understanding of adsorption, it is important to determine which elements of this approach are accurate representations of surface phenomena, and which could benefit from an alternative or modified approach. With respect to organic ligands, our ultimate goal is to develop structure–activity relationships for predicting the extent of adsorption based upon ligand structure and physical–chemical properties, (hydr)oxide composition and crystallographic structure, and the ionic composition of the aqueous medium. To reach this goal, more fundamental aspects of the adsorption process require close examination.

5–6 CHARACTERISTICS IMPORTANT FOR STRUCTURE–ACTIVITY RELATIONSHIPS APPLIED TO ADSORPTION

The number of organic ligands discovered in the natural environment is increasing. Acquiring information on individual adsorption behavior for various organic compounds is laborious. It is thus important to develop structure–activity relationships that can predict the adsorption behavior of previously unexplored organic ligands. In order to do this, we need to examine in detail characteristics of the ligand, surface and ionic medium likely to influence adsorption behavior.

5–6.1 Effect of Ligand Structure

There are three important factors to be considered in the comparison of organic ligand adsorption behavior: (i) the identity of the ligand donor atom, and the nature and relative position of ligand donor groups; (ii) substituents that affect the electronic structure of the compound through inductive and resonance effects; and (iii) substituents that impart hydrophobicity or impose steric constraints on surface complex formation.

The identity of the ligand donor atom determines the ionic versus covalent character of bond formed with metal ions. When bonding electrons are retained by the most electronegative atom, the bond is strongly ionic in character. The ionic contribution to bonding increases as the charge-to-radius ratio of the ligand and metal atom is increased. When bonding electrons are extensively shared between the ligand and the metal, the bond is strongly covalent in character. Heavier elements within each column of the periodic table more readily share electrons in this way. The ease with which electron distribution is distorted in an electric field is termed polarizability.

Nitrogen-bearing amino and pyridyl groups are electrically neutral, while O-bearing alcoholate and phenolate groups possess a negative charge. Nitrogen is more polarizable than O. As a consequence, the covalent contribution to bonding is greater for N-donor ligands, while the ionic contribution to bonding is greater for O-donor ligands.

The same donor atom can appear in different ligand structures and exhibit different metal-binding properties. The amine group of aniline, for example, complexes metal ions via the N sp^3 hybrid orbital, which can freely rotate relative to the aromatic ring until the optimum metal-amine bond arrangement has been reached. The N atom of pyridine, in contrast, complexes metal ions via an sp^2 hybrid orbital, which has lost electron density via resonance with the aromatic ring. This orbital points directly out from the plane of the aromatic ring, and experiences steric hindrance from neighboring ortho-substituents. Additionally, the degree of hybridization on the N (sp^3 vs. sp^2) and consequently the *s* character of the N influence compound basicity.

The presence of a second donor group capable of interacting with the surface is likely to result in the previously described chelate effect. In surface chelates, the stability of the surface metal-ligand complex is expected to be determined by the chelate ring size. Four-membered chelate rings are unstable because of steric and geometric constraints. Five-membered chelate rings are most stable; rings of larger size become progressively less stable because each linkage in the ring contributes additional rotational degrees of freedom, which affect the entropy of complex formation (decreases in entropy reduce stability). Relationships between chelate ring size, steric and geometric constraints, and entropic effects affecting adsorption have not been firmly established.

Substituents that do not serve as ligand donor groups can still exert a significant effect on adsorption. Bulky substituents can impede adsorption by blocking access to neighboring ligand donor groups. Electronic effects (inductive and resonance) also can be important; replacing a substituent with a more electron-withdrawing substituent is expected to lower both pK_a and $K^{intr.}{}_{ML}$ for ligand donor groups on the aromatic ring. These two quantities do not, however, change to the same extent; the coordinative requirements of protons and surface-bound metals are different, owing to differences in charge, ionic radius, and electronic configuration of the two Lewis acids. As a consequence, the competition between protons and surface-bound metals is altered by ring substituents. For one class of organic compounds, electron withdrawing substituents may favor surface-bound metals over protons and raise the extent of adsorption; for another class of organic ligands that coordinate protons and surface-bound metals in a different manner, the opposite may be true (Vasudevan, 1996).

Ligand-water interactions (relative to water-water interactions) are important because ligands lose part of their hydration shell when adsorption takes place. In the presence of hydrophobic substituents, ligands may experience expulsion from bulk solution that encourages adsorption. The effect of hydrophilic substituents is more difficult to evaluate, since most of the functional groups that interact strongly with solvent water molecules (via H bonding or dipole interactions) also interact favorably with surface-bound metals.

5–6.2 Effect of (Hydr)Oxide Structure

Differences in the adsorbent properties of (hydr)oxide minerals arise from (i) the identity and oxidation state of the constituent metal ion; (ii) the chemical composition of the mineral with respect to hydroxo (OH^-) and oxo (O^{2-}) ions; and (iii) the crystal structure of the oxide.

The above factors collectively define properties of the (hydr)oxide such as the dielectric constant, the Paulings bond strength, the metal ion radius and the degree of hydration of the surface. These properties are known to influence the protonation behavior of hydrated metal (hydr)oxide surfaces (Sverjensky, 1994) and are likely to influence the adsorption of organic ligands.

These factors also directly influence the specific surface characteristics and the geometry of the exposed surface. Trivalent and tetravalent metal ions (Ti(IV), Fe(III), and Al(III)) have a high charge to radius ratio, and as a result, the ionic contribution to bonding is high. The coordination of O-donor, anionic groups such as carboxylate, phenolate, and alcoholate ligands is favored over the more polarizable N-donor ligands. As discussed earlier, N-donor ligands are more polarizable than O-donor ligands, and therefore adsorb preferentially to (hydr)oxides comprised of transition metal ions with accessible *d* orbitals.

Hematite(α-Fe_2O_3), goethite (α-FeOOH), and lepidocrocite (γ-FeOOH) are common naturally occurring Fe oxides. The stacking of the lattice anions of hematite(O^{2-}) and goethite (OH^- and O^{2-}) are similar; however, as indicated by the stoichiometry with respect to the lattice anions, one-half of the octahedral sites are occupied by Fe(III) in the case of goethite, while two-thirds of the octahedral sites are occupied by Fe(III) in the case of hematite. On the other hand, goethite and lepidocrocite have the same molecular stoichiometry with respect to their lattice anions; however, the coordination polyhedra of the octahedral sites share faces in the case of goethite but not in the case of lepidocrocite. As a result, the geometry of the exposed surfaces of hematite, goethite, and lepidocrocite differ due to differences in the underlying crystal structure and in molecular stoichiometry. The geometry of exposed surfaces and other properties of the crystal structure, in turn, affect the manner in which surfaces can interact with water molecules and ligand solutes.

5–6.3 Effect of Medium Composition

The influence of pH and ionic strength on the adsorption behavior of organic ligands has been discussed earlier. Additionally, the presence of other adsorbing electrolyte ions can result in a competition between the ions and the organic ligand for available surface sites; this competition may alter the extent of ligand adsorption.

5–7 A METHODOLOGY FOR EXPLORING ADSORPTION

Most reports of ligand adsorption treat the subject as a modeling exercise: new surface complex formation stoichiometries and new values for $K^{intr.}{}_{ML}$ are put into a computer program until a reasonable representation of experimental measurements is obtained. We would like to go further by developing an understanding of the underlying phenomena that control stoichiometry and $K^{intr.}{}_{ML}$.

This understanding may allow us to predict adsorption behavior for organic ligands that have not yet been experimentally investigated.

Much of what we know about *dissolved* metal-ligand complexes was learned by making systematic changes to the structure of the organic ligand, by systematically comparing the properties of different metal ions, and by examining how aqueous medium characteristics affect complex formation. This same approach can be employed in the study of *surface* metal-ligand complexes.

5–7.1 Ligand Characteristics

As has already been mentioned, all carboxylic acids adsorb to some extent. Compounds possessing a single noncarboxylic acid Lewis base group typically adsorb more weakly, and other factors in the molecule must assist in order for adsorption to reach significant levels (e.g., 2,4-dinitrophenol).

The ability of Lewis base groups to facilitate adsorption can be readily determined by examining compounds possessing pairs of such groups. The first task in such a study is to determine the geometric arrangement necessary in order for cooperative interaction between Lewis base groups to occur. As an example, 2-aminophenol adsorbs to (hydr)oxide surfaces to a significant extent, but 3-aminophenol cannot, since the two Lewis base groups are too far apart to form a chelate ring (Stone et al., 1993). The second task in such a study is to systematically change the identity of the two Lewis bases. Thus catechol (possessing two –OH groups), 2-aminophenol (possessing one –OH and one NH_2 group) and 1,2-phenylenediamine (possessing two NH_2 groups) provide an interesting comparison: the ionic contribution to complex formation is expected to decrease in the order: catechol > 2-aminophenol > 1,2-phenylenediamine while the covalent contribution is expected to follow the reverse trend.

In a similar fashion, the identity of ring substituents that cannot serve as Lewis bases can be systematically varied. It is important to keep in mind that a number of important molecular characteristics change to some degree whenever ring substituents are varied; the challenge is to distinguish between these effects. Nitro ($-NO_2$) groups, for example, strongly lower the basicity of Lewis base groups and increase the molecular dipole moment through electronic effects, and increase hydrophobicity. Ethylene ($-CH_2CH_3$) groups, in contrast, have only a slight effect on basicity and dipole moment, but have a significant effect on hydrophobicity.

5–7.2 (Hydr)Oxide Characteristics

As has been discussed, (hydr)oxide minerals comprised of Si, Al, Fe, Ti, and Mn exhibit widely varying adsorbent properties. For each organic ligand under investigation, adsorption onto a variety of surfaces should be examined, reflecting differences in the identity and oxidation state of the metal center, the stoichiometry with respect to OH^- and O^{2-} groups, the bonding arrangement within the solid, and the crystal structure. Goethite (α-FeOOH) and diaspore (α-AlOOH), for example, make an interesting comparison; although the identity of the metal center is different, the metal oxidation state, the stoichiometry, and the crystal structure are the same. Goethite (α-FeOOH) and lepidocrocite (γ-FeOOH) share the same metal ion, oxidation state and stoichiometry, but possess important differences in bonding arrangement. Goethite (α-FeOOH) and hematite (α-Fe_2O_3) share the same metal ion and oxidation state, but differ in stoichiometry, bonding arrangement, and crystal structure.

5–7.3 Aqueous Medium Characteristics

Examining the effects of pH and ionic strength are important aspects of all adsorption studies. Effects arising from the chemical identity of electrolyte ions also should be considered. Dissolved cations may form ion pairs with anionic ligands, while dissolved anions may form ion pairs with cationic ligands, thereby altering adsorption behavior. Smectites, Mn (IV) oxides, and a number of other important (hydr)oxide minerals undergo cation exchange; the identity and amount of the bound cations may affect adsorption in unanticipated ways. Finally, competitive adsorption among dissolved species must be considered in any comprehensive treatment of adsorption.

The preceding paragraphs have listed a number of experiments that can be conducted to learn more about the underlying phenomena governing adsorption. The modeling exercise provides values for the reaction stoichiometry and for $K^{intr.}{}_{ML}$, which serve as a basis for detailed analysis and the development of structure–activity relationships. The identity of Lewis base groups and the electron-donating or electron-withdrawing nature of ring substituents can be compared with basicity and $K^{intr.}{}_{ML}$ and the identity of the metal center, mineral stoichiometry, bonding arrangement, and crystal structure of several (hydr)oxides can be compared with the extent of adsorption. Conclusions drawn from this kind of analysis can then, in turn, be tested in the laboratory and further refined.

5–8 RELEVANCE TO ECOSYSTEM HEALTH

Intentional or inadvertent releases of organic ligand pollutants into the natural environment adversely impacts ecosystem health. The potential toxicity and carcinogenicity of these pollutants necessitate an understanding of their environmental behavior. Adsorption is a key aspect of the environmental behavior of naturally-occurring organic ligands and organic ligand pollutants. The study of organic ligand adsorption provides insight into the transport and transformation of pollutants in the natural environment.

We have demonstrated that the identity of the ligand donor groups influences the extent of adsorption onto metal oxide surfaces. Likewise surface metal ions also show selectivity for certain ligand donor groups. In addition, the presence of other substituent groups on the aromatic ring also can increase or decrease the extent of adsorption. Electron donating or electron withdrawing substituents alter the competition between protons and the surface sites for binding the deprotonated ligand. Substituents that impart hydrophobicity are likely to raise the extent of adsorption as a result of ligand exclusion from bulk solution.

Careful examination of the phenomena responsible for adsorption, using the proposed methodology, can help us understand why the extent of adsorption changes when the ligand, the surface, or the ionic medium is changed. More importantly, studies of this nature will enable us to develop structure–activity relationships that can predict the adsorption behavior of previously unexplored organic ligands.

5–9 SUMMARY AND CONCLUSIONS

We have demonstrated that the identity of the ligand donor groups influences the extent of adsorption onto metal oxide surfaces. Likewise surface metal ions also show selectivity for certain ligand donor groups. In addition, the presence of other substituent groups on the aromatic ring also can increase or decrease the extent of adsorption. Electron donating or electron withdrawing substituents alter the competition between protons and the surface sites for binding the deprotonated ligand. Substituents that impart hydrophobicity are likely to raise the extent of adsorption as a result of ligand exclusion from bulk solution.

Careful examination of the phenomena responsible for adsorption, using the proposed methodology, can help us understand why the extent of adsorption changes when the ligand, the surface, or the ionic medium is changed. More importantly, studies of this nature will enable us to develop structure–activity relationships that can predict the adsorption behavior of previously unexplored organic ligands.

REFERENCES

Balistrieri, L.S., and J.W. Murray. 1987. The influence of the major ions of seawater on the adsorption of simple organic acids. Geochim. Cosmochim. Acta. 51:1151–1160.

Ballion, D., and N. Jaffrezic-Renault. 1985. Study of the uptake of inorganic ions and organic acids at the alpha-alumina-electrolyte interface in a colloidal system by radiochemical techniques and microelectrophoresis. J. Radioanalytical Nuclear Chem. 92:133–150.

Biber, M.V., and W. Stumm. 1994. An in-situ ATR-FTIR study: The surface coordination of salicylic acid on aluminum and iron (III) oxides. Environ. Sci. Technol. 28:763–768.

Bowers A.R., and Z. Huang. 1985. Adsorption characteristics of polyacetic amino acids onto hydrous gamma-Al_2O_3. J. Colloid Interface Sci. 105:1971–1985.

Cambier, P., and G. Sposito. 1991. Adsorption of citric acid by synthetic pseudoboehmite. Clays Clay Min. 39:369–374.

Chang, H.C., T.W. Healy, and E. Matijevic. 1983. Interactions of metal hydrous oxides with chelating agents: III. Adsorption onto spherical colloidal hematite particles. J. Colloid Interface Sci. 2:469–478.

Davis, J.A., and J.O. Leckie. 1978. Effect of adsorbed complexing ligands on trace metals uptake by hydrous oxides. Environ. Sci. Technol. 12:1309–1315.

Djafer, M., R.K. Khandal, and M. Terce. 1991. Interactions between different anions and the goethite surface as seen by different methods. Colloids Surf. 54:209–218.

Emsley, J. The elements. 1989. Clarendon, Oxford.

Giacomelli, C.E., M. J. Avena, and C.P. De Pauli. 1995. Aspartic acid adsorption onto TiO_2 particle surface. Experimental data and model calculations. Langmuir 11:3483–3490.

Hering, J.G., and W. Stumm. 1991. Fluorescence spectroscopic evidence for surface complex formation at the mineral–water interface: Elucidation of the mechanisms of ligand promoted dissolution. Langmuir 7:1567–1570.

Hochella, M.F. 1990. Atomic structure, microtopography, composition, and reactivity of mineral surfaces. p. 87. *In* M.F. Hochella and A.F. White (ed.) Mineral Water Interface Geochemistry Reviews in Mineralogy. Vol 23. Mineralogical Soc. of Am., Washington, DC.

Hug, S.J., and B. Sulzberger. 1994. In situ Fourier transform infrared spectroscopic evidence for the formation of several different surface complexes of oxalate on TiO_2 in the aqueous phase. Langmuir 10:3587–3597.

James, R.O, and G.A. Parks. 1982. Characterization of aqueous colloids by their electrical double-layer and intrinsic surface chemical properties. Surf. Colloid Sci. 12:119–216.

Kallay, N., and E. Matijevic. 1985. Adsorption at solid/solution interfaces: 1. Interpretation of surface complexation of oxalic and citric acids with hematite. Langmuir 1:195–205.

Karickhoff, S.W., D.S. Brown, and T.A. Scott. 1979. Sorption of hydrophobic compounds in natural sediments. Water Res. 13:241–248.

Kriegman-King, M.R., and M. Reinhard. 1992. Transformation of carbon tetrachloride in the presence of sulfide, biotite, and vermiculite. Environ Sci. Technol. 26:2198–2206.

Kummert, R., and W. Stumm. 1980. The surface complexation of organic acids on hydrous γ-Al_2O_3. J. Colloid Interface Sci. 75:373–385.

Kung, K.S., and M.B. McBride. 1989a. Coordination complexes of p-hydroxybenzoate on Fe oxides. Clays Clay Min. 37:333–340.

Kung, K.S., and M.B. McBride. 1989b. Adsorption of para-substituted benzoates on iron oxides. Soil Sci. Soc. Am. J. 53:1673–1678.

Laha, S., and R.G. Luthy. 1990. Oxidation of aniline and other primary aromatic amines by manganese dioxide. Environ. Sci. Technol. 24:363–373.

Lovgren, L. 1991. Complexation reactions of phthalic acid and aluminum (III) with the surface of goethite. Geochim. Cosmochim. Acta. 55:3639–3645.

Ludwig, L., and P.W. Schindler. 1995. Surface complexation of TiO_2: II. Ternary surface complexes: Coadsorption of Cu(II) and Organic Ligands (2,2'Bipyridyl, 8-Aminoquinoline, and o-phenylenediamine) onto TiO_2 (Anatase). J. Colloid Interface Sci. 169:291–299.

Lukenheimer, K., H. Fruhner, and F. Theil. 1993. Adsorption-catalyzed hydrolysis of sodium *n*-dodecyl sulfate at the solid/liquid interface. Colloids Surf. A. 76:289–294.

Machesky, M.L., B.L. Bischoff, and M.A. Anderson. 1989. Calorimetric investigation of anion adsorption onto goethite. Environ. Sci. Technol. 23:580–587.

McBride, M.B., and K.H. Kung. 1991. Adsorption of phenols and substituted phenols by iron oxides. Environ. Toxicol. Chem. 10:441–448.

McBride, M.B., and L.G. Wesselink. 1988. Chemisorption of catechol on gibbsite, boehmite, and noncrystalline alumina surfaces. Environ. Sci. Technol. 22:703–708.

McCarthy, J.F., and J.M. Zachara. 1989. Subsurface transport of contaminants. Environ. Sci. Technol. 23:496–502.

Means, J.C., S.G. Wood, J.J. Hassett, and W.L. Banwart. 1982. Sorption of amino- and carboxy- substituted polynuclear aromatic hydrocarbons by sediments and soils. Environ. Sci. Technol. 16:93–97.

Mesuere, K., and W. Fish. 1992. Chromate and oxalate adsorption on goethite: 1. Calibration of surface complexation models. Environ. Sci. Technol. 26:2357–2364.

Morel, F.M.M., and J.G. Hering. 1993. Principles and applications of aquatic chemistry. Wiley-Interscience, New York.

Murphy, E.M., J.M. Zachara, and S.C. Smith. 1990. Influence of mineral-bound humic substances on the sorption of hydrophobic organic compounds. Environ. Sci. Technol. 24:1507–1516.

Papelis, C., K.F. Hayes, and J.O. Leckie. 1988. HYRAQL: A program for the computation of chemical equilibria composition of aqueous batch systems including surface-complexation modeling of ion adsorption at the oxide/solution interface. Tech. Rep. 306. Stanford Univ., Stanford, CA.

Parfitt, R.L., V.C. Farmer, and J.D. Russell. 1977a. Adsorption on hydrous oxides: I. Oxalate and benzoate on goethite. J. Soil Sci. 28:29–39.

Parfitt, R.L., V.C. Farmer, and J.D. Russell. 1977b. Adsorption on hydrous oxides. II: Oxalate, benzoate, and phosphate on gibbsite. J. Soil Sci. 28:40–47.

Pope, C.G., E. Matijevic, and R.C. Patel. 1981. Adsorption of nicotinic, picolinic, and dipicolinic acids on monodispersed sols of α-Fe_2O_3 and $Cr(OH)_3$. J. Colloid Interface Sci. 80:74–83.

Rubio, J., and E. Matijevic. 1979. Interactions of metal hydrous oxides with chelating agents: 1. β-FeOOH-EDTA. J. Colloid Interface Sci. 68:408–421.

Schellenberg, K., C. Leuenberger, and R.P. Schwarzenbach. 1984. Sorption of chlorinated phenols by natural sediments and aquifer materials. Environ. Sci. Technol. 18:652–657.

Schindler, P.W., and W. Stumm. 1987. The surface chemistry of oxides, hydroxides, and oxide minerals. p. 83–110. *In* W. Stumm (ed.) Aquatic surface chemistry. John Wiley & Sons, New York.

Schwertmann, U., and R.M. Taylor. 1981. The significance of oxides for the surface properties of soils and the usefulness of synthetic oxides as models for their study. International Soc. Soil Sci. 60:62–66.

Sposito, G. 1989. The chemistry of soils. Oxford Univ. Press, New York.

Stone, A.T. 1987. Reductive dissolution of manganese (III/IV) oxides by substituted phenols. Environ. Sci. Technol. 21:979–988.

Stone, A.T., A. Torrents, J.M. Smolen, D. Vasudevan, and J. Hadley. 1993. Adsorption of organic compounds possessing ligand donor groups at the oxide/water interface. Environ. Sci. Technol. 27:895–909.

Stumm, W., and G. Furrer. 1987. The dissolution of oxides and aluminum silicates. Examples of surface-coordination controlled kinetics. p. 197–219. *In* W. Stumm (e.) Aquatic surface chemistry. John Wiley & Sons, New York.

Stumm, W, R. Kummert, and L. Sigg. 1980. A ligand exchange model for the adsorption of inorganic and organic ligands at hydrous oxide interfaces. Croat. Chem. Acta. 53:291–312.

Sverjensky, D.A. 1994. Zero-point-of-charge prediction from crystal chemistry and solvation theory. Geochim. Cosmochim. Acta. 58:3123–3129.

Tejedor-Tejedor, M.I., E.C. Yost, and M.A. Anderson, M. 1990. Characterization of benzoic and phenolic complexes at the goethite/aqueous solution interface using cylindrical internal reflection Fourier transform infrared spectroscopy: 1. Methodology. Langmuir 6:979–987.

Tejedor-Tejedor, M.I., E.C. Yost, and M.A. Anderson, M. 1992. Characterization of benzoic and phenolic complexes at the goethite/aqueous solution interface using cylindrical internal reflection Fourier transform infrared spectroscopy: 2. Bonding structures. Langmuir 8:525–533.

Tentorio, A., and L. Canova. 1989. Adsorption of alpha-amino acids on spherical TiO_2 particles. Colloids Surf. 39:311–319.

Thomas, F., J.Y. Bottero, and J.M. Caes. 1989. An experimental study of the adsorption mechanisms of aqueous organic acids on porous aluminas: 2. Electrochemical modeling of salicylate adsorption. Colloids Surf. 39:281–294.

Tipping, E., J.R. Giffith, and J. Hilton. 1983. The effect of adsorbed humic substances on the uptake of copper (II) by goethite. Croat. Chem. Acta. 56:613–621.

Torrents, A., and A.T. Stone. 1991. Hydrolysis of phenyl picolinate at the mineral/water interface. Environ. Sci. Technol. 25:143–149.

Torrents, A., and A.T. Stone. 1994. Oxide surface-catalyzed hydrolysis of carboxylate esters and phosphorothioate esters. Soil Sci. Soc. Am. J. 58:738–745.

Torres, R., M.A. Blesa, and E. Matijevic. 1990. Interactions of metal hydrous oxides with chelating agents: IX. Reductive dissolution of hematite and magnetite by aminocarboxylic acids. J. Colloid Interface Sci. 134:475–485.

Torres, R., N. Kallay, and E. Matijevic. 1988. Adsorption at solid/solution interfaces: 5. Surface complexation of iminodiacetic acid on hematite. Langmuir 4:706–710.

Westall, J.C., and H. Hohl. 1980. A comparison of electrostatic models for the oxide/solution interface. Adv. Colloid Interface Sci. 12:265–294.

Westall, J.C., J. Zachary, and F.M.M. Morel. 1976. MINEQL. Tech. Note. 18. Massachusetts Inst. of Technology, Cambridge.

Vasudevan, D. 1996. Adsorption of aromatic organic ligands possessing oxygen- and nitrogen- donor groups at the metal (hydr)oxide/water interface. Ph.D. diss. Johns Hopkins Univ., Baltimore, MD.

Verwey, E.J.W., and J.T.G. Overbeek. 1948. Theory of stability of lyophobic colloids. Elsevier, Amsterdam.

Zachara, J.M., C.C. Ainsworth, C.E. Cowan, and R.L. Schmidt. 1990. Sorption of aminonaphthalene and quinoline on amorphous silica. Environ. Sci. Technol. 24:118–126.

Zhang, Y., N. Kally, and E. Matijevic, E. Interaction of metal hydrous oxides with chelating agents: 7. Hematite-oxalic acid and -citric acid systems. Langmuir 1:201–206.

6 Organophosphorus Ester Hydrolysis Catalyzed by Dissolved Metals and Metal-Containing Surfaces[1]

Jean M. Smolen and Alan T. Stone

Johns Hopkins University
Baltimore, Maryland

6–1 INTRODUCTION

Vast amounts of pesticides are used on farmland soils each year. For example, 3500 tons of chlorpyrifos were used in 1982 (Green et al., 1987). These substances may persist in the environment for many years beyond the time required for pest control. Besides pesticides, other organic esters may be introduced inadvertently into the environment as plasticizers, gasoline and oil additives, flotation agents, and surfactants (Shen & Morgan, 1973). It is critical to understand the fate of phosphate esters from farmland and other sources in the natural environment because of their toxicity and prevalence. Important pathways that affect the fate of organic phosphate esters include biological degradation, oxidation–reduction, photolysis, adsorption, and hydrolysis. Hydrolysis is the major degradation pathway of these esters that occurs in surface and ground waters, which usually generates products of reduced toxicity.

Pollutants encounter a variety of organic and inorganic chemical constituents in natural waters, soils, and sediments that can alter pathways and rates of chemical reactions. Our research is specifically concerned with the possible role of dissolved metals and metal-containing surfaces in pesticide hydrolysis reactions. Evaluating the importance of metal catalysis requires that the following issues be addressed: (i) the susceptibility of pollutants to hydrolysis; (ii) metal speciation and abundance; (iii) the differences of catalytic properties of metal species; (iv) the principles governing metal-organic interactions.

[1] This work was supported by the Office of Exploratory Research of the U.S. Environmental Protection Agency (Grant R818894-01-1). The views and conclusions contained in this document are those of the authors and should not be interpreted as necessarily reflecting the official policies, either expressed or implied, of the U.S. Government.

6–2 HYDROLYSIS OF ORGANOPHOSPHATE ESTERS

Before addressing the role of metals in organophosphorus ester hydrolysis, it is important to examine the mechanisms of hydrolysis in the absence of potential catalysts. Hydrolysis of an organic compound (C) depends on specific acid- and base-catalyzed and neutral processes (Mabey & Mill, 1978):

$$-dC/dt = k_A[H^+][C] + k_B[OH^-][C] + k_N[C] = k_H[C] \quad [1]$$

where k_A, k_B, k_N, and k_H are the specific acid, base, neutral, and overall hydrolysis rate constants, respectively. Hydrolysis is dependent on temperature, pH, and concentration of the ester.

Common pesticides that may be hydrolyzed include carbamate, organophosphorus, and carboxylic acid ester derivatives. There are excellent reviews in the literature (Mabey & Mill, 1978, 1988) of current advances in understanding the hydrolysis of these and other organic chemicals.

The basic structure of organophosphorus ester pesticides is shown below:

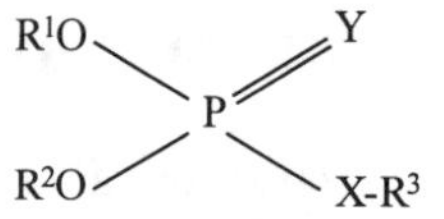

X and *Y* represent O or S atoms, R^1 and R^2 represent lower alkyl groups, and R^3 represents an alkyl or aromatic group that is sufficiently electron-withdrawing to impart biological activity to the molecule (Schmidt, 1975). According to Schrader's Rule (Schmidt, 1975), if the pK_a of the electron-withdrawing group is too acidic (<6), the XR^3 group is too electron-withdrawing and the compound hydrolyzes too quickly. If the pK_a is too basic (>8), the XR^3 group is strongly electron-donating and the compound is too stable to be biologically active as a pesticide. Therefore, a pK_a value of 6 to 8 favors relative stability towards hydrolysis and provides biological activity.

We are concerned with compounds in which alkyl and aromatic leaving groups are joined to the P atom via an ester (P–O–R^3) or thioester (P–S–R^3) linkage (Hassal, 1990). There are, however, pesticides in which alkyl and aromatic moieties are bound directly to the P (i.e., phosphonates). Figure 6–1 provides a guide to the nomenclature of phosphorus esters.

Hydrolysis by water and hydroxide ion at P and C electrophilic sites may be explained by Pearson's hard–soft rules (Pearson, 1966). Hard electrophiles such as the carbonyl C (C=O) and phosphoryl P (P=O) in these respective esters react with hard nucleophiles such as hydroxide, alkoxide, and imino groups. The C atoms of alkyl groups are soft electrophiles and are more likely to react with softer nucleophiles such as thiols, thioethers, and water (Schmidt, 1975). Hydroxide is a better nucleophile than water towards phosphate P by 10^8, but only by 10^4 for saturated C (Eto, 1979).

Among organophosphorus esters, the hydrolysis of phosphate esters (*X*,*Y* = O) is the best understood. The pH of the reaction solution has an effect on rates and mechanisms of phosphate ester hydrolysis. Alkaline hydrolysis of a dialkyl

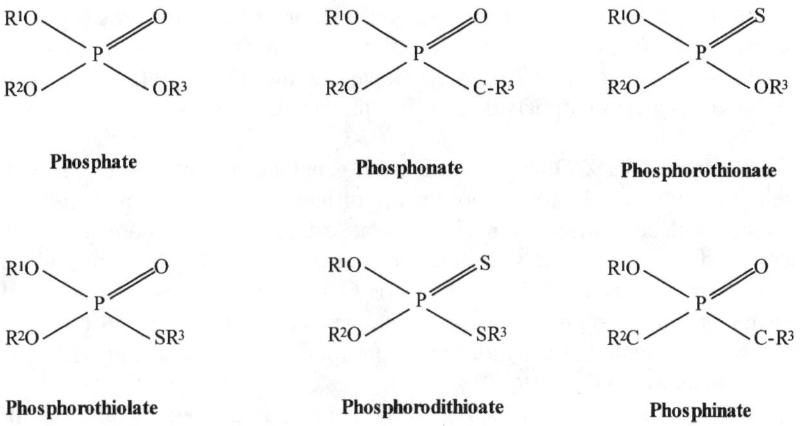

Fig. 6–1. Phorphorus ester structures (Shen & Morgan, 1973).

aryl phosphate proceeds via hydroxide attack on the electrophilic P atom, with fission of the aryl ester bond (Eto, 1979):

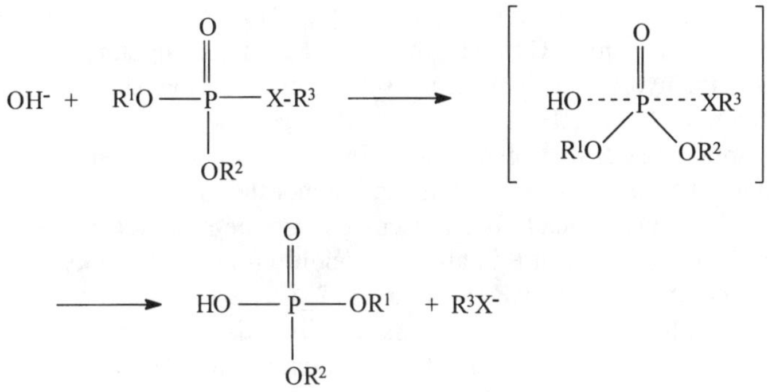

Whenever attack at P occurs, cleavage occurs at the ester linkage that yields the leaving group with the lowest pK_a (Eto, 1979).

Neutral and acid hydrolysis occurs with C–O bond fission (Eto, 1979):

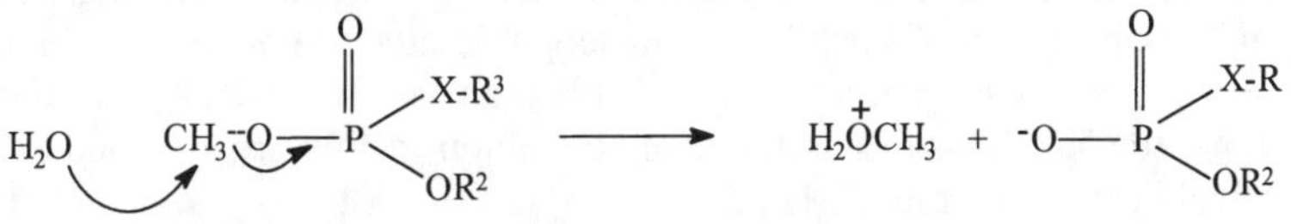

The most basic group is usually removed in this case.

6–2.1 Other Organophosphorus Esters

Replacing the *Y* O atom by a S atom converts an oxonate (phosphate) ester into a thionate ester. Sulfur is less electronegative than O and forms bonds that are more covalent in character. As a consequence, the P within thionate esters is less electrophilic, and thionate esters are more stable towards hydrolysis (Schmidt & Fest, 1982). For both insects and mammals, oxonates are more toxic than thionates (Schmidt, 1975). Insects have enzymes that rapidly convert thionates into oxonates; mammals do not. In this way, the insect *activates* the pesticide.

Replacing the *X* O atom by a S atom generates a thiolate ester. The electronic density on the P atom is not greatly influenced by this change, and hence it retains its electrophilicity; however, thiolate esters hydrolyze more quickly than corresponding oxonate esters because mercaptide ions (R_3S^-) are more acidic and better leaving groups than alkoxide ions (R_3O^-) (Schmidt & Fest, 1982). Another explanation for the high reactivity of thiolate esters is that the P–S bond has a lower bond strength due to a reduced π-contribution to the bond and is more susceptible to cleavage (Eto, 1979).

Phosphonate esters, which possess a direct P–C bond, hydrolyze more quickly than phosphate esters in neutral and alkaline solution, but hydrolyze more slowly than phosphate esters in acidic solution (Schmidt, 1975). This is due to the fact that there is no π-contribution to the P–C bond and the C has no lone pair of electrons to contribute to the P atom. For these reasons, the phosphonate P is more electrophilic than that of phosphates in which the lone pair of electrons from the ester O is donated to the P atom (Eto, 1979). As seen in Fig. 6–1, phosphinates have a second P–C bond, which makes these compounds more susceptible to alkaline hydrolysis than the phosphonates, and more stable in acidic conditions (Schmidt & Fest, 1982).

Phosphorus ester hydrolysis occurs in a series of steps, converting triesters into diesters, diesters into monoesters, and monoesters into inorganic phosphate. Triesters, which have already been discussed, are the most reactive of the three with hydrolysis occurring at P in alkaline conditions and at the alkyl C in neutral and acidic conditions (Shen & Morgan, 1973). Diesters are the least reactive toward nucleophiles. These compounds usually exist as anions and are stable towards nucleophilic attack, due to electrostatic repulsion. Hydrolysis at the alkyl C occurs in both acid and alkaline conditions. For monoesters, base hydrolysis occurs at the P atom, producing inorganic phosphate. Acid hydrolysis occurs via the alkyl C (Eto, 1979).

Ethyl esters are slightly more stable towards hydrolysis than methyl esters. Alkyl groups are electron-donating and reduce the electrophilic character of the P atom, making it less susceptible to nucleophilic attack. For example, methyl paraoxon (R^1, $R^2 = CH_3$) has a shorter half-life than paraoxon (R^1, $R^2 = CH_2CH_3$). At pH 5.0, the half-lives of parathion and parathion-methyl are 19.5 and 10.7 h, respectively (Faust & Gomaa, 1972).

6–3 METAL SPECIATION AND METAL-ORGANIC INTERACTIONS

In order for metals to affect hydrolysis reactions, they must either form complexes with organophosphorus esters or with the attacking nucleophile. In this section, the metallic elements encountered in the environment are listed, and their ability to form complexes is assessed and compared.

Table 6–1. Average crustal abundance of metallic elements within the earth's crust and pertinent chemical properties (Emsley, 1989; Stumm & Morgan, 1981; Smith et al., 1995).

Equilibrium reaction: $Me^{n+} + H_2O = MeOH^{(n-1)+} + H^+ \ {}^*K_a$			
Element	Average crustal abundance	Oxidation states	p^*K_a
Al	8.30	Al(III)	5.000
Fe	6.20	Fe(II, III)	9.4, 2.19
Ca	4.66	Ca(II)	12.7
Mg	2.76	Mg(II)	11.2
Ti	0.632	Ti(IV)	†
Mn	0.106	Mn(II, III, IV)	10.6, −0.4, †
Co	0.0029	Co(II, III)	9.7, 0.48
Ni	0.0099	Ni(II)	9.90
Cu	0.0068	Cu(II)	7.50
Zn	0.0072	Zn(II)	9.00

† Data not available.

6–3.1 Metal Abundance and Speciation

Metal abundance and speciation will influence the extent of metal complex formation and metal catalysis. In neutral and alkaline aquatic environments where the concentration of protons is low, metal species become the most abundant Lewis acids (electron acceptors) in the system. Under all pH conditions, metal ions have the advantage that they can simultaneously coordinate two or more ligand donor atoms of the same organic molecule, whereas protons can coordinate only one ligand donor atom (Hay, 1987). When the abundance of two metals are comparable, the metal that has the greater ability to form metal-ester complexes will usually have a greater effect on ester hydrolysis. When the abundance of a metal is low in comparison to others, it will only have a significant effect on ester hydrolysis if the ability to form metal-ester complexes is unusually high. Table 6–1 provides the average crustal abundance and other pertinent chemical properties for a list of environmentally significant metals.

6–3.2 Metal-Ligand Bonding

It is important to understand the principles that govern metal-ligand bonding in order to determine if metal-ligand complexes will form at environmental conditions. Ionic and covalent bonds may form between the metal and the ester, or between the metal and the nucleophile. Bonds that are ionic in character are held together by electrostatic forces; bond strength increases with increasing charge-to-radius ratio of the metal ion and the Lewis base. Bonds that are covalent in character are held together by the sharing of electrons within a molecular orbital; heavier elements within each column of the periodic table (especially those with *d*- and *f*-orbitals) share electrons more readily in this way. (This ability to share electrons is referred to as polarizability). In most instances, bonds are intermediate in character between these two extremes.

6–3.3 Metal-Hydroxide Complexes with Tri- and Tetra-Valent Metal Ions

Metal-hydroxide complexation also has a significant effect on the speciation and availability of metals under environmental conditions. Among Lewis bases, OH^- has a large charge-to-radius ratio and low polarizability. Tetravalent metal ions (Ti^{IV} and Mn^{IV}) and trivalent metal ions (Al^{III}, Mn^{III}, Fe^{III}, and Co^{III}) all exhibit high charge-to-radius ratios, and therefore form complexes with OH^- at relatively low pH values. Table 6–1 lists some important p^*K_a values, the pH where metal-hydroxo complexes become important species; however, the affinity of tetravalent and trivalent metal ions for OH^- is so great that they form sparingly-soluble (hydr)oxide solids that limit their solubility, especially under neutral and alkaline pH conditions.

6–3.4 Importance of Metal-Hydroxo Species as Nucleophiles

Formation of metal-hydroxide complexes will result in nucleophiles that may be important near neutral pH. $MeOH^{n+}$ is a weaker nucleophile than OH^-, but when dissolved metal concentrations are high, concentrations of $MeOH^{n+}$ may be several orders of magnitude higher than concentrations of OH^-, and therefore make up for the difference. The nucleophilicities of $MeOH^{n+}$ species are a function of the polarizability of the metal and its p^*K_a, the pH where the first hydroxo species [$MeOH^+$] is equal to the concentration of the free metal ion [Me^{2+}] (Table 6–1; Stone & Torrents, 1995).

6–3.4.1 Divalent Metal Ions

Among divalent metal ions, the tendency to form complexes with OH^- follows the Irving-Williams Series (Houghton, 1979): $Ca^{II} < Mg^{II} < Mn^{II} < Fe^{II} < Co^{II} < Ni^{II} < Cu^{II} > Zn^{II}$. Differences in charge-to-radius ratio and corresponding increases in the ionic contribution to bonding are primarily responsible for this trend. Increases in the covalent contribution to bonding from left to right in the periodic table also contribute to this trend; however, the selectivity trend of the Irving-Williams series may be modified for donor groups other than O (Evers et al., 1989). At neutral pH, the only metal-hydroxides present in any appreciable quantity are formed by Cu^{II} ($p^*K_a = 6.5$) and Pb^{II} ($p^*K_a = 7.2$; Morrow et al., 1992). Other divalent metal ions have higher p^*K_a values. Thus, reactions relying upon $MeOH^{n+}$ as the nucleophile for hydrolysis should be most pronounced in the presence of Cu^{II} and Pb^{II} at neutral pH.

6–3.5 Sensitivity of Organic Ligands

Among first-row transition metal ions, *K* values for metal–ligand complex formation differ by more than two orders-of-magnitude for oxalate, seven orders-of-magnitude for glycine, and eight orders-of-magnitude for mercaptoacetate (Fig. 6–2). To the right of Fe^{II} in the Irving-Williams series, Co^{II}, Ni^{II}, Cu^{II}, and Zn^{II} complexes with oxalic acid, glycine, and ethylenediamine, show a preference

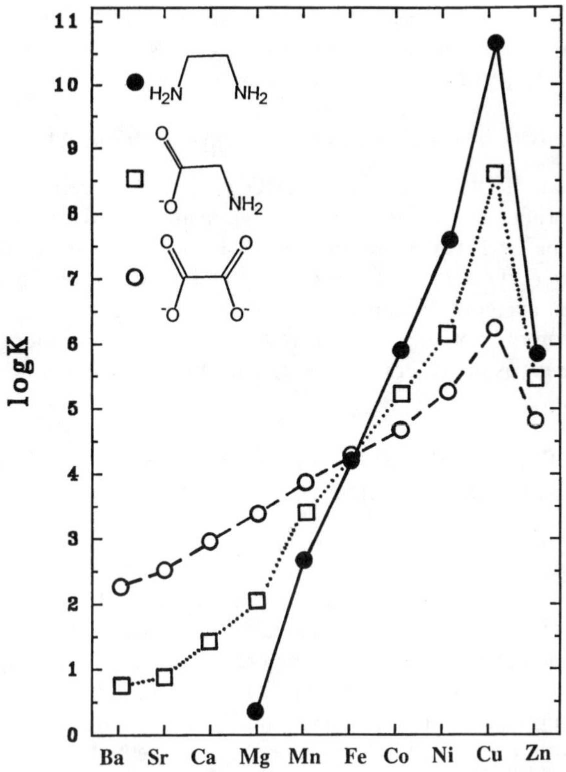

Fig. 6–2. Logarithms of the stability constants for the 1:1 complexes between Ba^{2+} through Zn^{2+} and the bidentate ligands oxalic acid, glycine ethylenediamine, mercaptoacetic acid, and mercaptoethylamine (adapted from Sigel & McCormick, 1970).

towards ligands with a N ligand donor atom. To the left of Fe^{II} in the series (Ca^{II}, Mg^{II}, and Mn^{II}), O is the preferred ligand donor atom (Sigel & McCormick, 1970). Softer donor atoms exhibit a wider range in log K_{ML} values, reflecting a greater sensitivity towards the identity of the divalent metal ion (Sigel & McCormick, 1970). Among divalent metal ions, Cu^{II} has the greatest ability to form complexes with all organic compounds, regardless of the nature of the donor atom. Thus, reactions relying solely upon metal complex formation should proceed most rapidly in the presence of Cu^{II}.

6–3.6 Ligand Donor Groups Within Organophosphorus Esters

Complex formation between metal ions and any portion of organophosphate ester molecules may have an effect on hydrolysis, even if the stability of the complex is low. The phosphate (P=O) O atom within phosphate esters, the thionate (P=S) S atom within thionate esters, and O, N, and S atoms linking the P atom to other substituents all have a capacity to complex metals. Organophosphorus esters often contain various auxiliary ligand groups. If suitably placed near the ester linkage, these ligand groups can facilitate complexation by forming five- or six-membered chelate rings involving the phosphate O, the thionate S, or O, N, or S atoms within the ester linkage.

6–3.6.1 Selectivity of Ligands Possessing Oxygen Donor Atoms

Alcoholate (RO^-) and phenolate (ArO^-) anions form complexes with metal ions that are primarily ionic in character, in a manner similar to hydroxide ions. For this reason, log K values are highest for those metal ions in Table 6–1 that possess low p^*K_a values. P=O and C=O O atoms bear only a partial negative charge, so the ionic contribution to bonding is low. Oxygen is a light element, so the covalent contribution to bonding is low. Therefore, these donor groups exhibit the same trends as the other O-donor ligands, but the extent of complex formation will be low.

6–3.6.2 Selectivity of Ligands Possessing Nitrogen and Sulfur Donor Atoms

Although N-donor ligands (i.e., amines, anilines, pyridines) exhibit an ionic contribution to binding, it is considerably weaker than that experienced by O-donor ligands. Nitrogen is a more polarizable atom than is O. The covalent contribution to binding is therefore more important. Because of this higher covalent contribution towards bonding, N-donor ligands exhibit a pronounced selectivity towards transition metal ions that possess polarizable *d*-orbital electrons. Because Fe^{III} is a transition metal, it binds the more polarizable N donor ligands far more strongly than Al^{III}. The log K for binding to NH_3 is similar to binding to a N donor ligand; log K for Al^{III} is 0.8, compared with 3.8 for Fe^{III} (Evers et al., 1989). In the case of sulfur-donor ligands, the covalent contribution to bonding is strongest, which results in a strong preference towards the heavier, more polarizable metals.

6–3.7 Acid–Base Equilibria

pH will have an effect on the extent of metal complexation. Since all metal ions complex with hydroxide ions to some degree, complex formation with organophosphate esters can be expected to diminish as the pH is increased. Some organophosphate ester ligand donor groups possess a pK_a within or above the pH range under consideration. In this case, competition between protons and metal ions is expected to occur, with metal-ligand complexation decreasing as the solution pH is decreased.

6–4 METAL-CATALYZED HYDROLYSIS OF PHOSPHORUS ESTERS

Several mechanisms have been postulated to explain the metal-catalyzed hydrolysis of organic esters (Suh & Chun, 1986): (i) The partial positive character of the electrophilic site can be enhanced by the coordination of a metal ion (complexation of P = S S atom or P = O O atom); (ii) coordination of a metal with the leaving group may enhance its leaving ability; (iii) metal ion coordination of the nucleophile (H_2O) can induce deprotonation, improving nucleophilicity ($MeOH^{n+}$); (iv) metal complexation may induce conformational changes that increase the reactivity of the ester or the nucleophile; (v) simultaneous complexation of the ester and nucleophile may increase their encounter frequency, thereby promoting reaction; and (vi) metal complexation may block inhibitory reverse paths by stabilizing hydrolysis products.

The literature provides detailed illustrations of these mechanisms. Mechanisms 1 through 3 are often the most important for organophosphorus hydrolysis reactions (Fig. 6–3).

6–4.1 Metal-Hydroxide Nucleophiles

A number of organophosphate triesters have been found to be susceptible to catalysis by $MeOH^{n+}$. The hydrolysis of isopropyl methylphosphono-fluoridate is catalyzed in the presence of Cu^{II}, Mn^{II}, and Mg^{II} (Epstein & Mosher, 1968). One possible mechanism is a push-pull interaction in which the metal acts as an electrophile interacting with electronegative O or F, resulting in a more electrophilic P site, and the OH^- as the attacking nucleophile at the P. Metal ion-hydroxide complexes ($CuOH^+$, $MnOH^+$, $MgOH^+$) appear to be more reactive nucleophiles than OH^- in this study (Epstein & Mosher, 1968).

A similar mechanism is proposed for the Zn-hydroxide catalysis of diphenyl p-nitrophenyl phosphate hydrolysis (Gellman et al., 1986). Zinc-bound hydroxide may be the attacking nucleophile or the Zn center may activate the phosphoryl P–O bond by coordination of the Zn center, followed by nucleophilic attack by free hydroxide. This complex mechanism must be called upon to explain the fact that the hydrolysis rate in the presence of OH^- is so much small-

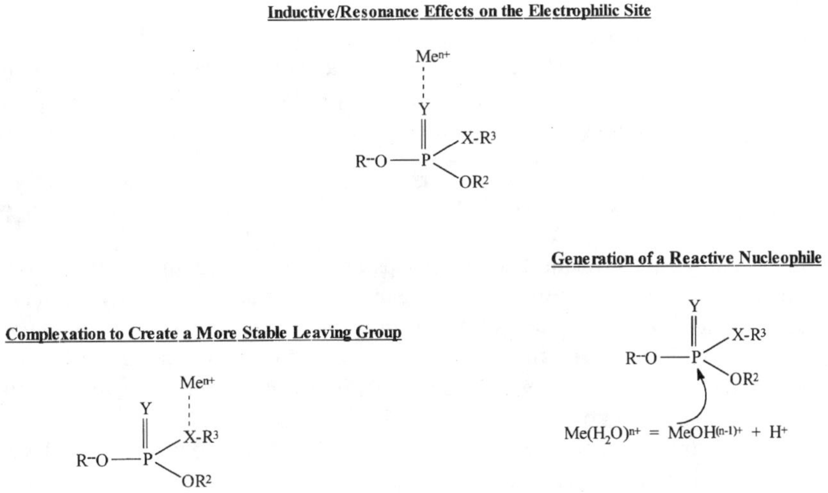

Fig. 6–3. Some potential mechanisms of metal catalysis (Suh & Chun, 1986).

er than in the presence of the Zn-complex. Simple coordination of Zn to the phosphate linkage is not strong enough mechanism for the high rate enhancements observed (Gellman et al., 1986).

6–4.2 Enhancement of Electrophilic Site

A number of studies have observed the Cu^{II}-catalyzed hydrolysis of phosphoro(thio)nate ester pesticides. Catalysis is believed to occur through coordination of the phosphate O or the thionate S. If the compound has a suitably placed auxiliary donor ligand, there is strong evidence that chelate formation occurs and that Type 1 catalysis occurs. Chelation in this manner can increase the extent of metal binding and increase the partial positive charge of the electrophilic site.

The hydrolysis of chlorpyrifos and diazinon, two thionate esters, is catalyzed by the presence of Cu^{II} at pH 5 to 6 (Mortland & Raman, 1967). The suggested mechanism involves bidentate chelation with Cu^{II} through the N atom in the ring structure and the thionate S on the phosphate linkage. The shifts in electron density can weaken the aryl-ester bond making it more susceptible to hydrolysis. Similarly, Cu^{II} has been found to catalyze the hydrolysis of chlorpyrifos and chlorpyrifos-methyl (Blanchet & St.-George, 1982). The proposed mechanism is the formation of a 6-membered chelate ring with the heterocyclic N and the S of the thionate moiety. This promotes hydrolysis by increasing the electrophilicity of the P atom in the complex intermediate. Ronnel and Zytron (Dow Chemical Co., Midland MI) are susceptible towards Cu-catalyzed hydrolysis, but not to the same extent as chlorpyrifos and diazinon (Mortland & Raman, 1967). Apparently the binding of Cu^{II} to the thionate S is sufficient to activate the P electrophilic site, despite the absence of an auxiliary donor atom.

In metal-free solutions, the oxonate ester paraoxon (P=O) hydrolyzes three times faster than the corresponding thionate ester parathion (P=S). This behavior is typical of most oxonate and thionate esters. In the presence of Cu^{II} catalysts, however, the order is reversed: parathion hydrolyzes four times faster than paraoxon (Ketelaar et al., 1956). It is therefore apparent that sensitivity towards metal catalysis depends very strongly upon the identity of the ligand donor atom within the ester, and that the relative rates of hydrolysis may change in the presence of metal catalysts.

6–4.3 Create a Better Leaving Group

When the pK_as of the leaving groups are high, exit of the leaving group often becomes the rate-limiting step in organophosphorus ester hydrolysis. In this situation, metal complexation of the leaving group during exit can catalyze hydrolysis. Cu^{II} catalyzes the hydrolysis of 2-(4(5)-imidazolyl) phenyl phosphate, a diester, by increasing the leaving ability of the phenolate group to promote hydrolysis (Benkovic & Dunikoski, 1971). The most probable mechanism involves neutralization of the charge on the phosphate group and/or induced strain on the P–O bond, leading to nucleophilic displacement by solvent on the P (Fig. 6–4).

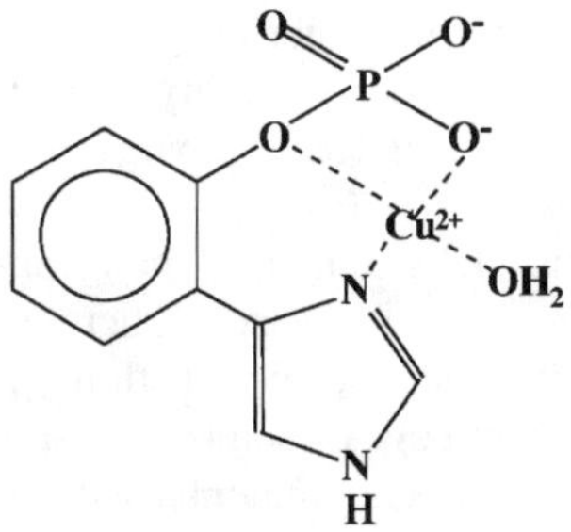

Fig. 6–4. Copper compexed with 2-(4(5)-imidazolyl) phenyl phosphate (Benkovic & Dunikoski, 1971).

6–5 PHOSPHORUS ESTER HYDROLYSIS ON METAL-CONTAINING SURFACES

In addition to soluble metals, organic pollutants may also encounter metal ions associated with surfaces. Clay and oxide mineral surface-catalyzed hydrolysis of phosphorus esters may be particularly important in the deep subsurface where organic matter is less abundant than immediately beneath the soil surface (Saltzman & Mingelgrin, 1984). Surface catalysis may arise from metal-ester or metal-hydroxide ion complex formation on the surface, in a manner analogous to solution catalysis. Surface catalysis may also arise from unique characteristics of mineral-water interfaces: ester adsorption caused by hydrophobic exclusion from bulk solution, the accumulation of protons or hydroxide ions near the surface (Stone, 1989), and changes in the dielectric constant and other properties of the solvent in the interfacial region (Nurnberg, 1974; Nurnberg & Wolff, 1980).

Silicates, clay minerals and oxide minerals serve as the predominant minerals in contact with water within soils and aquifers. Primary silicates are produced from the weathering of parent rock material and contain Na^{I}, Mg^{II}, K^{I}, Ca^{II}, Mn^{II}, Fe^{II}, as well as trace elements Co^{II}, Cu^{II}, and Zn^{II} (Sposito, 1989). Clay minerals are hydrated silicates of Al, Fe, and Mg (Birkeland, 1984). Aluminum, Fe, and Mn form the most important oxide minerals in soils. Oxide minerals are formed from the weathering of primary silicates or from the hydrolysis and desilication of clay minerals like smectite and kaolinite (Sposito, 1989).

Bronsted acidity is the ability of a species to donate protons (Voudrais & Reinhard, 1986). Clays and oxide minerals exhibit Bronsted acidity resulting from the dissociation of waters coordinated to exchangeable cations:

$$[M(H_2O)_X]^{n+} \Leftrightarrow [M(OH)(H_2O)_{x-1}]^{(n-1)+} + H^+$$

High Bronsted acidity surfaces are capable of promoting reactions that are normally acid-catalyzed; however, base-catalyzed reactions are inhibited by surface acidity of this kind.

Lewis acidity is the ability of a species to act as an electron pair acceptor. Clays and oxide minerals exhibit Lewis acidity resulting from metal cations adsorbed to the surface, held within ion exchange sites, or incorporated within the lattice of the solid. Metal cations in higher valence states (i.e., Fe^{III}, Al^{III}, Ti^{IV}) are particularly important sources of Lewis acidity. Lewis acidity depends upon the degree of soil hydration, and is strongest when the moisture content is low (Voudrais & Reinhard, 1986).

Unlike oxide minerals, clays contain exchangeable cations (Ca^{II}, Mg^{II}, Na^{I}, K^{I}, Al^{III}) which occupy the interlayer spaces, thereby neutralizing the negative electrical charge resulting from isomorphic substitution within the clay lattice and from broken bonds along clay edges. Isomorphic substitution of exchangeable cations is uncommon in 1:1 layer clay minerals such as kaolinite, but is common in 2:1 layer clay minerals such as smectite (Birkeland, 1984). Al^{III} substitutes for Si^{IV} in the tetrahedral sheet, and Fe^{II} substitutes for Mg^{II} and Al^{III} in the octahedral sheet (Birkeland, 1984).

Clays have been shown to exhibit a catalytic effect on the hydrolysis of phosphorothionate ester pesticides. The hydrolysis of parathion is catalyzed by kaolinite surfaces (Saltzman et al., 1974). Hydrolysis occurs via cleavage of the phosphate ester bond:

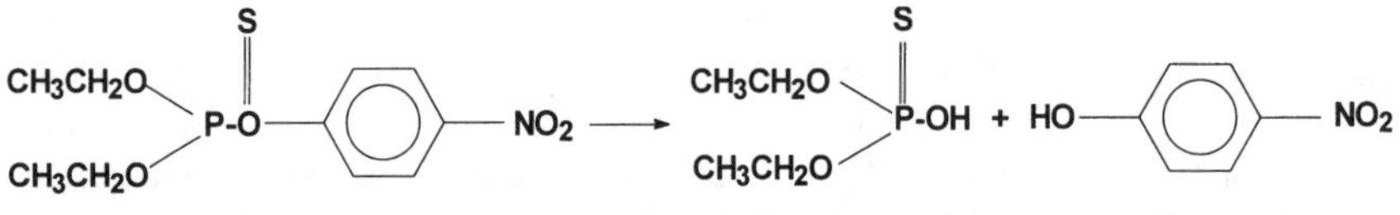

The presence of Ca^{II}-kaolinite results in the fastest hydrolysis rate followed by Na^{I}- and Al^{III}- kaolinite (Saltzman et al., 1974). If the moisture content of the clay exceeds that of a complete hydration layer, the catalytic effect is diminished and the rate of hydrolysis resembles that observed in homogeneous aqueous solution (Saltzman et al., 1974, 1976). It has been proposed that the waters associated with cations bound to the clay promote the hydrolysis of parathion (Mingelgrin et al., 1977). This implies that the hydration state of the cation has an effect on the rate of hydrolysis.

The hydrolysis of phosmet, another phosphorothionate ester, is catalyzed by montmorillonite (Sanchez-Camazano & Sanchez-Martin, 1983). X-ray diffraction and infrared (IR) spectroscopy have shown that phosmet adsorbs into the interlayer space of montmorillonite (Sanchez-Camazano & Sanchez-Martin, 1980). Chelate formation between the exchangeable cation of the clay, the thionate S, and the carbonyl O appears to be responsible for making the ester more susceptible to nucleophilic attack.

The hydrolysis of azinphos-methyl, a phosphorothionate ester pesticide with a S-methylene bridge, is catalyzed by Ca- and Cu-smectites. Interaction with the interlayer cations probably occurs through the thionate S and a N atom of the triazine ring, with adsorption into the interlayer space of the clays (Sanchez-Camazano & Sanchez-Martin, 1991).

Research examining the effect of surfaces on hydrolysis has demonstrated that the metal oxide surfaces can act as catalysts. Early work demonstrated that lanthanum hydroxide ($La(OH)_3$) gels catalyze the hydrolysis of methoxy-ethyl phosphate, amino-ethyl phosphate, hydroxy-ethyl phosphate, and 1-methoxy-2-

propyl phosphate through the cleavage of the P–O bond (Butcher & Westheimer, 1955).

Iron(III) oxide surfaces have been found to catalyze the hydrolysis of chlorpyrifos-methyl and parathion-methyl, although not to the same extent as Ti and Al oxides (Torrents & Stone, 1994). Both compounds contain thionate S atoms, whereas parathion-methyl contains a nitro group, and chlorpyrifos-methyl contains a heterocyclic N. In the presence of goethite, the hydrolysis of chlorpyrifos-methyl is faster than parathion-methyl. The reason may be due to the identity of the donor groups available for chelate formation. The heterocyclic N of chlorpyrifos-methyl will form a chelate, whereas the nitro group of parathion-methyl is not likely to do so. The hydrolysis of ronnel, which contains no N donor ligands, was not affected by the presence of goethite.

6–6 RELEVANCE TO ECOSYSTEM HEALTH

Organophosphorus pesticides come into contact with a wide variety of soil constituents. It has been demonstrated that metal species in pore waters and on naturally occurring surfaces can influence rates and pathways of degradation of these pollutants. Small changes in pesticide structure can have significant consequences on rates of degradation and susceptibility towards metal catalysis. Oxonate esters, which are the oxidized analogues of the commonly used thionate esters, are more toxic to mammals than the parent compound. The oxonate and thionate esters differ in their susceptibility towards hydrolysis in metal-free and metal-containing solutions. Noting these differences is important in evaluating the fate of these complex organic chemicals. Understanding rates and pathways of organophosphorus compounds can aid in the design of new pesticides that are less harmful to organisms other than the target pests, and those that are less persistent in the natural environment.

6–7 SUMMARY AND CONCLUSIONS

Organophosphorus esters are introduced into the environment as pesticides and industrial pollutants. These compounds may persist in the natural environment and contaminate water supplies. Hydrolysis is a major degradation pathway for many of these pollutants that usually results in products of reduced toxicity. We are interested in how constituents of the natural environment affect the fate of these hydrolyzable pollutants. Dissolved metals and metal-containing surfaces have been found to catalyze the hydrolysis of organophosphorus pesticides. Understanding the principles that affect metal-catalyzed hydrolysis such as metal speciation and complex formation is necessary in order to evaluate the effect of metals on hydrolysis reactions. Speciation of the nucleophile, the metal, and the organic ester determine mechanisms and rates of hydrolysis.

REFERENCES

Benkovic, S.J., and L.K. Dunikoski. 1971. An unusual rate enhancement in metal ion catalysis of phosphate transfer. J. Am. Chem. Soc. 93:1526–1527.

Birkeland, P.W. 1984. Soils and geomorphology. Oxford Univ., New York.

Blanchet, P.-F., and A. St.-George. 1982. Kinetics of chemical degradation of organophosphorus pesticides: Hydrolysis of chlorpyrifos and chlorpyrifos-methyl in the presence of copper(II). Pestic. Sci. 13:85–91.

Butcher, W.W., and F.H. Westheimer. 1955. The lanthanum hydroxide gel promoted hydrolysis of phosphate esters. J. Am. Chem. Soc. 77:2420–2424.

Emsley, J. 1989. The elements. Clarendon, Oxford, England.

Epstein, J., and W.A. Mosher. 1968. Magnesium ion catalysis of hydrolysis of isopropyl methylphosphonoflouridate. The charge effect in metal ion catalysis. J. Phys. Chem. 72:622–625.

Eto, M. 1979. Organophosphorus pesticides: Organic and biological chemistry. CRC Press, Boca Raton, FL.

Evers, A., R.D. Hancock, A.E. Martell, and R.J. Motekaitis. 1989. Metal ion recognition in ligands with negatively charged oxygen donor groups. Complexation of Fe(III), Ga(III), In(III), Al(III), and other highly charged metal ions. Inorg. Chem. 28:2189–2195.

Faust, S.D. and H.M. Gomaa. 1972. Chemical hydrolysis of some organic phosphorus and carbamate pesticides in aquatic environments. Environ. Lett. 3:171–201.

Gellman, S.H., R. Petter, and R. Breslow. 1986. Catalytic hydrolysis of a phosphate triester by tetracoordinated zinc complexes. J. Am. Chem. Soc. 108:2388–2394.

Green, M.B., G.S. Hartley, and T.F. West. 1987. Chemicals for crop improvement and pest management. Pergamon, Oxford, England.

Hassal, K.A. 1990. The biochemistry and uses of pesticides. VCH, New York.

Hay, R.W. 1987. Lewis acid catalysis and the reactions of coordinated ligands. p. 412–485. *In* G. Wilkinson et al. (ed.) Comprehensive coordination chemistry. Pergamon, England.

Houghton, R.P. 1979. Metal complexes in organic chemistry. Cambridge Univ., Cambridge, England.

Ketelaar, J.A.A., H.R. Gersmann, and M.M. Beck. 1956. Metal-catalyzed hydrolysis of thiophosphoric esters. Nature (London) 177:392–393.

Mabey, W., and T. Mill. 1978. Critical review of hydrolysis of organic compounds in water under environmental conditions. J. Phys. Chem. Ref. Data. 7:383–415.

Mill, T., and W. Mabey. 1988. Hydrolysis of organic chemicals. p. 72–111. *In* O. Hutzinger (ed.) The handbook of environmental chemistry: Reactions and processes. Springer-Verlag, Berlin.

Mingelgrin, U., S. Saltzman, and B. Yaron. 1977. A possible model for the surface-induced hydrolysis of organophosphorus pesticides on kaolinite clays. Soil Sci. Soc. Am. J. 41:519–523.

Morrow, J.R., L.A. Buttrey, and K.A. Berback. 1992. Transesterfication of a phosphate diester by divalent and trivalent metal ions. Inorg. Chem. 31:16–20.

Mortland, M.M., and K.V. Raman. 1967. Catalytic hydrolysis of some organic phosphate pesticides by copper(II). J. Agric. Food Chem. 15:163–167.

Nurnberg, H.W. 1974. The influence of double layer effects on chemical reactions at charged interfaces. p. 48–53. *In* U. Zimmerman and J. Dainty (ed.) Membrane transport in plants. Springer-Verlag, New York.

Nurnberg, H.W., and G. Wolff. 1980. Influence on homogeneous chemical reactions in the diffuse double layer. J. Electroanal. Interfacial Electrochem. 21:91–122.

Pearson, R.G. 1966. Acids and bases. Science (Washington, DC) 151:172–177.

Saltzman, S., and U. Mingelgrin. 1984. Nonbiological degradation of pesticides in the unsaturated zone. p. 153–161. *In* B. Yaron et al. (ed.) Pollutants in porous media: The unsaturated zone between soil surface and groundwater. Springer-Verlag, Berlin.

Saltzman, S., U. Mingelgrin, and B. Yaron. 1976. Role of water in the hydrolysis of parathion and methylparathion on kaolinite. J. Agric. Food Chem. 24:739–743.

Saltzman, S. B. Yaron, and U. Mingelgrin. 1974. The surface catalyzed hydrolysis of parathion on kaolinite. Soil Sci. Soc. Am. Proc. 38:231–234.

Sanchez-Camazano, M.S., and M.J.S. Sanchez-Martin. 1980. Interaction of phosmet with montmorillonite. Soil Sci. 129:115–118.

Sanchez-Camazano, M.S., and M.J.S. Sanchez-Martin. 1983. Montmorillonite-catalyzed hydrolysis of phosmet. Soil Sci. 136:89–93.

Sanchez-Camazano, M., and M.J. Sanchez-Martin. 1991. Hydrolysis of azinphosmethyl induced by the surface of smectites. Clays Clay Miner. 39:609–613.

Schmidt, K.J. 1975. Chemical aspects of organophosphate pesticides in view of the environment. p. 96–108. *In* F. Coulston and F. Korte (ed.) Environmental quality and safety, global aspects of chemistry, toxicology and technology as applied to the environment. Academic Press, New York.

Schmidt, K.J., and C. Fest. 1982. The chemistry of organophosphorus pesticides. Springer-Verlag, New York.

Shen, C.Y., and F.W. Morgan. 1973. Hydrolysis of phosphorus compounds. p. 241–263. *In* Environmental Phosphorus Handbook. Wiley-Interscience, New York.

Sigel, H., and D.B. McCormick. 1970. On the discriminating behavior of metal ions and ligands with regard to their biological significance. Acc. Chem. Res. 3:201–208.

Smith, R.M., A.E. Martell, and R.J. Motekaitis. 1995. NIST critically selected stability constants of metal complexes database. Version 2.0. NIST Standard Ref. Database 46. U.S. Dep. of Commerce, Gaithersburg, MD.

Sposito, G. 1989. The chemistry of soils. Oxford Univ., New York.

Stone, A.T. 1989. Enhanced rates of monophenyl terephthalate hydrolysis in aluminum oxide suspensions. Colloid Interface Sci. 127:429–441.

Stone, A.T., and A. Torrents. 1995. The role of dissolved metals and metal-containing surfaces in catalyzing the hydrolysis of organic pollutants. p. 271–295. *In* P.M. Huang (ed.) Environmental impact of soil component interactions. Lewis Publ., Chelsea, MI.

Stumm, W., and J.J. Morgan. 1981. Aquatic chemistry. Wiley Interscience, New York.

Suh, J., and K.H. Chun. 1986. Metal ion catalysis by blocking inhibitory reverse paths in the hydrolysis of 3-carboxyaspirin. J. Am. Chem. Soc. 108:3057–3063.

Torrents, A., and A.T. Stone. 1994. Oxide surface-catalyzed hydrolysis of carboxylate esters and phosphorothioate esters. Soil Sci. Soc. Am. J. 58:738–745.

Voudrais, E.A., and M. Reinhard. 1986. Abiotic organic reactions at mineral surfaces. p. 462–486. *In* J.A. Davis and K.F. Hayes (ed.) Geochemical processes at mineral surfaces. ACS Symp. Ser. Am. Chem. Soc., Washington, DC.

7 Impact of Chemical and Biochemical Reactions on Transport of Environmental Pollutants in Porous Media

Mark L. Brusseau

University of Arizona
Tucson, Arizona

7–1 INTRODUCTION

The transport and fate of contaminants or pollutants in subsurface systems has become one of the major research areas in the environmental–hydrological–earth sciences. This interest has been fomented by concerns associated with the effects of human activities on environmental pollution. There are numerous means by which humans may cause pollution. Some of the sources of pollution are the result of planned waste disposal activities, whereas other sources are the result of accidental releases. Just about all conceivable human activities result directly or indirectly in the production of waste materials. This has resulted in a major problem with regard to the existence of hazardous waste sites. It is estimated that the cost to clean up contaminated sites associated with industrial, commercial, energy production, and defense activities may be hundreds of billions of dollars in the USA alone.

An understanding of how contaminants move in the subsurface is required to address environmental pollution problems. For example, such knowledge is needed to conduct risk assessments, such as evaluating the probability of a chemical spill contaminating groundwater. Such knowledge is also required to develop and evaluate methods for cleaning up contaminated soils and aquifers. Just as importantly, knowledge of contaminant transport and fate is necessary to design *pollution-prevention* strategies. The purpose of this chapter is to briefly review, in a selective manner, how chemical and biochemical reactions influence transport. The chapter is based largely on a recent review of the subject (Brusseau, 1994).

7–2 BASIC CONCEPTS OF CONTAMINANT TRANSPORT IN POROUS MEDIA

Four processes that control the movement of contaminants in porous media are advection, dispersion, interphase mass transfer, and transformation reactions.

 Soil Chemistry and Ecosystem Health. Special Publication no. 52.

Advection, also referred to as convection, is the transport of dissolved matter (solute) by the movement of a fluid responding to a gradient of fluid potential. For example, a chemical dissolved in water will be carried along by the water as it flows through the soil. Dispersion represents spreading of solute about a mean position, such as the center of mass. Spreading at the microscopic scale is caused by movement through tortuous (nonlinear), nonuniformly sized pores and by molecular diffusion. Phase transfers, such as sorption, liquid–liquid partitioning, and volatilization, involve the transfer of matter in response to gradients of chemical potential or, more simply, to concentration gradients. Transformation reactions include any process by which the physicochemical nature of a contaminant is altered. Examples include biotransformation, radioactive decay, and hydrolysis.

The initial paradigm for transport of contaminants in porous media was based on assumptions that the porous medium was homogeneous and that the rates of inter-phase mass transfer and reaction were linear and essentially instantaneous. For discussion purposes, transport that follows this paradigm is considered to be ideal; however, it is well known that the subsurface is, in fact, heterogeneous and that many phase transfers and reactions are not instantaneous nor linear. Thus, contaminant transport usually deviates from that which is expected based on the original paradigm, especially at the field scale. Such transport can be considered as nonideal.

The equation used to describe the one-dimensional transport of dissolved reactive, sorbing contaminants following the assumptions associated with the original paradigm is:

$$R \frac{\partial C}{\partial t} = -v \frac{\partial C}{\partial x} + D \frac{\partial 2C}{\partial x^2} - \mu_l C - \mu_s S \qquad [1]$$

where x is a spatial coordinate, t is time, C is the concentration of contaminant in the fluid (e.g., water), v is the average linear velocity of the fluid in the pores of the medium, D^* is the dispersion coefficient, μ is a first-order transformation coefficient (l = solution phase and s = sorbed phase), and R is the retardation factor, defined as:

$$R = 1 + \frac{\rho}{\theta} K_d \qquad [2]$$

where K_d is a coefficient representing distribution of the contaminant between the solid and liquid phases (e.g., sorption), ρ is the bulk density of the soil, and θ is the volumetric fractional water content of the soil.

The dispersion coefficient is usually defined as:

$$D = \alpha v + \frac{D_0}{\tau} \qquad [3]$$

where α is the dispersivity, D_0 is fluid-phase diffusion coefficient, and τ is a factor accounting for the tortuosity of the porous medium. The first and second terms on the right hand side of Eq. [3] represent the contribution of mechanical mixing and molecular diffusion, respectively, to total dispersion. Mechanical mixing is caused by the tortuous nature of individual flow paths in the porous medium and by the existence of different pore sizes, both of which result in nonuniform flow velocities at the microscopic scale. Axial diffusion refers to the movement of solute in response to a concentration gradient.

Equation [1], known as the advection–dispersion equation, is by far the most widely used equation for describing the transport of dissolved matter in porous media. It is used in chemical engineering, petroleum engineering, and chromatography, in addition to earth sciences. The term on the left hand side of Eq. [1] represents the change in contaminant mass occurring at a specified location in response to transport and fate processes. The retardation factor represents the influence of phase transfer (sorption) on transport. The first term on the right hand side represents advective transport, while the second term represents dispersive transport. The third term represents a loss of contaminant from the system due to transformation reactions.

The movement of a contaminant from a point where it is first introduced to a point at which it is being monitored over time can be described by a breakthrough curve. The measurement and analysis of breakthrough curves is a widely used means of investigating contaminant transport in porous media. An example of a breakthrough curve, generated by solving Eq. [1] with $\mu = 0$ (no transformation), is shown in Fig. 7–1. To enhance our ability to analyze breakthrough curves, we use nondimensional parameters. The first is relative concentration

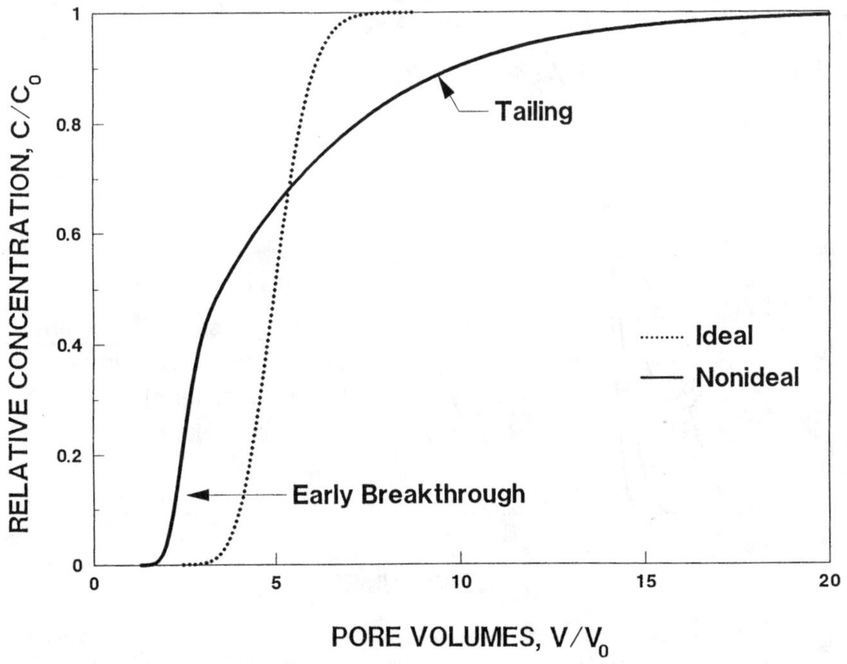

Fig. 7–1. Breakthrough curves for transport of a dissolved chemical through soil.

[C/C_0], which is the concentration of contaminant measured in the effluent [C] divided by the concentration in the influent [C_0]. The second is pore volume, which can be thought of as a nondimensional time parameter (real time divided by the average transit time of water) or as the volume of water discharged from soil (V) divided by the volume of water that can be held by the soil system (V_0). The third is the Peclet Number (P), which is a ratio of advective flux to dispersive flux.

The effect of dispersion on transport is illustrated in Fig. 7–2A, which shows breakthrough curves produced by varying the dispersion coefficient while keeping all other parameters the same. When the dispersion coefficient is 0 (no dispersion), the breakthrough curve is vertical. In this case, which is referred to as piston flow, the contaminant is transported only by advection. As the dispersion coefficient is increased, the breakthrough curve is rotated clockwise, showing greater and greater spreading. A sharp breakthrough curve with little spreading is characteristic of transport through homogeneous sandy soil. Breakthrough curves obtained for transport through soils with a wide particle size distribution usually exhibit greater dispersion.

Generally, the solid phase of soil is fixed in place and, therefore, contaminant molecules associated with soil particles are not transported (exceptions to this, such as transport facilitated by colloids, will not be considered herein). As a result, sorption reduces, or *retards*, the average transport velocity of a contaminant. The effect of sorption, or retardation, on transport is illustrated in Fig. 7–2B.

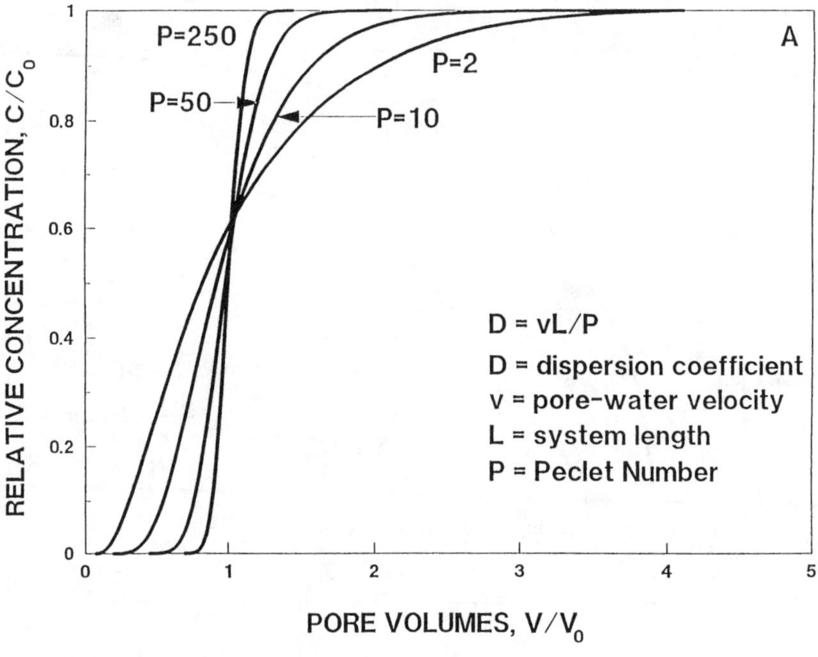

Fig. 7–2. (A) The effect of dispersion on contaminant transport through soil; (B) the effect of sorption and retardation on contaminant transport through soil; (C) the effect of transformation reactions on contaminant transport through soil.

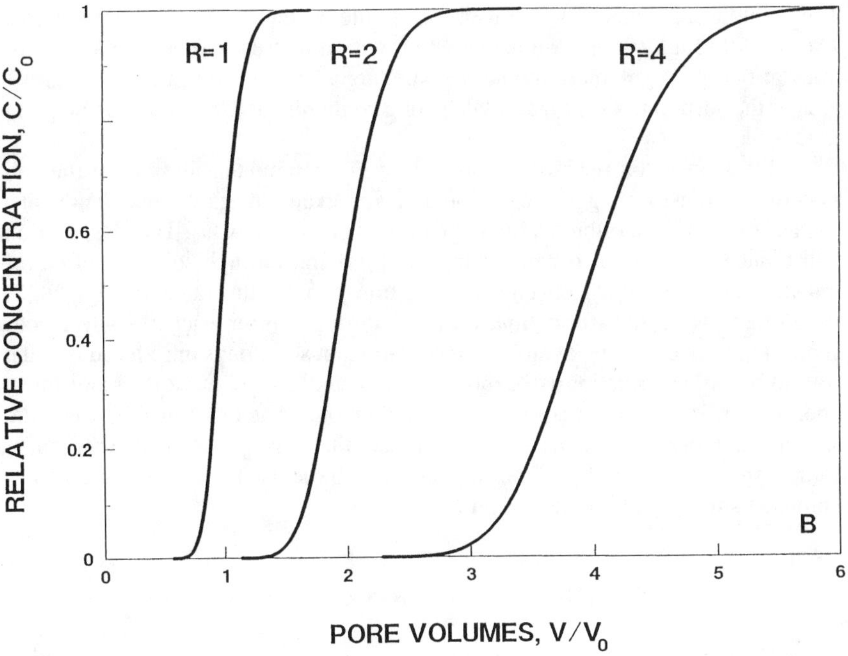

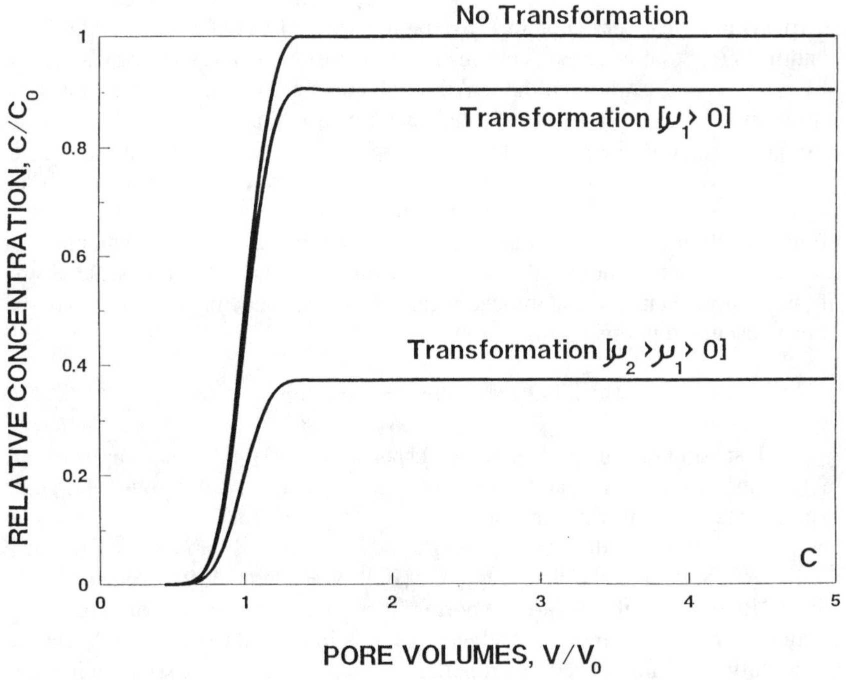

Fig. 7–2. Continued.

The breakthrough curve for a nonsorbing solute is centered at one pore volume. The breakthrough curve for a contaminant with a retardation factor of 2 is centered at two pore volumes, and so on. The larger the retardation factor (i.e., the greater the sorption), the longer it takes for a contaminant to move from one point to another.

Transformation reactions cause a loss of contaminant mass from the soil system by transforming the contaminant. For example, gasoline components, such as benzene, are often readily degraded by bacteria to CO_2. This degradation will reduce the mass of benzene while it is moving through soil. The effect of transformation reactions on contaminant transport is illustrated in Fig. 7–2C. When there is no transformation, the breakthrough curve reaches relative concentration of one, as long as sufficient contaminant solution is introduced into the soil. When transformation is occurring, the breakthrough curve does not reach one. Rather, it plateaus at some value less than one. The difference between the two curves represents the mass of contaminant that was transformed. Such transformation reactions are beneficial in that they can be used to help clean up contaminated sites (e.g., bioremediation).

7–3 NONIDEAL CONTAMINANT TRANSPORT

The discussion of contaminant transport presented in the preceding section was based on ideal conditions. It is recognized, however, that the transport of contaminants, especially at the field scale, is characteristically nonideal. Note the sharp, symmetrical curve obtained for the transport of a contaminant under ideal conditions (e.g., homogeneous porous medium, instantaneous sorption) as shown in Fig. 7–1. An illustration of a breakthrough curve exhibiting nonideal transport also is shown in Fig. 7–1. This curve is asymmetrical, and exhibits early breakthrough and tailing, i.e., an asymptotic approach to relative concentration of 1. This type of nonideal behavior can be caused several factors, such as physical–chemical heterogeneity, rate-limited diffusive mass transfer, and rate-limited sorption–desorption. It is possible, even probable, that more than one factor may contribute to nonideal transport of contaminants at the field scale. Some of the major chemical and biochemical factors influencing transport will be briefly discussed in this section.

7–3.1 Nonlinear Sorption

Most solute transport models, especially for field-scale applications, include the assumption that sorption is a linear process. Numerous laboratory experiments have shown that sorption is often linear for low-polarity organic compounds; however, this assumption should be evaluated for each case because nonlinear isotherms have been reported (Ball & Roberts, 1991a; Weber et al., 1992). Nonlinear isotherms are the norm rather than the exception for many polar organic, as well as for inorganic, chemicals. It is important to note that the nature of the isotherm is influenced by properties of the sorbent and the solution in addi-

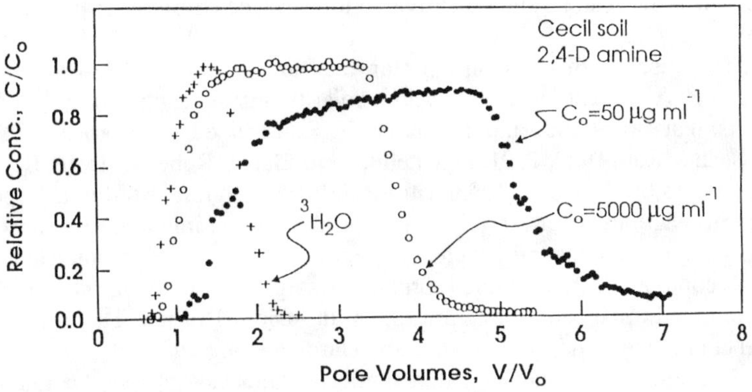

Fig. 7–3. The effect of nonlinear sorption on contaminant transport through soil; Cecil soil is a sandy soil. Adapted from Rao and Davidson (1979).

tion to those of the solute. Several equations exist for describing sorption isotherms (see Travis & Etnier, 1981; Kinniburgh, 1986 for reviews).

It is well known that nonlinear sorption can cause asymmetrical breakthrough curves and concentration-dependent retardation factors (Crittenden et al., 1986; Brusseau & Rao, 1989). This was clearly demonstrated by Rao and Davidson (1979), who reported breakthrough curves for transport of 2,4-dichlorophenoxyacetic acid through two soils (see Fig. 7–3). Retardation increased as the influent solute concentration decreased. In addition, the elution waves were much more spread out (exhibiting tailing) than were the injection waves.

There are several equations that may be used to represent nonlinear sorption isotherms. One of the most widely used is the Freundlich equation:

$$S = K_f\, C^n \qquad [4]$$

where K_f and n are the Freundlich magnitude and intensity parameters. Using Eq. [11], the resultant retardation factor is:

$$R = 1 + \frac{\rho}{\theta}\, K_f n\, C^{n-1} \qquad [5]$$

Inspection of Eq. [5] reveals that R is a function of concentration.

7–3.2 Rate-Limited Sorption

Most field-scale solute transport models include the so-called local equilibrium assumption, which specifies that interactions between the solute and the sorbent are so rapid in comparison to hydrodynamic residence time that the interactions can be considered instantaneous. Based on laboratory experiments, it well known that sorption–desorption of many contaminants by soils, sediments, and aquifer materials can be significantly rate limited (see Brusseau & Rao, 1989 for review).

The mechanisms responsible for rate-limited sorption–desorption have been a focus of recent research. For low-polarity organic compounds, the current view is that nonequilibrium is caused by constrained intrasorbent diffusion (Brusseau & Rao, 1989, 1991; Pignatello, 1990; Ball & Roberts, 1991b; Brusseau et al., 1991a,b; Brusseau, 1993; Hu et al., 1995). While rate-limited specific chemical reactions (chemisorption) are not likely to be important for most low-polarity organic compounds, such reactions can be important for ionic and polar organic compounds. Rate-limited sorption–desorption of inorganic chemicals has been reported to be caused by intrasorbent diffusion (Hodges & Johnson, 1987; Wood et al., 1990) and by chemisorption (Jardine et al., 1985).

Numerous experimental and theoretical studies have shown that rate-limited sorption–desorption can cause nonideal transport. This nonideality can take the form of asymmetrical breakthrough curves exhibiting early breakthrough and tailing, as well as decelerating contaminant plumes (temporally increasing *R* values). An example of the effect of rate-limited sorption–desorption on transport is presented in Fig. 7–4, which shows breakthrough curves for transport of several solutes through a sandy soil. Note that the breakthrough curves for the nonreactive tracers (^3H_2O and pentafluorobenzoate) are sharp and symmetrical, signifying ideal behavior. In contrast, the breakthrough curve for the sorbing organic solute (naphthalene) is asymmetrical. This suggests that the nonideality factor involves the sorption process.

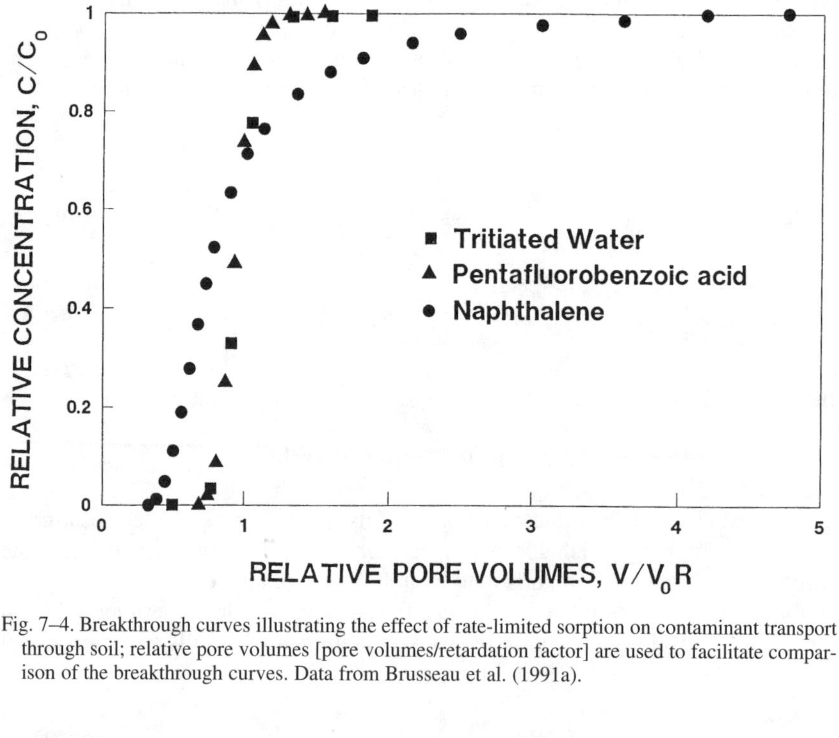

Fig. 7–4. Breakthrough curves illustrating the effect of rate-limited sorption on contaminant transport through soil; relative pore volumes [pore volumes/retardation factor] are used to facilitate comparison of the breakthrough curves. Data from Brusseau et al. (1991a).

Rate-limited sorption has been described by several equations. One often used equation is called the *one-site* model, wherein it is assumed that all sorption is rate-limited and governed by a single mechanism and associated rate coefficient. While the one-site model is mathematically expedient, it has been shown to be invalid for representing the sorption dynamics of many reactive solutes (Davidson & McDougal, 1973; Schwarzenbach & Westall, 1981; Brusseau & Rao, 1989; Brusseau et al., 1989).

Another widely used model is based on a two-domain description of rate-limited sorption. With this model, sorption is assumed to be governed by two sets of rate coefficients, each representing a separate sorption domain. It is often further assumed that the rate of sorption for one domain is so rapid that it can be considered instantaneous. The two-domain approach was developed by Giddings and Eyring (1955) for chromatographic systems, and was originally applied to soil systems by several researchers (Liestra & Dekkers, 1976; Selim et al., 1976; Cameron & Klute, 1977). This approach has been used successfully to simulate the sorption–desorption and, when coupled to the advective-dispersive equation, the transport of both organic and inorganic solutes in soils, sediments, and aquifer materials (see Brusseau & Rao, 1989 for review). This approach is very flexible in that the equations can be used to represent rate-limited sorption–desorption caused by chemisorption (Jardine et al., 1985) or by intrasorbent diffusion (Brusseau et al., 1991a).

One formulation of the two-domain approach is given by:

$$C \leftrightarrow S_1 \underset{k_2}{\overset{k_1}{\rightleftarrows}} S_2 \qquad [6]$$

where C is solution-phase concentration of solute [M L^{-3}], S_1 is the concentration of instantaneously sorbed solute [M M^{-1}], S_2 is the concentration of rate-limited sorbed solute [M M^{-1}], k_1 is the first-order sorption rate coefficient [T^{-1}], and k_2 is the first-order reverse sorption rate coefficient [T^{-1}]. The equation for sorption at equilibrium is:

$$S_T = S_1 + S_2 = F\, K_f\, C + (1 - F)\, K_f\, C \qquad [7]$$

where S_T is total sorbed-phase concentration [M M^{-1}], K_d is the sorption coefficient [L^3 M^{-1}], and F is the fraction of sorbent for which sorption is instantaneous.

The equations describing solute transport governed by steady-state, one-dimensional water flow, linear, rate-limited sorption are:

$$\theta \frac{\partial C}{\partial t} + \rho \frac{\partial S_1}{\partial t} + \rho \frac{\partial S_2}{\partial x} = -q \frac{\partial C}{\partial x} + \theta D \frac{\partial^2 C}{\partial x^2} \qquad [8]$$

$$\frac{\partial S_2}{\partial t} = k_1 S_1 - k_2 S_2 \quad (15) \qquad [9]$$

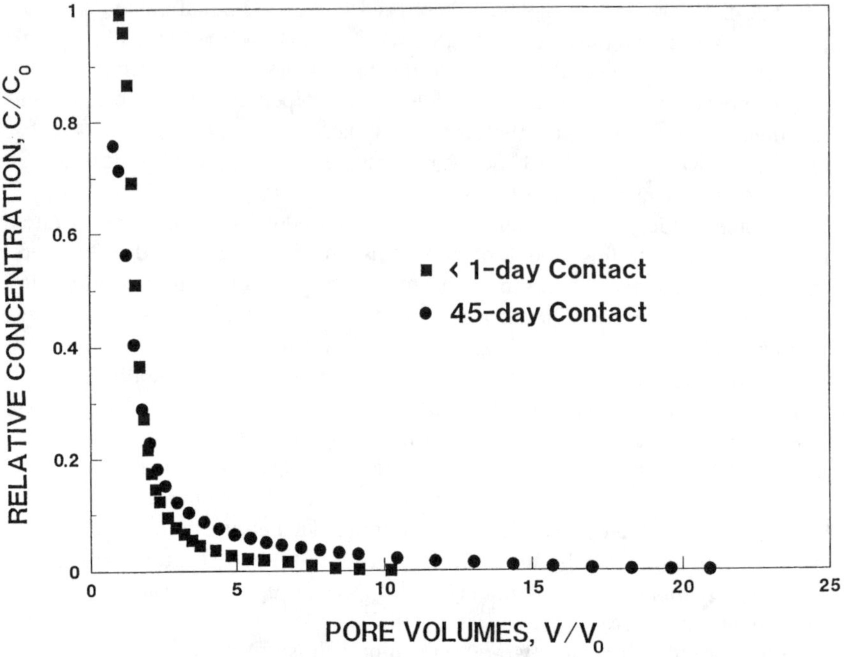

Fig. 7–5. Illustration of the effect of contact time on contaminant elution. For one experiment naphthalene was in contact with the sandy soil for a few hours prior to elution. In the other experiment, naphthalene was in contact with the soil for 45 d. Data from Brusseau et al. (1991a).

A phenomenon that, in many cases, is related to rate-limited sorption–desorption is so-called contaminant aging. Recent research has shown that contaminants that have been in contact with porous media for long times are much more resistant to desorption, extraction, and degradation. For example, contaminated samples taken from field sites exhibit solid/aqueous distribution ratios that are much larger than those measured or estimated based on spiking the porous media with the same contaminant (e.g., adding contaminant to uncontaminated sample) (Steinberg et al., 1987; Pignatello et al., 1990; Smith et al., 1990; Scribner et al., 1992). In addition, the magnitude of the desorption rate coefficients determined for previously contaminated media collected from the field have been shown to be much smaller than the values obtained for spiked samples (Steinberg et al., 1987; Connaughton et al., 1993). These field-based observations are supported by laboratory experiments wherein measured values of desorption rate coefficients have been observed to decrease with increasing time of contact prior to desorption (Karickhoff, 1980; McCall & Agin, 1985; Coates & Elzerman, 1986; Brusseau et al., 1991a).

The effect of aging on contaminant desorption is illustrated in Fig. 7–5. Here, the elution curve obtained for a case where the contaminant was in contact with the porous medium for several hours (usual experimental conditions) is compared with a case with a 45-d contact period. It is evident that the elution curve for the long-contact case exhibits greater tailing, reflecting a greater resistance to removal.

7–3.3 Transformation Reactions

Many contaminants may undergo various types of transformation reactions during transport. Examples of such reactions include radioactive decay, hydrolysis, and biodegradation. A recent review of mathematical models designed to simulate transport of solutes influenced by transformation reactions revealed that the vast majority were based on the use of a first-order transformation reaction, such as shown in Eq. [1] (Brusseau et al., 1992b). First-order kinetics, wherein the rate is a function only of solute concentration, accurately describes some transformation reactions (radioactive decay, pH-independent hydrolysis) but not others. Biodegradation is a good example of a transformation process that can be described as first order only for highly constrained conditions. Often, the rate of biodegradation of a contaminant will depend on many other factors (e.g., biomass growth, electron acceptor availability, sorption, environmental conditions) in addition to contaminant concentration. Contaminant transport under these conditions will not match that based on ideal transport (i.e., first-order reaction).

The effect of nonsteady-state microbial populations on contaminant transport is an example of transformation-related nonideality. This effect is illustrated in Fig. 7–6, which shows the transport of a pesticide, 2,4-D (2,4-dichlorophenoxyacetic acid), through soil packed in a column (Estrella et al., 1993). The soil

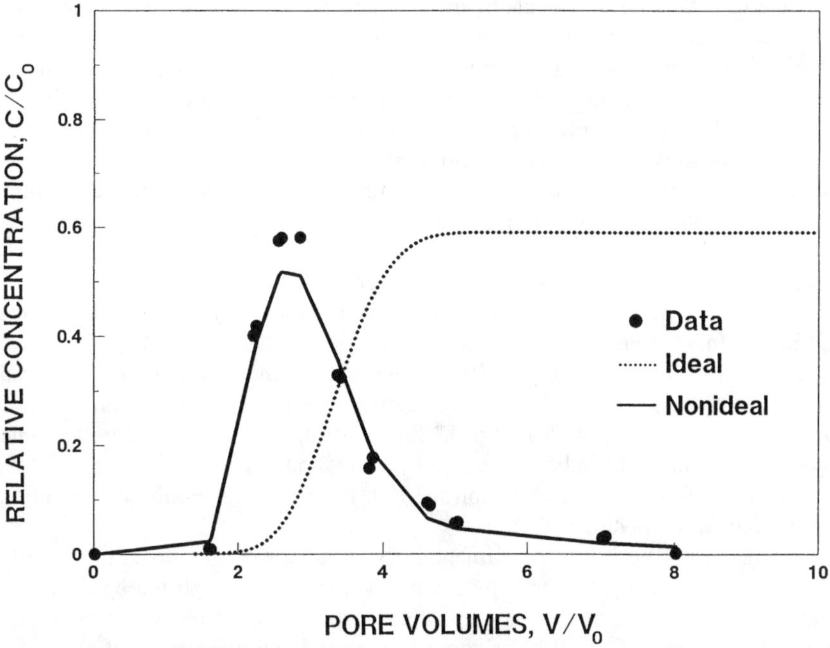

Fig. 7–6. Illustration of the effect of microbial growth on biodegradation and transport of 2,4-dichlorophenoxyacetic acid through a soil column partially saturated with water. The nonideal simulation accounts for rate-limited sorption and microbial growth, whereas the ideal curve is based on instantaneous sorption and a nongrowing microbial population. Adapted from Estrella et al. (1993).

contained a bacterial population capable of degrading the 2,4-D as demonstrated by separate microcosm experiments. A simulated breakthrough curve obtained for ideal conditions, including first-order biodegradation, also is shown in Fig. –6. Clearly, the biodegradation and transport of 2,4-D could not be simulated with the ideal model (Eq. [1]). One condition associated with the assumption of first-order biodegradation kinetics is that there is minimal microbial growth. This condition is inconsistent with the data, which show a steadily decreasing concentration of 2,4-D in the effluent to a point of about 8 pore volumes, whereupon 2,4-D was no longer detectable. It is important to note that the decline and disappearance of 2,4-D in the effluent occurred while the solution containing 2,4-D was still being injected into the column. This behavior indicates the occurrence of temporally nonuniform biological activity. A model that accounted for, among other factors, microbial growth could successfully simulate the data (see Fig. 7–6). Results similar to these have been reported by Chen et al. (1992).

7–3.4 Spatially Variable Chemical–Biological Properties

It is well known that physical properties of porous media are spatially variable in the subsurface. It is logical to expect chemical and biological properties to also be spatially variable. For example, it is likely that sorption coefficients may vary in space as a result of heterogeneous distributions of organic matter and clay content. Several field-scale investigations have shown that this is indeed the case (Brusseau, 1994). The impact of spatially variable sorption on transport has received little attention to date. Smith and Schwartz (1981), with a series of modeling exercises, evaluated the influence of spatially variable cation exchange capacity (i.e., variable retardation) on solute transport. They assumed that sorption was negatively correlated with hydraulic conductivity. Dispersion was observed to increase as the variability in retardation increased; however, they concluded that this effect would be of secondary importance in comparison to the effect of spatially variable hydraulic conductivity. Spatially variable sorption was proposed as one cause of the enhanced dispersion exhibited by lithium, in comparison to bromide, in a recent natural gradient experiment performed at Cape Cod, MA. In a theoretical analysis of the effect of coupled physical and chemical heterogeneity, Garabedian et al. (1988) observed enhanced dispersion compared to the uniform sorption case when a negative correlation between sorption and hydraulic conductivity was assumed. Similar results, based on theoretical and modeling exercises, have been reported by several other researchers (Bahr, 1986; Valocchi, 1989; Cvetkovic & Shapiro, 1990; Andricevic & Foufoula-Georgiou, 1991; Kabala & Sposito, 1991).

There have been few experimental investigations of the impact of spatially variable sorption on solute transport. One such study was reported by Brusseau and Zachara (1993), who investigated the transport of Co^{2+} in a column packed with layers of two media of differing hydraulic conductivities and sorption capacities. The sorption capacities of the two media differed by about a factor of 3. The asymmetrical breakthrough curve obtained for transport of a nonreactive tracer through the column demonstrated the effect of the physical heterogeneity on transport. The breakthrough curve obtained for transport of Co^{2+} was shifted to the left of a simulated curve obtained for ideal conditions (homogeneous, instantaneous sorption) and exhibited tailing (see Fig. 7–7). The comparison reveals that transport of Co^{2+} through the heterogeneous porous medium was significantly nonideal. The optimized curve obtained for the case of hydraulic-conductivity variability, sorption-capacity variability, rate-limited mass transfer between the two layers, and rate-limited sorption–desorption matched the experimental data quite well. In contrast, a simulated curve for the case of homogeneous sorption, hydraulic-conductivity variability, and rate-limited mass transfer and sorption did not match the data well. The influence of sorption variability had to be considered to accurately simulate the transport of Co^{2+}. The impact of spatially variable sorption on solute transport was shown also by Brusseau (1991) in an analysis of data reported by Starr et al. (1985), who investigated the transport of Sr in a flow cell containing layers of two media with different conductivity and sorption properties.

It also is possible that microbial activity is spatially variable, either due to heterogeneous population distributions or to spatially variability of factors controlling activity, or to some combination thereof. The impact of spatially variable transformation rates on solute transport was examined by Xu and Brusseau (1996). Results of the analysis indicate that spatial variation of the transformation coefficient reduces the global rate of contaminant mass loss, as compared with the case of a spatially constant coefficient. This suggests that the lower range of transformation-coefficient values somewhat controls the transport behavior.

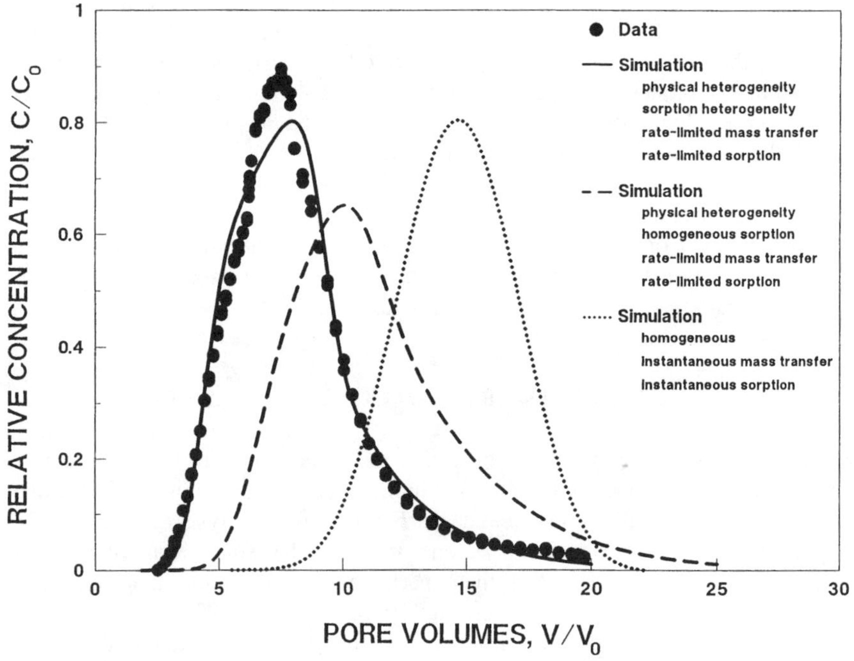

Fig. 7–7. Illustration of the effect of porous-medium heterogeneity and rate-limited sorption–desorption on transport of Co. Adapted from Brusseau and Zachara (1993).

7–3.5 Multiple-Factor Nonideality

The discussion above focused on several factors that can cause nonideal transport. Most research (experimental and modeling) is focused on investigating a single process, which is necessary to develop an in-depth understanding of that process; however, the transport and fate of contaminants is often influenced by more than one process, and it is possible that nonideal behavior may be caused by more than one factor. The investigation of such systems, sometimes referred to as coupled-process systems, is a burgeoning field of research.

The transport, in heterogeneous porous media, of reactive contaminants undergoing rate-limited sorption is one example of a coupled-process system. In this case, it is of interest to understand the relative contributions of rate-limited sorption and physical heterogeneity to observed nonideal transport (e.g., early breakthrough and tailing). Brusseau and colleagues (Brusseau et al., 1989; Brusseau, 1991; Brusseau & Zachara, 1993; Hu & Brusseau, 1996) have investigated a number of systems wherein transport was influenced by rate-limited sorption and physical heterogeneity and conclude that the relative contribution of these two factors to nonideal transport is controlled, in part, by contaminant distribution potential and characteristic reaction times.

The effects of interactions between sorption and biodegradation, two rate-controlled processes, on transport is another example of nonideal transport influenced by two or more factors or processes. Recent research performed with batch (i.e., nonflowing) systems has suggested that the rate of biodegradation can be controlled by the rate of desorption or mass transfer (Rijnaarts et al., 1990; Robinson et al., 1990; Scow & Alexander, 1992), depending on the relative time scales of the two processes. When the rate of desorption or mass transfer is slower than that of biodegradation, the transport of the contaminant will not follow the behavior expected based on ideal conditions. This has been shown by Brusseau and colleagues, who have investigated the influence of rate-limited sorption–desorption (Angley et al., 1992; Estrella et al., 1993) and heterogeneous porous media combined with rate-limited sorption–desorption (Brusseau et al., 1992a) on the transport of contaminants undergoing biodegradation. In such cases, the bioavailability of the contaminant to the bacteria is constrained by mass transfer processes associated with sorption or structured soil.

7–4 SUMMARY AND CONCLUSIONS

Many factors can influence the transport and fate of contaminants in porous media. It is important to remember that in many cases more than one factor may be important especially at the field scale. For example, analysis of several field-scale contaminant transport experiments revealed that some form of nonideal transport was observed for all of the experiments, and that more than one factor was responsible in each case (Brusseau, 1992; Brusseau, 1994). Knowledge of which factors may be controlling transport will allow one to better evaluate how far a contaminant may move and how long it may take to reach a given point. Such information is critical to evaluating or assessing the risk posed by a contaminant to human health and to other components of the ecosystem. Knowledge of how contaminants move through porous media also is critical to successful development and evaluation of plans for cleaning up contaminated sites.

REFERENCES

Andricevic, R., and E. Foufoula-Georgiou. 1991. Modeling kinetic non-equilibrium using the first two moments of the residence time distribution. Stochastic Hydrol. Hydraul. 5:155–171.

Angley, J.T., M.L. Brusseau, W.L. Miller, and J.J. Delfino. 1992. Nonequilibrium sorption and aerobic biodegradation of dissolved alkylbenzenes during transport in aquifer material: Column experiments and evaluation of a coupled-process model. Environ. Sci. Technol. 26(7):1404–1410.

Bahr, J.M. 1986. Applicability of the local equilibrium assumption in mathematical models for groundwater transport of reacting solutes. Ph.D. diss., Stanford Univ., Palo Alto, CA.

Ball, W.P., and P.V. Roberts. 1991a. Long-term sorption of halogenated organic chemicals by aquifer material: 1. Equilibrium. Environ. Sci. Technol. 25(7):1223–1236.

Ball, W.P., and P.V. Roberts. 1991b. Long-term sorption of halogenated organic chemicals by aquifer material: 2. Intraparticle diffusion. Environ. Sci. Technol. 25(7):1237–1249.

Brusseau, M.L. 1991. Application of a multiprocess nonequilibrium sorption model to solute transport in a stratified porous medium. Water Resour. Res. 27(4):589–595.

Brusseau, M.L. 1992. Transport of rate-limited sorbing solutes in heterogeneous porous media: Application of a one-dimensional multi-factor nonideality model to field data. Water Resour. Res. 28(9):2485–2497.

Brusseau, M.L. 1993. Using QSAR to evaluate phenomenological models for sorption of organic compounds by soil. Environ. Toxicol. Chem. 12(10):1835–1846.

Brusseau, M.L. 1994. Transport of reactive contaminants in porous media. Rev. Geophys. 32(3):285–314.

Brusseau, M.L., R.E. Jessup, and P.S.C. Rao. 1989. Modeling the transport of solutes influenced by multi-process nonequilibrium. Water Resour. Res. 25(9):1971–1988.

Brusseau, M.L., R.E. Jessup, and P.S.C. Rao. 1991a. Nonequilibrium sorption of organic chemicals: Elucidation of rate-limiting processes. Environ. Sci. Technol. 25(1):134–142.

Brusseau, M.L., R.E. Jessup, and P.S.C. Rao. 1992a. Modeling solute transport influenced by multiprocess nonequilibrium and transformation reactions. Water Resour. Res. 28(1):175–182.

Brusseau, M.L., and P.S.C. Rao. 1989. Sorption nonideality during organic contaminant transport in porous media. CRC Crit. Rev. Environ. Contr. 19:33–99.

Brusseau, M.L., and P.S.C. Rao. 1991. The influence of sorbate structure on nonequilibrium sorption of organic compounds. Environ. Sci. Technol. 25(8):1501–1506.

Brusseau, M.L., P.S.C. Rao, and C.A. Bellin. 1992b. Modeling coupled processes in porous media: Sorption, transformation, and transport of organic solutes. p. 147–184. *In* Advances in soil science. CRC Press, Boca Ratan, FL.

Brusseau, M.L., A.L. Wood, and P.S.C. Rao. 1991b. The influence of organic cosolvents on the sorption kinetics of hydrophobic organic chemicals. Environ. Sci. Technol. 25(5):903–910.

Brusseau, M.L., and J. Zachara. 1993. Transport of Co^{2+} in a physically and chemically heterogeneous porous medium. Environ. Sci. Technol. 27(9):1937–1939.

Cameron, D.R., and A. Klute. 1997. Convective–dispersive solute transport with a combined equilibrium and kinetic adsorption model. Water Resour. Res. 13:183.

Chen, Y.M., L.M. Abriola, P.J.J. Alvarez, P.J. Anid, and T.M. Vogel. 1992. Modeling transport and biodegradation of benzene and toluene in sandy aquifer material: Comparisons with experimental measurements. Water Resour. Res. 28(7):1833–1847.

Coates, J.T., and A.W. Elzerman. 1986. Desorption kinetics for selected PCB congeners from river sediments. J. Contam. Hydrol. 1:191–201.

Connaughton, D.F., J.R. Stedinger, L.W. Lion, and M.L. Shuler. 1993. Description of time-varying desorption kinetics: Release of naphthalene from contaminated soils. Environ. Sci. Technol. 27(11):2397–2403.

Crittenden, J.C., N.J. Hutzler, D.G. Geyer, J.L. Oravitz, and G. Friedman. 1986. Transport of organic compounds with saturated groundwater flow: Model development and parameter sensitivity. Water Resour. Res. 22(3):271–284.

Cvetkovic, V.D., and A.M. Shapiro. 1990. Mass arrival of sorptive solute in heterogeneous porous media. Water Resour. Res. 26(9):2057–2067.

Davidson, J.M., and J.R. McDougal. 1973. Experimental and predicted movement of three herbicides in a water-saturated soil. J. Environ. Qual. 2:428–433.

Estrella, M.R., M.L. Brusseau, R.S. Maier, I.L. Pepper, P.J. Wierenga, and R.M. Miller. 1993. Biodegradation, sorption, and transport of 2,4-dichlorophenoxyacetic acid (2,4-D) in a saturated and unsaturated soil. Appl. Environ. Microbiol. 59(12):4266–4273.

Garabedian, S.P., L.W. Gelhar, and M.A. Celia. 1988. Large-scale dispersive transport in aquifers: Field experiments and reactive transport theory. Rep. 315. Dep. of Civil Eng., Massachusetts Inst. of Technol., Cambridge.

Giddings, J.C., and H. Eyring. 1955. A molecular dynamic theory of chromatography. J. Phys. Chem. 59:416–421.

Hodges, S.C., and G.C. Johnson. 1987. Kinetics of sulfate adsorption and desorption by Cecil soil using miscible displacement. Soil Sci. Soc. Am. J. 51:323–331.

Hu, Q., and M.L. Brusseau. 1996. Transport of rate-limited sorbing solutes in an aggregated porous medium: A multiprocess non-ideality approach. J. Contamin. Hydrol. 24:53–73.

Hu, Q., X. Wang, and M.L. Brusseau. 1995. Quantitative structure–activity relationships for evaluating the influence of sorbate structure on sorption of organic compounds by soil. Environ. Toxicol. Chem. 14(7):1133–1140.

Jardine, P.M., L.W. Zelazny, and J.C. Parker. 1985. Mechanisms of aluminum adsorption on clay minerals and peat. Soil Sci. Soc. Am. J. 49:862–867.

Kabala, Z.J., and G. Sposito. 1991. A stochastic model of reactive solute transport with time- varying velocity in a heterogeneous aquifer. Water Resour Res. 27(3):341–350.

Karickhoff, S.W. 1980. Sorption kinetics of hydrophobic pollutants in natural sediments. p. 193–205. *In* R.A. Baker (ed.) Contaminants and sediments. Ann Arbor Sci., Ann Arbor, MI.

Kinniburgh, D.G. 1986. General purpose adsorption isotherms. Environ. Sci. Technol. 20(9):895–904.

Leistra, M., and W.A. Dekkers. 1976. Computed effects of adsorption kinetics on pesticide movement in soils. J. Soil Sci. 28:340–350.

McCall, P.J., and G.L. Agin. 1985. Desorption kinetics of picloram as affected by residence time in soil. Environ. Toxicol. Chem. 4:37–45.

Pignatello, J.J. 1990. Slowly reversible sorption of aliphatic hydrocarbons in soils: II. Mechanistic aspects. Environ. Toxicol. Chem. 9:1117–1126.

Pignatello, J.J., C.R. Frink, P.A. Marin, and E.X. Droste. 1990. Field-observed ethylene dibromide in an aquifer after two decades. J. Contam. Hydrol. 5:195–210.

Rao, P.S.C., and J.M. Davidson. 1979. Adsorption and movement of selected pesticides at high concentrations in soils. Water Res. 13:375–342.

Rijnaarts, H.M., A. Bachmann, J.C. Jumelet, and A.J.B. Zehnder. 1990. Effect of desorption and intraparticle mass transfer on the aerobic biomineralization of α-hexachlorocyclohexane in a contaminated calcareous soil. Environ. Sci. Technol. 24:1349–1354.

Robinson, K.G., W.S. Farmer, and J.T. Novak. 1990. Availability of sorbed toluene in soils for biodegradation by acclimated bacteria. Wat. Res. 24:345–350.

Schwarzenbach, R.P., and J. Westall. 1981. Transport of nonpolar organic compounds from surface water to groundwater: Laboratory sorption studies. Environ. Sci. Technol. 15:1360–1369.

Scow, K., and M. Alexander. 1992. Effect of diffusion on the kinetics of biodegradation: Experimental results with synthetic aggregates. Soil Sci. Soc. Am. J. 56:128–134.

Scribner, S.L., T.R. Benzing, S. Sun, and S.A. Boyd. 1992. Desorption and bioavailability of aged simazine residues in soil from a continuous corn field. J. Environ. Qual. 21:1115–1121.

Selim, H.M., J.M. Davidson, and R.S. Mansell. 1976. Evaluation of a two-site adsorption–desorption model for describing solute transport in soils. p. 444–448. *In* Proc. Summer Computer Simulation Conf., Washington, DC. 12–14 July 1976. Simulations Councils, La Jolla, CA.

Smith, J.A., C.T. Chiou, J.A. Kammer, and D.E. Kile. 1990. Effect of soil moisture on the sorption of trichloroethene vapor to vadose-zone soil at Picatinny Arsenal, New Jersey. Environ. Sci. Technol. 24:676–682.

Smith, L., and F.W. Schwartz. 1981. Mass transport: 2. Analysis of uncertainty in prediction. Water Resour. Res. 17(2):351–369.

Starr, R.C., R.W. Gillham, and E.A. Sudicky. 1985. Experimental investigation of solute transport in stratified porous media: 2. The reactive case. Water Resour. Res. 21:1043–1050.

Steinberg, S.M., J.J. Pignatello, and B.L. Sawhney. 1987. Persistence of 1,2-dibromoethane in soils: Entrapment in intraparticle micropores. Environ. Sci. Technol. 21:1201–1210.

Travis, C.C., and E.L. Etnier. 1981. A survey of sorption relationships for reactive solutes in soil. J. Environ. Qual. 10:8–17.

Valocchi, A.J. 1989. Spatial moment analysis of the transport of kinetically adsorbing solutes through stratified aquifers. Water Resour. Res. 25(2):273–279.

Weber, W., J. McGinley, and L.E. Katz. 1992. A distributed reactivity model for sorption by soils and sediments: 1. Conceptual basis and equilibrium assessments. Environ. Sci. Technol. 26(10):1955–1962.

Wood, W.W., T.F. Kraemer, and P.P. Hearn. 1990. Intragranular diffusion: An important mechanism influencing solute transport in clastic aquifers? Science (Washington, DC) 247:1569–1572.

Xu, L., and M.L. Brusseau. 1996. Semi-analytical solution for solute transport in porous media with multiple spatially variable reaction processes. Water Resour. Res. 32(7):1985–1991.

8 Soil–Root Interface: Biological and Biochemical Processes

Horst Marschner

Institute of Plant Nutrition
University Hohenheim
Stuttgart, Germany

8–1 INTRODUCTION

Conditions at the soil–root interface, the rhizosphere, differ in many respects from those in the bulk soil. Roots can take up ions or water preferentially, leading to the depletion or accumulation of ions in the rhizosphere. Roots also release H^+ or HCO_3^- (and CO_2) which changes the pH. Additionally, consumption or release of O_2 by roots may alter the redox potential. Low molecular weight (LMW) root exudates may mobilize sparingly soluble mineral nutrients either directly, or indirectly by providing the energy for microbial activity in the rhizosphere. For a comprehensive review, refer to Marschner (1995).

These root-induced changes lead to gradients both in radial and longitudinal direction along an individual root (Fig. 8–1). These gradients are of crucial importance for the acquisition of mineral nutrients by plants, and are determined by soil chemical and physical factors, as well as by plant factors such as species and nutritional status of the plants, and by microbial activity in the rhizosphere. In plant roots colonized by mycorrhizal fungi, the soil–root interface is ill-defined, and the biological and biochemical processes are more complex than in nonmycorrhizal roots. Mycorrhizal colonization increases the spatial availability of several mineral nutrients and might also protect roots from metal toxicity and pathogens. In this chapter, representative examples are given for various biological and biochemical processes occurring at the soil–root interface.

8–2 ROOT-INDUCED CHANGES IN THE RHIZOSPHERE

8–2.1 Nutrient Concentration

The concentration of a particular ion in the rhizosphere can be lower, similar, or higher than in the bulk soil, depending on the concentration in the bulk soil solution, the rate of delivery of the ion to the root surface, and its uptake rate by the roots. For mineral nutrients of low concentration in soil solution compared

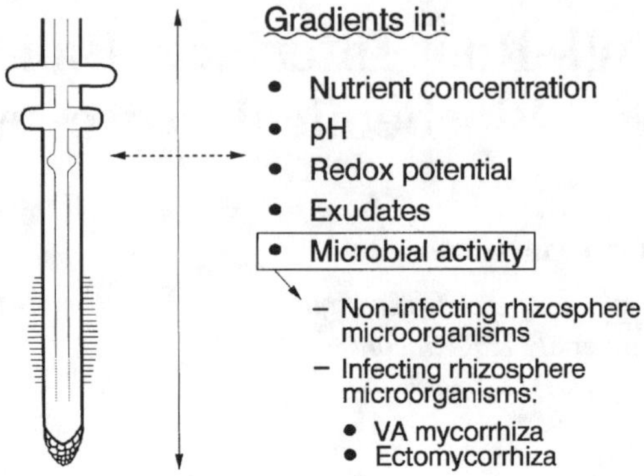

Fig. 8–1. Gradients at the root–soil interface (rhizosphere).

with plant demand such as P and K, typical depletion profiles develop around the roots (Kuchenbuch & Jungk, 1984; Jungk & Claassen, 1986). In soils low in K, depletion of K in the rhizosphere soil solution can enhance the release of interlayer K with concomitant transformation of trioctahedral mica into vermiculite, i.e., enhance weathering of clay minerals (Hinsinger & Jaillard, 1993).

On the other hand, for nutrients of relatively high concentration in soil solution compared with root uptake, accumulation in the rhizosphere has been shown, for example, for Ca and Mg in barley (*Hordeum vulgare* L.;Youssef & Chino, 1987) or to the precipitation of $CaSO_4$ in the rhizosphere of azalea (*Azalea indica* L.; Jungk, 1991). Rates of ion and water uptake also differ along the root axis that has important implications not only for the ion concentrations in the rhizosphere but also for ion competition and selectivity in uptake. For example, in nutrient solutions Mg uptake is usually markedly depressed by K. In soil-grown plants, however, preferential uptake of K and, thus, depletion of K in the rhizosphere, particularly in basal root zones, enhances Mg uptake in these root zones (Seggewiss & Jungk, 1988). Thus, different gradients of individual ions along the root axis can overcome limitations in mineral nutrient acquisition caused by competition for uptake sites.

8–2.2 pH and Redox Potential

The pH of the rhizosphere may differ from that of the bulk soil by up to two units, depending on plant and soil factors. Of the plant factors, the most important ones responsible for root-induced changes in rhizosphere pH are imbalance in cation-anion uptake ratio (C/A) and corresponding differences in net excretion of H^+ and HCO_3^- (or OH^-); excretion of organic acids and, indirectly, microbial acid production from root release of organic C (sugars in particular); and enhanced CO_2 production. As a rule, the most common of these plant factors is imbalance in cation–anion uptake ratio. The form of N supply (NH_4^+; NO_3^-; N_2 fixation) has the most prominent influence on rhizosphere pH (Table 8–1). Nitrate supply is usually correlated with a higher rate of HCO_3^- net release (or H^+ consumption) than net excretion of H^+, and with NH_4^+ supply the reverse is the case.

Table 8–1. Effect of the N form applied to a sandy loam soil (pH 6.8) on rhizosphere pH and nutrient uptake per meter root length in bean (*Phaseolus vulgaris* L.) plants (Source: Thomson et al., 1993).

N form applied	Rhizosphere pH (H_2O)	Uptake P	K	Fe	Mn	Zn	Cu
		$\mu g\ m^{-1}$ root length					
Ca $(NO_3)_2$	6.6	815	2026	68	23	11	2.7
$(NH_4)_2SO_4$†	4.5	1818	1756	184	37	21	3.7

† Nitrification inhibitor.

In neutral or alkaline soils, rhizosphere acidification in NH_4^+-fed plants can enhance mobilization of sparingly soluble Ca phosphates or micronutrients and, thus, their uptake rates per unit root length (Table 8–1). On the other hand, on acid soils the pH increase induced by NO_3^- supply enhances P uptake, most likely by HCO_3^- mediated desorption of P from Fe and Al oxides (Jungk et al., 1993).

For legumes in which N requirement is met by symbiotic N_2 fixation rather than NO_3^- nutrition, uptake of cations is much greater than anions and thus, the rhizosphere is acidified. The capacity of legumes to use P from rock phosphate, or from other sparingly soluble Ca phosphates, is therefore higher in N_2 fixing as compared with NO_3^--fed plants (Aguilar & van Diest, 1981).

Striking differences in rhizosphere pH also exist between plant species growing in the same soil and supplied with NO_3^-–N. Typically, such plant species as chickpea (*Cicer arietinum* L.), white mustard (*Sinapis alba* L.), and buckwheat (*Fagopyrum esculentum* Moench.), have a very low rhizosphere pH compared, for example, with that of wheat (*Triticum aestivum* L.), sorghum [*Sorghum bicolor* (L.) Moench.], or maize (*Zea mays* L.; Marschner & Römheld, 1983). These differences in rhizosphere pH mainly reflect differences in the cation–anion uptake ratio (Table 8–2). Accordingly, usage of sparingly soluble Ca phosphates such as rock phosphate is low in maize compared with buckwheat (Bekele et al., 1983). Rhizosphere acidification also might be achieved by root exudation of organic acids (explained later) and this plays an important role in the adaptation of wild plants to calcareous soils (Tyler & Ström, 1995).

In plants grown in, and adapted to, acid soils, the pH in apical root zones is often considerably higher than in basal zones and in the bulk soil (Fig. 8–2). Such a higher rhizosphere pH at the root tip might decrease both, the exchangeable Al

Table 8–2. Dry matter yield and cation–anion (C–A) uptake pattern in maize and buckwheat supplied with triple superphosphate (TPS) and rock phosphate (RP; Source: Bekele et al., 1983).

Plant species	P source	Dry weight	Meq g^{-1} dry matter Cations	Anions	C–A	Final soil pH
		g pot^{-1}				
Maize	TPS	30.0	943	979	−36	6.4
	RP	2.6	2350	2760	−410	6.4
Buckwheat	TPS	31.0	2058	1213	+845	4.9
	RP	27.0	2467	1486	+979	5.1

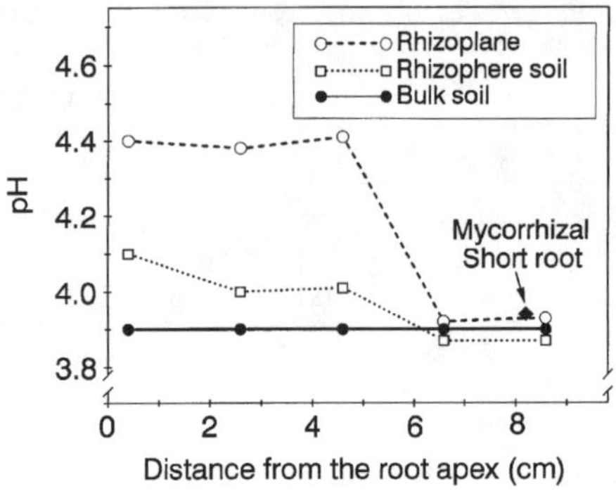

Fig. 8–2. pH pattern in bulk soil and in the rhizosphere along the axes of nonmycorrhizal roots of 80-yr-old Norway spruce in September of 1986 near Heidelberg (Marschner, 1991).

and the release of Al from the solid phase into the rhizo–soil solution and thereby contribute to detoxification of Al at the root tip.

Even in aerated soils, anaerobic microsites do occur and they vary in location and time. Such microsites are most likely much more abundant in the rhizosphere than in the bulk soil (Fischer et al., 1989) and are particularly important for the acquisition of Mn and for gaseous losses of N (N_2; N_2O, and others). As O_2 consumption (roots plus rhizosphere microorganisms) is higher in the rhizosphere compared with the bulk soil, in poorly aerated soils the risk of losses of N by denitrification—or incomplete nitrification—is higher in planted than in unplanted soil (Prade & Trolldenier, 1989) and increases as the organic C input from roots into the rhizosphere increases (Bakken, 1988).

In plants adapted to waterlogging and to submerged soils (e.g., lowland rice), high redox potentials in the rhizosphere are maintained by O_2 transport from the shoot through aerenchyma to the roots and release of O_2 into the rhizosphere. This oxidation of the rhizosphere is essential to decrease the phytotoxic concentrations of organic solutes (e.g., volatile monocarboxylic acids) and Fe^{2+}, Mn^{2+}, and H_2S present in the bulk soil solution. Root-induced oxidation of Fe^{2+} (and Mn^{2+}) is associated with production of H^+ and, thus, decrease in rhizosphere pH (Fig. 8–3). As NH_4^+ is the dominant form of N taken up by roots in submerged soils, the corresponding increase in net excretion of H^+ in response to the excess uptake of cations compared with anions contributes to the rhizosphere acidification. This acidification enables the plants to cover most of the P demand from the acid soluble fraction of soil-P (Saleque & Kirk, 1995) and also may enhance net uptake of CO_2 by the roots (Begg et al., 1994). In wetland species, CO_2 transport from the roots through the aerenchyma into the leaves might serve as an endogenous source of CO_2 for photosynthesis (Constable & Longstreth, 1994).

In submerged soils, at a very low redox potential large amounts of methane (CH_4) are formed, for example, from acetic acid. Wetland fields are a major source of CH_4 emissions and of concern in relation to global warming. Methane emissions are higher in planted than in unplanted paddy fields, as root exudates increase the CH_4 production (Kimura et al., 1991), and the aerenchyma in rice

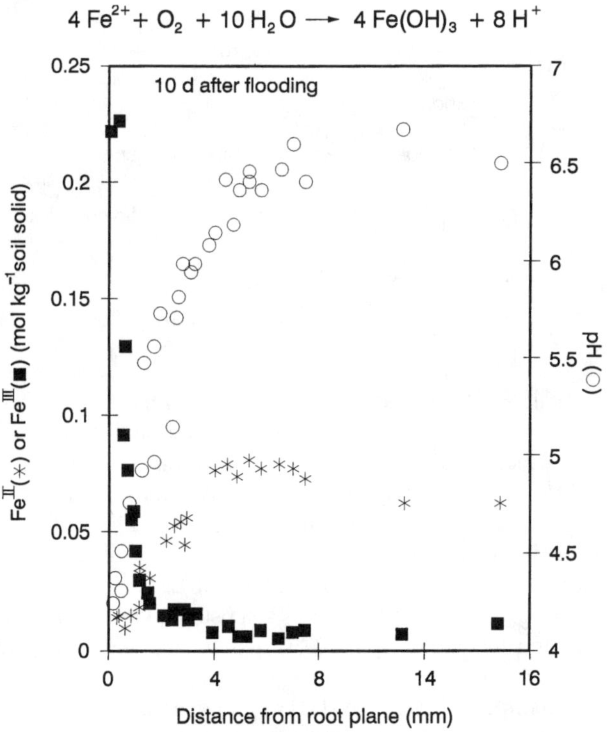

Fig. 8–3. Root-induced Fe oxidation and pH changes in the rhizosphere of lowland rice (Begg et al., 1994).

(*Oryza sativa* L.) roots enhances its transport into the shoot and subsequent release into the atmosphere (Nouchi et al., 1990). On the other hand, wetland grass species such as salt meadow cordgrass [*Spartina patens* (Aiton) Muhlenb.], a perennial species of salt marshes, may decrease rather than enhance CH_4 emissions, presumably due to decomposition of CH_4 by microbial methane monooxygenase in the rhizosphere (Kludze & DeLaune, 1994).

8–3 RHIZODEPOSITION AND ROOT EXUDATES

8–3.1 General

On average, in annual species between 30 and 60% of the net photosynthetic C is allocated to the roots, and of this C an appreciable portion (4–70%) is released as organic C (**rhizodeposition**) into the rhizosphere (Lynch & Whipps, 1990; Liljeroth et al., 1994). In soil-grown crop plant species, over the whole vegetation period more than twice as much organic C was released into the rhizosphere (rhizodeposition) as was retained in the root system at harvest (Sauerbeck et al., 1981). The major components of rhizodeposition are low and high molec-

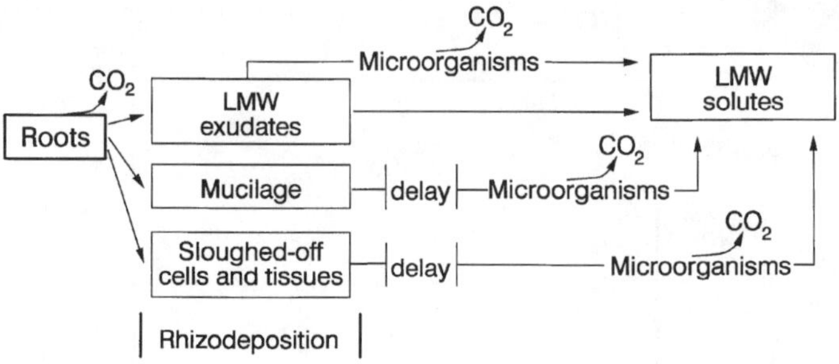

Fig. 8–4. Model of C flux in the rhizosphere; LMW, low molecular weight (modified from Warembourg & Billes, 1979).

ular weight organic solutes and sloughed-off cells and tissues (Fig. 8–4). The low molecular weight (LMW) exudates include organic acids, phenolics and phytosiderophores, which are directly able to mobilize mineral nutrients or detoxify metals (e.g., Al) in the rhizosphere. These LMW exudates may, however, also mobilize heavy metals such as Pb and Cd (Mench et al., 1988) thus contributing substantially to differences in heavy metal uptake by different plant species (Mench & Martin, 1991). The total root release of LMW solutes might be higher than usually measured as net release as a substantial proportion of sugars and particularly amino acids are recaptured by an active uptake process (Jones et al., 1994).

Of the high molecular weight (HMW) root exudates, mucilage, which consists mainly of polysaccharides such as polygalacturonic acids, may bind polyvalent metal cations and thereby protect root apical meristems from Al toxicity (Horst et al., 1990; Marschner, 1991) or heavy metal toxicity (Morel et al., 1986).

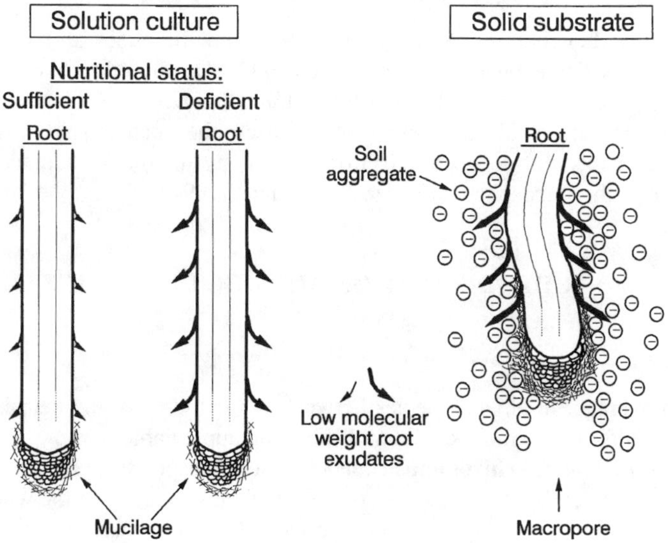

Fig. 8–5. Schematic presentation of root exudation as affected by mineral nutrient deficiency and mechanical impedance (Marschner, 1995).

Table 8–3. Influence of mechanical impedance (MI) on growth and root exudation of axenic maize at the 5 to 6 leaf stage (Source: Boeuf-Tremblay et al., 1995).

Treatment†	Leaf area	Dry weight		Root exudates‡
		Shoot	Root	
	cm^2	g		$mg\ g^{-1}$ root dry weight
−MI	246	1.17	0.33	26.2
+MI	133	0.59	0.36	82.1

† −MI, nutrient solution; +MI, glass beads + nutrient solution.
‡ Soluble plus insoluble.

Sloughed-off cells and tissues are primarily a C substrate for rhizosphere microorganisms, but can become effective for mineral nutrient mobilization and detoxifyers of metal cations as metabolites of microbial activity (Fig. 8–4). Rhizodeposition includes mineral nutrients previously taken up by the plant. In wheat, over the whole growing period, between 18 and 33% of the total N in the plants is transferred to the rhizosphere (Janzen, 1990).

Rhizodeposition is increased by various forms of stress such as anaerobiosis, drought, mechanical impedance, and mineral nutrient deficiency (Fig. 8–5). Some examples of this are given in the following sections.

8–3.2 Soil Physical and Chemical Factors, Rhizosphere Microorganisms

Root exudation is much higher from roots growing in solid substrates like quartz sand, glass beads, or soils than those in nutrient solution (Table 8–3). Shoot dry weight is often more depressed than root dry weight, but root length is usually strongly negatively related to the substrate bulk density. Accordingly, allocation of photosynthates to roots is absolutely or relatively enhanced by mechanical impedance and the consumption of photosynthates for respiration and exudation per unit root length might increase by a factor of two in soils with high as compared with low bulk density (Sauerbeck & Helal, 1986).

Increased root exudation as a result of mechanical impedance has important implications not only for the dynamics of mineral nutrients and heavy metals in the rhizosphere but also for the ability of plants to tolerate high Al concentrations (Table 8–4). Whereas in nutrient solution a concentration of only 74 μM Al great-

Table 8–4. Influence of Al on root length and mineral element concentration in apical (0–5 mm) root zones of soybean grown in nutrient solution or in sand culture percolated with nutrient solution (Source: Horst et al., 1990).

Substrate	Root length	Mineral element concentration in root tips		
		Al	Ca	Mg
	cm $plant^{-1}$	$mg\ g^{-1}$ dry weight		
Nutrient solution				
Control (−Al)	189	<0.1	0.69	1.37
+74 μM Al	39	3.9	0.36	0.47
Sand culture				
Control (−Al)	114	<0.1	1.56	1.39
+741 μM Al	50	0.9	1.22	1.02

ly inhibited root elongation, the same concentration in the percolating nutrient solution in sand culture was without inhibitory effect (Horst et al., 1990). Even increasing the Al concentration about 10-fold (741 μ*M*), in plants grown in sand culture the inhibitory effects on root elongation were less severe than those at 74 μ*M* in water culture. As indicated by the mineral element concentration of the root tips, suppression of Al uptake and the correspondingly lesser depression in Ca and Mg concentration in the root tips were presumably the responsible factors for the higher Al tolerance of the roots growing in the solid substrate. This effect was most likely brought about by higher root exudation and a corresponding decrease in the concentration of the toxic monomeric Al species in the rhizosphere.

High root exudation rates of HMW or LMW solutes, either inherent or in response to Al stress, are key factors for higher Al tolerance of certain plant species or genotypes of a species. In natural grassland in South Africa, on acid soils the dominance of the unpalatable grass [*Aristida junciformis* Trin et Rupr.] is most likely related to its high Al tolerance brought about by an unusual high production of root cap mucilage (Johnson & Bennet, 1991). The importance of mucilage for Al tolerance has been shown in wheat genotypes where a tolerant genotype produced several times more mucilage than a sensitive genotype (Puthota et al., 1991). In wheat genotypes a close relationship also has been established between Al tolerance and excretion of organic acids. In tolerant genotypes excretion of malic acid was enhanced 5 to 10 times by as little as 10 μ*M* Al and continued linearly during 24 h (Fig. 8–6). These enhanced exudation rates of organic acids were confined to the apical 3 to 4 mm (Delhaize et al., 1993).

Rhizodeposition of HMW and LMW solutes supports a higher microbial population in the rhizosphere compared with the bulk soil, and different plant species bear a different rhizosphere microflora, both in number and physiological

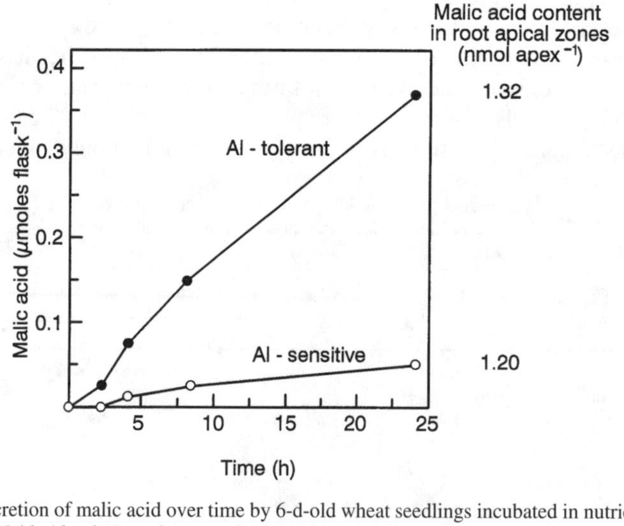

Fig. 8–6. Excretion of malic acid over time by 6-d-old wheat seedlings incubated in nutrient solutions with 50 μ*M* Al. Aluminum-tolerant and Al-sensitive near-isogenic lines (Delhaize et al., 1993).

Table 8–5. Partitioning of ^{14}C after 48h pulse labeling with $^{14}CO_2$ of shoots of perennial ryegrass (*Lolium perenne* L.) seedlings inoculated with different microorganisms (Source: Meharg & Killham, 1995).

	^{14}C		
Microorganisms	Root exudates	Roots	Shoots
	%		
Uninoculated	1.0	27.8	71.2
Aspergillus niger	3.4	29.3	67.3
Penicillium rubium	9.5	20.3	70.3
Fusarium oxysporum	25.3	19.2	55.5
Penicillium notatum	33.8	27.3	39.8

characteristics (Kloepper et al., 1991). This also is true for a given plant species for different zones of roots, for example, for sheathed and bare root zones of C_4 grasses (Gochnauer et al., 1989).

Rhizosphere microorganisms also increase rhizodeposition, particularly the LMW fraction (Schönwitz & Ziegler, 1982). The extent of enhanced root exudation strongly depends on the species or strains of microorganisms (Table 8–5) and may vary by a factor of 3 to 33. This enhancing effect of rhizosphere microorganisms on LMW root exudates may be attributed to lowering the proportion of reabsorption (retrieval) by the roots (Jones & Darrah, 1993) or an increase in efflux rates via toxin production (Meharg & Killham, 1995), or both these factors.

In fast growing roots there is usually a steep gradient of rhizoplane and rhizosphere microorganisms from apical to basal zones along the root axis (Fig. 8–7). In maize, for example, of the total root surface area, bacteria surface cover is about 4% in apical zones, 7% in the root hair zone, and might rise up to 20% in basal zones (Schönwitz & Ziegler, 1986). This gradient of microbial population along the root axis has important implications for the efficiency of root exudates released in response to deficiency of mineral nutrients (explained later).

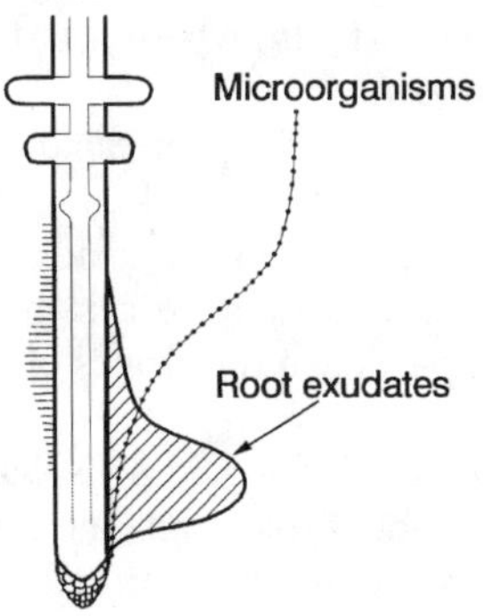

Fig. 8–7. Schematic presentation of spatial separation of low molecular weight (LMW) root exudates (e.g., phytosiderophores, organic acids) and microbial activity in the rhizosphere of soil-grown plants.

Table 8–6. Effect of Zn nutritional status of cotton and wheat plants on root exudation and Zn mobilization from a Zn-loaded resin by root exudates (Source: Marschner et al., 1990).

Zinc nutritional status (Zn supply)	Root exudates			Zn mobilization by root exudates
	Amino acids	Sugars	Phenolics	
	µg g^{-1} root dry weight h^{-1}			µmol g^{-1} root dry weight 4 h^{-1}
Cotton: +Zn	32	250	78	0.6
−Zn	110	500	107	0.6
Wheat: +Zn	21	315	34	0.4
−Zn	48	615	80	4.9

8–3.3 Nutritional Status of Plants

When plants are nutrient deficient, the amounts of LMW root exudates often increase and the composition of these exudates is altered, both are often of key importance for enhanced mobilization of mineral nutrients in the rhizosphere. Under K deficiency in maize the amounts of exudates increase and the proportion of sugars of organic acids is shifted in favour of organic acids (Kraffczyk et al.,1984). Under Zn deficiency in both dicots and monocots (grasses), the amounts of amino acids, sugars, and phenolics in root exudates increase, but the increase in Zn-mobilizing root exudates is confined to grasses such as wheat (Table 8–6).

Enhanced root exudation of organic acids is often observed under P deficiency in dicots and in legumes in particular. In alfalfa (*Medicago sativa* L.) even under latent P deficiency when the total dry weight has not yet been depressed, root exudation of citric acid increases about two-fold (Lipton et al., 1987). Legume species respond quite differently to P deficiency in terms of increase in root exudation of organic acids. For example, exudation is very high in chickpea and peanut (*Arachis hypogaea* L.), but low in soybean [*Glycine max* (L.) Merr.] and moderate in pigeon pea [*Cajanus cajan* (L.) Huth; Ohwaki & Hirata, 1992). The relatively low exudation of organic acids in pigeon pea is in apparent contrast to the outstanding capacity of this species for P acquisition from Alfisols (Ae et al., 1990). This high capacity of pigeon pea to use P from sparingly soluble Fe phosphates is probably related to the presence of a particular organic acid in root exudates, namely piscidic acid (*p*-hydroxybenzyl tartaric acid), which is a strong chelator for Fe^{III} (Ae et al., 1990).

In rape (*Brassica napus* L.), rhizosphere acidification in P deficient plants might be related to a high cation-anion uptake ratio (i.e., H^+ net excretion), as well as to enhanced net excretion of organic acids (Table 8–7). This enhanced acid excretion is confined to the apical root zones and coincides with higher contents of malate and citrate in the apical root zones of P deficient plants (Hoffland et al., 1992).

In P deficient roots, the activity of acid phosphatases also is increased (Fig. 8–8). Acid phosphatases are ectoenzymes and their activity is particularly high in apical root zones (Dinkelaker & Marschner, 1992), mainly associated with mucilage (Eltrop, 1993). In view of the high proportion of organic P (P_{org}) in most soils and also in the bulk soil solution of P deficient soils (Shand et al., 1994; Macklon et al., 1994), enhanced hydrolysis of P_{org} in the rhizosphere and at the

Table 8–7. Organic acids in exudates from different root zones of rape *(Brassica napus* L.) plants grown for 7 d without or with P (Source: Hoffland et al., 1989).

Phosphorus supply	Root zone	Organic acids in exudates: Malic	Citric
		nmol cm^{-1} root 2 h^{-1}	
−P	apical	0.87	0.27
	basal	0.20	0.13
+P	apical	0.15	0.06
	basal	0.03	0.03

rhizoplane by root-borne acid phosphatases might play an important role in P acquisition of plants from P deficient soils.

Under Fe deficiency, root responses differ between plant species in a distinct manner and can be classified into Strategy I and Strategy II (Römheld & Marschner, 1990). Strategy I is typical for dicots and nongraminaceous monocots and characterized by increased reducing capacity of roots and enhanced net excretion of H^+ and in many cases also enhanced release of reducing or chelating compounds, mainly phenolics (Fig. 8–9). These root responses are confined to

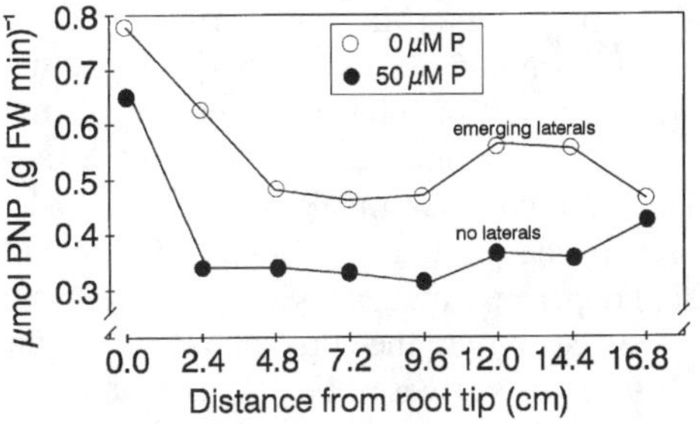

Fig. 8–8. Surface phosphatase activity (*p*-NPPase) along a white, nonsuberized long root of Norway spruce seedlings from nutrient solution culture and with two levels of P supply (Eltrop, 1993).

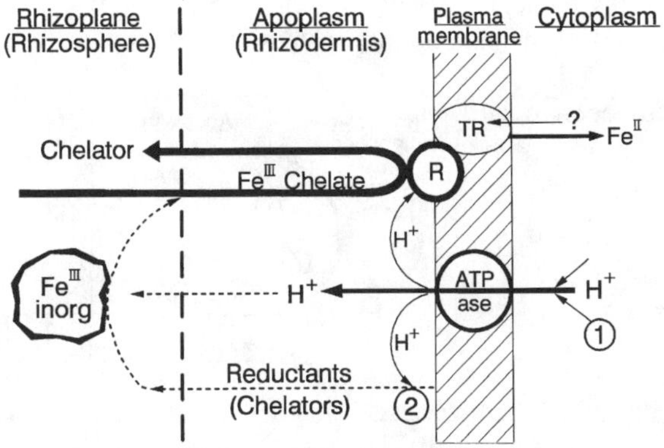

Fig. 8–9. Model for root responses to Fe deficiency in dicots and nongraminaceous monocots; Strategy I: (R) inducible reductase; [TR] transporter or channel for Fe^{II}?; (1) stimulated proton efflux pump; (2) increased release of reductants–chelators (modified from Marschner et al., 1986; Römheld, 1987).

Table 8–8. Effect of Fe deficiency in cucumber (Strategy 1) on proton excretion (pH) reducing capacity of the roots and Fe uptake rate (Source: Römheld & Kramer, 1983; Römheld & Marschner, 1990).

Fe nutritional status (Preculture)	Chlorophyll	H^+ excretion (solution pH)	Reducing capacity	Fe uptake
	mg m^{-1} dry weight		μmol Fe (II) g^{-1} root dry weight 4 h^{-1}	μmol g^{-1} root dry weight 4 h^{-1}
+Fe†	12.2	6.2	3.2	0.03
–Fe	7.8	4.8	96.8	2.6

† Supply of 1 × 10^{-6} *M* FeEDDHA, pH 6.2.

the apical zones of growing roots and repressed within 1 or 2 d after resupply of Fe. An example for the Fe deficiency-induced enhancement of H^+ excretion and of reducing capacity of roots, and the corresponding increase in uptake rates of Fe are shown in Table 8–8.

Strategy II is confined to graminaceous species (grasses) and characterized by an Fe deficiency-induced enhanced release of nonproteinogenic amino acids, so called phytosiderophores (Takagi et al., 1984). The release is confined to apical root zones, and in most plant species follows a distinct diurnal rhythm, and is rapidly repressed by resupply of Fe (Marschner, 1995).

Phytosiderophores form complexes of high stability with Fe^{III}, and these complexes are transferred across the plasma membrane of root cells by a highly specific, constitutive transporter into the cytoplasm of Strategy II plants (Fig. 8–10). Strategy I plants not only lack the release of phytosiderophores, but also the specific transporter in the plasma membrane. Phytosiderophores also form complexes with zinc, copper and manganese, but the transporter in the plasma membrane has only a low affinity to the corresponding complexes (Marschner et al., 1989; Ma et al., 1993). The sources of Fe^{III} for formation of Fe^{III} PS are either sparingly soluble inorganic Fe hydroxides in the rhizosphere and at the rhizoplane, or soluble microbial and synthetic Fe^{III} chelates (Fe^{III}-L) and corresponding ligand exchange $Fe^{III}\text{-L} + PS \rightarrow Fe^{III}\,PS + L$ (Yehuda et al., 1996).

Although this phytosiderophore system in Strategy II plants has features resembling those of of the siderophore system in microorganisms, its affinity to

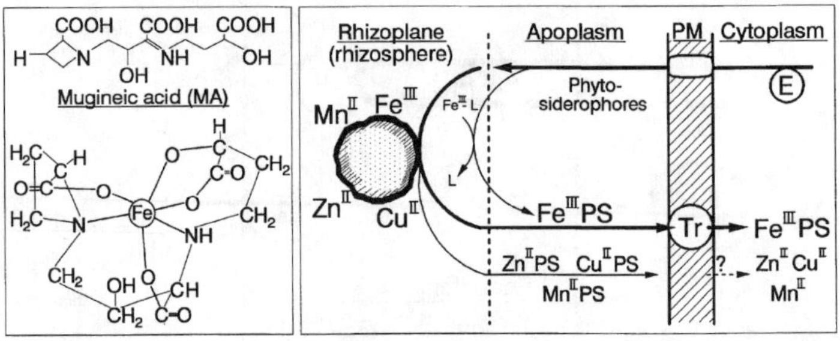

Fig. 8–10. Model for root responses to Fe deficiency in graminaceous species. Strategy II: (E) enhanced synthesis and release of phytosiderophores; (TR) translocator for Fe^{III} phytosiderophores in the plasma membrane; structure of the phytosiderophore mugineic acid and its corresponding Fe^{III} chelate (Fe^{III} PS); Fe^{III}-L = synthetic or microbial Fe^{III} chelate (modified from Marschner, 1995).

Table 8–9. Iron mobilization from a calcareous soil by Fe^{3+} chelators and uptake rates of Fe^{2+} supplied as $^{59}Fe^{3+}$ chelates by Fe deficient barley plants (Source: Römheld & Marschner, 1990).

Chelator 10^{-5} *M*	Mobilization	Uptake rate
	nmol g^{-1} soil 12 h^{-1}	nmol g^{-1} root dry weight 4 h^{-1}
Phytosiderophore (HMA)	23.6	3456.00
Microbial (Desferal)	19.2	1.21

phytosiderophores is orders of magnitude higher than for siderophores with a similarly high complex stability such as desferal or ferrioxiamine B (Crowley et al., 1992; Marschner & Römheld, 1994; see also Table 8–9).

In graminaceous species, under Zn deficiency release of phytosiderophores also is enhanced (Zhang et al., 1989; Walter et al., 1994). It is not yet clear whether this indicates a common control mechanism of the biosynthetic pathway by both, Fe and Zn, or a Zn deficiency-induced impairment of Fe metabolism at a cellular level leading to the Fe deficiency-root responses in Strategy II plants (Cakmak et al., 1994).

8–4 ROLE OF ROOT EXUDATES FOR MINERAL NUTRIENT ACQUISTITION

8–4.1 General

Low molecular weight root exudates such as organic acids or phytosiderophores can readily be collected and quantified in nutrient solution experiments. As previously mentioned, however, root exudation is favored by the presence of solid substrates. Factors such as a mechanical impedance, bulk soil density, confinement of release to certain root zones, enhancement of exudation by rhizosphere microorganisms, as well as utilization of root exudates such as phytosiderophores as C source for rhizosphere microorganisms (von Wirén et al., 1995) have to be considered. The proposed importance of root exudates for the mineral acquisition of soil-grown plants has therefore to be confirmed. In the following section examples of experiments are presented that show this importance.

8–4.2 Enhancement of Mineral Nutrient Mobilization in the Rhizosphere

A high local exudation rate of organic acids or phytosiderophores, as well as H^+, has ecological advantages. In well-buffered soils a pH decline can only be achieved by high flux densities of H^+ and organic acids. Furthermore, sites of localized low pH in the rhizosphere may inhibit bacterial growth and thereby prevent, or at least restrict, microbial degradation of exudates (Hoffland et al., 1989). The confinement of enhanced release of phytosiderophores to only a few hours per day, and to apical root zones with the typically lower microbial activity is another means to ensure high efficiency in mobilization of sparingly soluble mineral nutrients by root exudates (Römheld, 1991).

Table 8–10. Soil pH and concentrations of citrate and micronnutrients in bulk and rhizosphere soil of lupine (*Lupinus albus* L.) grown in a P deficient soil (23% $CaCO_3$; Source: Dinkelaker et al., 1989).

	Bulk soil	Rhizosphere soil (Proteoid roots)
pH (H_2O)	7.5	4.8
Citrate, μmol g^{-1} soil	<0.05	47.7
DTPA extractable, μmol kg^{-1}		
Fe	34	251
Mn	44	222
Zn	2.8	16.8

This principle of localized high exudation rates is almost perfectly obtained in plants forming root clusters, such as members of the family Proteaceae, tree and shrub species widely distributed on nutrient-poor soils in Mediterranean climates of Australia and South Africa (Dinkelaker et al., 1995). In annual legume species too such as white lupin (*Lupinus albus* L.) formation of root clusters (proteoid roots) is common. Root clusters are bottlebrush-like rootlets with determinate growth that form along lateral roots. They are induced primarily under P deficiency. In white lupin, citric acid is the dominant solute in the exudate of the root clusters. Within these root clusters, an intensive chemical extraction of a limited soil volume is made possible by the root exudates that would otherwise diffuse into a larger soil volume with corresponding dilution. In natural stands, members of the Proteaceae such as banksia (*Banksia integrifolia* L.) form a dense mat of root clusters beneath the litter layer, and in this mat the concentration of H^+, reductants and chelating compounds (mainly organic acids) is much higher than in the bulk soil (Grierson & Attiwill, 1989).

In white lupin, the high local citric acid excretion acidifies the rhizosphere pH even in calcareous soils (Table 8–10) and mobilizes sparingly soluble Ca phosphate by dissolution and subsequent formation of sparingly soluble Ca citrate in the rhizosphere (Dinkelaker et al., 1989). In the rhizo–soil solution of the root clusters the concentration of P, Fe, and Al is increased (Gerke et al., 1994a) as well as in the rhizosphere soil the extractable Fe, Mn, and Zn (Table 8–10). When white lupins and other root cluster-forming plant species are supplied with soluble P, root cluster formation is suppressed and, accordingly, also citric acid excretion. Plant concentrations of P can remain similar but those of Mn and Zn can decrease drastically (Marschner, 1995) and Fe deficiency chlorosis may be induced as is the case in the Proteaceae banksia (*Banksia ericifolia* L.; Handreck, 1991).

When grown in mixed culture in a P deficient soil supplied with rock phosphate, wheat can profit from the P mobilized in the rhizosphere of white lupin, provided the root systems of both plant species can intertwine (Horst & Waschkies, 1987). Thus, white lupin is able not only to mobilize sparingly soluble phosphates for its own demand but also provide additional P for other plant species grown in mixed culture.

The amount of citric acid released into the rhizosphere of white lupin (Table 8–10) accounted for 1 g plant^{-1}, or 23% of the net photosynthetic C after 13 wk of growth. This appears to be a high cost for P acquisition; however, in view of the benefits (mobilization not only of P but also other mineral nutrients)

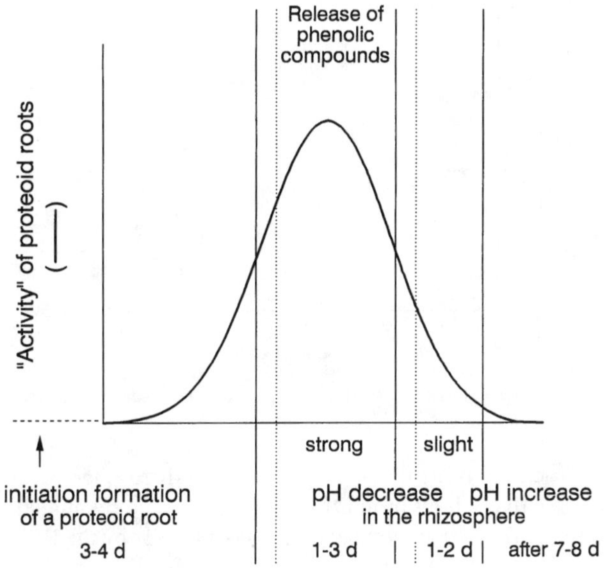

Fig. 8–11. Schematic presentation of the time-course of the *activity* of a proteoid root in *Hakea undulata* as reflected by rhizosphere acidification and release rate of phenolic compounds. (Dinkelaker et al., 1995).

and the costs of alternative strategies (increase in root surface area, or mycorrhizal associations (discussed later), this strategy of plants forming root clusters and excreting large amounts of organic acids seems to be quite efficient. Furthermore, citric acid excretion of white lupin in acid soils (e.g., Oxisols), mobilizes P also from sparingly soluble Fe and Al phosphates by chelation of Fe and Al and by ligand exchange (Gerke et al., 1994b).

In Proteaceae the activity of proteoid roots is confined to only a few days after initiation (Fig. 8–11). In hakea (*Hakea undulata* R. Br.), malic acid was the dominant organic acid excreted by the proteoid roots (data not shown), and this excretion was associated with enhanced release of phenolics and a decrease in rhizosphere pH (Dinkelaker et al., 1995). With such an intensive extraction of a small soil volume the depletion of mineral nutrients must be quite fast and, thus, only a short-term activity of these roots is needed. The endogenous mechanism of regulation of this short-term root exudation, however, is not yet understood. In white lupin the enhanced exudation of citrate (Table 8–10) is associated with higher citrate biosynthesis, PEP carboxylase activity and CO_2 dark fixation in the proteoid roots (Johnson et al., 1994).

The activity of acid phosphatases is much higher in the rhizosphere than in the bulk soil (Fig. 8–12). This gradient in enzyme activity can derive from root-borne phosphatases (ectoenzymes) and rhizosphere microorganisms. The differences in acid phosphatase activity between the three plant species can therefore be the result of different population densities of rhizosphere microorganisms, as well as differences in the P nutritional status of the plants (Section 3.3).

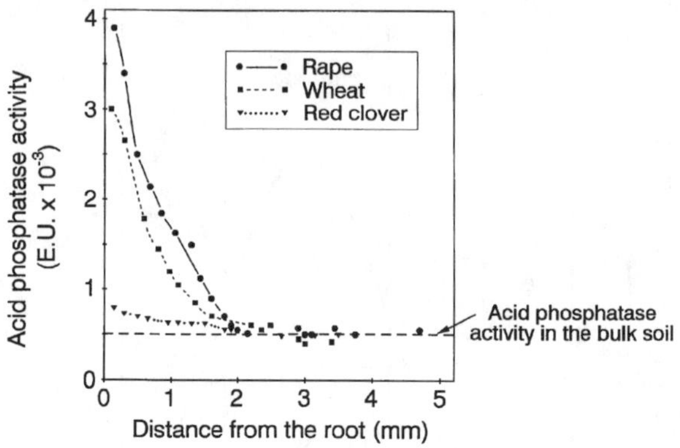

Fig. 8–12. Acid phosphatase activity in the rhizosphere of different plant species grown in a silt loam soil (Tarafdar & Jungk, 1987).

The high activity of acid phosphatases in the rhizosphere is in good agreement with the relatively high proportion of P_{org} taken up by annual crop species by depletion of the rhizosphere that can be accounted for by the P_{org} fraction, particularly in low P soils. For the depletion of the P_{org} fraction, values of about 50% have been obtained (Helal & Sauerbeck, 1989) or even more (Tarafdar & Jungk, 1987). In the rhizosphere of Norway spruce [*Picea abies* (L.) Karst.; Fig. 8–13] the depletion of P was even confined to the P_{org} fraction and closely correlated with the acid phosphatase activity. In this case of ectomycorrhizal tree root systems (Section 5.2) the phosphatase activity might derive from both, roots and fungal mycelium.

The ecological importance of the Fe deficiency induced root responses (Strategy I and II) for Fe acquisition from calcareous soils and for genotypic differences in sensitivity to *lime-induced chlorosis* is well established (for a review see Marschner & Römheld, 1994). In Strategy I plants a close positive correlation exists between the extent to which Fe deficiency induces enhanced reducing capacity of roots and net excretion of H^+ on the one hand, and the resistance of plants to Fe deficiency on calcareous soils on the other. This is true for different

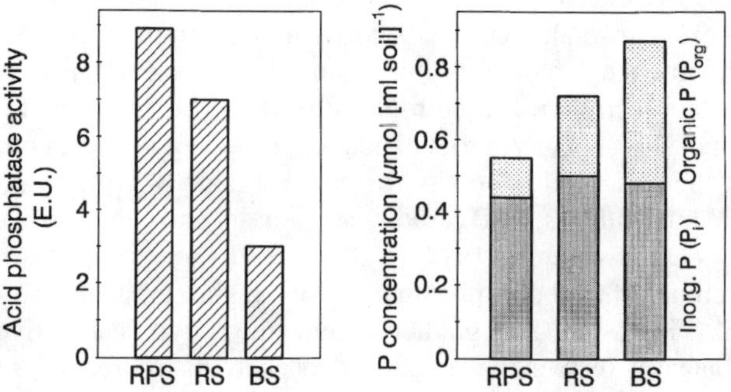

Fig. 8–13. Acid phosphatase activity and P concentration (P_i; P_{org}) in the H_2O extractable fraction of rhizoplane (RPS) and rhizosphere (RS) soil or bulk soil (BS) in 80-yr-old Norway spruce. (Häussling & Marschner, 1989).

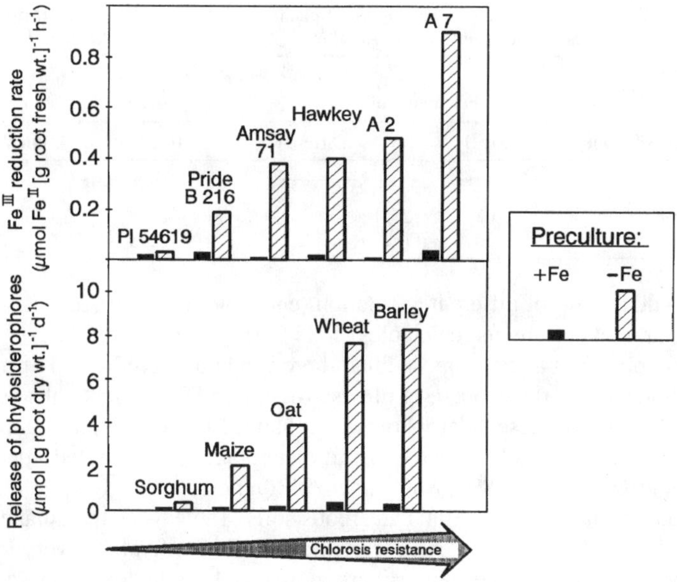

Fig. 8–14. Relationship between resistance against *lime-induced* chlorosis under field conditions and reducing capacity of roots of soybean cultivars (upper) and release of phytosiderophores of graminaceous species (lower) under controlled conditions (compiled from Römheld, 1987; Römheld & Marschner, 1990).

plant species as well as genotypes within a given species such as soybean (Fig. 8–14). Regardless of plant species and genotype, the effectivity of root responses in Fe acquisition in Strategy I plants is markedly depressed by the pH buffering effect of high HCO_3^- concentrations in the rhizosphere (Chaney, 1984; Romera et al., 1991). Accordingly, poor soil aeration combined with low soil temperature is a major factor responsible for Fe deficiency chlorosis on poorly aerated calcareous soils.

In Strategy II plants, enhanced release of phytosiderophores under Fe deficiency is much less affected by the rhizosphere pH, but the efficiency of the mechanism in Fe acquisition in soil-grown plants has been questioned because phytosiderophores might either form complexes with other micronutrients, or be metabolized by rhizosphere microorganisms or both (Crowley et al., 1987; Darrah, 1993); however, as shown in Table 8–11, under nonaxenic conditions, Fe deficient wheat plants also mobilize much more Fe in the rhizosphere of a calcareous soil and also take up much more Fe although a direct root-soil contact was prevented by confining the root systems to nylon bags filled with quartz sand.

The ecological importance of Strategy II also is reflected in a close positive correlation between the extent of enhanced release of phytosiderophores under Fe deficiency and resistance of plants to Fe deficiency chlorosis on calcareous soils (Fig. 8–14). Similar positive correlations have been found between the Fe deficiency-induced enhanced release of phytosiderophores and the natural distribu-

Table 8–11. Effect of Fe nutritional status of wheat plants on Fe mobilization in the rhizosphere and Fe uptake from a calcareous soil (Source: Awad et al., 1994).

	Fe content of plants			DTPA extractable Fe distance from rhizoplane		
Preculture of plants	Day 0	Day 4	Difference	0–2 mm	2–4 mm	4–6 mm
	μg pot^{-1}			mg kg^{-1} soil		
+Fe	205	255	+50	2.69	2.18	2.05
–Fe	50	182	+132	3.38	2.69	1.95

tion of wild grasses of different adaptation (calcicoles or calcifuges) on acid (calcifuges) and on calcareous (calcicoles) soils (Gries & Runge, 1992).

Although it may be supposed that there is a higher production of microbial siderophores in the rhizosphere of plants growing in Fe deficient calcareous soil, the contribution of these siderophores to enhanced Fe acquisition is most likely of minor importance, at least in terms of direct uptake of the ferrated siderophores (Alexander & Zuberer, 1993; Marschner & Römheld, 1994). Despite their similar capacity to mobilize Fe from calcareous soils (Table 8–9), the uptake rates of Fe from Fe^{III} siderophores such as ferrioxiamine B (Desferal) are very low compared with the plant-borne phytosiderophores such as hydroxy-mugineic acid. Nevertheless, siderophores, or siderophore-producing microorganisms in the rhizosphere might provide higher levels of freshly precipitated amorphous Fe hydroxides at the rhizoplane and thus, in the long run, serve as a readily accessible source of Fe for root exudates (Marschner & Römheld, 1994). In addition, in Strategy II plants, Fe^{III} siderophores might also serve as direct source of Fe via ligand exchange (Fig. 8–10).

There is increasing evidence that enhanced release of phytosiderophores of the mugineic acid family that also occurs under Zn deficiency in Strategy II plants (Section 3.3) plays an important role in genotypical differences in Zn efficiency when grown on Zn deficient soils (Table 8–12). The higher Zn efficiency of Aroona compared with Durati when grown on calcareous soils was related to higher Zn deficiency-induced release of phytosiderophores in Aroona compared with Durati when grown in nutrient solutions. The general importance of this Zn deficiency-induced enhanced release of phytosiderophores for efficiency of Zn acquisition from calcareous soils has now been established for a range of graminaceous species (Cakmak et al., 1994). Within the same species (e.g., bread wheat), however, phytosiderophore release of different cultivars does not relate well with zinc efficiency (Erenoglu et al., 1996).

Table 8–12. Grain yield and phytosiderophore (PS) release of two wheat cultivars in response to Zn deficiency [Source: field experiments on calcareous soils (Graham et al., 1992) and from laboratory studies with 18-d-old plants (Cakmak et al., 1994).

	Grain yield		PS release	
Cultivar	–Zn	+Zn	–Zn	+Zn
	t ha^{-1}		μmol per 60 plants 4 h^{-1}	
Aroona	1.21	1.42	6.9	0.5
Durati	0.45	1.12	1.8	0.5

As shown in this section, enhanced root exudation, for example, of organic acids or phytosiderophores under deficiency of a given mineral nutrient (e.g., P or Fe) is a mechanism of limited specificity in terms of mobilization of mineral nutrients in the rhizosphere. This limited specificity might be of ecological advantage under many conditions (e.g., low availability of both, Fe and Zn and calcareous soils) but might have negative side effects under others. A representative example of negative side effects is the risk of Mn toxicity in Strategy I plants grown on calcareous soils. In flax, the Mn concentration in the shoots was poorly related to the extractable Mn in soils, but inversely related to the extractable Fe (Moraghan & Freeman, 1978). In calcareous soils low in Fe but high in extractable Mn, Mn toxicity occurred in flax and could be prevented by application of Fe chelates to the soil (Moraghan, 1979). In white lupin, Mn acquisition and shoot concentrations are primarily related to the formation of root clusters and, thus, P supply. In P deficient soils therefore Mn toxicity might occur in white lupin (Moraghan, 1991).

8–4.3 Other Mobilizing Effects of Root Exudates

The enhanced mobilization of sparingly soluble mineral nutrients by root exudates is, as a rule, beneficial for plant growth and yield and also can improve plant quality, at least in vegetative plant organs. Negative side effects as described above, however, or the losses of N by denitrification in the rhizosphere, deserve attention. Root exudates also might affect the soil-plant transfer of organic chemicals such as polychlorinated dibenzo-p-dioxins (PCDD) and dibenzo-furans (PCDF). These extremly hydrophobic substances are firmly bound particularly to the organic soil matrix, and uptake by roots and transport to shoots is usually extremely small. Accordingly, shoot concentrations of various plant species are usually below 1 ng I-TEq kg^{-1} dry matter, even when grown on soils with up to 6000 ng I-TEq kg^{-1} (Hülster & Marschner, 1993). Exceptional high PCDD–PCDF concentrations have been found, however, in leaves and fruits of zucchini (*Cucurbita pepo* L.). When grown on soils with 148 ng I-TEq kg^{-1}, the concentrations in fruits were about 20 ng I-TEq kg^{-1} dry weight, in both the peel and inner parts of fruits, and about 100 times higher than in the inner parts (or flesh) of other fruits such as cucumber (*Cucumis sativus* L.; Table 8–13). Higher PCDD–PCDF concentrations in the peel of fruits like cucumber, apple (*Malus silvestris* Mill.) or pear (*Pyrus domestica* Medik.) are of atmogenic origin (Müller et al., 1993; Hülster & Marschner, 1994).

Collecting root exudates from zucchini and other plant species like tomato (*Lycopersicon esculentum* Mill.) has revealed, that root exudates of zucchini had a marked capacity to mobilize PCDD–PCDF from contaminated soils (Fig. 8–15). A similar capacity for PCDD–PCDF mobilization has been found in xylem and phloem exudates and leaf press sap of this plant species. These substances are obviously ubiquitous in zucchini tissues including root exudates, and facilitate mobility in the rhizosphere, uptake by roots and distribution of PCDD–PCDF in shoot organs. The chemical nature of these substances is not yet known. This example demonstrates the limitations in defining *critical* contents of chemicals in soils in terms of transfer into the food chain.

Table 8–13. PCDD–PCDF concentrations in vegetable fruits of plants grown in a soil contaminated with 148 ng I-TEQ kg^{-1} (I-TEq = International Toxicity Equivalents; Source: Hülster et al., 1994; Hülster et al., unpublished data).†

Plant species	Fruit part	PCDD/PCDF concentration
		ng I-TEQ kg^{-1} dry weight
Cucumber	Outer part (peel)	2.4
Cucumis sativus L.	Inner part	0.2
Zucchini	Outer part (peel)	~20
Cucurbita pepo L.	Inner part	~20

† Based on the relative toxicity of the individual congener and homologue groups compared with the most toxic congener 2, 3, 7, 8-TCDD (which is set as 1.0) toxicity equivalent factors have been developed. The concentrations of the individual congeners and homologues are multiplied by a corresponding factor (e.g., 0.4) and the sum of all I-TEqs are expressed as a single value.

8–4.4 Root Exudates as Carbon Source for Rhizosphere Microorganisms

Since roots act as a source of organic C, the population density of microorganisms, especially bacteria, is considerably higher in the rhizosphere than in the bulk soil (Table 8–14), particularly in basal root zones (Fig. 8–7). Despite the high supply of C rhizosphere microorganisms are often limited by N. Lower N concentrations in plants grown under elevated atmospheric CO_2 levels are therefore attributed at least in part to higher root exudation of organic C and associated increased sequestration of soil N by rhizosphere microorganisms (Rouhier et al., 1994). Therefore, as a rule, numbers of rhizosphere bacteria increase with supply of fertilizer N (Table 8–14; Liljeroth et al., 1990). However, the proportion of denitrifyers often increases and, thus, also the potential for losses of gaseous N from the rhizosphere, particularly in poorly aerated soils under condi-

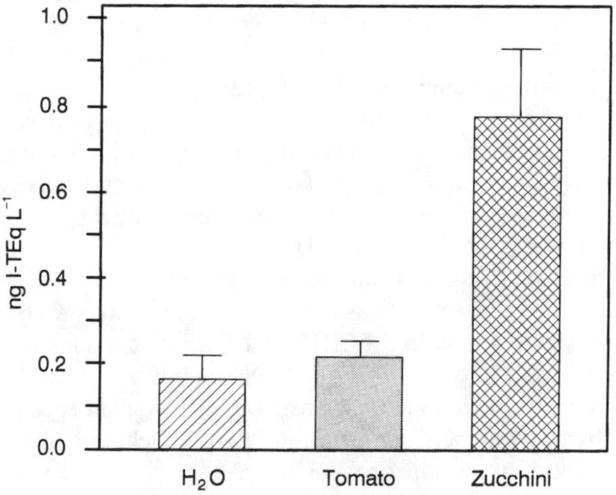

Fig. 8–15. Polychlorinated dibenzo-p-dioxin (PCDD) and dibenzo-furan (PCDF) concentrations extracted from a contaminated soil by H_2O or root exudates of tomato or zucchini (Hülster & Marschner, 1994).

Table 8–14. Effect of N-fertilization (mineral N) on total number of bacteria and on number of denitrifying bacteria in the bulk soil and rhizosphere soil of wheat (Source: Jagnow & Söchtig, 1983).

Number of bacteria × 10^6 (g soil)$^{-1}$					
		Rhizosphere soil			
Bulk soil (0 N levels)		N_0		N_{120}	
Total	Denitrifying	Total	Denitrifying	Total	Denitrifying
6.4	0.21	46.6	6.7	121.0	43.6

tions where root exudation is high such as in plants suffering from K deficiency or where roots are infected with pathogens.

When a large supply of organic C from roots is associated with a low O_2 partial pressure in the rhizosphere, high microbial activity can increase plant availability of Mn (Marschner, 1988). In aerated soils of high pH such an increase in Mn reduction is mainly microbial-mediated and can be considered as beneficial for Mn nutrition of plants, but often becomes critical in poorly aerated low pH soils, when excessive amounts of reduced Mn are formed and cause Mn toxicity in plants. In the rhizosphere of lowland rice, enhanced root exudation, for example as result of K or P deficiency, can impair Fe oxidation in the rhizosphere and, thus increase the risk of Fe toxicity in plants (Ottow et al., 1982).

There are many speculations to account for the marked microbial production of organic acids in the rhizosphere as a result of root exudation of carbohydrates. The relationship is considered of particular importance in the context of bacteria being capable of dissolving sparingly soluble Ca phosphates. The proportion of these acid-producing bacteria in the rhizosphere of crop plants might vary between 5 and 32% of the total rhizosphere bacteria (Domey, 1992), and this capability can readily be demonstrated in pure cultures of many bacteria species and strains. So far, however, the evidence is poor that such a mechanism plays an important role in the rhizosphere. Many of these bacteria also produce phytohormones and thereby alter root morphology and root metabolism. It can be assumed that, in general, the major mechanism of Plant Growth Promoting Rhizosphere Microorganisms (PGPRM) is based on production of phytohormones, or on the suppression of minor pathogens. Depending on the species, strains and isolates, rhizosphere microorganisms can enhance or depress plant growth (Table 8–15). The absence of marked effects on plant growth following inoculation with a mix-

Table 8–15. Influence of a soil inoculum with rhizosphere microflora and of pure cultures of rhizosphere bacteria on root morphology of maize seedlings (Source: Schönwitz & Ziegler, 1986).

Treatment	Dry weight		Primary root length	Lateral roots	Adventitious roots
	Shoot	Roots			
	— mg plant^{-1} —		mm	—— no. plant^{-1} ——	
Sterile	33.4	28.2	122	31	1.0
Soil inoculum	33.3	21.4	97	31	1.0
Pure culture I†	19.3	16.4	86	17	4.9
Pure culture II†	42.4	37.8	137	25	5.0

† Bacterial colonies isolated from rhizosphere microflora.

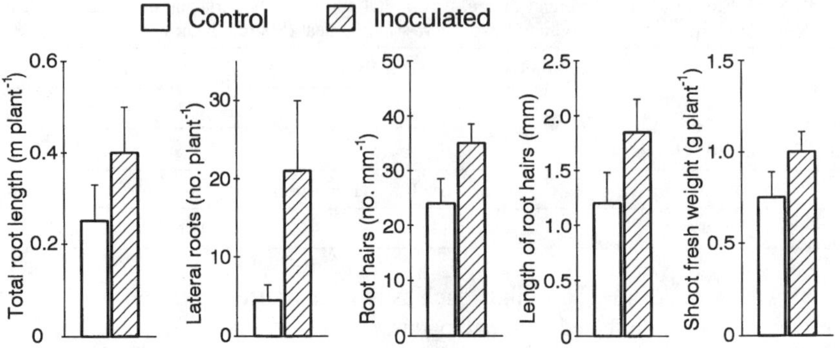

Fig. 8–16. Effect of inoculation with *Azospirillum brasilense* Cd on root and shoot growth of soil-grown wheat plants (Martin et al., 1989).

ture of rhizosphere microflora (soil inoculum) might therefore often be the result of one rhizosphere microorganism nullifying the effects of another.

Diazotrophic (N_2 fixing) associative rhizosphere bacteria present a particular example of relationships between plant roots and microorganisms (Zuberer, 1990). Several genera of diazotrophic bacteria such as *Azospirillum* sp. or *Enterobacter* sp. are quite common in the rhizosphere, at the rhizoplane, and even in the apoplasm of the root cortex particularly of C_4 graminaceous species (Boddey & Döbereiner, 1988). Compared with the free-living diazotrophic bacteria, the associative bacteria have better access to root exudates as an energy substrate for nitrogenase activity. Establishment of rhizosphere associations of diazotrophic bacteria depends not only on the amount but also on the composition of root exudates. For example, *Azospirillum* sp. prefer malate as substrate (Alexander & Zuberer, 1989). In many instances, these associations are rather host specific (Zuberer, 1990), and in *Panicum maximum* L. of 24 ecotypes tested, the proportion of N in the plants derived from biological N_2 fixation varied between zero and 39% (Miranda et al., 1990). Many diazotrophic bacteria produce and also secrete phytohormones like auxins and cytokinins (Jagnow et al., 1991) and thereby also considerably affect root growth and morphology (Fig. 8–16). Similar effects on root morphology are achieved by application of auxins.

Over the last decade, many field experiments have been conducted on the effect of inoculation with diazotrophic bacteria on growth and yield of temperate and tropical crop species. The results are highly variable, and yield increases of up to 20–40% are reported (Jagnow, 1990; Okon et al., 1994). Whether the yield increases are attributable to N_2 fixation or hormonal effects, or both, remains unclear. In tendency, hormonal effects dominate in temperate climates in C_3 species, and N_2 fixation in tropical C_4 species grown in soils low in N (Marschner, 1995).

8–4.5 Root Exudates as Signals for Microorganisms

Root exudates can also act for microorganisms as *signals* in the recognition of the host (Fig. 8–17). In many instances, the active components are flavonoids,

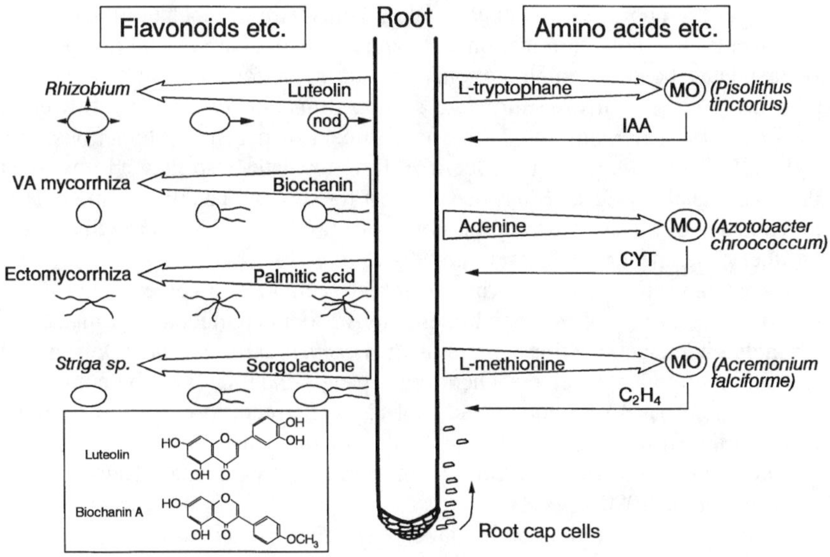

Fig. 8–17. Possible role of certain low-molecular-weight root exudates as *signals* or as sources (precursors) for phytohormone production for microorganisms (MO) in the rhizosphere (Marschner, 1995).

but can include betaines, as for example in lucerne (*Medicago sativa* L.; Phillips et al., 1994). For such signal function only very low concentrations are required, for chemotaxis of rhizobia, for example, a concentration of 10^{-9} *M* luteolin is sufficient (Bauer & Caetano-Anolles, 1990). The concentration of flavonoids needed as signal substances in root exudates of legumes also might be modulated by the form of N supply (Table 8–16). Compared with NO_3^- supply, NH_4^+ strongly enhances exudation of the isoflavonoids 2-hydroxygenistein and genistein. These isoflavonoids are putative signal molecules for *Rhizobium* and, thus, the N form supplied might affect nodulation. In agreement with this, the well-known inhibitory effect of N supply on nodulation in legumes such as pea (*Pisum sativum* L.) is primarily caused by NO_3^- whereas NH_4^+ supply even as high as 1 m*M* stimulates nodulation compared with plants without N supply (Waterer et al., 1992).

Table 8–16. Exudation of isoflavanoids by roots of lupine (*Lupinus albus* L.) supplied with NO_3^-–N or NH_4^+–N (Source: Wojtaszek et al., 1993).

	Exudation	
N supply	2 ‘Hydroxygenistein	Genistein
m*M*	——— μg plant^{-1} ———	
0	3	12
0.5 NO_3^-	–	5
9.0 NO_3^-	–	–
0.5 NH_4^+	37	48
10.0 NH_4^+	36	27

Other flavonoids in root exudates (e.g., quercetin or biochanin) act as signal molecules for spore germination and hyphal growth of arbuscular mycorrhiza. Root exudates, however, might not only act as a signal for establishment of symbiotic interactions, but also stimulate seed germination of parasitic flowering plants such as *Striga* and, thus, the establishment of parasitic interactions with sorghum [*Sorghum bicolor* (L.) Moench]. Root exudates also play an important role in allelopathic effects as has been shown for certain phenolic compounds in root exudates of quack grass (*Agropyron repens*) inhibiting root growth of various other plant species (Schulz et al., 1994).

Root cap cells and root cap mucilage also seem to be involved in the establishment of specific root-microbial interactions. Root cap mucilage of maize has a strongly chemotactic action on strains of *Azospirillum lipoferum.* During root penetration through the soil, detached root cap cells and mucilage come into contact with more basal root zones, and root cap cells seem to carry host-specific traits into the rhizosphere for the establishment of a characteristic rhizosphere bacterial flora, and to suppress certain soil-borne root pathogens (Hawes, 1990; Gochnauer et al., 1990).

Phytohormone production by rhizosphere microorganisms is often markedly stimulated by precursors released in root exudates (Fig. 8–17). Amino acids such as L-tryptophan and L-methionine are particularly important in this respect (Arshad & Frankenberger, 1990, 1991). The examples cited in this section demonstrate that in the evaluation of the importance of root exudates in chemical and microbial processes in the rhizosphere specific effects of particular components acting as signals for microorganisms or precursors of phytohormones also have to be considered in addition to the direct effects such as mobilization of mineral elements and C source for microorganisms.

8–5 MYCORRHIZAL ASSOCIATIONS

8–5.1 General

Mycorrhizas are the most widespread associations between microorganisms and higher plants. Roots of most soil-grown plants are mycorrhizal, and on a global scale about 80% of the Gymnosperms are mycorrhizal (Wilcox, 1991). As a rule, the fungus is strongly or wholly dependent on the higher plant, whereas the plant may or may not benefit. Mycorrhizal associations are therefore mutualistic, neutral, or parasitic, but as a rule, mutualism dominates.

8–5.2 Endo- (Arbuscular) and Ectomycorrhizae

There are two major mycorrhizal groups according to how the fungal mycelium relates to the root structure (Fig. 8–18). In endomycorrhizae the fungi live within cortical cells and also grow intercellularly, and extends its extraradical hyphae up to several centimeters into the surrounding soil. Of the various types the vesicular-arbuscular mycorrhizae (VAM) are the best known and most widely distributed. As not all endomycorrhizal fungi form vesicles, instead of

VAM the term arbuscular mycorrhizae (AM) has recently become the preferred term.

Ectomycorrhizae (ECM) occur mainly on roots of woody plants and are characterized by an interwoven mantle of hyphae around the roots (fungal sheath), and hyphae that penetrate the intercellular space of the cortex (Fig. 8–18). From the fungal sheath, hyphae, or rhizomorphs extend well into the surrounding soil. In many forest tree species of the northern hemisphere, both AM and ECM occur simultaneously. On a global scale, ECM is more abundant in boreal and temperate forests with a distinct surface humus horizon, and in N-limited ecosystems, whereas AM is more abundant in warmer climates with drier soils, in pasture land and deciduous forests with high turnover of organic material, and where P is limited (Read, 1991).

In mycorrhizal roots, a substantial proportion of the net photosynthates allocated to the roots is required for fungal growth and maintenance. For example, in AM plants *root* respiration may be 20 to 30% higher than in nonmycorrhizal plants, and estimations on the additional C demand are in the range of 10 to 20% of the net photosynthesis. In forest stands between 5 and 30% of the net photosynthates are allocated to the ECM (Söderström, 1992). This additional C demand is presumably one of the reasons that in mycorrhizal plants host root growth is less enhanced or even depressed compared with shoot growth, leading to a typically higher shoot–root dry weight ratio in mycorrhizal compared with nonmycorrhizal plants. Competition for photosynthates, as well as hormonal effects, might be responsible for alterations of root growth and morphology. Such changes include a decrease in total root length with a simultaneous increase in the number of lateral roots per unit root length, or per plant, associated with more rapid decline in activities of apical root meristems (Berta et al., 1990). In poplar, average longevity of AM roots was much lower than of nonmycorrhizal roots (Fig. 8–19). Therefore, higher respiration rates of mycorrhizal root systems also might be caused by higher root turnover.

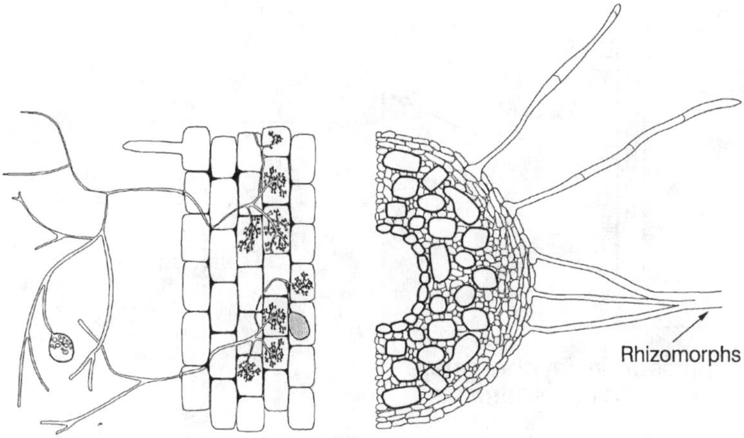

Fig. 8–18. Schematic presentation of the main structural features of the vesicular–arbuscular (VA) mycorrhizae (left) and the ecto-(ECM) mycorrhizae (right).

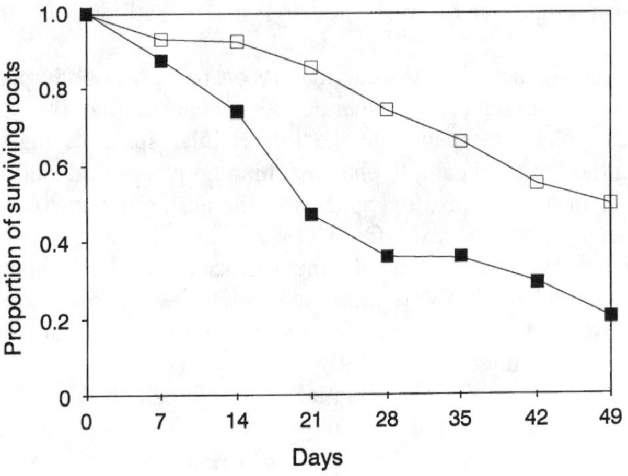

Fig. 8–19. Survival of mycorrhizal (■ *Glomus mosseae*) and nonmycorrhizal (□) roots of poplar (*Populus generosa*) in a sandy loam soil (Hooker et al., 1995).

8–5.3 Mycorrhizosphere, Hyphosphere

In mycorrhizal plants the soil–root interface is altered in many aspects, and a *mycorrhizosphere* is formed (Linderman, 1988) which is an ill-defined compartment (Fig. 8–20). As most soil-grown plants are mycorrhizal, the occurrence

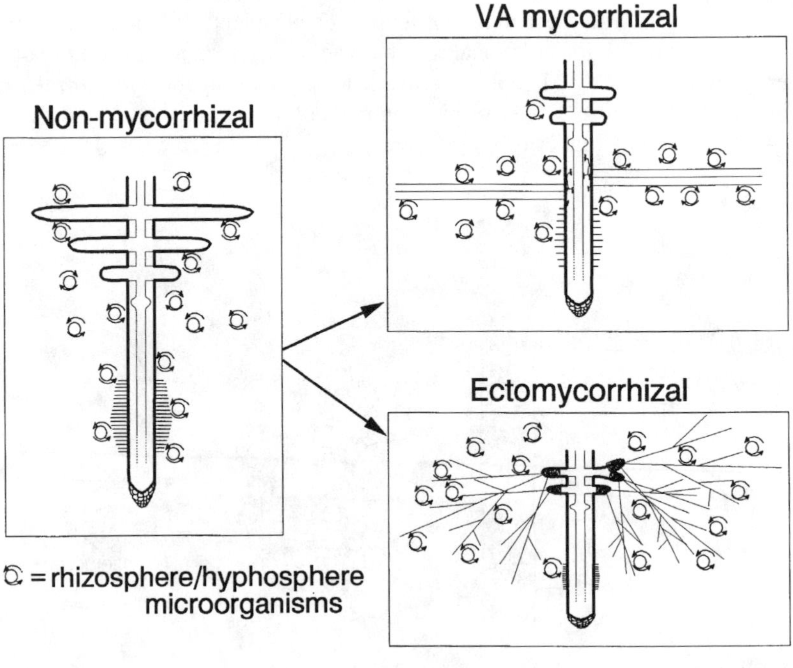

Fig. 8–20. Schematic presentation of effects of mycorrhizal colonization on root morphology and noninfecting rhizosphere microorganisms (Marschner, 1995).

Table 8–17. Bacteria and actinomycetes in the rhizosphere soil of guineagrass (*Panicum maximum* Jacq.; Source: Secilla & Bagyaraj, 1987).

	Rhizosphere populations (cfu g^{-1} soil)†		
Treatment	Bacteria × 10^6	N_2 fixer × 10^5	Actinomycetes × 10^4
Control (−VAM)	14.7	12.4	13.4
Glomus fasciculatum	41.9	42.0	26.1
Gigaspora margarita	34.0	87.9	17.6
Acaulospora laevis	8.0	10.6	28.6

† cfu, colony forming units.

of a mycorrhizosphere might be the rule, rather than the exception. In addition, around the extraradical hyphae and mycelia a new interface with the soil is formed, the *hyphosphere* (Linderman, 1988). These extraradical hyphae and mycelia may be responsible not only for altering microbial distribution and activity but also for providing substrate for soil fauna such as collembola feeding on mycorrhizal hyphae (McGonigle & Fitter, 1988). The soil structure also may be changed by AM extraradical hyphae (Kothari et al., 1991b; Andrade et al., 1995) by binding microaggregates into stable macroaggregates by intermeshing hyphae or the production of extracellular polysaccharides.

Mycorrhizal colonization not only alters the number of rhizosphere microorganisms but also the spectrum. This alteration might even differ between mycorrhizal species (Table 8–17). The low numbers of rhizosphere bacteria in nonmycorrhizal plants are caused by P limitation and the corresponding poor plant growth.

Since rhizosphere microorganisms can modify root morphology and activity (Section 4.5), the acquisition of mineral nutrients can in turn be affected, and in some cases also plant health (Section 5.4). An example of the alterations in Mn dynamics in the rhizosphere is shown in Table 8–18. Mycorrhizal colonization did not affect shoot dry weight but slightly inhibited root dry weight and root length in particular (Posta et al., 1994). In the rhizosphere soil, total microbial population was higher but the number of Mn reducers and effectivity of root exudates for Mn reduction lower in mycorrhizal plants (Table 8–18). Of the total rhizosphere microorganisms, 22% were able to reduce MnO_2 in nonmycorrhizal plants but only 2% in mycorrhizal plants. In accordance with this shift, Mn concentrations in the shoot dry matter were higher in nonmycorrhizal plants (115 mg kg^{-1}) than in mycorrhizal plants (89 mg kg^{-1}).

Table 8–18. Effect of arbuscular mycorrhiza (*Glomus mosseae*) on growth and rhizosphere microorganisms in 7-wk-old maize plants.

	Dry weight		Rhizosphere microorganisms†		
Treatment	Shoot	Roots	Total × 10^7	Mn reducers × 10^5	MnO_2 solubilized by root exudates
	— g $plant^{-1}$ —				mg L^{-1} 24 h^{-1}
−AM	19.8	5.2	3.93	88.0	12.1
+AM	20.6	4.2	8.43	19.2	4.9

† Numbers are expressed per gram of soil; Posta et al., 1994.

8–5.4 Role for Acquisition of Mineral Nutrients

There are various ways by which mycorrhizal colonization might affect mineral nutrient acquisition of the host plant. In principle, three possibilities exist:

1. Root growth and activity are not affected and extraradical mycorrhizal hyphae add surface area and increase the spatial availability of mineral nutrients.
2. Mycorrhizal hyphae supply mineral nutrients such as P, N, or Zn, which are growth-limiting factors in nonmycorrhizal plants. If it is assumed that photosynthetic capacity (source) is unlimited, not only shoot but also root growth will be enhanced and, thus, also acquisition of mineral nutrients (and water) by the roots.
3. Mycorrhizae are either ineffective in delivering mineral nutrients, or mineral nutrients are not growth-limiting in nonmycorrhizal plants; under conditions of source limitation, mycorrhizae depress root growth by sink competition. Harmful effects on shoot growth are then to be expected if the extraradical mycelium is not fully able to compensate for the root's functions in uptake of mineral nutrients and water.

As a rule, the most prominent influence of mycorrhizal colonization of roots on mineral nutrient acquisition can be attributed to the increase in absorbing surface area and, thus, soil volume delivering mineral nutrients to the plants (Marschner & Dell, 1994). Accordingly, the importance of mycorrhizal colonization for host plant mineral nutrition is higher on nutrient-poor than on nutrient-rich sites. It also is higher for acquisition of mineral nutrients of low mobility in the soil (e.g., P; NH_4^+–N, Zn) than of high mobility (e.g., Ca, NO_3^-–N), and high-

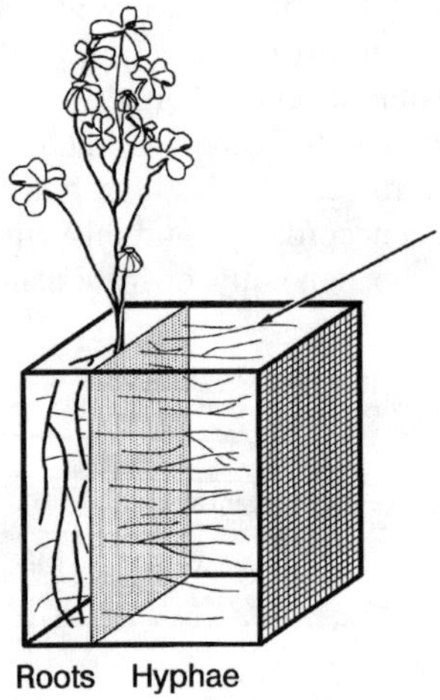

Estimated contribution of extraradical hyphae (% of shoot content)

	Clover	Maize
P	79	20
Zn	50	25
Cu	60	25

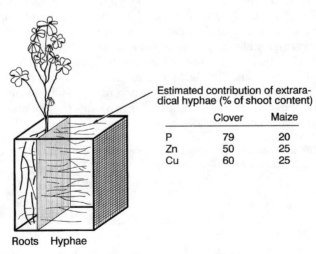

Fig. 8–21. Contribution of extraradical hyphae (*Glomus mosseae*) to the uptake of phosphorus, zinc and copper in white clover and maize plants in compartmented boxes (compiled data of Kothari et al., 1991a; Li et al., 1991b).

er for plants that have a coarse and poorly branched root system such as many legume species or cassava (*Manihot esculenta* Crantz), than for many graminaceous species. For example, in white clover (*Trifolium repens* L.), the extension of the depletion zone of P in the rhizosphere might increase from <5 mm in nonmycorrhizal plants to several centimeters in mycorrhizal plants (Li et al., 1991a, b). Accordingly, P uptake per unit root length is usually several times higher in mycorrhizal than in nonmycorrhizal plants.

The capacity of the extraradical hyphae for uptake and delivering mineral nutrients such as P, Zn, and Cu is quite high as shown for red clover (*Trifolium pratense* L.) and maize in Fig. 8–21. There is good evidence that not only ECM but also AM fungi might contribute substantially to the N acquisition of plants supplied with NH_4^-–N (George et al., 1992; Frey & Schuepp, 1994).

In addition to uptake of mineral nutrients, mycorrhizal hyphae also might be involved in plant-to-plant exchange of nutrients. At forest sites, for ECM such a nutrient exchange had been discussed between overstory trees and tree seedlings (Griffiths et al., 1991). In mixed stands, e.g., pastures, extraradical hyphae of AM can act as bridges between individual plants and also of different species, thus, offering the potential for nutrient exchange. Such exchange, however, is very slow and of limited ecological importance, for example, for P, but might become important for N under certain circumstances (Newman et al., 1992). In mixed stands with legumes, AM hyphal connections also exist between legumes and nonlegumes, but only insignificant amounts of N fixed by the legumes are directly transported via the AM hyphae to the nonlegumes (Bethlenfalvay et al., 1991).

Figure 8–22 summarizes the potential role of mycorrhizal colonization in mineral nutrient acquisition. For AM much more detailed data are available because of its role in annual (crop) species and also because it is relatively easy to carry out model experiments under controlled environmental conditions. In addition to the potential contribution shown for AM, in ECM at least some of the fungal species can provide the host plant additional access to sparingly soluble mineral nutrients, for example, by excretion of organic acids. These acids are presumably contributing factors in enhancing weathering of mica in the substrate of ectomycorrhizal compared with nonmycorrhizal pine plants (*Pinus sylvestris* L.; Leyval & Berthelin, 1991). Some ECM such as *Paxillus involutus* release large amounts of oxalic acid, which dissolves sparingly soluble Ca phosphates and provides the host plant such as eucalyptus (*Eucalyptus globulas* L.) with additional P and presumably also other mineral nutrients mobilized by acidification of a calcareous soil (Lapeyrie et al., 1990). Many ECM fungi also release acid phosphatases and thereby enhance mobilization of P_{org} in the rhizosphere. Excretion of acid phosphatases is, however, not confined to ECM but also a property of the extraradical hyphae of some, but not all (Joner & Jakobsen, 1995), AM fungi that therefore use P_{org} from soils very effectively (Tarafdar & Marschner, 1994), particularly in association with other phosphatase-producing fungi such as *Aspergillus fumigatus* (Table 8–19). Compared with inoculation with only one fungus, dual inoculation was more effective in increasing plant growth and useage of P_{org}. The enhancing effect of inoculation on Zn concentration in the shoots was, however, confined to the AM fungus.

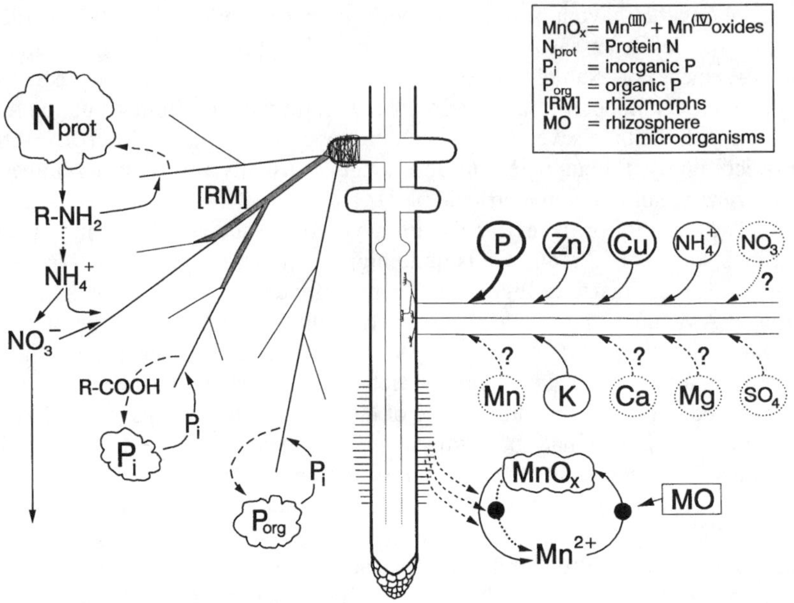

Fig. 8–22. Schematic presentation of components of the nutrient dynamics in and acquisition from the *hyphosphere* of endo-(VA) mycorrhizal roots and of additional components found in ectomycorrhizal (ECM) roots (Marschner, 1995).

So far, excretion of acid proteinases has not been shown for AM and seems to be a peculiarity of some ericoid and ECM fungi (Fig. 8–22). Since the host plants themselves have little or no access to complex organic N such as protein N, their fungal associate might provide access to the host plant to this resource (Table 8–20). Pine seedlings in association with *Suillus bovinus* can readily use N from protein sources, similar to plants provided with NH_4^+–N. The capacity to excrete acid proteinases and, thus, use protein-N is, however, only expressed in some ECM fungi (Finlay et al., 1992) but not, for example, in *Pisolithus tinctorius* (Table 8–20). Use of complex organic N might have important ecological advantages in N limited forest ecosystems of the northern hemisphere by shortening and tightening the N cycle, thereby minimizing N losses, from soil by

Table 8–19. Effect of *Aspergillus fumigatus* and *Glomus mosseae* on growth and use of Na-phytate (200 mg P kg^{-1} soil) in wheat grown in a Cambisol (source: Tarafdar & Marschner, 1995).

	Dry weight		Mineral element concentration in shoot dry matter				
Treatment	Shoot	Root	P	K	Zn	Fe	P org in Rhizo-soil at harvest
	— g plant^{-1}—		— mg g^{-1}—		— µg g^{-1} —		mg kg^{-1}
Steril	5.0	1.8	0.8	8.4	42	67	652
+*A. fumigatus*	5.5	2.2	1.3	9.4	45	37	604
+*G. mosseae*	5.5	2.4	1.6	9.5	81	68	584
+*A. fumigatus* +*G. mosseae*	6.9	2.5	1.8	9.9	93	72	558

Table 8–20. Nitrogen Content in Pinus contorta seedlings either nonmycorrhizal or infected with *Suillus bovinus* and *Pisolithus tinctorlus* and supplied with NH^+_4 or protein as source of N (Source: Abuzinadah et al., 1986).

Treatment	N content†	
	NH^+_4–N	Protein–N
	mg plant^{-1}	
Nonmycorrhizal	3.66	1.14
Suillus bovinus	4.05	3.20
Pisolithus tinctorius	3.27	1.30

† Seed content and starter N: 1.6 mg N flask^{-1}.

leaching, and by gaseous losses, and simultaneously decreasing competition for N from other soil microorganisms (Vogt et al., 1991).

8–5.5 Protection Against Metal Toxicity and Root Pathogens

A relatively large number of ECM fungi are effective in increasing heavy metal tolerance of their host plants. This increase is brought about mainly by a high retention capacity of the fungal mycelium, or its exudates, for heavy metals (Turnau et al., 1993; Denny & Ridge, 1995). An example for this is shown for Zn in *Paxillus involutus* in Table 8–21. Accordingly, concentration and content of Zn in young mycorrhizal pine plants inoculated with *Paxillus involutus* is much lower than in nonmycorrhizal plants. In contrast, and despite of similar fungal biomass, another ECM fungus (*Thelephora terrestris*) hardly retains Zn in its structures and even further increases the Zn content in the host plant. This and other examples demonstrate that it is not possible to make a general statement that ECM fungi increase the heavy metal tolerance of plants.

Some ECM fungi may also increase the Al tolerance of the host plant as shown in Table 8–22 for pine seedlings. Plant growth is reduced even at 50 μM Al in the substrate, and this is associated with marked increase in Al concentration in the needles. In contrast, in ECM plants even at 200 μM Al growth is not affected and Al concentration in the needles remain relatively low. The protecting effect of ECM against Al toxicity is brought about by a combination of improved P nutrition and Al binding in fungal structures (Martin et al., 1994) or release of Al complexing solutes (organic acids, phenolics). Binding and complexation of Al decreases the concentration of monomeric, toxic Al in the rhi-

Table 8–21. Shoot and root contents of Zn in nonmycorrhizal and ectomycorrhizal seedlings of *Pinus sylvestris* supplied with high Zn concentration (Source: Colpaert & van Assche, 1992).

Treatment (fungus)	Shoot dry weight	Zinc content			Fungal biomass % of short roots
		Shoot	Short roots	Shoot	
	g plant^{-1}	µg g^{-1} dry weight		mg plant^{-1}	
Nonmycorrhizal	16.2	197	273	3.19	--
Paxiius involutus	14.3	106	708	1.52	54
Theiephora terr.	16.2	240	309	3.89	66

Table 8–22. Effect of mycorrhizal inoculation (*Pisolithus tinctorius*)and solution Al on growth and Al content in needles of pitch pine (Pinus rigida L.) seedlings (Source: Cumming & Weinstein, 1990).

	Dry weight				Al concentration in needles	
	Roots		Shoots			
Treatment	–Myc	+Myc.	–Myc.	+Myc.	–Myc.	+Myc.
μ*M* Al	mg plant^{-1}				μg g^{-1} dry weight	
0	251	262	442	537	90	50
50	218	259	404	570	365	100
200	139	224	237	522	375	175

zosphere soil solution and enable host roots to maintain growth even in acid mineral soils.

In contrast to ECM, the role of AM in metal tolerance of plants is not clear. Although in soils polluted with heavy metals the AM population might become more tolerant to these heavy metals (Weissenhorn et al., 1994), the reported results on increase in host plant heavy metal tolerance are contradictory (Schuepp et al., 1987; Galli et al., 1995; Weissenhorn et al., 1995). Various reasons for this are possible. Root apical zones are the principal sites of metal toxicity, and in plants colonized by ECM most of the apical root zones are enclosed by the fungal mycelium, in contrast to the plants colonized by AM (Fig. 8–20). Furthermore, the fungal biomass is much lower in AM fungi, and they also lack the capacity to excrete organic acids and other potential chelators for heavy metals and Al. A simultaneous decrease in heavy metal concentrations in AM plants with an increase in tolerance of heavy metals by these plants is often an indirect effect resulting from enhanced growth, for example, by improved P nutrition and, thus, dilution of heavy metals in the shoot dry matter. In the absence of growth enhancing effects, AM might further increase the Zn concentration in the host plant even at high external supply (Symeonidis, 1990), or the Cd concentrations by delivery of Cd via the extraradical mycelium (Guo et al., 1995).

There is a long list of examples on suppression of soil-borne fungal and bacterial root pathogens by mycorrhizae. An example for ECM is shown in Table 8–23. Without root pathogens, the ECM fungus has no enhancing effect on host plant growth. In the presence of the pathogen (*Fusarium oxysporum*), host plant growth is severely depressed and seedling mortality high. Inoculation with the ECM fungus effectively suppresses the harmful effect of the root pathogen on host plant growth. This suppression is achieved by the production of oxalic acid by *Paxillus involutus*, and this production is enhanced by root exudates of red pine (*Pinus resinosa* Ait; Duchesne et al., 1989).

The list of examples on suppression of root pathogens by mycorrhizae is particularly long for AM fungi. This suppression can be attributed either to improved host plant nutrition (Perrin, 1990) or changes in root exudation, rhizosphere microflora, or host root physiology in terms of induced resistance (Dehne & Schönbeck, 1979; Benhamou et al., 1994). In the northern wheat belt of Australia, wheat root infection with common root rot (*Bipolaris sarokiniania*) has been found to be inversely related to root colonization by AM. Low root colonization was caused by low AM infection potential in soils due to long fallow

Table 8–23. Suppression of root pathogens in *Pinus resinosa* seedlings by *Paxillus involutus* (Source: Chakravarty et al., 1991).

Treatment	Seeling mortality	Length	
		Shoot	Root
	%	cm plant^{-1}	
Control	0	3.0	2.3
+*Paxillus involutus*	0	3.0	2.5
+*Fusarium oxysporum*	50	1.5	0.6
+*P. involutus* + *F. oxysp*	20	2.5	1.5

Table 8–24. Effect of VAM (*Gl. mosseae*) on shoot growth and number of rhizoplane bacteria in rapevine cuttings (source: Waschkies et al., 1994).

Soil ± innoculation	Shoot	Roots	VAM infection	Bacteria no. g^{-1} root fresh weight	
				Total × 10^7	*Pseudomonas fluorescens* ×10^5
	g dry weight	g fresh weight	%		
Control					
−VAM	6.3	10.1	33	3.2	0.18
+VAM	6.2	12.5	31	3.7	0.16
Soil sickness					
−VAM	1.3	3.6	21	4.4	5.88
+VAM	2.3	7.8	34	3.2	0.71

periods or a high proportion of nonhost plants in the crop rotation (Thompson & Wildermuth, 1989). Higher AM infection potential in soils is often found in low input compared with high input management practices and mainly the result of higher proportion of weeds in fields with low input management (Kurle & Pfleger, 1994).

Suppressing effects of AM are also evident in cases of *soil sickness* or *replant disease* where minor pathogens or deleterious soil microorganisms may harm root growth or activity (Section 3.2). An example for such suppressing effect is shown in Table 8–24. The growth of grapevine (*Vitis vinifera* L.) seedlings was poor on soils with replant disease but could be markedly improved by inoculation with VAM, which also raised the level of mycorrhizal root colonization. Suppression of *Pseudomonas fluorescence* by VAM inoculation was presumably the main factor responsible for improvement of plant growth in the soil with replant disease. Soil sterilization was, however, more effective than VAM inoculation and restored plant growth to the level in the control soil (Waschkies et al., 1994).

8–6 SUMMARY AND CONCLUSIONS

Conditions at the soil-root interface (rhizosphere) differ in many respects from those in the bulk soil. Roots preferentially take up either particular nutrients or water, leading to depletion or accumulation of nutrients in the rhizosphere. Depending on the plant species, form of N supply and nutritional status of plants, the rhizosphere pH can markedly differ from the bulk soil pH. Changes in rhizosphere pH may considerably affect mineral element uptake via mobilization and immobilization. Roots also consume or release O_2 and thereby alter the redox potential and the availability of certain mineral elements.

Depending on the soil chemical and physical conditions, species, age, and nutritional status of plants, a varied proportion of photosynthetic C is released into the rhizosphere as high- or low-molecular-weight organic solutes. These solutes (e.g., organic acids) either directly mobilize mineral elements, or provide an energy substrate for microbial activity in the rhizosphere. Plants have developed various strategies to increase the effectivity of these root exudates by short-term release of high amounts at localized zones of the root system. Root exudates also may act as precursors for microbial phytohormone production and as signal substances for microorganisms and thereby modulate composition of the rhizosphere microflora. This in turn may affect root morphology, activity and mineral nutrient acquisition as well as uptake of heavy metals.

On a global basis, about 80% of the plants are mycorrhizal. In roots colonized by endomycorrhizas (AM) or ectomycorrhizas (ECM), the soil-root interface is ill-defined, and the biological and biochemical processes more complex than in nonmycorrhizal roots. A mycorrhizosphere, or hyphosphere, is formed extending the soil volume accesible for mineral nutrient acquisition. Mycorrhizal hyphae can deliver a high proportion of the host plant demand of mineral nutrients like phosphorus and zinc. In general, the importance of mycorrhizal associations for the host plant nutrition is much higher in nutrient-poor than in nutrient-rich soils. In some instances root colonization with ECM also can protect roots against metal toxicity by metal binding in the fungal tissue. For a given C investment into belowground parts, mycorrhizal associations ensure a higher long-term stability, in which protection against root pathogens may also be an important component, whereas root-induced changes in the rhizosphere increase flexibility and specificity of plant responses to adverse soil chemical conditions.

REFERENCES

Abuzinadah, R.A., R.D. Finlay, and D.J. Read. 1986. The role of proteins in the nitrogen nutrition of ectomycorrhizal plants: II. Utilization of proteins by mycorrhizal plants of *Pinus contorta*. New Phytol. 103:495–506.

Ae, N., J. Arihara, K. Okada, T. Yoshihara, and C. Johansen. 1990. Phosphorus uptake by pigeon pea and its role in cropping systems of the Indian subcontinent. Science (Washington, DC) 248:477–480.

Aguilar S., A., and A. van Diest, A. 1981. Rock-phosphate mobilization induced by the alkaline uptake pattern of legumes utilizing symbiotically fixed nitrogen. Plant Soil 61:27–42.

Alexander, D.B., and D.A. Zuberer. 1989. ^{15}N fixation by bacteria associated with maize roots at a low partial O_2 pressure. Appl. Environ. Microbiol. 55:1748–1753.

Alexander, D.B., and D.A. Zuberer. 1993. Response by iron-efficient and inefficient oat cultivars to inoculation with siderophore-producing bacteria in a calcareous soil. Biol. Fertil. Soils 16:118–124.

Andrade, G., R. Azcón, and G.J. Bethlenfalvay. 1995. A rhizobacterium modifies plant and soil responses to the mycorrhizal fungus *Glomus mosseae*. Appl. Soil Ecol. 2:195–202.

Arshad, M., and W.T. Frankenberger, Jr. 1990. Response of *Zea mays* and *Lycopersicon esculentum* to the ethylene precursors, L-methionine and L-ethionine applied to soil. Plant Soil 122:219–227.

Arshad, M., and W.T. Frankenberger, Jr. 1991. Microbial production of plant hormones. Plant Soil 133:1–8.

Awad, F., V. Römheld, and H. Marschner. 1994. Effect of root exudates on mobilization in the rhizosphere and uptake of iron by wheat plants. Plant Soil 165:213–218.

Bakken, L.R. 1988. Denitrification under different cultivated plants: Effects of soil moisture tension, nitrate concentration, and photosynthetic activity. Biol. Fertil. Soils 6:271–278.

Bauer, W.D., and G. Caetano-Anolles. 1990. Chemotaxis, induced gene expression and competitiveness in the rhizosphere. Plant Soil 129:45–52.

Begg, C.B.M., G.J.D. Kirk, A.F. Mackenzie, and H.-U. Neue. 1994. Root-induced iron oxidation and pH changes in the lowland rice rhizosphere. New Phytol. 128:469–477.

Bekele, T., B.J. Cino, P.A.I. Ehlert, A.A. van der Mass, and A. van Diest. 1983. An evaluation of plant-borne factors promoting the solubilization of alkaline rock phosphates. Plant Soil 75:361–378.

Benhamou, N., J.A. Fortin, C. Hamel, M. St-Arnaud, and A. Shatilla. 1994. Resistance responses of mycorrhizal R_i T-DNA-transformed carrot roots to infection by *Fusarium oxysporum* f. sp. *chrysanthemi*. Phytopathology 84:958–968.

Berta, G., A. Fusconi, A. Trotta, and S. Scannerini. 1990. Morphogenetic modifications induced by the mycorrhizal fungus *Glomus* strain E_3 in the root system of *Allium porrum* L. New Phytol. 114:207–215.

Bethlenfalvay, G.J., M.G. Reyes-Solis, S.B. Camel, and R. Ferrera-Cerrato. 1991. Nutrient transfer between the root zones of soybean and maize plants connected by a common mycorrhizal mycelium. Physiol. Plant. 82:423–432.

Boddey, R.M., and J. Döbereiner. 1988. Nitrogen fixation associated with grasses and cereals: Recent results and perspectives for future research. Plant Soil 108:53–65.

Boeuf-Tremblay, V., S. Plantureux, and A. Guckert. 1995. Influence of mechanical impedance on root exudation of maize seedlings at two development stages. Plant Soil 172:279–287.

Cakmak, I., K.Y. Gülüt, H. Marschner, and R.D. Graham. 1994. Effect of zinc and iron deficiency on phytosiderophore release in wheat genotypes differing in zinc efficiency. J. Plant Nutr. 17:1–17.

Chakravarty, C., R.L. Peterson, and B.E. Ellis. 1991. Interaction between the ectomycorrhizal fungus *Paxillus involutus*, damping-off fungi and *Pinus resinosa* seedlings. J. Phytopathol. 132:207–218.

Chaney, R.L. 1984. Diagnostic practices to identify iron deficiency in higher plants. J. Plant Nutr. 7:47–67.

Colpaert, J.V., J.A. van Assche. 1992. Zinc toxicity in ectomycorrhizal *Pinus sylvestris*. Plant Soil 143:201–211.

Constable, J.V.H., and D.J. Longstreth. 1994. Aerenchyma carbon dioxide can be assimilated in *Thypha latifolia* L. leaves. Plant Physiol. 106:1065–1072.

Crowley, D.E., C.P.P. Reid, and P.J. Szaniszlo. 1987. Microbial siderophores as iron source for plants. p. 375–386. *In* G. Winkelmann et al. (ed.) Iron transport in microbes, plants and animals. Verlag Chemie, Weinheim.

Crowley, D.E., V. Römheld, H. Marschner, and P.J. Szaniszlo. 1992. Root-microbial effects on plant iron uptake from siderophores and phytosiderophores. Plant Soil 142:1–7.

Cumming, J.R., and L.H. Weinstein. 1990. Aluminum-mycorrhizal interactions in the physiology of pitch pine seedlings. Plant Soil 125:7–18.

Darrah, P.R. 1993. The rhizosphere and plant nutrition: A quantitative approach. p. 3–22. *In* N.J. Barrow (ed.) Plant nutrition: From genetic engineering to field practice. Kluwer Academic Publ., Dordrecht, the Netherlands.

Dehne, H.-W., und F. Schönbeck. 1979. Untersuchungen zum Einfluß der endotrophen Mykorrhiza auf Pflanzenkrankheiten: II. Phenolstoffwechsel und Lignifizierung. Phytopathol. Z. 95:210–216.

Delhaize, E., P.R. Ryan, and P.I. Randall. 1993. Aluminum tolerance in wheat (*Triticum aestivum* L.): II. Aluminum-stimulated excretion of malic acid from root apices. Plant Physiol. 103: 695–702.

Denny, H.J., and I. Ridge. 1995. Fungal slime and its role in the mycorrhizal amelioration of zinc toxicity to higher plants. New Phytol. 130:251–257.

Dinkelaker, B., C. Hengeler, and H. Marschner. 1995. Distribution and function of proteoid roots and other root clusters. Bot. Acta 108:183–200.

Dinkelaker, B., and H. Marschner. 1992. In vivo demonstration of acid phosphatase activity in the rhizosphere of soil-grown plants. Plant Soil 144:199–205.

Dinkelaker, B., V. Römheld, and H. Marschner. 1989. Citric acid excretion and precipitation of calcium citrate in the rhizosphere of white lupin (*Lupinus albus* L.). Plant Cell Environ. 12:285–292.

Domey, S. 1992. Vorkommen phosphatmobilisierender Bakterien in der Rhizosphäre landwirtschaftlicher Kulturpflanzen bei mittlerer bis hoher Phosphor-Versorgung des Bodens. Zentralbl. Mikrobiol. 147:270–276.

Duchesne, L.C., B.E. Ellis, and R.L. Peterson. 1989. Disease suppression by the ectomycorrhizal fungus *Paxillus involutus*: Contribution of oxalic acid. Can. J. Bot. 67:2726–2730.

Eltrop, L. 1993. Role of ectomycorrhiza in the mineral nutrition of Norway spruce (*Picea abies* L. Karst.). Ph.D. diss. Univ. of Hohenheim, Hohenheim, Germany.

Erenoglu, B., I. Cakmak, H. Marschner, V. Römheld, S. Eker, H. Daghan, M. Kalayci, and H. Ekiz. 1996. Phytosiderophore release does not relate well with zinc efficiency in different bread wheat genotypes. J. Plant Nutr. 19:569–580.

Finlay, R.D., A. Frostegard, and A.M. Sonnerfeldt. 1992. Utilization of organic and inorganic nitrogen sources by ectomycorrhizal fungi in pure culture and in symbiosis with *Pinus contorta* Dougl. ex Loud. New Phytol. 120:105–115.

Fischer, W., H. Felssa, and G. Schaller. 1989. pH values and redox potentials in microsites of the rhizosphere. Z. Pflanzenernähr. Bodenk. 152:191–195.

Frey, B., and H. Schüepp. 1994. Acquisition of nitrogen by external hyphae of arbuscular mycorrhizal fungi associated with *Zea mays* L. New Phytol. 124:221–230.

Galli, U., H. Schüepp, and C. Brunold. 1995. Thiols of Cu-treated maize plants inoculated with the arbuscular-mycorrhizal fungus *Glomus intraradices*. Physiol. Plant. 94:247–253.

George, E., K.-U.Häussler, D. Vetterlein, E. Gorgus, and H. Marschner. 1992. Water and nutrient translocation by hyphae of *Glomus mosseae*. Can. J. Bot. 70:2130–2137.

Gerke, J., W. Römer, and A. Jungk. 1994a. The excretion of citric and malic acid by proteoid roots of *Lupinus albus* L.: Effects on soil solution concentrations of phosphate, iron, and aluminum in the proteoid rhizosphere in samples of an oxisol and a luvisol. Z. Pflanzenernähr. Bodenk. 157:189–294.

Gerke, J., W. Römer, and U. Meyer. 1994b. Die Nutzung von Bodenphosphaten durch chemische Phosphatmobilisierung von Leguminosen. Mitt. Dtsch. Bodenk. Ges. 73:43–46.

Gochnauer, M.B., M.E. McCully, and H. Labbé. 1989. Different populations of bacteria associated with sheathed and bare regions of roots of field-grown maize. Plant Soil 114:107–120.

Gochnauer, M.B., L.J. Sealey, and M.E. McCully. 1990. Do detached root-cap cells influence bacteria associated with maize roots? Plant Cell Environ. 13:793–801.

Graham, R.D., J.S. Ascher, and S.C. Hynes. 1992. Selecting zinc-efficient cereal genotypes for soils low in zinc status. Plant Soil 146:241–250.

Grierson, P.F., and P.M. Attiwill. 1989. Chemical characteristics of the proteoid root mat of *Banksia integrifolia* L. Aust. J. Bot. 37:137–143.

Gries, D., and M. Runge. 1992. The ecological significance of iron mobilization in wild grasses. J. Plant Nutr. 15:1727–1737.

Griffiths, R.P., M.A. Castellano, and B.A. Caldwell. 1991. Hyphal mats formed by two ectomycorrhizal fungi and their association with Douglas-fir seedlings: A case study. Plant Soil 134:255–259.

Guo, Y., E. George, and H. Marschner. 1996. Contribution of an arbuscular mycorrhizal fungus to the uptake of cadmium and nickel in bean and maize plants. Plant Soil 184:195–205.

Handreck, K.A. 1991. Interactions between iron and phosphorus in the nutrition of *Banksia ericifolia* L. f. var. *ericifolia* (Proteaceae) in soil-less potting media. Aust. J. Bot. 39:373–384.

Häussling, M., and H. Marschner. 1989. Organic and inorganic soil phosphates and acid phosphatase activity in the rhizosphere of 80-year-old Norway spruce (*Picea abies* (L.) Karst.) trees. Biol. Fertil. Soils 8:128–133.

Hawes, M.C. 1990. Living plant cells released from the root cap: a regulator of microbial populations in the rhizosphere. Plant Soil 129:19–27.

Helal, H.M., and D. Sauerbeck. 1989. Input and turnover of plant carbon in the rhizosphere. Z. Pflanzenernähr. Bodenk. 152:211–216.

Hinsinger, P., and B. Jaillard. 1993. Root-induced release of interlayer potassium and vermiculitization of phlogopite as related to potassium depletion in the rhizosphere of ryegrass. J. Soil Sci. 44:525–534.

Hoffland, E., G.R. Findenegg, and J.A. Nelemans. 1989. Solubilization of rock phosphate by rape: II. Local root exudation or organic acids as a response to P-starvation. Plant Soil 113:161–165.

Hoffland, E., R. van den Boogaard, J.A. Nelemans, and G.R. Findenegg. 1992. Biosynthesis and root exudation of citric and malic acid in phosphate-starved rape plants. New Phytol. 122:675–680.

Hooker, J.E., K.E. Black, R.L. Perry, and D. Atkinson. 1995. Arbuscular mycorrhizal fungi induced alteration to root longevity of poplar. Plant Soil 172:327–329.

Horst, W.J., F. Klotz, and P. Szulkiewicz. 1990. Mechanical impedance increases aluminium tolerance of soybean (*Glycine max*) roots. Plant Soil 124:227–231.

Horst, W.J., and Ch. Waschkies. 1987. Phosphatversorgung von Sommerweizen (*Triticum aestivum* L.) in Mischkultur mit Weisser Lupine (*Lupinus albus* L.). Z. Pflanzenernähr. Bodenk. 150:1–8.

Hülster, A., and H. Marschner. 1993. Transfer of PCDD/PCDF from contaminated soils to food and fodder crop plants. Chemosphere 27(1–3):439–446.

Hülster, A., and H. Marschner. 1994. PCDD/PCDF-Transfer in Zucchini und Tomaten. Angew. Ökologie 8:579–589.

Hülster, A., J.F. Müller, and H. Marschner. 1994. Soil-plant transfer of polychlorinated dibenzo-*p*-dioxins and dibenzofurans to vegetables of the cucumber family (*Cucurbitaceae*). Environ. Sci. Technol. 28:1110–1115.

Jagnow, G. 1990. Differences between cereal crop cultivars in root-associated nitrogen fixation, possible causes of variable yield response to seed inoculation. Plant Soil 123:255–259.

Jagnow, G., G. Höflich, and K.-H. Hoffmann. 1991. Inoculation of non-symbiotic rhizosphere bacteria: Possibilities of increasing and stabilizing yields. Angew. Botanik 65:97–126.

Jagnow, G., and H. Söchtig. 1983. Nitrogen losses from the soil to the atmosphere and to ground water–possible ways of limiting them: A survey. p. 68–87. *In* Plant Research and Development 17. Institute for Scientific Cooperation, Federal Republic of Germany.

Janzen, H.H. 1990. Deposition of nitrogen into the rhizosphere by wheat roots. Soil Biol. Biochem. 22:1155–1160.

Johnson, J.F., D.L.Allan, and C.P. Vance. 1994. Phosphorus stress-induced proteoid roots show altered metabolism in *Lupinus albus.* Plant Physiol. 104:657–665.

Johnson, P.A., and R.J. Bennet. 1991. Aluminium tolerance of root cap cells. J. Plant Physiol. 137:760–762.

Joner, E.J., and I. Jakobsen. 1995. Uptake of ^{32}P from labelled organic matter by mycorrhizal and non-mycorrhizal subterranean clover (*Trifolium subterraneum* L.). Plant Soil 172:221–227.

Jones, D.L., and P.R. Darrah. 1993. Re-absorption of organic compounds by roots of *Zea mays* L. and its consequences in the rhizosphere: II. Experimental and model evidence for simultaneous exudation and re-absorption of soluble C compounds. Plant Soil 153:47–59.

Jones, D.L., A.C. Edwards, K. Donachie, and P.R. Darrah. 1994. Role of proteinaceous amino acids released in root exudates in nutrient acquisition from the rhizosphere. Plant Soil 158:183–192.

Jungk, A. 1991. Dynamics of nutrient movement at the soil–root interface. p. 455–481. *In* J. Waisel et al. (ed.) Plant roots. The hidden half. Marcel Dekker, New York.

Jungk, A., and N. Claassen. 1986. Availability of phosphate and potassium as the result of interactions between root and soil in the rhizosphere. Z. Pflanzenernähr. Bodenk. 149:411–427.

Jungk, A., B. Seeling, and J. Gerke. 1993. Mobilization of different phosphate fractions in the rhizosphere. Plant Soil 155/156:91–94.

Kimura, M., H. Murakami, and H. Wada. 1991. CO_2, H_2, and CH_4 production in rice rhizosphere. Soil Sci. Plant Nutr. (Tokyo) 38:55–60.

Kloepper, J.W., R. Rodriguez-Kabana, J.A. McInroy, and D.J. Collins. 1991. Analysis of populations and physiological characterisation of microorganisms in rhizosphere of plants with antagonistic properties to phytopathogenic nematodes. Plant Soil 136:95–102.

Kludze, H.K., and R.D. DeLaune. 1994. Methane emissions and growth of *Spartina patens* in response to soil redox intensity. Soil Sci. Soc. Am. J. 58:1838–1845.

Kothari, S.K., H. Marschner, and V. Römheld. 1991a. Contribution of the VA mycorrhizal hyphae in acquisition of phosphorus and zinc by maize grown in a calcareous soil. Plant Soil 131:177–185.

Kothari, S.K., H. Marschner, and V. Römheld. 1991b. Effect of a vesicular–arbuscular mycorrhizal fungus and rhizosphere microorganisms on manganese reduction in the rhizosphere and manganese concentrations in maize (*Zea mays* L.). New Phytol. 117:649–655.

Kraffczyk, I., G.Trolldenier, and H. Beringer. 1984. Soluble root exudates of maize: Influence of potassium supply and rhizosphere microorganisms. Soil Biol. Biochem. 16:315–322.

Kuchenbuch, R., and A. Jungk. 1984. Wirkung der Kaliumdüngung auf die Kaliumverfügbarkeit in der Rhizosphäre von Raps. Z. Pflanzenernähr. Bodenk. 147:435–448.

Kurle, J.E., and F.L. Pfleger. 1994. Arbuscular mycorrhizal fungus spore populations respond to conversions between low-input and conventional management practices in a corn–soybean rotation. Agron. J. 86:467–475.

Lapeyrie, F., C. Picatto, J. Gerard, and J. Dexheimer. 1990. T.E.M. study of intracellular and extracellular calcium oxalate accumulation by ectomycorrhizal fungi in pure culture or in association with *Eucalyptus* seedlings. Symbiosis 9:163–166.

Leyval, C., and J. Berthelin. 1991. Weathering of a mica by roots and rhizosphere microorganisms of pine. Soil Sci. Soc. Am. J. 55:1009–1016.

Li, X.-L., E. George, and H. Marschner. 1991a. Extension of the phosphorus depletion zone in VA-mycorrhizal white clover in a calcareous soil. Plant Soil 136:41–48.

Li, X.-L., H. Marschner, and E. George. 1991b. Acquisition of phosphorus and copper by VA-mycorrhizal hyphae and root-to-shoot transport in white clover. Plant Soil 136:49–57.

Liljeroth, E., P. Kuikman, and J.A. Van Veen. 1994. Carbon translocation to the rhizosphere of maize and wheat and influence on the turnover of native soil organic matter at different soil nitrogen levels. Plant Soil 161:233–240.

Liljeroth, E., G.C. Schelling, and Van Veen, J.A. 1990. Influence of different application rates of nitrogen to soil on rhizosphere bacterial. Neth. J. Agric. Sci. 38:355–364.

Linderman, R.G. 1988. Mycorrhizal interactions with the rhizosphere microflora: The mycorrhizosphere effect. Phytopathology 78:366–371.

Lipton, D.S., R.W. Blanchar, and D.G. Blevins. 1987. Citrate, malate, and succinate concentration in exudates from P-sufficient and P-stressed *Medicago sativa* L. seedlings. Plant Physiol. 85:315–317.

Lynch, J.M., and J.M. Whipps. 1990. Substrate flow in th rhizosphere. Plant Soil 129:1–10.

Ma, J.F., G. Kusano, S. Kimura, and K. Nomoto. 1993. Specific recognition of mugineic acid-ferric complex by barley roots. Phytochemistry 34:599–603.

Macklon, A.E.S., L.A. Mackie-Dawson, A. Sim, C.A. Shand, and A. Lilly. 1994. Soil P resources, plant growth and rooting characteristics in nutrient poor upland grassland. Plant Soil 163:257–266.

Marschner, H. 1988. Mechanism of manganese acquisition by roots from soils. p. 191–204. *In* R.D. Graham et al. (ed.) Manganese in soils and plants. Kluwer Academic Publ., Dordrecht, the Netherlands.

Marschner, H. 1991. Mechanism of adaptation of plants to acid soils. Plant Soil 134:1–20.

Marschner, H. 1995. Mineral nutrition of higher plants. 2nd ed. Academic Press, London.

Marschner, H., and B. Dell. 1994. Nutrient uptake in mycorrhizal symbiosis. Plant Soil 159:89–102.

Marschner, H., and V. Römheld, 1983. *In vivo* measurement of root-induced pH changes at the soil-root interface: Effect of plant species and nitrogen source. Z. Pflanzenphysiol. 111:241–251.

Marschner, H., and V. Römheld. 1994. Strategies of plants for acquisition of iron. Plant Soil 165:261–274.

Marschner, H., V. Römheld, and M. Kissel. 1986. Different strategies in higher plants in mobilization and uptake of iron. J. Plant Nutr. 9:695–713.

Marschner, H., M. Treeby, and V. Römheld. 1989. Role of root-induced changes in the rhizosphere for iron acquisition in higher plants. Z. Pflanzenernähr. Bodenk. 152:197–204.

Marschner, H., V. Römheld, and F.S. Zhang. 1990. Mobilization of mineral nutrients in the rhizosphere by root exudates. p. 158–163. *In* Trans. 14th Int. Cong. of Soil Science, Kyoto, Japan. 12–18 Aug. 1990. Vol. II. ISSS, Vienna, Austria.

Martin, F., P. Rubini, and I. Kottke. 1994. Aluminium polyphosphate complexes in the mycorrhizal basidiomycete *Laccaria bicolor*: A ^{27}Al-nuclear magnetic resonance study. Planta 194:241–246.

Martin, P., A. Glatzle, W. Kolb, H. Omay, and W. Schmidt. 1989. N_2-fixing bacteria in the rhizosphere: Quantification and hormonal effects on root development. Z. Pflanzenernähr. Bodenk. 152:237–245.

McGongile, T.P., and A.H. Fitter. 1988. Ecological consequences of arthropod grazing on VA mycorrhizal fungi. Proc. R. Soc. Edin. 94B:25–32.

Meharg, A.A., and K. Killham. 1995. Loss of exudates from the roots of perennial ryegrass inoculated with a range of micro-organisms. Plant Soil 170:345–349.

Mench, M., and E. Martin. 1991. Mobilization of cadmium and other metals from two soils by root exudates of *Zea mays* L., *Nicotiana tabacum* L. and *Nicotiana rustica* L. Plant Soil 132:187–196.

Mench, M., J.L. Morel, A. Guckert, and B. Guillet. 1988. Metal binding with root exudates of low molecular weight. J. Soil Sci. 39:521–527.

Miranda, C.H.B., S. Urquiaga, and R.M. Boddey. 1990. Selection of ecotypes of *Panicum maximum* for associated biological nitrogen fixation using the ^{15}N isotope dilution technique. Soil Biol. Biochem. 22:657–663.

Moraghan, J.T. 1979. Manganese toxicity in flax growing on certain calcareous soils low in available iron. Soil Sci. Soc. Am. J. 43:1177–1180.

Moraghan, J.T. 1991. The growth of white lupin on a Calcaquoll. Soil Sci. Soc. Am. J. 55:1353–1357.

Moraghan, J.T., and T.J. Freeman. 1978. Influence of Fe EDDHA on growth and manganese accumulation in flax. Soil Sci. Soc. Am. J. 42:445–460.

Morel, J.L., M. Mench, and A. Guckert. 1986. Measurement of Pb^{2+}, Cu^{2+} and Cd^{2+} binding with mucilage exudates from maize (*Zea mays* L.) roots. Biol. Fertil. Soils 2:29–34.

Müller, J.F., A. Hülster, O. Päpke, M. Ball, and H. Marschner. 1993. Transfer pathways of PCDD/PCDF to fruits. Chemosphere 27:195–201.

Newman, E.I., W.R. Eason, D.M. Eissenstat, and M.I.R.F. Ramos. 1992. Interactions between plants: The role of mycorrhizae. Mykorrhiza 1:47–53.

Nouchi, I., S. Mariko, and K. Aoki. 1990. Mechanisms of methane transport from the rhizosphere to the atmosphere through rice plants. Plant Physiol. 94:59–66.

Ohwaki, Y., and H. Hirata. 1992. Differences in carboxylic acid exudation among P-starved leguminous crops in relation to carboxylic acid contents in plant tissues and phospholipid level in roots. Soil Sci. Plant Nutr. (Tokyo) 38:235–243.

Okon, Y., and C.A. Labandera-Gonzales. 1994. Agronomic applications of *Azospirillum*: An evaluation of 20 years worldwide field inoculation. Siol Biol. Biochem. 26:1591–1601.

Ottow, J.C.G., G. Benckiser, S. Santiago, and I. Watanabe. 1982. Iron toxicity of wetland rice (*Oriza sativa* L.) as a multiple nutritional stress. p. 454–460. *In* A. Scaife (ed.) Proc. of the 9th Int. Plant Nutrition Colloquium, Warwick, England. 22–27 Aug. 1982. Commonwealth Agric. Bur. Farriham House, Slough, England.

Perrin, R. 1990. Interactions between mycorrhizae and diseases caused by soil-borne fungi. Soil Use Manage. 6:189–195.

Phillips, D.S., F.D. Dakora, E. Sande, C.M. Joseph, and J. Zon. 1994. Synthesis, release, and transmission of alfalfa signals to rhizobial symbionts. Plant Soil 161:69–80.

Posta, K., H. Marschner, and V. Römheld. 1994. Manganese reduction in the rhizosphere of mycorrhizal and non-mycorrhizal maize. Mycorrhiza 5:119–124.

Prade, K., and G. Trolldenier. 1989. Further evidence concerning the importance of soil-air-filled porosity, soil organic matter and plants for denitrification. Z. Pflanzenernähr. Bodenk. 152:391–393.

Puthota, V., R. Cruz-Ortega, J. Johnson, and J. Ownby. 1991. An ultrastructural study of the inhibition of mucilage reaction in the wheat root cap by aluminium. p. 779–787. *In* R.J. Wright et al. (ed.) Plant–soil interactions at low pH. Kluwer Academic Publ., Dordrecht, the Netherlands.

Read, D.J. 1991. Mycorrhizas in ecosystems. Experientia 47:376–391.

Romera, R.J., E. Alcantara, and M.D. de la Guardia. 1991. Characterization of the tolerance of iron chlorosis in different peach rootstocks grown in nutrient solution. Plant Soil 130:121–125.

Römheld, V. 1987. Different strategies for iron acquisition in higher plants. What's new in plant physiology. Physiol. Plant. 70:231–234.

Römheld, V. 1991. The role of phytosiderophores in acquisition of iron and other micronutrients in graminaceous species: An ecological approach. Plant Soil 130:127–134.

Römheld, V., and D. Kramer. 1983. Relationship between proton efflux and rhizodermal transfer cells induced by iron deficiency. Z. Pflanzenphysiol. 113:73–83.

Römheld, V., and H. Marschner. 1990. Genotypical differences among graminaceous species in release of phytosiderophores and uptake of iron phytosiderophores. Plant Soil 123:147–153.

Rouhier, H., G. Billès, A. El Kohen, M. Mousseau, and P. Bottner. 1994. Effect of elevated CO_2 on carbon and nitrogen distribution within a tree (*Castanea sativa* Mill.): Soil system. Plant Soil 162:281–292.

Saleque, M.A., and G.J.D. Kirk. 1995. Root-induced solubilization of phosphate in the rhizosphere of lowland rice. New Phytol. 129:325–336.

Sauerbeck, D., and H.M. Helal. 1986. Plant root development and photosynthate consumption depending on soil compaction. p. 948–949. *In* Trans. 13th Congr. Int. Soil Sci. Soc., Hamburg. 13–20 Aug. 1986. ISSS, Vienna, Austria.

Sauerbeck, D., S. Nonnen, and J.L. Allard. 1981. Assimilateverbrauch und -umsatz im Wurzelraum in Abhängigkeit von Pflanzenart und -anzucht. Landwirtsch. Forsch. Sonderh. 37:207–216.

Schönwitz, R., and H. Ziegler. 1982. Exudation of water-soluble vitamins and of some carbohydrates by intact roots of maize seedlings (*Zea mays* L.) into a mineral nutrient solution. Z. Pflanzenphysiol. 107:7–14.

Schönwitz, R., and H. Ziegler. 1986. Quantitative and qualitative aspects of a developing rhizosphere microflora and hydroponically grown maize seedlings. Z. Pflanzenernähr. Bodenk. 149:623–634.

Schüepp, H., B. Dehn, und H. Sticher. 1987. Interaktionen zwischen VA-Mykorrhizen und Schwermetallbelastungen. Angew. Bot. 61:85–96.

Schulz, M., A. Friebe, P. Kück, M. Seipel, and H. Schnabl. 1994. Allelopathic effects of living quackgrass (*Agropyron repens* L.). Identification of inhibitory allelochemicals exuded from rhizome borne roots. Angew. Bot. 68:195–200.

Secilia, J., and D.J. Bagyaraj. 1987. Bacteria and actinomycetes associated with pot cultures of vesicular–arbuscular mycorrhizas. Can. J. Microbiol. 33:1069–1073.

Seggewiss, B., und A. Jungk. 1988. Einfluß der Kaliumdynamik im wurzelnahen Boden auf die Magnesiumaufnahme von Pflanzen. Z. Pflanzenern. Bodenk. 151:91–96.

Shand, C.A., A.E.S. Macklon, A.C. Edwards, and S. Smith. 1994. Inorganic and organic P in soil solutions from three upland soils: I. Effect of soil solution extraction conditions, soil type and season. Plant Soil 159:255–264.

Söderström, B. 1992. The ecological potential of the ectomycorrhizal mycelium. p. 77–83. *In* D.J. Read et al. (ed.) Mycorrhizas in ecosystems. CAB Int., Wallingford, England.

Symeonidis, L. 1990. Tolerance of *Festuca rubra* L. to zinc in relation to mycorrhizal infection. Biol. Metals 3:204–207.

Takagi, S., K. Nomoto, and T. Takemoto. 1984. Physiological aspect of mugineic acid, a possible phytosiderophore of graminaceous plants. J. Plant Nutr. 7:469–477.

Tarafdar, J.C., and A. Jungk. 1987. Phosphatase activity in the rhizosphere and its relation to the depletion of soil organic phosphorus. Biol. Fertil. Soils 3:199–204.

Tarafdar, J.C., and H. Marschner. 1994. Phosphatase activity in the rhizosphere of VA-mycorrhizal wheat supplied with inorganic and organic phosphorus. Soil Biology Biochem. 26:387–395.

Tarafdar, J.C., and H. Marschner. 1995. Dual inoculation with *Aspergillus fumigatus* and *Glomus mosseae* enhances biomass production and nutrient uptake in wheat (*Triticum aestivum* L.) supplied with organic phosphorus as Na-phytate. Plant Soil 173:97–102.

Thompson, J.P., and G.B. Wildermuth. 1989. Colonization of crop and pasture species with vesicular–arbuscular mycorrhizal fungi and a negative correlation with root infection by *Bipolaris sorokiniana.* Can. J. Bot. 69:687–693.

Thomson, C.J., H. Marschner, and V. Römheld. 1993. Effect of nitrogen fertilizer form on pH of the bulk soil and rhizosphere, and on the growth, phosphorus, and micronutrient uptake of bean. J. Plant Nutr. 16:493–506.

Turnau, K., I. Kottke, and F. Oberwinkler. 1993. *Paxillus involutus–Pinus sylvestris* mycorrhizae from heavily polluted forest: I. Elemental localization using electron energy loss spectroscopy and imaging. Bot. Acta 106:213–219.

Tyler, G., and L. Ström. 1995. Differing organic acid exudation pattern explains calcifuge and acidifuge behaviour of plants. Ann. Bot. 75:75–78.

Vogt, K.A., D.A. Publicover, and D.J. Vogt. 1991. A critique of the role of ectomycorrhizas in forest ecology. Agric. Ecosyst. Environ. 35:171–190.

von Wirén, N., V. Römheld, T. Shioiri, and H. Marschner. 1995. Competition between micro-organisms and roots of barley and sorghum for iron accumulated in the root apoplasm. New Phytol. 130:511–521.

Walter, A., V. Römheld, H. Marschner, and S. Mori. 1994. Is the release of phytosiderophores in zinc-deficient wheat plants a response to impaired iron utilization? Physiol. Plant. 92:493–500.

Warembourg, F.R., and G. Billes. 1979. Estimation carbon transfers in the plant rhizosphere. p. 183–196. *In* J.L. Harley and R. Scott-Russell (ed.) The soil–root interface. Academic Press, London.

Waschkies, C., A. Schropp, and H. Marschner. 1994. Relations between grapevine replant disease and root colonization of grapevine (*Vitis* sp.) by fluorescent pseudomonads and endomycorrhizal fungi. Plant Soil 162:219–227.

Waterer, J.G., J.K. Vessey, and C.D. Raper, Jr. 1992. Stimulation of nodulation in field pears (*Pisum sativum*) by low concentrations of ammonium in hydroponic culture. Physiol. Plant. 86:215–220.

Weissenhorn, I., A. Glashoff, C. Leyval, and J. Berthelin. 1994. Differential tolerance to Cd and Zn of arbuscular mycorrhizal (AM) fungal spores isolated from heavy metal-polluted and unpolluted soils. Plant Soil 167:189–196.

Weissenhorn, I., C. Leyval, and J. Berthelin. 1995. Bioavailability of heavy metals and abundance of arbuscular mycorrhiza in a soil polluted by atmospheric deposition from a smelter. Biol. Fertil. Soils 19:22–28.

Wilcox, H.E. 1991. Mycorrhizae. p. 731–765. *In* Y. Waisel et al. (ed.) The plant root: The hidden half. Marcel Dekker, New York.

Wojtaszek, P., M. Stobiecki, and K. Gulewicz. 1993. Role of nitrogen and plant growth regulators in the exudation and accumulation of isoflavenoids by roots of intact white lupin (*Lupinus albus* L.) plants. J. Plant Physiol. 142:689–694.

Yehuda, Z., M. Shenker, V. Römheld, H. Marschner, Y. Hadar, and Y. Chen. 1996. The role of ligand exchange in the uptake of iron from microbial siderophores by gramineous plants. Plant Physiol. 112:1273–1280.

Youssef, R.A., and M. Chino. 1987. Studies on the behavior of nutrientes in the rhizosphere: I. Establishment of a new rhizobox system to study nutrient status in the rhizosphere. J. Plant Nutr. 10:1185–1195

Zhang, J., and W.J. Davies. 1989. Sequential response of whole plant water relations to prolonged soil drying and the involvement of xylem sap ABA in the regulation of stomatal behaviour of sunflower plants. New Phytol. 113:167–174.

Zhang, D.V., V. Römheld, and H. Marschner. 1989. Effect of zinc deficiency in wheat on the release of zinc and iron mobilizing root exudate. Z. Pflanzenernahr. Bodenk. 152:205–210.

Zuberer, D.A. 1990. Soil and rhizosphere aspects of N_2-fixing plant-microbe associations. p. 317–368. *In* J.M. Lynch (ed.) The rhizosphere. John Wiley & Sons, Chichester, England.

9 Soil–Root Interface: Physicochemical Processes

M. J. McLaughlin

CSIRO Land and Water
Glen Osmond, Australia

E. Smolders and R. Merckx

Laboratory for Soil Fertility and Soil Biology
Katholieke Universität
Leuven, Belgium

9–1 INTRODUCTION

The physicochemical reactions between plant root cells, microorganisms and the solution and solid phases of soil are critical in determining the environmental impacts of contaminants on ecosystem health. The term *rhizosphere* was first coined by Hiltner (1904) to describe the region in soil at the soil–root interface, as part of his work examining the relationship between bacteria and roots of leguminous plants. Since then the term has assumed a more general meaning related to the Greek derivation of the word *rhiza* meaning root and *sphere* meaning field of action or influence. The actual physical dimensions of the rhizosphere vary among plants and among soils. Soil up to several centimeters from the root is often termed *rhizosphere* soil (Campbell & Greaves, 1990), although the zone is more usually restricted to a few millimeters from the root. While no discrete zones exist within the rhizosphere, a number of classifications have been proposed to divide the soil–root interface into component parts (Lynch, 1983). The endorhizosphere is comprised of the root itself (Fig. 9–1) and includes the stele, root cortex, endo-, and epidermis (rhizoplane) and the root cap. The ectorhizosphere includes the region of soil close to the root where root hairs are active, where plant exudates, sloughed root cells, plant mucigel and mucilage, and plant-induced microbial activity significantly alter soil physical and chemical reactions compared with bulk soil. Figure 9–2 summarizes some of the major features of the rhizosphere with regard to contaminant chemistry.

Much of the research on physicochemical reactions of contaminants in soil has focused on the use of well-defined model systems simulating bulk soil characteristics. The chemistry of contaminants in the rhizosphere is often inferred from a knowledge of conditions in the rhizosphere and the results of controlled

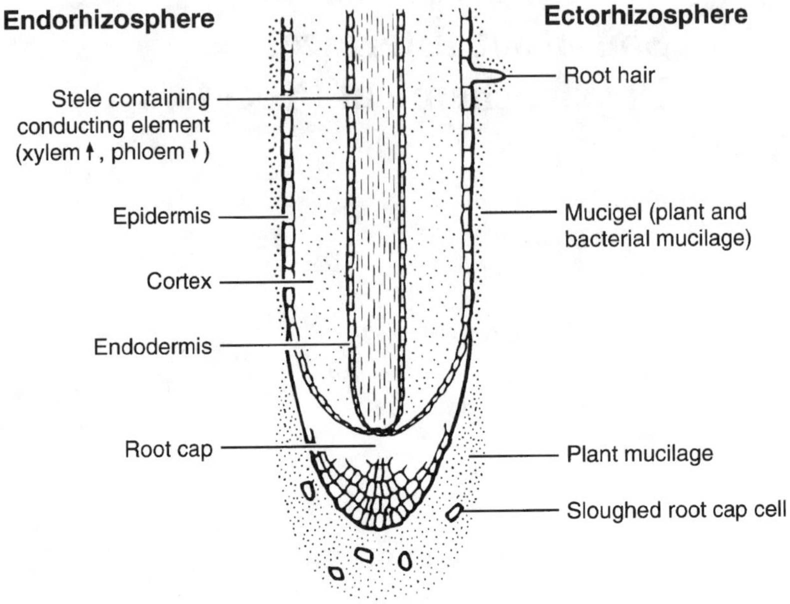

Fig. 9–1. The structure of the rhizosphere (from Lynch, 1983).

laboratory studies, often in the absence of plants. Studies attempting to assess the impact of rhizosphere processes on contaminant chemistry in vivo are, unfortunately, few. With some recent exceptions, studies of physicochemical reactions specific to the rhizosphere have been mostly limited to the nutrient elements C, N, and P. In attempting to review this topic therefore, we have drawn from infor-

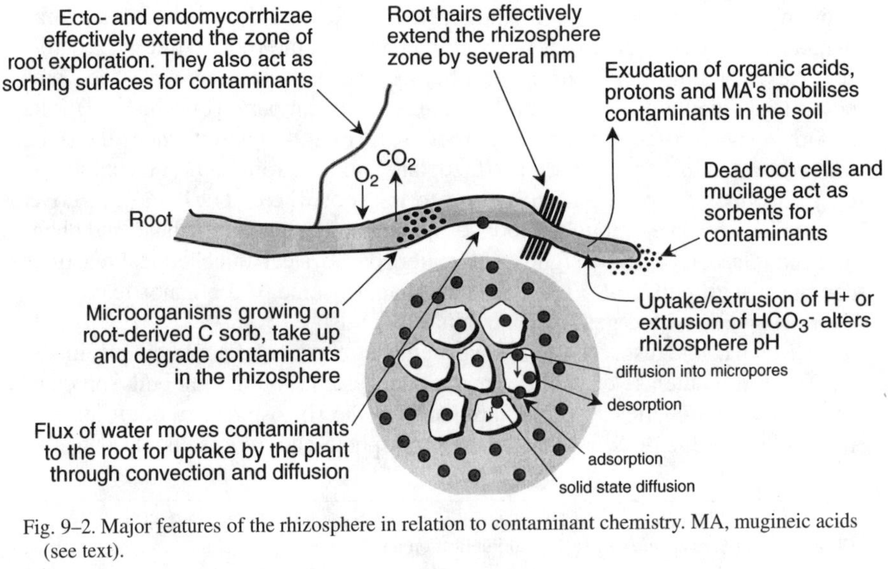

Fig. 9–2. Major features of the rhizosphere in relation to contaminant chemistry. MA, mugineic acids (see text).

mation gathered for macronutrient elements and apply this to potential reactions for contaminants in soil. Of necessity, we will focus our discussion on some of the more persistent contaminants in soil systems, i.e., heavy metals and radionuclides. The increasing use of fertilizers and crop protection chemicals in agriculture is posing a number of soil and food contamination issues. Heavy metals in fertilizers, principally Cd, are slowly accumulating in many soils and monitoring of food Cd concentrations has alerted authorities in some countries to potential problems. Radionuclides, principally Cs have recently attracted renewed interest in Europe due to the emission of radionuclides into the atmosphere from the Chernobyl incident. ^{137}Cs has a long half life (30.2 y) and is mobile and biologically available in terrestrial and aquatic systems. Reactions of organic chemicals in the rhizosphere have not been considered as part of this review.

9–2 SOIL PROCESSES

9–2.1 Reactions of the Soil Solid Phase with Contaminants: Adsorption–Desorption of Metal Contaminants

The solid phase in soil has a great capacity to adsorb many contaminants. The adsorption characteristics of soils are due to layer silicate clays, oxides and hydroxide minerals, and organic matter, all of which exhibit either permanent or variable surface charge characteristics. Permanent negative charge in layer silicate clays develops due to isomorphous substitution of Al for Si and Mg for Al in the crystal lattice. Variable charge exists on oxide–hydroxide surfaces and on soil organic matter, where net surface charge is related to solution pH, ionic strength, and solution composition. Some of these reactions have been described earlier in more detail by Hayes and Traina (1998, this publication), Fellows et al. (1998, this publication), and Vasudevan and Stone (1998, this publication). Several excellent reviews also have been published on the reactions of ions with soil surfaces (Barrow, 1987; Schindler & Sposito, 1991; McBride, 1989, 1991). Few studies have examined these reactions in the rhizosphere, with most examining bulk soil reactions and some modifying conditions to simulate expected rhizosphere conditions (e.g., increased concentrations of organic acids). Only a brief review of these reactions is given here with emphasis on the metal contaminants and some discussion of the potential impacts of rhizosphere conditions on these processes.

Sorption of metal contaminants to soils has been extensively studied. Sorption can either be electrostatic (outer sphere complexes) or involve chemisorption to the surface (specific adsorption or inner sphere complexes). Electrostatic adsorption is nonspecific in that bonding depends only on the charge and hydration properties of the cation (McBride, 1991).

$$Cd^{2+} + 2Na - Clay \Leftrightarrow Cd^{2+} - Clay + 2Na^{+} \quad [1]$$

With these sorption reactions there is no change in surface charge and sorption per unit weight is dependent on the particle surface area and charge density, which varies with mineralogy (Tiller et al., 1984; Barrow, 1993). The selectivity coefficient for Na–Cd exchange can therefore be written

$$K_{\mathrm{Na}}^{\mathrm{Cd}} = \left(\frac{(m_{\mathrm{Na}})^2 \bullet (N_{\mathrm{Cd}})}{(m_{\mathrm{Cd}}) \bullet (N_{\mathrm{Na}})^2} \right) \qquad [2]$$

where m_{Na} and m_{Cd} are the activities in solution of these ions and N_{Na} and N_{Cd} are the adsorbed mole fractions of metal. At low pH it may be observed that alkaline earth metals are bound with similar energy as transition metal (contaminant) ions if charge and size are similar (El-Sayed et al., 1970), with selectivity values for metal ions $K^{x}{}_{y}$ close to unity. At higher pH values, however, the strength of bonding of heavy metals to clay surfaces is much greater than alkaline earth metal ions and the selectivity coefficient is often significantly greater than unity. In an early series of studies by Tiller and coworkers (Tiller & Hodgson, 1962; Tiller et al., 1963; Hodgson et al., 1964) this *specific adsorption* was attributed to metal hydrolysis and adsorption of the hydrolyzed species. They studied Co and Zn adsorption by soils in the presence of excess Ca (to minimize electrostatic sorption) and found the specifically sorbed metal was bound so strongly to the clay surface that mineral dissolution was necessary to liberate the metal. They postulated that some of the adsorbed metal had migrated into the crystal structure. The hypothesis that hydrolysis controls sorption reactions for metals has gained popularity (Farrah & Pickering, 1977; Gerth & Brummer, 1983; Barrow, 1993), with the recognition that the relative affinity of surfaces for metals can be related to their tendency to hydrolyze (Kinniburgh et al., 1976; McBride, 1991). For example, Kinniburgh et al. (1976) determined the affinity series for amorphous Fe hydroxide as

$$Pb^{2+} > Cu^{2+} > Zn^{2+} > Ni^{2+} > Cd^{2+} > Co^{2+} > Sr^{2+} > Mg^{2+} \qquad [3]$$

which is an extension of the original Irving–Williams series for the stability of metal complexes (Irving & Williams, 1948). It should be noted, however, that Kinniburgh et al.'s data were generated at fairly high solution metal concentrations (0.125 m*M*), several orders of magnitude higher than metal concentrations normally found in soil solutions.

Specific sorption or inner-sphere complexation occurs on oxide–hydroxide surfaces and at the edges of aluminosilicate clays (Schindler & Sposito, 1991). There also is evidence that metals can form inner-sphere complexes with soil organic matter. These surfaces exhibit variable charge in response to solution compositional changes (pH, ionic strength, and cation and anion composition) and the adsorbed metal ion forms a covalent bond with the surface (McBride, 1991). Schindler and Sposito (1991) have described the general behavior of hydroxide–oxide surfaces, where surface hydroxyls can be protonated or deprotonated depending on solution conditions, and metal binding is through a deprotonation of the surface.

$$S - OH + M^{z+} \rightarrow S - OM^{(z-1)+} + H^{+} \qquad [4]$$

$$2S - OH + M^{z+} \rightarrow (S - O)_2 M^{(z-2)+} + 2H^+ \qquad [5]$$

Direct evidence for the formation of such bonds has been obtained from electron spin resonance and ultra violet spectroscopic studies of adsorbed species (Bleam & McBride, 1986). The participation of many oxides in metal retention would not be predicted on the basis of the charge characteristics of pure oxide–hydroxide surfaces, as some have a net positive charge at normal soil pH values, however, the impure nature of most natural oxides in soil, where adsorbed organic matter and other anionic ligands cause structural distortion of the oxide and may impart a net negative charge to the surface, means that oxides in soil can be important constituents determining sorption of metals (Kwong & Huang, 1979; Xue & Huang, 1995; Helmke & Naidu, 1996). Hydrous oxides of Fe and Mn were identified almost 30 yr ago by Jenne (1968) as being the most important surfaces controlling heavy metal sorption by soils, and much work subsequently has supported this hypothesis (Loganathan & Burau, 1973; Kinniburgh et al., 1976; Lion et al., 1982; Tiller et al., 1984).

Organic matter in soil also acts as a variably charged surface due to deprotonation of carboxylic, phenolic, and enolic groups and often exhibits a high affinity for metal contaminants (Swift & McLaren, 1991). Although not discussed here, soil organic matter also acts as the major sorbent for many hydrophobic contaminants (e.g., pesticides, polyaromatic hydrocarbons, and others).

The sorption of metals in most soils is highly pH dependent, due to the effects of pH on variable charge sorption sites and the tendency for metallic contaminants to hydrolyze in solution. With the exception of metals that form oxyanions (e.g., Cr and As), metal retention is low at low pH and as solution pH increases metal sorption increases dramatically, the so-called pH *sorption edge* (Fig. 9–3). Given the dramatic changes in solution pH that can occur in the rhizosphere (see below), plant root activity is likely to have a marked effect on metal availability through changes in solution pH.

Despite being an alkali metal ion, ^{137}Cs also is strongly sorbed to the soil solid phase. The solid–liquid distribution coefficient K_D varies from about 50 mL g^{-1} for light textured soils to > 10 000 mL g^{-1} for loam and clay soils, however, once in solution, ^{137}Cs is readily available for root uptake, most probably because of its chemical resemblance to K (Smolders & Shaw, 1995). ^{137}Cs activity in soil solution is never controlled by (co)precipitation because ^{137}Cs is soluble in water and, as a radiocontaminant, only present at trace concentrations. Instead, sorption is interpreted in terms of ion exchange laws applied to the qualitative findings that Cs is selectively sorbed by nonexpanding 2:1 minerals such as illite (Sawhney, 1972; Francis & Brinkley, 1976). It is generally accepted that specific sorption occurs at the frayed edge sites of illitic particles. These sites are quantitatively occupied in soils by K and NH_4^+ and to a very limited extent by Ca and Mg. Although the exchange capacity of the frayed edge sites is only a small percentage of the soil cation-exchange capacity (CEC), the majority of the ^{137}Cs is sorbed on these sites because the Cs–K or Cs–NH_4^+ selectivity coefficients on these sites are more than two orders of magnitude larger than those on the planar

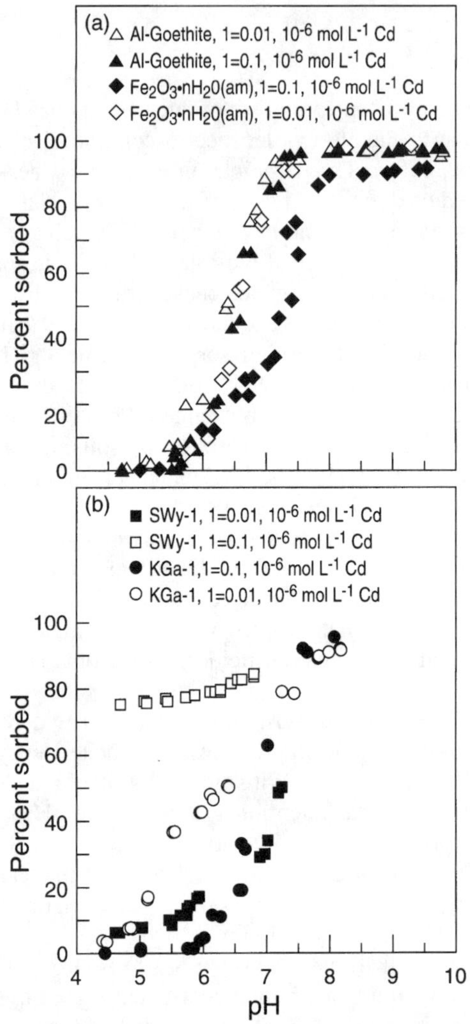

Fig. 9–3. Effect of pH on the sorption of Cd to (a) Fe–Al oxide surfaces and (b) aluminosilicate mineral surfaces. Background electrolyte = $NaClO_4$. SWy = Wyoming bentonite, KGa = kaolinite (from Zachara et al., 1992).

exchange sites of clay minerals (Cremers et al., 1988; Sweeck et al., 1994). The ion exchange properties of the frayed edge sites can be studied by masking the regular exchange sites of the soil (i.e., planar exchange sites of clay minerals and the humic acids). This is traditionally performed with large cations such as silver–thiourea, which have a high selectivity for the regular exchange sites of the soil, and that are sterically excluded from the frayed edge sites (Cremers et al., 1988).

9–2.2 Factors Affecting Adsorption–Desorption of Contaminants in the Rhizosphere

9–2.2.1 Solution pH

As discussed in Section 9–2.1.1, soil solution pH has a dramatic effect on total concentrations of contaminants in the soil solution, due to the effect of pH on surface charge properties of the soil solid phases and hydrolysis of metal cations. Hence, at low solution pH, concentrations of cationic contaminants in solution increase, e.g., Cd, Zn, Cu, and others, while anionic contaminants, e.g., CrO_4^{2-}, AsO_4^{3-} and F^- will become firmly adsorbed. Changes in solution pH also may lead to dissolution of solid phase contaminants, e.g., Pb particulates. Protons have only very limited effect on ^{137}Cs sorption: the H–K selectivity coefficient on the frayed edge sites is about one (De Preter, 1990).

Changes in solution pH also may lead to protonation or deprotonation of soluble organic ligands in solution, thus affecting the interaction between contaminants and soil surfaces (see Section 9–2.2.3).

9–2.2.2 Inorganic Ligands

A number of inorganic ligands may complex with metal contaminants and lead to changes in the charge of the ion, which therefore affects retention by the soil. Chloride is perhaps the best example, as it can often exist in high concentrations in the soil solution and is known to form complexes with a number of contaminant metal ions (Hahne & Kroonjte, 1973). For example, chloro-complexation of Cd has been demonstrated to increase Cd mobility through soil, through desorption of Cd from the solid phase and maintenance of a higher equilibrium solution Cd concentration (Doner, 1978; Egozy 1980; Boekhold et al., 1993).

Sulfate also can form ion pairs with some metal contaminants reducing the charge on the metal cation, e.g., $CdSO_4^0$. Effects of SO_4^{2-} on Cd sorption are not clear as SO_4^{2-} itself is a sorbing species in many soils and sorbed SO_4^{2-} imparts a greater negative charge to the sorbing surface (Benjamin & Leckie, 1982; Gessa et al., 1984). Garcia-Miragaya and Page (1976) found that increasing SO_4^{2-} concentrations in solution moderately reduced Cd sorption by soil. Benjamin and Leckie (1982) found that increasing SO_4^{2-} concentrations in solution also reduced Cd sorption by model oxides (amorphous Fe oxide, silica, and gibbsite) and had no effect on Cd sorption by lepidocrocite. Hoins et al. (1993) found increasing concentrations of SO_4^{2-} enhanced Cd sorption by goethite. The effects of SO_4^{2-} on Cd sorption will therefore depend on the nature of the sorbing surfaces present in soil surfaces, which are able to effectively sorb sulfate and decrease surface charge (i.e., more negative) and are likely to increase Cd retention, overcoming any effects of ion pairing.

The addition of P as phosphate to solution appears to enhance metal retention by soil. This effect may be related to precipitation of metal phosphates if P is added to the same solution as the metal ions, or may be due to enhanced sorption as evidenced by increased metal retention of soils pretreated with P where precipitation of metal phosphates is not expected (Kuo & McNeal, 1984).

In alkaline soils, hydroxide and bicarbonate also can form complexes with a range of metal contaminants (Ritchie & Sposito, 1995), but activities of metal ions in solution at alkaline pH values are generally low, due to increased retention by the soil solid phase. Indeed, the hydrolyzed metal species (MOH^+) is often regarded as being preferentially sorbed by soil surfaces over the aquo free metal ion (see Section 9–2.1).

9–2.2.3 Organic Ligands

The exudation of soluble organic compounds by plants is perhaps one of the more important processes likely to affect retention of contaminants by soil in the rhizosphere. Soluble organics can affect retention of metal ions on mineral surfaces through:

1. Competition for sorption sites or blocking of sorption sites on the mineral colloid (Levy & Francis, 1976),
2. Increased retention due to formation of a strongly sorbed metal-ligand complex (Elliott & Huang, 1979; Chubin & Street, 1981; Naas & Horowitz, 1986),
3. Decreased retention due to strong complexation of metal in solution in a nonsorbing form, lowered free metal activity leading to desorption of surface-bound metal (Elliott & Denneny, 1982; Inskeep & Baham, 1983; Neal & Sposito, 1986),
4. Increased retention through increases in surface negative charge, dependent on the solution pH and point of zero charge (PZC) of the surface (Chairidchai & Ritchie, 1990),
5. Increased retention through increased surface area and charge due to structural distortion of minerals during formation (Xue & Huang, 1995).
6. Dissolution of clay minerals (Huang & Keller, 1972; Boyle & Voigt, 1973) leading to reduction in retention of contaminants.

A recent review of some of the above processes is given in Harter and Naidu (1996). Most of the above studies have examined the effects of organics on metal retention by soil or model minerals in the absence of plants. The ecological and environmental effects of the exudation of soluble organics by plant roots depends to a large extent on the composition of the exudates involved, their persistence in rhizosphere soil and plant uptake characteristics for the contaminant in question. This is discussed further in Section 9–3.3.

9–2.2.4 Soil Solution Ionic Strength and Cation Concentrations

The effect of solution composition on metal retention by soils has implications for the rhizosphere environment. Increasing salt concentration in solution affects metal retention by soils through effects on the surface potential, effect of the cation in competition for sorption sites, effect of ionic strength on the metal activity coefficient and the effect of ionic strength on dissociation constants (Barrow, 1987). There has been no thorough study of the degree to which the ionic strength of the rhizosphere soil solution differs from that of bulk soil. It could be argued that, with the uptake of nutrient ions by the plant, it is likely that the ionic strength of the soil solution adjacent to the root will be low. If it is assumed that plants transpire approximately 250 mL water for each gram of dry matter produced and contain approximately 300 $cmol^+$ ions kg^{-1} dry matter, then if the concentrations of ions in the bulk soil solution is <1.2 $cmol^+ L^{-1}$, the rhizosphere soil solution will be of lower ionic strength than that of the bulk soil; however, exudation of organics and the accumulation (through mass flow) of certain ions (e.g., Ca) and organic material at the root surface in excess of plant demand (see Section 9–2.5.1) could increase solution ionic strength (divalent ions have a larger effect on ionic strength than monovalent ions).

Competitive effects of cations in reducing metal sorption are commonly observed. For Cd sorption, certain divalent cations (e.g., Ca^{2+}, Co^{2+}, Cu^{2+}, Ni^{2+} Pb^{2+}, and Zn^{2+}) competitively inhibit Cd retention (Christensen, 1987; Homann & Zasoski, 1987). Many authors have reported greater sorption in the presence of Na^+ than Ca^{2+} ions (Garcia-Miragaya & Page, 1976; Christensen, 1984; Boekhold et al., 1993). The lower sorption in the Ca^{2+} medium was attributed to the competition for sorption sites between Ca^{2+} and Cd^{2+} ions. The accumulation of Ca in rhizosphere soil solution may be important due to these competitive effects on sorption, as well as due to competition for binding on soluble organic ligands (Hamon et al., 1995).

The major competing cations in ^{137}Cs sorption are K and NH_4^+. According to De Preter (1990), the K_D of ^{137}Cs in soil depends on K and NH_4 concentrations m_K and m_{NH4} as

$$K_D = \frac{RIP}{m_K + 5m_{NH_4}} \qquad [6]$$

in which RIP is the radiocaesium interception potential, a soil characteristic that can be measured experimentally (Sweeck et al., 1994). The RIP is a measure of the frayed edge site capacity and the Cs–K selectivity coefficient on these sites. NH_4^+ has been found to be five times more competitive with Cs than is K. Stable Cs in soil also can reduce the radiocaesium sorption by occupying the sites with the highest affinity; however, naturally occurring concentrations of stable Cs are generally below the threshold concentrations level at which ^{137}Cs sorption decreases (Coughtrey & Thorne, 1983; Wauters, 1994). The divalent cations Ca and Mg also have limited effect on ^{137}Cs sorption because of their low selectivity for the frayed edge sites. In highly organic soils (i.e., organic matter content >80 %), significant fractions of ^{137}Cs can be present in the organic matter phase (Valcke & Cremers, 1994). No selectivity differences have been found among K, Cs, and NH_4 on these sites, which exhibit readily reversible exchange properties (Stevenson, 1982; Sweeck et al., 1994; Valcke & Cremers, 1994). In these soils, protons and divalent cations can be expected to provide significant competition for ^{137}Cs sorption.

9–2.2.5 Solution Concentration of Contaminant

Early work on retention of heavy metals by soils used unrealistically high solution concentrations (μ*M* up to m*M*) of metals (John, 1972; Bittel & Miller,

1974; Cavallero & McBride, 1978; Milberg et al., 1978). Concentrations of most contaminant metals in soil solution are at the pM to sub μM level. As pointed out by Hendrickson and Corey (1981), the selectivity for metal sorption is highly dependent on the solution concentration of metal. The greater the surface coverage by metal, which is dependent on the equilibrium solution concentration of metal, the lower the strength of metal retention by the surface. These authors reanalyzed a number of the earlier studies of Cd adsorption by soils, to determine the impact of surface coverage of exchange sites on the relative binding of Cd by the surface, measured as the solid–liquid distribution coefficient (K_d) from a modified Langmuir plot of the adsorption data (Fig. 9–4). Similar results have been demonstrated for Cu, Pb, and Zn (Hendrickson & Corey, 1981).

9–2.2.6 Time of Contact

Adsorption of metal contaminants is rarely fully reversible, with desorption often occurring much more slowly than adsorption (Tiller et al., 1963; McLaren et al., 1983). Thus, *hysteresis* is often noted in metal adsorption–desorption experiments (Fig. 9–5). This characteristic has been attributed to migration of metal ions into the crystal structure or surface (Tiller et al., 1963; Brummer et al., 1988) or to the activation energy required for desorption (McBride, 1989). Thus, these hypotheses indicate that all sorption should be reversible, so that the apparent hysteresis is due to the rate of desorption being much slower than the rate of adsorption. The implication of this is that once a contaminant ion is added to soil, it will slowly become less available for either plant uptake or for desorption and/or leaching to water supplies. Various contaminant ions have been found to have different degrees of *reversibility* in terms of reaction with soil surfaces (Barrow, 1987; Brummer et al., 1988).

In some situations, the slow desorption of metals could be due to the formation of solid solutions, where the metal is occluded in the solid matrix by dis-

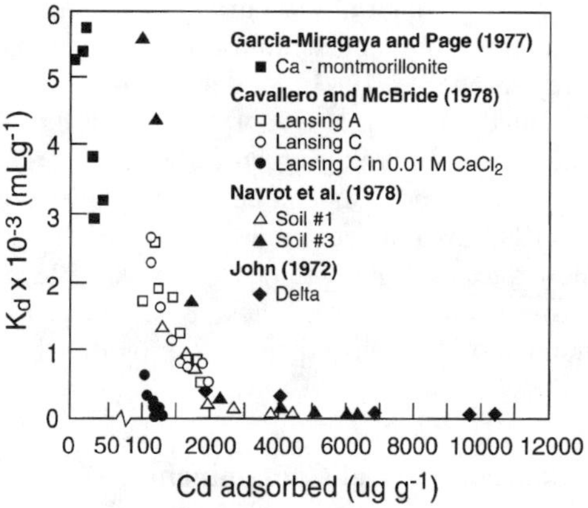

Fig. 9–4. Effect of surface coverage of soil surfaces with Cd (Cd loading) on apparent selectivity coefficients (K_d) of the surface for Cd (from Hendrickson & Corey, 1981).

solution and crystallization at the solid surface. For example, Davis et al. (1987) and Papadopoulos and Rowell (1988) found that Cd was incorporated into the structure of calcite under conditions where the solutions were undersaturated with respect to $CdCO_3$; however, evidence that the formation of solid solutions in soils controls metal solubility is mostly speculative (McBride, 1989).

Similar to data for hysteresis of metal sorption–desorption in soils, desorption studies of ^{137}Cs using concentrated salt solutions (e.g., 1 *M* KCl or 1 *M* NH_4Cl) reveal a nondesorbable (*fixed*) fraction of ^{137}Cs. More recently, this desorption methodology has been questioned because high salt concentrations are thought to promote fixation. New desorption methodologies have been developed based on the infinite sink principle (Wauters et al., 1994). In this method, ^{137}Cs is desorbed using an *accumulator adsorbent*. The accumulator can be an ion exchange resin or the so-called giese granulate (ammonium copper hexacyanoferrate impregnated on an inert carrier). The amount of ^{137}Cs desorbed by this method is generally greater than with the concentrated salt solutions. Even so, the *fixed* fraction can be as large as 90% and this fraction increases with ageing time

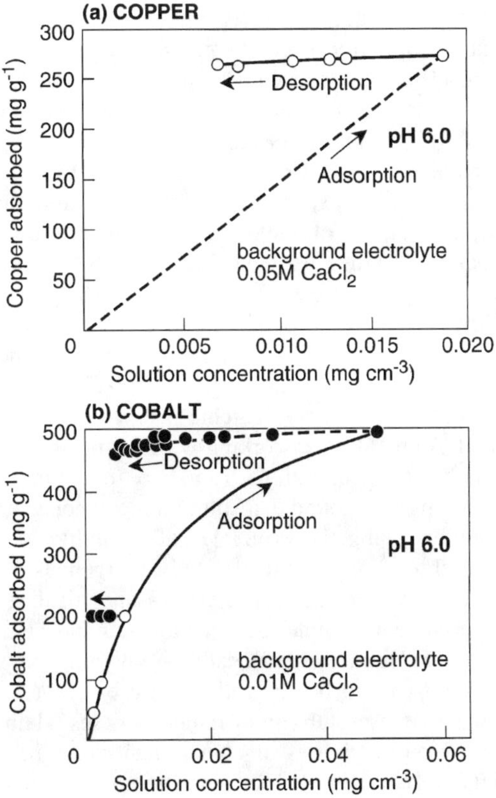

Fig. 9–5. Hysteresis of adsorption–desorption curves for copper and cobalt (from McLaren et al., 1983).

(adsorption time) and with increasing concentrations of divalent ions present during adsorption (Wauters, 1994). The sorption and desorption kinetics have been described in several models (Comans, 1990; Wauters, 1994) in which up to three pools are distinguished in the solid, instantaneous equilibrium sites, slow sorption–desorption sites and irreversibly sorbed (fixed) sites. The fixed fraction varies considerably between soils and, to date, it is not known which soil factor is behind the fixation process. The fixation of ^{137}Cs in soil is an important process for long-term reduction in availability and can explain the gradual reduction in food contamination after a nuclear accident. As an example, the ^{137}Cs activities in skim milk powder in Austria have been found to decrease exponentially between 1.5 and 7 yr after the Chernobyl accident with an effective half life in the range 1.5 to 2 yr (Mück, 1995).

9–2.3 Precipitation and Dissolution Reactions

Lindsay and coworkers (Lindsay, 1979; Lindsay & Norvell, 1969; Ma & Lindsay, 1990) postulate that solubility of metals such as Zn may be controlled by dissolution and precipitation of minerals such as franklinite ($ZnOFe_2O_3$). Others contend that precipitation and dissolution reactions are not the major determinants of metal solubility in most soils, because metal concentrations in solution are seldom sufficient to exceed the solubility product of any solid phase so that adsorption–desorption reactions control metal solubility in most soils (Brummer et al., 1983; McBride, 1989; Tiller, 1996).

Evidence for the control of Cd and Pb solubility by solid phase dissolution–precipitation derives from experiments with soils equilibrated in the laboratory with high concentrations of metal in solution. For example, Santillan-Medrano and Jurinak (1975) investigated Cd and Pb solubility in three soils, with Cd or Pb salts added to achieve equilibrium concentrations in solution varying from 10^{-7} to 10^{-3} M for both metals. Similarly, Cavallaro and McBride (1978) and Papadopoulos and Rowell (1988) used equilibrium Cd concentrations varying from 10^{-6} to 10^{-2} M. These concentrations are well above those found in most soils, even highly contaminated soils (Ritchie & Sposito, 1995). Indeed, recent work by Jopony and Young (1994), working with a range of highly contaminated soils (total Cd and Pb concentrations from 1 to 1638 and 330 to 38 178 mg kg^{-1}, respectively) in England, found that solution metal concentrations could not be adequately described using the principles of solubility of discrete mineral phases. Most of the above studies did not study indigenous soil solutions, but dilute salt extracts of soil. With improvements in analytical sensitivity and the ability to now measure free uncomplexed metal ion activities in true soil solutions (Berggren, 1990; Fitch & Helmke, 1989), data are emerging that support the view that metal solubility appears to be controlled more by ion exchange and complexation phenomena than by solubility of mineral phases (Helmke et al., 1998). In rhizosphere soils where activities of contaminants may be reduced through plant and microbial uptake, and where concentrations of complexing ligands may be relatively high, we contend that it is unlikely that dissolution/precipitation reactions play a major role in controlling metal contaminant chemistry.

9–2.4 Redox Reactions

Oxidation–reduction (redox) reactions are important in controlling the chemical speciation of a number of contaminant elements, notably As, Cr, and Se. Redox reactions also are important in controlling the form and reactivity of Fe and Mn oxides in soil, which are major sorbing surfaces for metal contaminants. Furthermore, reduction of sulfate to sulfide in anaerobic environments also may affect metal contaminant solubility through the precipitation of highly insoluble metal sulfides (Alloway, 1995). The general equation for redox reactions is shown in Eq. [7] (Lumsden & Evans, 1995).

$$\text{oxidant} + ne^- \Leftrightarrow \text{reductant} \qquad [7]$$

Electrons are produced in soils as a result of microbial respiration of organic matter, and when O_2 is absent other elements may act as electron acceptors. The redox status of soil is measured as the millivolt difference in potential (Eh) between a platinum electrode and a reference electrode (e.g., calomel electrode; Lumsden & Evans, 1995). High Eh values indicate oxidizing conditions and low values reducing conditions. Redox status also is sometimes reported as pE values, which is a hypothetical or equivalent electron activity expressed as a negative logarithm of the electron activity. pE is related to Eh as follows:

$$\text{Eh (mV)} = 59.2 \text{ pE } (T = 25°C) \qquad [8]$$

Metals such as As, Cr, and Se exist in more than one oxidation state and therefore redox reactions are critical in controlling their mobility and toxicity in soil. Inorganic As exists in two main oxidation states, arsenite—As^{III} (H_3AsO_3) and arsenate—As^V ($H_2AsO_4^-$ and $HAsO_4^{2-}$). Arsenate behaves analogously to P in soils, being strongly sorbed by oxides and to a lesser extent aluminosilicate clays. Manganese oxides are capable of oxidizing As^{III} to As^V in soil (Oscarson et al., 1983). Arsenite is a weak acid and only dissociates at high pH (pK = 9.29), so is less firmly bound by soil surfaces than As^V. Reducing conditions may therefore release As into solution due to reduction of Fe oxides liberating sorbed As^V, and reduction of As^V to As^{III}. Various organic As forms also occur, e.g., mono- and dimethyl arsenic acid (O'Neill, 1995).

Chromium exists mainly as either soluble Cr^{VI} species (CrO_4^{2-} and $HCrO_4^-$) or as insoluble precipitates or adsorbed Cr^{III} (Cr^{3+}) in soils (Katz & Salem, 1994). Virtually no inorganic Cr^{III} is found in soluble form in soil (Massechelyn & Patrick, 1994). Cr^{VI} is much more toxic than Cr^{III} to both plants and animals, so that redox reactions of this element are important in determining environmental impact. Similar to As^{III}, soluble Cr^{III} may be oxidized to Cr^{VI} by Mn oxides (Bartlett & James, 1979), but precipitated or organically complexed Cr^{III} forms are less prone to oxidation (James & Bartlett, 1983).

Selenium can exist in a wide range of oxidation states (inorganic Se^0, Se^{II}, Se^{III}, Se^{IV}, and Se^{VI}) as well as organic-Se compounds. Massechelyn and Patrick, (1994) recently summarized the critical redox potentials for transformation of some contaminant ions in wetland soils (Fig. 9–6).

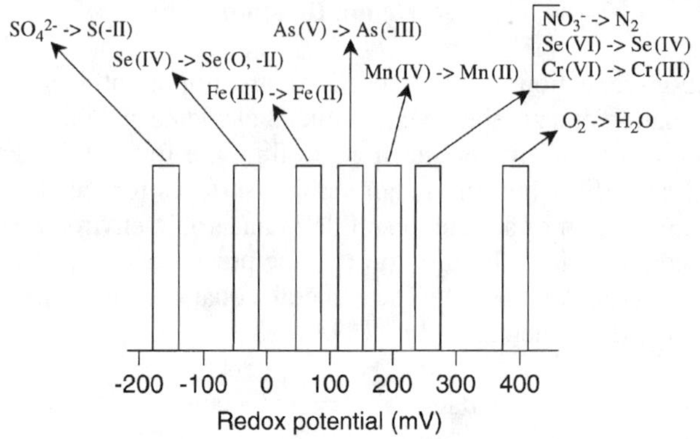

Fig. 9–6. Critical redox potentials for some contaminant ions (from Massechelyn & Patrick, 1994).

There has been little study of how soil redox potential changes in the rhizosphere could affect contaminant chemistry, but the creation of an oxidized zone adjacent to the plant root in wetland soils has been identified as one process affecting As and Zn chemistry in soil (Section 9–3.4).

9–2.5 Soil Solution in the Rhizosphere

9–2.5.1 Solution Composition

The soil solution is the medium through which contaminants are taken up, either actively or passively, by plant roots and microbial cells. Several reviews have covered the composition and behavior of the soil solution (Adams, 1974; Stevenson & Ardakani, 1972; Sposito, 1989; Ritchie & Sposito, 1995). Soil factors (adsorption–desorption, precipitation–dissolution, or oxidation–reduction) that control solution composition have been described above. In the study of contaminant concentrations in soil solution, few studies have specifically investigated behavior in the rhizosphere.

If the uptake of an ion from soil solution by roots differs from that transported with water flow to the roots (mass flow), its concentration will change at the soil–root interface. Typically, P and K are depleted in the rhizosphere and Ca accumulates (Brewster & Tinker, 1970). The concentration change creates a diffusive gradient around the roots. For some ions that are strongly depleted in the rhizosphere, the resulting diffusive flux can be consistently larger than the contribution from mass flow and is the limiting factor for total ion uptake from soil.

Using autoradiographs, concentration gradients in the rhizosphere have been visually demonstrated for ^{32}P, ^{86}Rb, ^{45}Ca, $^{35}SO_4$, ^{99}Mo, and ^{65}Zn (references in Barber, 1984). Also, the analysis of soil sampled at different distances away from single roots or root mats has demonstrated a concentration gradient (Gahoonia & Nielsen, 1992). As far as we are aware, no such concentration changes have been visually demonstrated by autoradiography or other techniques for soil contaminants such as Cd, Pb, ^{137}Cs, or other contaminants. Hamon et al. (1995) measured Cd and Zn concentrations in soil solution during growth of radish (*Raphanus sativus* L.) and equated the change in solution composition to

Table 9–1. Prevalent species of trace elements in soil solution (from Ritchie & Sposito, 1995).

Cation	pH 3.5–6.0	pH 6.0–8.5
Al^{3+}	Al^{3+}, organic, AlF^{2+}, Al-hydroxy species	$Al(OH)_4^-$, organic
Cd^{2+}	Cd^{2+}, $CdCl^+$, $CdSO_4^0$	Cd^{2+}, $CdCl^+$, $CdSO_4^0$
Cr^{3+}	Cr^{3+}, $CrOH^{2+}$	$Cr(OH)_4^-$,
Cu^{2+}	Organic, Cu^{2+}	Cu-hydroxy species, $CuCO_3^0$, organic
Mn^{2+}	Mn^{2+}, $MnSO_4^0$, organic	Mn^{2+}, $MnSO_4^0$, $MnCO_3^0$
Fe^{3+}	Organic, Fe-hydroxy species	Organic, Fe-hydroxy species
Ni^{2+}	Ni^{2+}, $NiSO_4^0$, organic	Ni^{2+}, $NiHCO_3^+$, $NiCO_3^0$
Pb^{2+}	Pb^{2+}, organic, $PbSO_4^0$	Pb hydroxy and carbonate species, organic
Zn^{2+}	Zn^{2+}, organic, $ZnSO_4^0$	Zn^{2+}, organic, Zn hydroxy and carbonate species

a change from nonrhizosphere to rhizosphere conditions as roots explored all the soil in the pots. They found reductions in concentrations of Cd and Zn compared with initial soil solution composition. Calcium concentrations in the *rhizosphere* soil solution were found to be low however, which is unusual as Ca is usually found to accumulate around roots in the rhizosphere (see Section 9–3.1). This may have been because Hamon et al. (1995) used a progressive nutrient addition technique to maintain plant growth, and nutrient additions may have not matched plant demand. The soil solution in this experiment was collected from intensively rooted soil using a water displacement technique (Lorenz et al., 1994). This technique is reported by the authors to mainly sample soil pores occupied by plant roots but will most probably sample more than the solution at the soil–root interface alone. Hence, we are left with uncertainty about trace metal concentration at the soil–root interface. Transport modeling is, however, a tool to speculate about this composition (see Section 9–3.1).

Accumulation or depletion of ions in the rhizosphere can influence contaminant availability in several ways. The concentration of the contaminant itself at the root surface obviously affects availability since the uptake rate generally obeys first order kinetics at low solution activities. The availability of contaminants in the rhizosphere also can be altered by other ions that may interact with root uptake, soil–solution distribution or complexation of the contaminant.

9–2.5.2 Speciation–Complexation of Contaminants in Solution

Interest in the speciation of micronutrient metals (Cu, Co, and Zn) in soil solution is over 30 yr old (Hodgson et al., 1965), but during the last 15 yr, investigation of contaminant behavior in soil solution has increasingly focused on this aspect. Since Hodgson et al. (1965) demonstrated that a large proportion of Cu in soil solution is complexed with organic ligands, many studies have sought to determine the speciation of metal contaminants in solution (Tills & Alloway, 1983; Fujii et al., 1983; Brown et al., 1984; Hirsch & Banin, 1990; Holm et al., 1995, Helmke et al., 1998). The predominant species of metal ions in soil solution are shown in Table 9–1.

The study of metal speciation in soil solutions has been encouraged by the so-called *free metal ion* hypothesis in environmental toxicology (Lund, 1990). This states that the toxicity or bioavailability of a metal is related to the activity of the free aquo ion. The principle is widely used in aquatic toxicology (Borgmann, 1983) and is gaining popularity in studies of soil–plant relationships (Parker et al., 1995). Evidence that the free metal ion activity controls plant uptake of metals from solution derives from studies almost 40 yr ago where organic chelating agents were added to nutrient solutions and plant uptake of the metal was found to decrease markedly (DeKock & Mitchell, 1957). As discussed by Parker et al. (1995) in a review of this topic, the absorption of most metals is an energetically favored process as plant cells have negative transmembrane potentials. Complexation reduces the positive charge (or may even reverse charge) on the metal ion as well as increasing the size of the ion, both of which may mitigate against uptake by the root.

Some evidence is now emerging, however, that the *free metal ion* hypothesis may not be valid in all situations. Taylor and Foy (1985), Checkai et al. (1987b), Bell et al. (1991), and Laurie et al. (1991) noted differences in plant uptake of metals at the same metal activity in solution when different chelators were used in solution culture studies. For example, Bell et al. (1991) found that critical Fe^{3+} activities in solution for growth of barley (*Hordeum vulgare* L.) depended on the chelate being used. Also, given the same chelate, total Fe concentration in solution also affected plant uptake of Fe. The authors indicated that either kinetic limitations to chelate dissociation or uptake of the intact chelate could explain these data. Römheld and Marschner (1981) found that maize (*Zea mays* L.) could absorb intact FeEDDHA (Fe-ethylenediamine-di(o-hydroxyphenyl acetic acid)) at breaks in the endodermis where lateral roots budded. Taylor and Foy (1985) could induce Cu toxicity in wheat (*Triticum aestivum* L.) by increasing CuEDTA (Cu-ethylenediaminetetraacetic acid) concentrations while maintaining solution Cu^{2+} activity constant, which also suggests uptake of the intact chelate, or better buffering of free Cu^{2+} activities at the site of uptake leading to enhanced Cu levels in the plant. This latter hypothesis requires the assumption that there is a large diffusive limitation to Cu uptake in the unstirred layer adjacent to the root, or in the root apoplast. Laurie et al. (1991) diagrammatically presented the possible reactions of complexed metals at the soil–root interface (Fig. 9–7).

Recently Phinney and Bruland (1994) examined the uptake of Cu, Cd, and Pb by marine diatoms (*Thalassiosira weissflogii*) in relation to complexation by 8-hydroxyquinoline (8-HQ), diethyldithiocarbamate (DDC), EDTA, and sulfoxine (SOx). While these workers did not distinguish between metal actually absorbed across the cell membrane and metal adsorbed to the cell surface, their data indicate (Fig. 9–8) that organic ligands forming uncharged complexes with metals (DDC, 8-HQ) increase metal uptake dramatically in comparison to either free metal only or negatively charged metal organic complexes (EDTA, Sox).

As metal adsorbed to cell wall components is likely to be greatest with the free metal only in solution, these data suggest that uncharged metal complexes can indeed cross cell membranes; however, further mechanistic studies examining the uptake of metals in relation to size and charge of metal-organic complexes are needed to resolve this issue.

There also is evidence that metal complexes with inorganic ligands such as Cl^- may also be taken up by plants (Smolders & McLaughlin, 1996a,b). In these

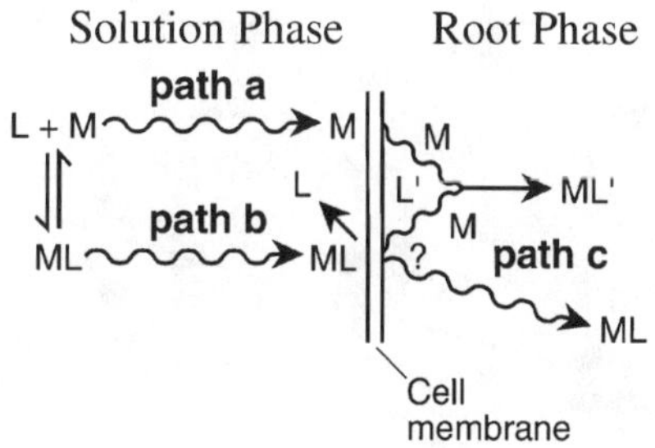

Fig. 9–7. Conceptual model proposed by Laurie et al. (1991) for potential uptake by plants of metals ions (M) complexed by organic ligands (L).

experiments, using both solutions unbuffered and buffered with regard to Cd^{2+} activity, Cd uptake by Swiss chard [*Beta vulgaris* (L.) Koch] was found to be related not only to Cd^{2+} activity in solution, but also to activities of $CdCl_n^{2-n}$ complexes, formed as a result of increasing Cl concentrations in solution (Fig. 9–9).

The explanation offered for these results was that either the $CdCl_n^{2-n}$ complexes were taken up directly by the root, or that the presence of labile $CdCl_n^{2-n}$ complexes allowed greater mobility (and hence buffering) of Cd^{2+} in the unstirred liquid layer adjacent to the root, or in the root apoplast, to sites of ion uptake. The hypothetical *efficiency of uptake* of the chloro-complexes was less than for Cd^{2+}, as expected; however, the results help to explain the large increase in Cd uptake by field crops noted under conditions of Cl salinity (Li et al., 1994; McLaughlin et al., 1994). McLaughlin and Tiller (1994) found that chloro complexes constituted the dominant species of inorganic Cd in soil solutions from irrigated horticultural soils in south Australia. Chloro- and sulfato-complexes accounted for up to 75 and 20% respectively, of the inorganic Cd in soil solution. Indeed, given the

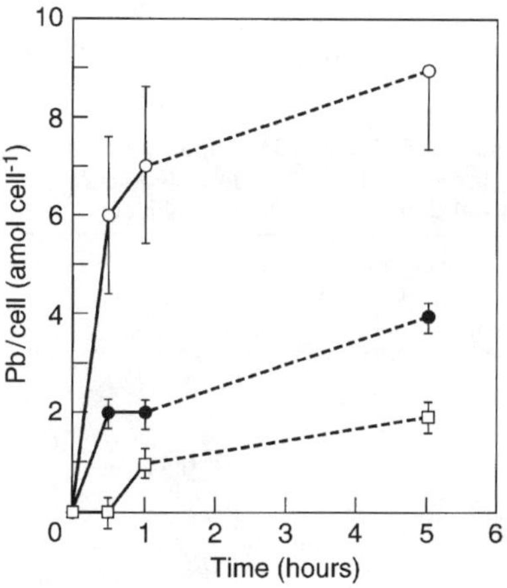

Fig. 9–8. Lead associated with cells of diatoms (*Thalassiosira weissflogii*) as a function of time in relation to free Pb^{2+} (•), Pb-complexed by dithiodicarbamate (○) and EDTA (□). Total Pb concentrations in all solutions = 5 n*M* , (from Phinney & Bruland, 1994).

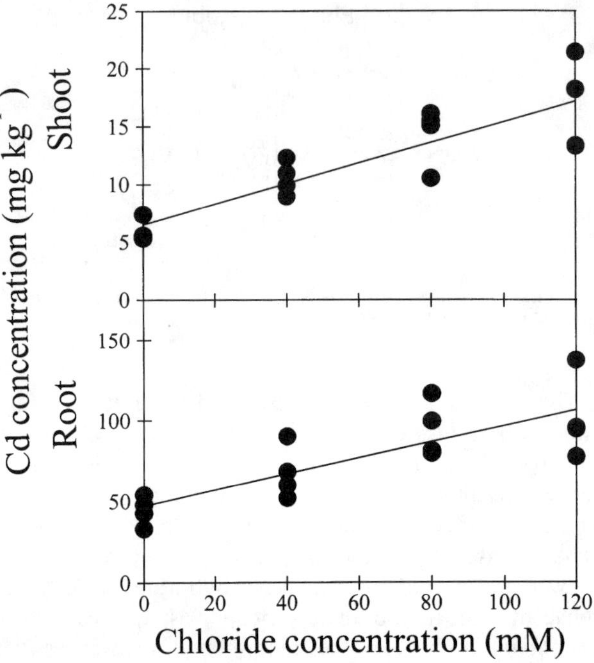

Fig. 9–9. Influence of chloride concentrations in solution on uptake of Cd by Swiss chard from solution culture. Solution ionic strength was constant across all treatments (compensated using $Ca(NO_3)_2$) and solution Cd^{2+} activity (4 n*M*) was held constant across Cl treatments using a chelex resin buffering method (from Smolders & McLaughlin, 1996b).

composition of saturation extracts of a wide range of saline soils in USA shown by Jurinak and Suarez (1990), it is likely that Cl^- and SO_4^{2-} complexation of Cd^{2+} is significant in many saline soils (Table 9–2). Recent results indicate that complexation of Cd^{2+} by SO_4^{2-} in solution culture, unbuffered with respect to Cd^{2+} activity, has no impact on Cd uptake (McLaughlin et al., 1996, unpublished data).

Table 9–2. Median concentrations of Cl^- and SO_4^{2-} anions in well waters, river waters and saturation extracts of a range of salt-affected soils in USA (from Jurinak & Suarez 1990) and impact on speciation of inorganic Cd in soil solution (from McLaughlin et al., 1996).

Ligand	Number of samples analysed	Median concentration	Percentage inorganic Cd complexed by ligand at median concentration
		m*M*	%
Cl^-			
Saturated extract	139	34.8	65
Well water	115	2.5	17
River water	58	1.5	11
SO_4^{2-}			
Saturated extract	134	29.4	50
Well water	23	3.6	23
River water	58	4.1	25

This may result either due to uptake of the uncharged $CdSO_4^0$ complex or due to the presence of labile $CdSO_4^0$ complexes allowing greater mobility (and hence buffering) of Cd^{2+} in the unstirred liquid layer adjacent to the root or in the root apoplast, as for $CdCl_n^{2-n}$ complexes.

Regardless whether it is the free ion which is taken up by the plant, or uptake is via some complexed form, it is evident that the relationship between free metal activity in solution and plant metal uptake is not as close a relationship as previously assumed. Thus, improvements in methodology for speciation of contaminant forms in soil solutions will provide useful information on the potential for transfer of contaminants to the food chain only if parallel information is gathered on the phytoavailability of contaminants in both free and complexed forms.

9–2.5.3 Nonequilibrium Situations

In the rhizosphere, where plants and microorganisms rapidly remove ions from solution and exude substances into soil, it is unlikely that any chemical or physicochemical reaction proceeds to equilibrium. Despite this scenario, much of the work in soil chemistry, both pertaining to the solid and solution phases, has approached the study of contaminant behavior in soil assuming reactions proceed to equilibrium. Recently, progress has been made in the description of the kinetics of adsorption/desorption processes (Sparks & Suarez, 1991; Zhang & Sparks, 1989), although it is important to note the time scales over which these studies are performed, as some reactions may occur in timescales measured in ns (Zhang & Sparks, 1989) while others may take days, weeks or months (Brummer et al., 1988). In terms of plant uptake of contaminants, reactions that have time scales measured in hours or days will be critical in terms of reaction kinetics controlling availability for uptake.

Similarly, the kinetics of reactions in soil solution are important in terms of describing contaminant behavior in the rhizosphere. Rate constants for complexation of metal contaminants by inorganic ligands are large, i.e., reactions are fast (Margerum et al., 1978; Ritchie & Sposito, 1995), while on the basis of limited data, those for metal humate–fulvate complexes are smaller, i.e., reactions are slower (Lavigne et al., 1987; Langford & Gutzman, 1992). Differences in the uptake by barley of metals complexed by EDTA and DTPA were explained on the basis of dissociation kinetics of metal complexes with these two ligands (Laurie et al., 1991). Little is known of the kinetics of complexation of contaminants in the rhizosphere with root-derived organic ligands and this would be a fruitful topic for further research.

9–3 PLANT PROCESSES AFFECTING PHYSICAL CHEMISTRY OF THE RHIZOSPHERE

Plants affect contaminant chemistry in soil at the soil–root interface through a number of processes:

1. Plant uptake may reduce ion activity and desorb contaminants from surfaces, or convective flow of solution to the root may move additional contaminant to the rhizosphere, leading to sorption,
2. Plant-induced changes in solution chemistry can affect sorption, e.g., pH, ionic strength, macronutrient cation concentrations (e.g., Ca);
3. Plants excrete organic ligands that may increase or decrease the total concentration of contaminant ions in solution depending on whether free activity is well buffered or poorly buffered, respectively;
4. Living or dead plant material in the rhizosphere can act as new sorbing surfaces for contaminants, and
5. Microbial activity, stimulated by plants, can also affect contaminant behavior by the above processes.

These processes will be discussed in the following sections.

9–3.1 Root Growth and Uptake of Water and Solutes: Mass Flow and Diffusion

There are three principle pathways for transport of contaminants from the soil to plant roots: mass flow, diffusion, and transport within symbiotic microorganisms. The last mechanism will be discussed later in Section 9–4.2. Mass flow is the convective transport of solutes in water from soil to the root surface. This flow is created by the uptake of water by plant roots, required for transpiration by plant shoots. Diffusive transport occurs when the plant takes up more (or less) contaminant from solution than is supplied by convective flow. Where plant uptake is greater than the convective flux, concentrations of ions in solution adjacent to the root are depleted and a diffusion gradient is created from the soil to the root surface (Brewster & Tinker, 1970).

Knowledge of the actual concentration of contaminants at the soil–root interface is not merely an academic issue. If contaminant transport to the roots by mass flow cannot match the uptake by the roots, the contaminant concentration at the root surface will be decreased and the diffusive flux will control soil availability. In this scenario, it can be speculated that uptake can be reduced by factors reducing the effective root surface area of the plant and by those reducing the diffusive mobility of the contaminant through soil (e.g., a *reduction* in root exudates, see Section 9–3.3). On the other hand, root uptake of the contaminant can be similar to or even smaller than the amount transported by mass flow. In this scenario, the root uptake characteristics (e.g., rate of uptake per unit root length) are rate limiting and it can be speculated that those factors reducing the uptake kinetics (e.g., addition of ions competing for uptake) are most appropriate to reduce soil–plant transfer of the contaminant.

The composition of the soil solution at the soil–root interface has a key role in predictions of plant uptake of nutrient or toxic elements from soil (Nye & Tinker, 1977; Barber, 1984). Where the soil solution composition is unknown, but plant uptake of the contaminant and behavior of the contaminant in soil (e.g., K_d, diffusion coefficient) are known, solute transport–plant uptake models provide a tool to speculate on solution composition at the soil–root interface. In these models, the ion concentration at the soil–root interface is the driving variable for root uptake and is calculated from the balance between root uptake characteristics and supply by mass flow and diffusion. Under well defined experimental conditions, these models are relatively successful in predicting uptake of nutrients such as P and K for which uptake is limited by mobility through soil. In contrast, predictions for nutrients such as Ca and Mg (for which root uptake characteristics control uptake) are often subjected to a wide error (Rengel, 1993). As far as we are aware, similar mechanistic modeling efforts for contaminants have only been made for Cd and ^{137}Cs (Mullins et al., 1986; Kirk & Staunton, 1989). Under conditions representative of agricultural soils (i.e., not grossly contaminated urban or industrial soils), both ions are predicted to deplete in the rhizosphere. These predictions will be critically evaluated here.

An example of calculated concentration gradients of nutrients around roots of spinach (*Spinacia oleracea* L.) plants is given in Fig. 9–10.

The model calculations are based on uptake rates measured in solution culture and chemical characteristics of a soil and soil solution representative of agricultural topsoils. Model assumptions and calculation methods are described elsewhere (Smolders, 1993) and are basically the same as those in the well-known approach of Barber (1984). Model parameters are given in the caption. The model predictions show a steep depletion of P, a small K depletion and a small Ca accumulation around roots after 3 d of uptake. Such profiles are similar to other model calculations and observations (Brewster & Tinker, 1970; Barber, 1984). Using the same model assumptions, concentration gradients of Cd and ^{137}Cs were calculated (Fig. 9–11).

These calculations are based on the soil characteristics and the transfer coefficients (solution–plant concentration ratio) of Cd and ^{137}Cs derived in solution culture (Smolders et al., 1996, unpublished data). For Cd, this was related to concentration of the free Cd^{2+} ion in an EDTA-buffered nutrient solution.

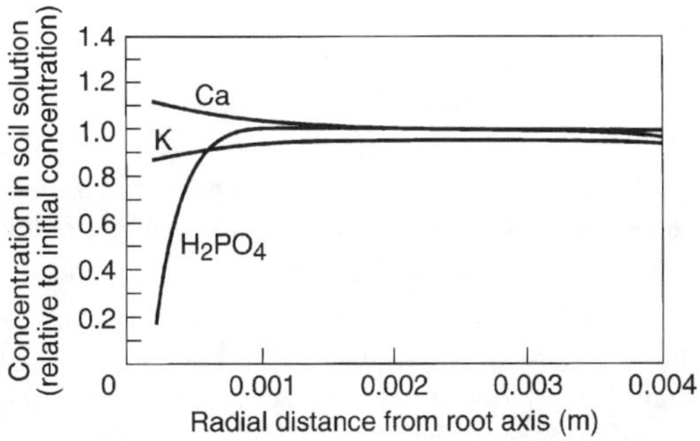

Fig. 9–10. Simulated concentration profile of Ca, K, and H_2PO_4 around a spinach root growing in a loamy sand after 3 d of uptake. Solute transport calculation method as in Smolders (1993). Simulation parameters: effective diffusion coefficient in soil, 1.6×10^{-9} cm^2 s^{-1} (P), 3.4×10^{-7} cm^2 s^{-1} (K), 5.8×10^{-8} cm^2 s^{-1} (Ca); buffer power in soil, 218 (P), 2.3 (K), 5.4 (Ca); initial concentration, 0.01 m*M* (P), 3.3 m*M* (K), 5.8 m*M* (Ca); volumetric moisture content, 0.22; root density, 2 cm^{-2}; root radius, 0.02 cm; water flux at root surface, 7.4×10^{-7} cm s^{-1}; root uptake as Michaelis-Menten kinetics per root area (Barber, 1984) with V_{max}, 4.4 nmol m^{-2} s^{-1} (P and Ca), 65 nmol m^{-2} s^{-1} (K), K_m, 0.002 m*M* (P), 0.014 m*M* (K), and 0.039 m*M* (Ca) and c_{min} 0.0002 m*M* (P), and 0 (Ca and K).

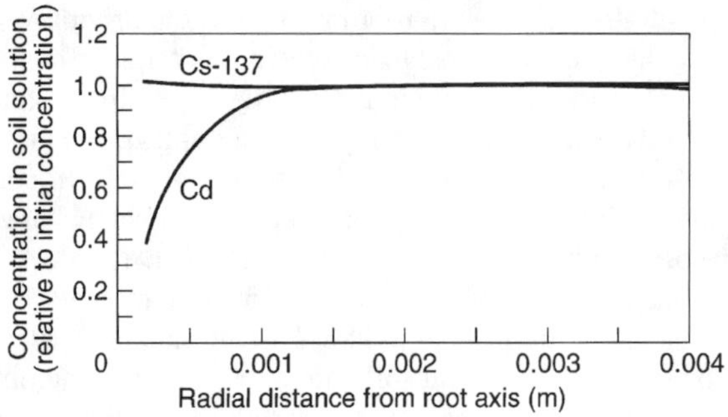

Fig. 9–11. Simulated concentration profile of Cd and ^{137}Cs around a spinach root growing in a loamy sand after 3 d of uptake. Solute transport calculation and parameter as for previous figure except for the following: effective diffusion coefficient in soil, 1.6×10^{-9} cm^2 s^{-1} (^{137}Cs), 8.6×10^{-9} cm^2 s^{-1} (Cd); buffer power in soil, 500 (^{137}Cs), 34 (Cd); initial concentration, 5 kBq L^{-1} (^{137}Cs), 60 n*M* (Cd); root uptake of Cd as Michaelis-Menten kinetics per root area with V_{max}, 0.39 nmol m^{-2} s^{-1} (Cd), K_m, 76 n*M* and c_{min} ,0. Uptake rate of ^{137}Cs per root area proportional to solution activity (in Bq m^{-3}) with a proportionality factor of 1.78×10^{-9} m s^{-1}.

Cadmium concentrations are predicted to decrease in the rhizosphere, but depletion is not severe even at the soil–root interface. Indirect evidence that Cd is likely to be depleted in the rhizosphere was also recently given by Hamon (1995). In these experiments, nutrient and metal concentrations in solution and water transpired by the (radish) plants were monitored over time. Comparison of amounts of elements transported to the root via mass flow to those taken up by the plant indicated that Ca and Mg would have accumulated in the rhizosphere, while Cd, K, P, and Zn were depleted in the rhizosphere (Table 9–3).

^{137}Cs is predicted to accumulate slightly in the rhizosphere (Fig. 9–11). This prediction may be surprising since ^{137}Cs is strongly sorbed by soils (Section 9–2.1) and the estimated effective diffusion coefficient of ^{137}Cs in soil is about equal to that of $H_2PO_4^-$ (1.6×10^{-9} cm^2 s^{-1}), which is strongly depleted in the rhizosphere; however, this difference between $H_2PO_4^-$ and ^{137}Cs is due to a strong difference in root uptake kinetics as observed in solution culture experiments.

The value of the ^{137}Cs uptake rate was derived from experiments at optimum nutrient supply in solution culture; however, under conditions of K deficiency, the uptake rate of ^{137}Cs may be considerably higher than in the calculations above. Recently, Smolders and Shaw (1995) found in solution culture experiments with wheat that the ^{137}Cs uptake rate is more than 100-fold higher at micromolar K concentrations than at millimolar K concentrations. Nevertheless, even where K concentrations were very low, no depletion of ^{137}Cs activity in the rhizosphere was predicted (Smolders et al., 1996, unpublished data) because: (i) ^{137}Cs activity in solution is instantaneously buffered by a large adsorbed pool and (ii) ^{137}Cs uptake rate is reduced when its activity in solution is reduced.

It may therefore be concluded that the characteristic of low mobility in soil, as in the case of strongly sorbed contaminants, is not necessarily related to a depletion in the rhizosphere because the root uptake rate may be very low.

Some of the *classical* model assumptions may be critical for modeling contaminant movement in the rhizosphere. First of all, the soil parameters are gener-

Table 9–3. Ratio of convective flow of elements to actual uptake into leaves and tubers of radish plants grown in a sandy topsoil (pH 6.8, total soil Cd and Zn concentrations, 2.8 and 150 mg kg^{-1}, respectively. Soil solution concentrations of elements determined by a displacement method (from Hamon, 1995).

Element	Ratio
Ca	6.79
Mg	5.79
K	0.22
Cd	0.42
Zn	0.39
P	0.05

ally derived from bulk soil properties. As described in this review, conditions in the rhizosphere are markedly different than those in bulk soil and modelers should be aware of this. More complicated nutrient uptake models account for rhizosphere conditions such as gradients in soil pH (Saleque & Kirk, 1995), ion gradients that influence the buffer power of each ion (Bouldin, 1989; Silberbush et al., 1993), the presence of organic acids (Hoffland, 1992) and exudates in general (Nye, 1984). Secondly, the root uptake characteristics can be different in the rhizosphere from those measured in nutrient solutions. Furthermore, the concentration of ions affecting uptake, either through direct competition or through influences on membrane permeability, can be lower or higher at the root surface than in the bulk soil. As an example, Mg uptake from acid soil is influenced by Mg/Al, Mg/Ca, and Mg/K concentration ratios at the soil–root interface rather than by Mg concentration alone (Rengel & Robinson, 1990).

As far as we are aware, no model has incorporated these inter-ionic effects on root uptake characteristics. It is unlikely that plants have developed element-specific uptake mechanisms for nonessential elements. It can be argued therefore that for many contaminants, which are not essential plant nutrients, interionic effects are even *more* important than for nutrients; however, contaminant ions may enter the root through channels or mechanisms designed for nutrient elements, so that by *substitution* an effective demand for the contaminant is created. The depletion of Cd in nutrient solutions by plants in excess of the amounts entering the plant through the transpirational demand is a case in point.

Examples of inter-ionic effects on contaminant uptake are that the uptake of ^{137}Cs strongly depends on the K concentration (Shaw et al., 1992) and the uptake of Cd varies with the solution activity of Zn^{2+} (McKenna et al., 1993) and H^+ (Hatch et al., 1988). An illustration for competition reactions in the rhizosphere is given for the K: ^{137}Cs pair in Table 9–4. The composition of the bulk soil solution was insufficient to explain a considerable reduction in ^{137}Cs levels in wheat plants by increased K supply to the soil. The predictions were, however, improved if they were based on rhizosphere concentrations of K (Table 9–4). The improvement of the prediction is basically explained because K concentrations have only little effect on ^{137}Cs uptake between 0.25 and 1 m*M* K whereas below 0.25 m*M* K, ^{137}Cs uptake increases sharply with decreasing K concentrations.

Further criticisms of model calculations of contaminant mobility in soils concern the value of the effective diffusion coefficient D_e in soil. Traditionally in most nutrient uptake models, D_e is calculated as follows (Nye & Tinker, 1977):

Table 9–4. The ^{137}Cs activity in the shoots of young wheat plants grown in ^{137}Cs contaminated soil at different K supply. The predicted activities are based on root uptake characteristics measured in solution culture (^{137}Cs uptake at various K concentrations) and ^{137}Cs and K concentrations in soil solution in either the bulk soil or at the soil–root interface (modified from Smolders et al., 1996b).

	Bulk soil		Soil–root interface		
K dose	[K] (μ*M*)	Predicted ^{137}Cs content	[K] (μ*M*)	Predicted ^{137}Cs content	Observations
mmol pot^{-1}		cpm g^{-1}		cpm g^{-1}	
0	170	345	44	4444	6191
2	810	244	776	249	755

$$D_e = D_l\, \theta\, f/b \qquad [9]$$

in which D_l represents the diffusion coefficient in water, θ is the volumetric moisture content, *f* is the tortuosity measured by diffusion of a nonadsorbing ion such as Cl^- and *b* is the volumetric buffer power, equal to dC/dC_l. *C* and C_l are ion concentration in soil and soil solution respectively, both expressed on a unit volume basis. This formulation is based on the assumption that solute moves through the solution phase of the soil and that instantaneous equilibrium exists between the solid phase and solution.

Mullins and Sommers (1986) measured D_e values for Cd and Zn in four soils treated with sewage sludge and compared these observations with D_e values calculated by the Nye and Tinker formula. The *observed* D_e values were derived from a cationic resin paper method in which the amount of Cd and Zn, collected on the resin paper, was determined after a given contact time between the paper and soil. The *calculated* D_e values were based on soil solution Cd and Zn concentrations and DTPA-extractable Cd and Zn concentrations in soil. Observed D_e values were between 5 and 35 (Cd) and between 23 and 344 (Zn) times lower than calculated values (based on DTPA-extractable metal contents). A similar discrepancy also was found for Zn by Warncke and Barber (1973). The selection of either an observed or a calculated D_e value gives very different predictions for Cd availability to roots (Fig. 9–12).

Using the calculated D_e value, uptake per root area was calculated to be 0.56 μmol Cd m^{-2} after 3 d of uptake. With the observed D_e value, only 0.15 μmol Cd m^{-2} was calculated. Mullins et al. (1986) used the observed D_e values to predict Cd uptake from soil and proposed that uptake is not markedly affected by root uptake characteristics. Most probably, root uptake characteristics would be more critical in predictions based on the calculated D_e values. No convincing explanation could be given for the discrepancy between both methods to derive D_e (Mullins & Sommers, 1986). On the one hand, the calculated D_e value could overestimate mobility because it neglects slow desorption that may control mobility. On the other hand, the methods for measuring D_e can have shortcomings. In the case of the methods of Mullins and Sommers (1986), it can be questioned if the cation exchange resin provides an infinite sink for some contaminants (e.g., Cd) and if a 1-h soil–resin paper contact is sufficiently long to allow the diffusive flux to become dominant.

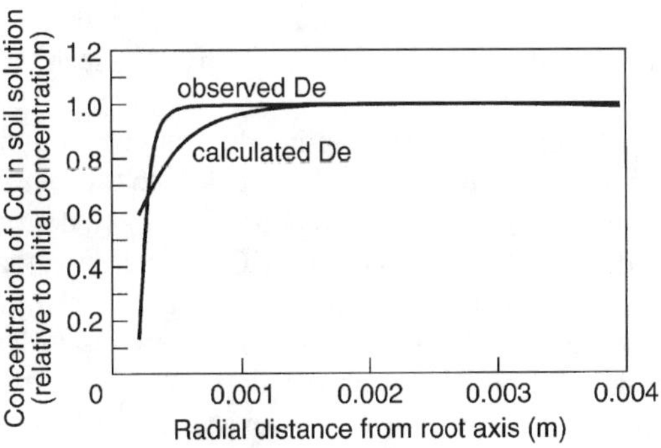

Fig. 9–12. Simulated concentration profile of Cd around a spinach root growing in a silt loam soil with characteristics given by Mullins and Sommers (1986, Chalmers soil at zero sludge application). Root uptake characteristics as for previous figure and two values of the effective diffusion coefficient D_e: either a value (7.3×10^{-9} cm^2 s^{-1}) calculated according to Nye and Tinker (1977) with a buffer power 108 (expressed on a DTPA-Cd content in the soil, see text) or the value measured in a diffusion cell (2.8×10^{-10} cm^2 s^{-1}). Initial Cd concentration, 18 n*M*.

Finally, it is evident that some contaminants (e.g., Cd) are taken from soil in excess of the amounts supplied to the root surface by mass flow, although it is not known if the mode of entry to the root is similar to that for nutrient elements, i.e., ion channels or active transport systems (ATPase or H^+-coupled transport systems). This highlights the need to understand uptake mechanisms at the root surface, which at present are poorly understood for metal contaminants, and even for micronutrient metals such as Zn (Kochian, 1993). It also is important that the extent and activity of uptake zones along the root be characterized, because significant heterogeneity in uptake for nutrient elements has been demonstrated (Marschner, 1995). Nutrient uptake models are, as yet, not sophisticated enough to account for these factors, and this should be considered when interpreting output data from such models.

In conclusion, availability of contaminants in the rhizosphere can be affected in various ways because of the presence of the living root. The root acts as a sink for ions in solution. The different soil solution composition in the rhizosphere compared with bulk soil affects sorption–desorption, complexation, and root uptake characteristics of contaminants. Furthermore, the contaminant itself can be either accumulated or depleted around the root depending on root uptake and soil transport factors. As information on contaminant concentrations at the soil–root interface is scarce and difficult to model, more experimental and modeling work should be carried out in this area to understand the mechanisms controlling the soil-plant transfer of contaminants.

9–3.2 Changes in Solution pH and Composition at the Soil–Root Interface

The pH in the rhizosphere generally differs from that in the bulk soil. The pH changes have been demonstrated experimentally and are often larger than one unit. A review on rhizosphere pH in relation to mineral nutrition has been given by Marschner et al. (1986) and is also discussed by Brusseau (1998, this publication).

Rhizosphere pH changes are predominantly brought about by net excretion of H^+ or HCO_3^- by the roots. This excretion is related to the cation–anion uptake ratio of the plant. The form of N supply (NO_3^- or NH_4^+) has a dominant role in that ratio. Roots supplied with NO_3^- have a higher $HCO^-{}_3$ excretion than H^+ excretion whereas the reverse occurs with NH_4^+ supply. Nitrogen uptake through N_2-fixation leads to rhizosphere acidification. Genotypical effects in HCO_3^- release in NO_3^--fed plants are large. Iron or phosphorus deficiency cause rhizosphere acidification in many plants. In most grasses however, Fe deficiency does not result in strong rhizosphere acidification as these plants have a different strategy to acquire Fe (Marschner, 1995).

pH changes in the rhizosphere are not uniform over the whole root system. In NO_3^--fed plants, rhizosphere alkalinization is most prominent along the main root axes whereas on lateral roots no pH changes or even a pH decrease are found (Marschner & Römheld, 1983). A local pH decrease around the root tips in combination with a pH increase around the other root parts has been demonstrated for NO_3^- fed rape (*Brassica napus* L.) as a result of P-deficiency (Hoffland et al., 1989). In NH_4^+-fed plants, rhizosphere acidification is more or less uniformly distributed over the whole root system whereas Fe deficiency causes an intense, but local, acidification around the root tip (Marschner et al., 1986).

The consequence of local pH changes around the root on contaminant availability in the rhizosphere remains speculative, to our knowledge. Local pH changes may influence contaminant sorption (Section 9–2.1.1) and root uptake characteristics. While Jarvis et al. (1976) found in solution culture that Cd uptake by plants *increased* as solution pH increased, the observation that Cd availability in soil *decreases* with increasing soil pH (Chaney & Hornick, 1978) suggests that rhizosphere acidification also will result in an increase in Cd uptake. As an example, NH_4^+ supply has been shown to increase Cd uptake in wheat grain (Williams & David, 1976), which could be ascribed to rhizosphere acidification in response to NH_4^+ supply. Eriksson (1990) and Willaert and Verloo (1992) also demonstrated in glasshouse experiments that the more acidifying N fertilizers, such as ammonium sulfate, increased Cd uptake over those having an alkaline effect in soil (e.g., Ca nitrate). Similar results have been obtained by Reuss et al. (1978) who found in pot trials that banded superphosphate increased Cd uptake by plants in comparison to banded diammonium phosphate (DAP). They suggested the effect was due to pH changes in the fertilizer band. Such effects will depend on the pH-buffering capacity of the soil and the rates of N fertilizer applied. Under field conditions, rates of soil acidification due to N fertilization are often less marked than under glasshouse conditions where soil temperature, moisture and root activity are optimized.

Impacts of solution composition on contaminant chemistry are more difficult to predict, as little information exists on actual concentrations of ions at the soil–root interface for most plant species and environmental conditions. It is known that Ca concentrations in solution tend to be higher at the soil–root interface, so that contaminants whose retention by soil may be affected by Ca con-

Table 9–5. The effect of increasing Ca concentration in soil solution on Cd and Zn complexation, expressed as percentage of the total concentration present as the free divalent ion (from Hamon et al., 1995).

Ca concentration	% Cd^{2+}	% Zn^{2+}
mg L^{-1}		
42	14	8
42	ND	11
292	45	50
292	47	59
1042	97	83
1042	100	88

centrations (e.g., Cd) may be less strongly retained in the rhizosphere. The interaction of Ca with soluble organic ligands also may alter contaminant behavior. For example, it is known that Ca competes with metals for binding sites on organic ligands. Recently, Hamon et al. (1995) demonstrated the impact of solution Ca concentrations on the speciation of Cd and Zn in soil solution. Isolated soil solutions were spiked with Ca and the speciation of Cd and Zn between free and complexed forms in solution assessed using an ion exchange procedure (Holm et al., 1995). Increasing concentrations of Ca in solution significantly reduced the percentage of total Cd and Zn in solution associated with organic ligands (Table 9–5).

Thus, while concentrations of organic ligands in solution may be high in the rhizosphere, leading to complexation of metals, concentrations of Ca also may be high which will tend to reduce complexation.

9–3.3 Impacts of Root Exudates on Contaminant Chemistry

9–3.3.1 Nature and Amounts of Root Exudates in the Rhizosphere

Plant roots excrete a wide variety of organic compounds into the rhizosphere. Whipps (1990) recently reviewed this topic and noted that up to 70% of the C transferred from shoots to roots could be *lost* as exudates in the soil (excludes root respiration). While many early studies examined root exudation under sterile conditions in order to exclude microbial contributions to exudation and respiration, Barber and Martin (1976) found that losses of C from roots are much greater under conditions where microorganisms are present (i.e., nonsterile conditions).

A number of low molecular weight organic acids have been identified in the rhizosphere of plants (Rovira, 1965; Rovira & McDougal, 1967; Hale et al., 1971; Smith, 1976; Huang & Violante, 1986). These can interact with soil colloids in a number of ways to either desorb or promote adsorption of contaminants in the rhizosphere as outlined in Section 9–2.2.3. A summary of the major source of root-derived material in the rhizosphere is given in Table 9–6 and details of some of the more important acids are given in Table 9–7.

Table 9–6. Root material that could affect contaminant behaviour in the rhizosphere (from Uren & Reisenauer, 1988).

Product	Compounds
Root exudates	
Diffusates	Sugars, organic acids, amino acids, water, inorganic ions, O_2, riboflavin, and others
Excretions	Carbon dioxide, bicarbonate ions, protons, electrons, ethylene, and others
Secretions	Mucilage, protons, electrons, enzymes, siderophores, allelopathic compounds, and others
Root debris	Root-cap cells, cell contents

9–3.3.2 Effect of Exudates on Metal Chemistry in the Rhizosphere

Seasonal changes in Cu, Mn, Zn, and Co concentrations in rhizosphere soil were related to the presence of complexing agents of biological origin (Linehan et al., 1989; Nielsen, 1976). Exudates of various kinds isolated from axenically grown plants have been proven to complex metals in vitro (Morel et al., 1986; Mench et al., 1987, 1988; Gries et al., 1995) and have also been demonstrated to be able to extract Cd and other elements from soils (Mench & Martin, 1991; Jones & Darrah, 1994). For example, Mench and Martin (1991) compared the metal solubilizing ability of axenic root exudates of two species. They found that the solubility of Cd, Cu, Mn, and Ni in soil was generally enhanced in the presence of root exudates of either maize or tobacco (*Nicotiana tabacum* L.; Fig. 9–13).

Root-derived compounds, as evidenced by ^{14}C-labeling, have been shown to be able to complex Co, Mn, and Zn (Table 9–8) during plant growth (Merckx et al., 1986). Similarly, Hamon et al. (1995) found that both Cd and Zn were complexed in rhizosphere soil solutions after radish growth (Table 9–9); however, in this experiment Ca concentrations in solution were low due to plant growth depleting concentrations in solution. This allowed greater complexation of both Zn and Cd (see Section 9–2.2.4) than would be expected in the rhizosphere where Ca concentrations are usually higher than in the bulk soil. Helmke et al. (1998) using a new Donnan equilibrium method to measure free Cd^{2+} activities in solution, clearly demonstrated that the activity of Cd^{2+} in sludge-amended soils was strongly dependent on Ca and Mg concentrations in solution. These data indicate that complexation of Cd by both solid and liquid phases in soil is dependent on Ca activity in solution, with Ca displacing free Cd^{2+} from both soil surfaces and from organic ligands in solution.

In these studies, as in many others, the question whether the complexing agents in the rhizosphere are released from living roots, mediated by soil microflora or merely derived from dead root cells, cannot be answered adequately; however, in some cases, evidence for a pure root origin of complexing agents has been obtained from axenically grown plants as is the case for Fe-deficient plants with Strategy II (Marschner et al., 1989). The question of the origin (plant or microbial) of complexing agents in the rhizosphere may not just be an academic question, as attempts to control contaminant behavior through changes in complexation will need to know if microbial or plant systems require (genetic) modification.

Table 9–7. A selection of organic acids occurring in soil solution (adapted from Huang & Violante, 1986).

Organic acid	Concentration in soil solutions ($\times 10^{-5}$ *M*)	Occurrence in nature	References
Acetic	265–570	Microbial metabolites. Accumulates if microbial respiration is anaerobic. Commonly found in root exudates of many grasses and herbaceous plants and in green manures. Volatile but can be adsorbed by soil clays.	Vancura, 1964; Stevenson, 1967; Rovira & McDougal, 1967; Rao & Mikkelson, 1977
Amino	8–60	Building blocks of all plant protein. Amnipresent. Concentration in soils that are under cultivation is higher than in fallow fields.	Stevenson & Ardankani, 1972
Benzoic	7.5 or less	Accumulates if microbial action is intense. The hydroxy derivative (p-hydroxy-benzoic acid) is more common and inhibits plant growth and seed germination. p-hydroxy benzoic acid has been identified in Podzol B horizons.	Shorey, 1913; Webley et al., 1963; Whitehead, 1964; Wang et al., 1967; Davies, 1971
Citric	1.4	Important intermediate in the tricarboxylic acid cycle of all higher plants and organisms. Identified in root exudates of mustard plants. Present in high concentrations in many fruits and leaves. Produced by soil fungi.	Stevenson, 1967, 1982; Bruckert, 1970; Boyle & Voigt, 1973; Förstner, 1981
Formic	250–435	Produced by bacteria in rhizosphere. Has been isolated from root exudate of corn. Volatile.	Stevenson, 1967
Malic, tartaric, malonic	100–400	Excreted by roots of many cereals and solanaceous crops.	Stevenson, 1967; Stevenson & Ardakani 1972; Robert & Berthelin, 1986
Mugineic acids	?	Produced in response to micronutrient stress.	Treeby et al., 1989; Zhang et al., 1989
Oxalic	6.2	Produced during lysis of microbial cells. May make up 50% of the dry weight of the leaves of plants. Commonly present in root exudates of cereals. Nonvolatile.	Stevenson, 1967; Bruckert, 1970
Tannic acid related compounds (gallic acid, tannins and other phenolic acids)	5–30	Comon in tea leaves and barks of trees. Related compounds isolated in root exudates of musturd. Common in waters near tanneries. If the phenolic acids in senescent leaves are polymerized and condensed, tannic acid is produced. Tannins believed to be building blocks of humus because of the simlarity and stability of the functional groups.	Coulson et al., 1960; Davies, 1971; Ladd & Butler, 1975; Förster, 1981; Stevenson, 1982

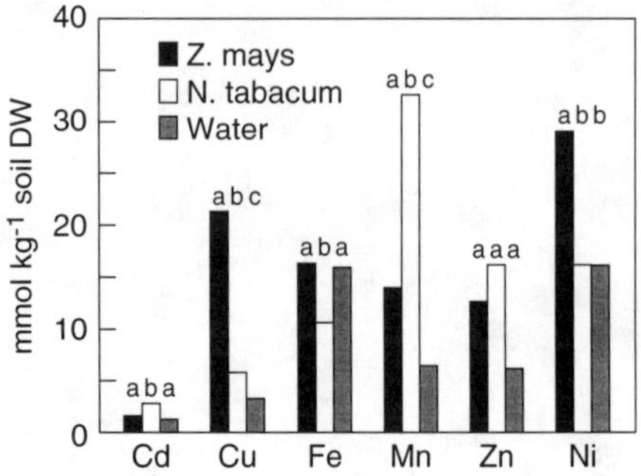

Fig. 9–13. Metals extracted from a sandy acidic topsoil by water and by root exudates of maize and tobacco. Root exudates were collected under axenic conditions and were used to extract metals at a concentration of 150 μg C g^{-1} soil. Iron concentrations are scaled (divided by 10) for presentation (from Mench & Martin, 1991).

9–3.3.3 Effects of Exudates on Plant Uptake of Contaminants

A clear and mechanistic picture on how complexation affects nutrient or contaminant availability remains the subject of much speculation. Nevertheless, the concept seems to have had an impact in the design of many soil fertility tests, very often relying on extracting solutions containing one or other chelating agent

Table 9–8. Percentage of metals complexed after addition to solutions containing water extracts from soils growing maize and wheat, compared with an unplanted control soil (from Merckx et al., 1986).

Sample origin	Co	Mn	Zn
Fallow	6.4	0.2	1.9
4 wk maize	58.6	3.5	11.8
6 wk maize	60.6	5.9	15.8
4 wk wheat	22.7	0.4	3.7
6 wk wheat	30.5	0.8	15.2

Table 9–9. Percentages of total Cd and Zn in soil solution present as the free ion in relation to days of growth of radish plants. Note the soil solution at Day 0 is assumed to be equivalent to bulk soil solution and that at 30 d rhizosphere soil solution, due to root exploration throughout the pot (from Hamon et al., 1995).

Days of radish growth	% Cd^{2+}	% Zn^{2+}
0	95	100
14	15	17
16	11	3
18	15	8
20	25	21
22	31	11
24	50	0
26	36	11
28	28	0
30	28	13

to mimic root activity, e.g., diethylenetriaminepentaacetic acid (DTPA; Lindsay & Norvell, 1969). Moreover, a number of studies have provided indirect evidence for plant-induced modifications of nutrient availabilities. Differences in exudation characteristics have been linked with abilities of given plant species to survive in specific environments, for example the low root exudation of phytosiderophores from calcifuge species and their consequent inability to grow in alkaline soils due to insufficient mobilization of Fe (Ström et al., 1994). Classical examples in this context are derived from the work on Fe-efficient versus Fe-inefficient species (Marschner, 1995). In many cases the distinction between adequate and inadequate Fe-uptake resides in the ability of the cultivar to produce and exude a phytosiderophore (or phytochelator for elements other than Fe) as for example demonstrated in oats (Mench & Fargues, 1994) or suggested for clover (Wei et al., 1994). Similarly, Mench and Martin (1991) demonstrated a clear link between the bioavailability of Cd to different plants and the complexation capacity of the corresponding exudates for Cd.

The phytochelators exuded in response to Fe-deficiency are apparently not specific as they also are reported to complex Cu, Mn, and Zn (Treeby et al., 1989) whereas the compounds released in response to Zn-deficiency also may mobilize Fe (Zhang et al., 1989); however, it remains to be seen how these complexing agents operate in situ amidst a myriad of other complexing agents likely to be formed as a result of enhanced microbial activity in the rhizosphere. Nevertheless, it can be anticipated that the final result will be that a larger fraction of the nutrients or contaminants will be in a complexed (and usually soluble) form in the rhizosphere as compared with the bulk soil.

A clear understanding of the interaction of organic chelators with soil solid surfaces, element transport through the soil and uptake by the plant root has yet to be developed. Early explanations of the impact of chelators on metal uptake by plants relied only on their ability to increase metal mobility through soil (Lindsay, 1974; Halvorson & Lindsay, 1977); however, the situation is more complex than this and it is necessary to consider both the impacts on plant uptake characteristics as well as impacts on soil/solution equilibria for the metal in question.

In hydroponics, the presence of complexing agents leads to a decrease in free metal activity and a concomitant decrease in uptake rate, which suggests that plant uptake is correlated with the activity of the free, uncomplexed ion in solution (DeKock, 1956; DeKock & Mitchell, 1957; Parker et al., 1995). A good example of this was given by Checkai et al. (1987a,b), who maintained free metal activities constant in a resin-buffered hydroponic solution while total concentrations were increased as a result of additions of selected chelating agents. Plant uptake of Cd was not significantly increased despite solution Cd concentrations being increased 50-fold; however, for Cu an increase in uptake (at constant Cu^{2+} activity) was measured as the amount of metal chelate in solution increased, although the increase in plant uptake was less than the increase in total concentration of metal (free + complexed; Checkai et al., 1987b). As discussed earlier (Section 9–2.5.2), reasons for the imperfect fit of the *free metal model* may reside in various ill-understood phenomena, such as the uptake of intact metal-chelates by plant root cells as demonstrated for Fe by Römheld and Marschner (1981), or alleviation of diffusion limitations to uptake within or at the root surface.

In soils, the picture is far more complex due to solid–liquid reactions involving sorption–desorption, precipitation–solution, and oxidation–reduction. In general, interactions with the solid matrix lead to a reduced mobility of contaminants in soil, but chelates may mitigate these reactions considerably (Section 9–2.2.3) depending on the charge of the contaminant-chelate complex and surface charge characteristics of the soil. Solubilization of otherwise insoluble compounds by root exudates has been repeatedly demonstrated for rock phosphates (Hoffland et al., 1989; Hoffland, 1992) and manganese oxides (Bromfield, 1958; Godo & Reisenauer, 1980; Jauregui & Reisenauer, 1982), apart from the enhancement of nutrient desorption from soil particles discussed earlier (Jones & Darrah, 1994). Nevertheless, whether such processes also will lead to an increased uptake of contaminants by the plant depends on root uptake characteristics for the contaminant in question. For example, it is appropriate to make a distinction between the situation where plant demand exceeds supply by convection of a given element from soil and the case where convection matches or surpasses plant demand (see Section 9–3.1). As there are few data in the literature to indicate how complexation by organics in the rhizosphere affects uptake of various contaminants by plants, we can only speculate as to the possible outcomes.

Taking the first situation, where the movement of the contaminant in the convective flow of water to the root from soil is insufficient to meet plant uptake, complexing agents will, in line with common knowledge, enhance transport of the element to the root surface and result in *increased* uptake. This is particularly important in the case where kinetic limitations exist for desorption or dissolution of the element as can be envisaged for Fe, for example. In this case one could even speculate that, in view of the desorption facilitated by the chelating agents, higher concentrations, even of free metal, may occur at the root surface. Theoretically, the situation where equilibria between solid and liquid phase are instantaneous also exists. In this situation, exudates with complexing properties will also lead to improved transport to the root surface and a *higher* uptake by the plant.

Secondly, we must consider the situation where root uptake of the contaminant is less than that supplied by convective flow, which can be considered typical for elements for which no selective uptake should exist. For some contaminants, the uptake rate can indeed be very slow (see the example of ^{137}Cs, Fig. 9–11). Following the same train of logic as above, in the situation where no kinetic limitations exist for the desorption or dissolution of the element in soil, we can envisage that complexation will *not* reduce uptake since a decrease in free metal concentration at the root surface is unlikely given a large buffer capacity in the soil. This raises an interesting anomaly, in that exudation of organic compounds into the rhizosphere has been linked to plant tolerance to Al toxicity in acidic soils (Delhaize et al., 1993). Given the large amount of exchangeable-Al and the presence of relatively soluble amorphous Al phases in most acidic soils, it is unlikely that exudation of organics could significantly affect equilibrium Al activities in soil solution to any great extent; however, Uren (1989) discussed a possible mechanism involving an interaction between both root exudates and plant vigor, which could explain how release of root exudates may result in increased tolerance to Al as follows. Rates of diffusion of Al through the root cap and mucilage to meristematic cells (where Al injury occurs to retard root growth) may be affected by organic exudates. If roots divide and elongate at a rate faster than the rate of diffusion of Al to the meristem, which can be reduced by complexation with organics, injury may be avoided.

In conclusion, it appears that complexing agents may enhance uptake predominantly in situations where plant demand exceeds the supply from the soil. Balancing plant processes against soil processes remains essential for a proper judgement on the final outcome of any soil–plant transfer process. It is clear from the above that the role of exudates also must be viewed in this perspective.

9–3.4 Effects of Plant Roots on Redox Reactions of Contaminants in the Rhizosphere

Data on gradients in the redox potential around the roots are generally lacking. Consumption of O_2 by roots and rhizosphere biomass will most likely reduce the redox potential in the rhizosphere; however, in one of the few studies reporting redox potentials in the rhizosphere zone of aerobic soils, Blanchar and Lipton (1986) found no gradients in redox potential around alfalfa (*Medicago sativa* L.) roots.

In contrast, in wetlands, it has been known for some time that steep gradients in redox potentials develop around plant roots as a result of O_2 release from the roots (Schreiner & Sullivan, 1910). This process is reflected in a precipitation of FeOOH (*iron plaque*) on the roots (Jeffrey, 1961; Otte et al., 1989; Kirk & Bajita, 1995). It has been demonstrated with the salt-marsh plant (*Aster tripolium*) that these iron plaques are enriched in Zn and Cu compared with the surrounding soil (Otte et al., 1989). Recently, it has been reported that Zn accumulates in the rhizosphere of rice (*Oryza sativa* L.), in line with the zone of oxidation of Fe^{2+} to Fe^{3+} adjacent to the root (Kirk & Bajita, 1995). Otte et al. (1989) reported that Zn concentration in red roots (with iron plaque) was higher than in white roots and short-term uptake studies demonstrated a positive effect of the Fe concentration on the root surface, up to a certain level, on Zn uptake into the xylem fluid. Above this level the Fe coating reduced Zn uptake by the plant, due possibly to complete coating of the root surface by iron hydroxides and blocking of absorption sites. Otte et al. (1989) speculated that Fe coatings on roots could enhance uptake of heavy metal contaminants by plants, but no data were given to support this.

Arsenic has been found to accumulate in the rhizosphere of this same plant and other plant species (Otte et al., 1991, 1995). Arsenic is mobilized in reduced conditions due to reduction of Fe and Mn oxides and reduction of As(V) to As(III). In the rhizosphere of these plants, As was immobilized owing to the oxidation to As(V) and adsorption to FeOOH (Otte et al., 1991). It has been shown in incubation experiments that As uptake from solution is increased with increasing Fe concentration on the roots, but most of the As is likely to be retained on the root surface in this situation. Concentrations of As in the plaque on the roots of wetland species in polluted environments were found to be highly correlated to concentrations of Fe (Otte et al., 1995).

9–4 PHYSICOCHEMICAL PROCESSES AFFECTED BY MICROBIAL ACTIVITY

9–4.1 Impact of Microbial Activity on Contaminant Chemistry

In the rhizosphere, it is difficult to separate the impacts on contaminant chemistry of microbial activity from those of plant root activity. Microorganisms act in a similar fashion to plant roots in that they can accumulate contaminants through uptake and adsorption, as well as mobilizing contaminants through the action of microbial exudates, diffusates and excretions. As shown by Rovira and Davey (1974), bacteria in particular are abundant in the rhizosphere (Fig. 9–14) and because of their high surface area to volume ratio, they have a large capacity to sorb metals (Beveridge, 1988).

Both gram-positive and gram-negative bacteria have a net negative surface charge in the pH range of 5 to 8, and metals bind strongly to phosphoryl and carboxyl groups in the cell wall (Beveridge et al., 1995). Furthermore, the polysaccharide gel surrounding bacteria can also bind metals (Geesey et al., 1968) and may be implicated in tolerance to high metal concentrations by some bacteria (Bitton & Freihofer, 1978). This adsorption of metals to bacterial cell walls or mucilages may account for a large proportion of the metal taken up from solution by microorganisms. For example, Surowitz et al. (1984) found that up to 95% of the Cd removed from solution by *Bacillus subtilis* was attributable to binding on the cell wall and membrane, with just over 5% found in the soluble fraction within the cell; however, specific transport mechanisms for transport of metals across bacterial cell membranes have been identified for some metals (Hughes & Poole, 1989; Lepp, 1992).

Much of the information with regard to uptake of metal contaminants by microorganisms has been generated in vitro. Most data have been collected for design and development of wastewater treatment facilities using microbial growth as a means to remove both nutrients and metals from contaminated waters (Norberg & Rydin, 1984; Lepp, 1992). The effect of free-living rhizosphere flora on metal uptake by plants is not known, with little data collected for indigenous soil microflora and even less for rhizosphere flora.

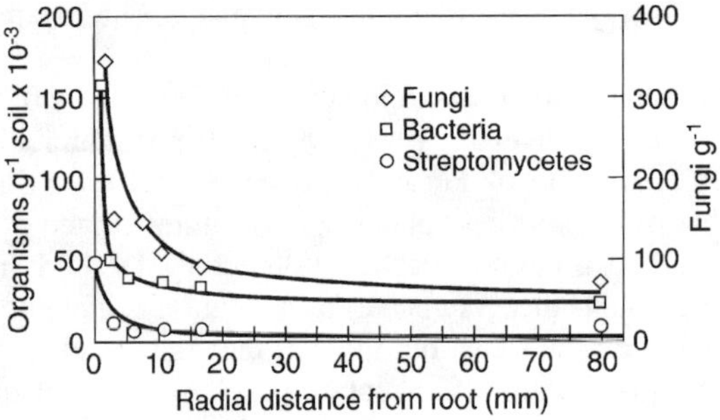

Fig. 9–14. Distribution of organisms with distance from the roots of 18-d old lupin (*Lupinus angustifolius* L.) seedlings (from Rovira & Davey, 1974).

9–4.2 Effect of Mycorrhizal Infection on Contaminant Uptake by Plants

The growth and morphology of vesicular arbuscular mycorrhizae and their impact in terms of plant nutrition have been outlined in Marschner (1998, this publication).

Mycorrhizae also modify uptake of contaminants by plants in polluted soils; however, it is important to draw a distinction between the effect of mycorrhizae on uptake of contaminants from soils mildly polluted by trace concentrations of contaminants (e.g., Cd from phosphatic fertilizers) and those grossly polluted by sewage sludges or point sources.

At low levels of metal pollution in soil, infection by mycorrhizae increases uptake of metal contaminants from soil (Cooper & Tinker, 1978; Rogers & Williams, 1986; Li et al., 1991). Presumably, the mycorrhizae in these situations are acting in a similar way to their enhancement of nutrient uptake, i.e., extension of the effective rooting volume explored by the root:mycelial mass. Hence, elements that are likely to be limited by diffusion through soil to the root are most likely to be enhanced by mycorrhizal infection, e.g., Cd, Cu, Pb, Zn, and Ni. This could potentially lead to problems where low levels of metal pollution lead to concentrations of contaminants in crops, which could be of concern from a human health viewpoint (e.g., Cd).

In highly polluted soils, there is contradictory evidence regarding the effect of mycorrhizal infection on metal uptake by plants. For example, Killham and Firestone (1983) reported that colonization of roots of bunchgrass (*Erharta calycina*) by mycorrhizae enhanced uptake of Cu, Ni, Pb, and Zn and enhanced toxicity of these elements. Other authors have reported that mycorrhizal infection inhibits metal uptake by plants in polluted soils and affords some protection to the plant from metal toxicity (Dueck et al., 1986; Heggo et al., 1990). This has been attributed to a number of factors: binding of metal in the external mycelium, or binding to cell walls or membranes in mycelia within the host plant. Metal binding appears to be ecotype related, with mycorrhizae from polluted soils having a greater ability to bind metals than those from non-contaminated soils. Galli et al. (1994) recently reviewed the various factors responsible for binding of metals by mycorrhizal fungi. They concluded that metal binding is predominantly related to sorption to cell wall components such as chitin, melanin, cellulose, and cellulose derivatives, rather than due to coprecipitation within polyphosphate granules, as previously suggested.

9–5 SUMMARY AND CONCLUSIONS

The soil–root interface is a zone of intense biological activity which alters markedly the physical and chemical processes affecting contaminant behavior in soils. Much of the study of physicochemical processes in soils has concentrated on bulk soil conditions, with little regard for the interactions in the rhizosphere. While some studies of nutrient dynamics in the rhizosphere have been undertaken, there are few studies for contaminants that pose ecological or human health risks.

Plant growth induces significant fluxes of water and solutes to the root surface, significantly altering the composition of the soil solution in the rhizosphere. The depletion or accumulation of nutrient ions at the root surface have been investigated, with depletion shown for P, Mo, Zn, and other nutrients recorded, while others such as Ca tend to accumulate at the root surface. Few studies exist for contaminant ions.

In addition to the direct effects of root uptake of water and solutes on rhizosphere chemistry, plants excrete many ions and compounds either passively or actively into soil adjacent to the root, e.g., protons, organic ligands, enzymes, and others. These compounds and ions alter conditions in the rhizosphere sufficiently to markedly affect adsorption and retention of contaminants. Examination of the effect of root exudates or secretions on uptake of contaminant ions must consider not only soil chemistry, but must include some assessment of plant demand and uptake pathways. For example, the effects of organic ligands on metal uptake by plants depend in part on the kinetics of uptake in relation to the kinetics of desorption or dissolution in soil, with either increases or decreases in uptake possible as ligand concentrations increase.

Microbial activity is markedly enhanced in rhizosphere compared with bulk soil. Microorganisms in the rhizosphere can act as sinks for pollutants, effect the degradation of some organic pollutants, or mobilize pollutants due to excretion of extracellular complexing agents. Infection of plants by mycorrhizae may increase or decrease phytoavailability of metal species and offer some protection from toxicity in metal-rich soils.

The lack of knowledge of contaminant dynamics in the rhizosphere suggests that more emphasis should be placed on integrating the study of soil chemical and physical properties in relation to root growth. Predictions of contaminant behavior in the rhizosphere on the basis of models assuming equilibrium and/or homogenous conditions could be improved by consideration of kinetics of reactions and heterogeneity of the soil-root environment.

ACKNOWLEDGMENTS

The senior author thanks the Horticultural Research and Development Corporation for support while preparing this review article.

REFERENCES

Adams, F. 1974. Soil solution. p. 441–481. *In* E.W. Carson (ed.) The plant root and its environment. Univ. of Virginia Press, Charlottesville.

Alloway, B.J. 1995. Soil processes and the behavior of metals. p. 11–37. *In* B.J. Alloway (ed.) Heavy metals in soils. Blackie Academic & Professional, London.

Barber, D.A., and J.K. Martin. 1976. The release of organic substances by cereal roots into soil. New Phytol. 76:69–80.

Barber, S.A. 1984. Soil nutrient bioavailability. A mechanistic approach. John Wiley & Sons, New York.

Barrow, N.J. 1987. Reactions with variable charge soils. Martinus Nijhoff, Dordrecht, the Netherlands.

Barrow, N.J. 1993. Mechanisms of retention of zinc with soil and soil components. p. 15–31. *In* A.D. Robson (ed.) Zinc in soils and plants. Kluwer Academic Publ., Dordrecht, the Netherlands.

Bartlett, R., and B.R. James. 1979. Behavior of chromium in soils: III. Oxidation. J. Environ. Qual. 8:31–35.

Bell, P.F., R.L. Chaney, and J.S. Angle 1991. Free metal activity and total metal concentrations as indices of micronutrient availability to barley *(Hordeum vulgare* L. 'Klages'). Plant Soil 130:51–62.

Benjamin, M.M., and J.O. Leckie. 1982. Effects of complexation by Cl, SO_4, and S_2O_3 on adsorption behavior of Cd on oxide surfaces. Environ. Sci. Technol. 16:162–170.

Berggren, D. 1990. Speciation of cadmium (II) using donnan dialysis and differential-pulse anodic stripping voltammetry in a flow-injection system. Int. J. Environ. Anal. Chem. 41:133–148.

Beveridge, T.J. 1988. The bacterial surface: General considerations towards design and function. Can. J. Microbiol. 34:363–372.

Beveridge, T.J., S. Schultze-Lam, and J.B. Thompson. 1995. Detection of anionic sites on bacterial cell walls, their ability to bind toxic heavy metals and form sedimentable flocs and their contribution to mineralization in natural freshwater environments. p. 183–205. *In* H.E. Allen et al. (ed.) Metal speciation and contamination of soil. Lewis Publ., Boca Raton, FL.

Bittel, J.E., and R.J. Miller. 1974. Lead, cadmium and calcium selectivity coefficients on a montmorillonite, illite and kaolinite. J. Environ. Qual. 3:250–253.

Bitton, G., and V. Freihofer. 1978. Influence of extracellular polysaccharide on the toxicity of copper and cadmium toward *Klebsiella aerogenes*. Microb. Ecol. 4:119–125.

Blanchar, R.W., and D.S. Lipton. 1986. The pe and pH in alfalfa seedling rhizospheres. Agron. J. 78:216–218.

Bleam, W.F., and M.B. McBride. 1986. The chemistry of adsorbed Cu(II) and Mn(II) on aqueous titanium dioxide suspensions. J. Colloid. Interface Sci. 103:124–132.

Boekhold, A.E., E.J.M. Temminghoff, and S.E.A.T.M. van der Zee. 1993. Influence of electrolyte composition and pH on cadmium sorption by an acid soil. J. Soil Sci. 44:85–96.

Borgmann, U. 1983. Metal speciation and toxicity of free metals ions to aquatic biota. p. 47–72. *In* J.O. Nriagu (ed.) Aquatic toxicology. John Wiley & Sons, New York.

Bouldin, D.R. 1989. A multiple ion uptake model. J. Soil Sci. 40:309–319.

Boyle, J.R., and G.K. Voigt. 1973. Biological weathering of silicate minerals. Implications for tree nutrition and soil genesis. Plant Soil 38:191–201.

Brewster, J.L., and P.B. Tinker. 1970. Nutrient cation flows around plant roots. Soil Sci. Soc. Am. Proc. 34:421–426.

Bromfield, S.M. 1958. The solution of γ-MnO_2 by substances released from soil and from the roots of oats and vetch in relation to manganese availability. Plant Soil 10:147–160.

Brown, R.M., C.J. Pickford, and W.L. Davison. 1984. Speciation of metals in soils. Int. J. Environ. Anal. Chem. 18:135–141.

Bruckert, S. 1970. Influence des composés organiques solubles sur la pédogenese en milieu acide: II. Expériences de laboratoire. Ann. Agron. 21:725–757.

Brummer, G.W., J. Gerth, and K.G. Tiller. 1988. Reaction kinetics of the adsorption and desorption of nickel, zinc and cadmium by goethite: II. Adsorption and diffusion of metals. J. Soil Sci. 39:37–52.

Brummer, G.W., K.G. Tiller, U. Herms, and P.M. Clayton. 1983. Adsorption–desorption and/or precipitation-dissolution processes of zinc in soils. Geoderma 31:337–354.

Campbell, R., and M.P. Greaves. 1990. Anatomy and community structure of the rhizosphere. p. 11–34. *In* J.M. Lynch (ed.) The rhizosphere. John Wiley & Sons, New York.

Cavallaro, N., and M.B. McBride. 1978. Copper and cadmium adsorption characteristics of selected acid and calcareous soils. Soil. Sci. Soc. Am. J. 42:550–556.

Chairidchai, P., and G.S.P. Ritchie. 1990. Zinc adsorption by a lateritic soil in the presence of organic ligands. Soil Sci. Soc. Am. J. 54:1242–1248.

Chaney, R.L., and S.B. Hornick. 1978. Accumulation and effects of cadmium on crops. p. 125–140. *In* Proc. of the Int. Cadmium Conf., 1st, San Francisco. 1977. Metals Bulletin, London.

Checkai, R.T., R.B. Corey, and P.A. Helmke. 1987b. Effects of ionic and complexed metal concentrations on plant uptake of cadmium and micronutrient metals from solution. Plant Soil 99:335–345.

Checkai, R.T., L.L. Hendrickson, R.B. Corey, and P.A. Helmke. 1987a. A method for controlling the activities of free metal, hydrogen and phosphate ions in hydroponic solutions using ion exchange and chelating resins. Plant Soil 99:321–334.

Christensen, T.H. 1984. Cadmium soil sorption at low concentrations: II. Reversibility, effect of changes in solute composition, and effect of soil aging. Water Air Soil Pollut. 21:115–125.

Christensen, T.H. 1987. Cadmium soil sorption at low concentrations: V. Evidence of competition by other heavy metals. Water Air Soil Pollut. 34:293–303.

Chubin, R.G., and J.J. Street. 1981. Adsorption of cadmium on soil constituents in the presence of complexing ligands. J. Environ. Qual. 10:225–228.

Comans, R.N.J. 1990. Sorption of cadmium and cesium at mineral/water interfaces. Reversibility and its implications for environmental mobility. Ph.D. diss. Rijksuniversiteit Utrecht, the Netherlands.

Cooper, K.M., and P.B. Tinker. 1978. Translocation and transfer of nutrients in vesicular–arbuscular mycorrhizas: II. Uptake and translocation of phosphorus, zinc and sulphur. New Phytol. 73:901–912.

Coughtrey, P., and M. Thorne. 1983. Radionuclide distribution and transport in terrestrial and aquatic ecosystems. A critical review of data. A.A. Balkema Publ., Rotterdam, the Netherlands.

Coulson, C.B., R.I. Davies, and D.A. Lewis. 1960. Polyphenols in plant, humus and soil: I. Polyphenols of leaves, litter and superficial humus from mull and mor sites. J. Soil Sci. 11:20–29.

Cremers, A., E. Elsen, P. De Preter, and A. Maes. 1988. Quantitative analysis of radiocaesium retention in soils. Nature (London) 335:247–249.

Davies, R.J. 1971. Relation of polyphenols to decomposition of organic matter and to pedogenic processes. Soil Sci. 111:80–85.

Davis, J.A., C.C. Fuller, and A.D. Cook. 1987. A model for trace metal sorption processes at the calcite surface: Adsorption of Cd^{2+} and subsequent solid solution formation. Geochim. Cosmochim. Acta 51:1477–1490.

DeKock, P.C. 1956. Heavy metal toxicity and iron chlorosis. Ann. Bot. 20:133–141.

DeKock, P.C., and R.L. Mitchell. 1957. Uptake of chelated metals by plants. Soil Sci. 84:55–62.

Delhaize, E., P.R. Ryan, and P.J. Randall 1993. Aluminum tolerance in wheat (*Triticum aestivum* L.): II. Aluminum-stimulated excretion of malic acid from root apices. Plant Physiol. 103:695–702.

De Preter, P. 1990. Radiocesium retention in aquatic, terrestrial and urban environments: A quantitative and unifying analysis. Dissertations de Agricultura no. 190, Katholieke Universiteit Leuven, Belgium.

Doner, H.E. 1978. Chloride as a factor in mobilities of Ni(II), Cu(II) and Cd(II) in soil. Soil Sci. Soc. Am. J. 42:882–885.

Dueck, Th.A., P. Visser, W.H.O. Ernst, and H. Schat. 1986. Vesicular–arbuscular mycorrhizae decrease zinc-toxicity to grasses growing in zinc-polluted soil. Soil Biol. Biochem. 18:331–333.

Egozy, Y. 1980. Adsorption of cadmium and cobalt on montmorillonite as a function of solution composition. Clays Clay Min. 28:311–318.

El-Sayed, M.H., R.G. Burau, and K.L. Babcock. 1970. Thermodynamics of copper (II)–calcium exchange on bentonite clay. Soil Sci. Soc. Am. Proc. 34:397–400.

Elliott, H.A., and C.M. Denneny. 1982. Soil adsorption of cadmium from solutions containing organic ligands. J. Environ. Qual. 11:658–662.

Elliott, H.A., and C.P. Huang. 1979. The adsorption characteristics of Cu(II) in the presence of chelating agents. J. Colloid Interface Sci. 70:29–45.

Eriksson, J.E. 1990. Effects of nitrogen-containing fertilizers on solubility and plant uptake of cadmium. Water Air Soil Pollut. 49:355–368.

Farrah, H., and W.F. Pickering. 1977. The sorption of lead and cadmium species by clay minerals. Aust. J. Chem. 30:1417–1422.

Fitch, A., and P.A. Helmke. 1989. Donnan equilibrium/graphite furnace atomic absorption estimates of soil extract complexation capacities. Anal. Chem. 61:1295–1298.

Förstner, U. 1981. Metal transfer between solid and aqueous phases. p. 197–270. *In* U. Förstner and G.T.W. Wittman (ed.) Metal pollution in the aquatic environment. Springer-Verlag, New York.

Francis, C., and F.S. Brinkley. 1976. Preferential adsorption of ^{137}Cs to micaceous minerals in contaminated freshwater sediment. Nature (London) 260:511–513.

Fujii, R., L.L. Hendrickson, and R.B. Corey. 1983. Ionic activities of trace metals in sludge-amended soils. Sci. Total Environ. 28:179–190.

Gahoonia, T.S., and N.E. Nielsen. 1992. The effect of root-induced pH changes on the depletion of inorganic and organic phosphorus in the rhizosphere. Plant Soil 143:185–191.

Galli, U., H. Schuepp, and C. Brunold. 1994. Heavy metal binding by mycorrhizal fungi. Physiol. Plant. 92:364–368.

Garcia-Miragaya, J., and A.L. Page. 1976. Influence of ionic strength and inorganic complex formation on the sorption of trace amounts of Cd by montmorillonite. Soil Sci. Soc. Am. J. 40:658–663.

Garcia-Miragaya, J., and A.L. Page. 1977. Influence of exchangeable cations on the sorption of trace amounts of cadmium by montmorillonite. Soil Sci. Soc. Am. J. 41:718–721.

Geesey, G.G., L. Lang, J.G. Jolley, M.R. Hankins, T. Iawoka, and P.R. Griffiths. 1968. Binding of metal ions by extracellular polymers of biofilm bacteria. Water Sci. Technol. 20:161–165.

Gerth, J., and G. Brummer. 1983. Adsorption and immobilization of nickel, zinc and cadmium by goethite (a-FeOOH). Fresenius J. Anal. Chem. 316:616–620.

Gessa, C., M.L. de Cherchi, P. Melis, G. Micera, and S.L. Strina Erre 1984. Anion-induced metal binding in amorphous aluminum hydroxide. Colloids Surf. 11:109–117.

Godo, G.H., and H.M. Reisenauer 1980. Plant effects on soil manganese availability. Soil Sci. Soc. Am. J. 44:993–995.

Gries, D., S. Brunn, D.E. Crowley, and D.R. Parker 1995. Phytosiderophore release in relation to micronutrient metal deficiencies in barley. Plant Soil 172:299–308.

Hahne, H.C.H., and W. Kroontje. 1973. Significance of pH and chloride concentration on behavior of heavy metal pollutants, mercury (II), cadmium (II), zinc (II) and lead (II). J. Environ. Qual. 2:444–450.

Hale, M.G., C.L. Foy, and F.J. Shay. 1971. Factors affecting root exudation. Adv. Agron. 23:89–109.

Halvorson, A.D., and W.L. Lindsay 1977. The critical Zn^{2+} concentration for corn and the nonabsorption of chelated zinc. Soil Sci. Soc. Am. J. 41:531–534.

Hamon, R.E. 1995. Identification of factors governing cadmium and zinc bioavailability in polluted soils. Ph.D. diss. Univ. of Nottingham, England.

Hamon, R.E., S.E. Lorenz, P.E. Holm, T.H. Christensen, and S.P. McGrath. 1995. Changes in trace metal species and other components of the rhizosphere during growth of radish. Plant Cell Environ. 18:749–756.

Harter, R.D., and R. Naidu. 1996 Role of metal-organic complexation in metal sorption by soils. Adv. Agron. 55:219–264.

Hatch, D.J., L.H.P. Jones, and R.G. Burau. 1988. The effect of pH on the uptake of cadmium by four plant species grown in flowing solution culture. Plant Soil 105:121–126.

Heggo, A., J.S. Angle, and R.L. Chaney. 1990. Effects of vesicular–arbuscular mycorrhizal fungi on heavy metal uptake by soybeans. Soil Biol. Biochem. 22:865–869.

Helmke, P.A., and R. Naidu. 1996. Fate of contaminants in the soil environment: Metal contaminants. p. 69–93. *In* R. Naidu et al. (ed.) Contaminants and the soil environment in the Australasia-Pacific region. Kluwer Academic Publ., Dordrecht, the Netherlands.

Helmke, P.A., A.B. Salam, and Y. Li. 1998. Measurement and behavior of indigenous levels of the free, hydrated cations of Cu, Zn and Cd in the soil–water system. *In* R. Prost (ed.) Proc. of the Int. Conf. on the Biogeochemistry of Trace Elements in the Environment, 3rd. INRA, Versailles, France. (in press).

Hendrikson, L.L., and R.B. Corey. 1981. Effect of equilibrium metal concentrations on apparent selectivity coefficients of soil complexes. Soil Sci. 131:163–171.

Hiltner, L. 1904. Uber neuere Erfahrungen und Proleme auf dem Gebeit der Bodenbakteriologie und unter besonderer Berucksichtigung der Grundungung und Brache. Arb. Dsch. Landwirt. Ges. 98:59–78.

Hirsch, D., and A. Banin. 1990. Cadmium speciation in soil solutions. J. Environ. Qual. 19:366–372.

Hodgson, J.F., H.R. Geering, and W.A. Norvell. 1965. Micronutrient cation complexes in soil solution: Partition between complexed and uncomplexed forms by solvent extraction. Soil. Sci. Soc. Am. Proc. 29:665–669.

Hodgson, J.F., K.G. Tiller, and M. Fellows. 1964. The role of hydrolysis in the reaction of metals with soil forming materials. Soil Sci. Soc. Am. Proc. 28:42–46.

Hoffland, E. 1992. Quantitative evaluation of the role of organic acid exudation in the mobilization of rock phosphate by rape. Plant Soil 140:279–289.

Hoffland, E., G.R. Findenegg, and J.A. Nelemans 1989. Solubilization of rock phosphate by rape: II. Local root exudation of organic acids as a response to P-starvation. Plant Soil 113:161–165.

Hoins, U., L. Charlet, and H. Sticher. 1993. Ligand effect on the adsorption of heavy metals: The sulfate–cadmium–goethite case. Water Air Soil Pollut. 68:241–255.

Holm, P.E., T.H. Christensen, J.C. Tjell, and S.P. McGrath. 1995. Speciation of cadmium and zinc with application to soil solutions. J. Environ. Qual. 24:183–190.

Homann, P.S., and R.J. Zasoski. 1987. Solution composition effects on cadmium sorption by forest soil profiles. J. Environ. Qual. 16:429–433.

Huang, P.M., and A. Violante. 1986. Influence of organic acids on crystallization and surface properties of precipitation products of aluminum. p. 159–221. *In* P.M. Huang and M. Schnitzer (ed.) Interactions of soils minerals with natural organics and microbes. SSSA, Madison, WI.

Huang, W.H., and W.D. Keller. 1972. Dissolution of clay minerals in dilute organic acids at room temperature. Am. Mineral. 56:1082–1095.

Hughes, M.N., and R.K. Poole. 1989. Metal mimicry and metal limitation in studies of metal-microbe interactions. p. 1–17. *In* R.K. Poole and G.M. Gadd (ed.) Metal-microbe interactions. IRL Press, Oxford.

Inskeep, W.P., and J. Baham. 1983. Competitive complexation of Cd(II) and Cu(II) by water-soluble organic ligands and Na-montmorillonite. Soil Sci. Soc. Am. J. 47:1109–1115.

Irving, H., and R.J.P. Williams. 1948. Order of stability of metal complexes. Nature (London) 162:746–747.

James, B.R., and R.J. Bartlett. 1983. Behavior of chromium in soils: VI. Interactions between oxidation–reduction and organic complexation. J. Environ. Qual. 12:173–176.

Jarvis, S.C., L.H.P. Jones, and M.J. Hopper. 1976. Cadmium uptake from solution by plants and its transport from roots to shoots. Plant Soil 44:179–191.

Jauregui, M.A., and H.M. Reisenauer. 1982. Dissolution of oxides of manganese and iron by root exudate components. Soil Sci. Soc. Am. J. 46:314–317.

Jeffrey, J.W.O. 1961. Defining the state of reduction of a paddy soil. J. Soil Sci. 12:172–179.

Jenne, E.A. 1968. Controls on Mn, Fe, Co, Ni, Cu and Zn concentrations in soils and water: The significant role of hydrous Mn and Fe oxides. Adv. Chem. Serv. 73:337–387.

John, M.K. 1972. Cadmium adsorption maxima of soils as measured by the Langmuir isotherm. Can. J. Soil Sci. 52:343–350.

Jones, D.L., and P.R. Darrah. 1994. Role of root derived organic acids in the mobilization of nutrients from the rhizosphere. Plant Soil 166:247–257.

Jopony, M., and S.D. Young. 1994. The soil:solution equilibria of lead and cadmium in polluted soils. Eur. J. Soil Sci. 45:59–70.

Jurinak, J.J., and D.L. Suarez. 1990. The chemistry of salt-affected soils and waters. p. 42–63. *In* K.K. Tanji (ed.) Agricultural salinity assessment and management. Am. Soc. of Civil Eng., New York.

Katz, S.A., and H. Salem. 1994. The biological and environmental chemistry of chromium. VCH Publ., New York.

Killham, K., and M.K. Firestone. 1983. Vesicular arbuscular mycorrhizal mediation of grass response to acid and heavy metal deposition. Plant Soil 72:39–48.

Kinniburgh, D.G., M.L. Jackson, and J.K. Syers. 1976. Adsorption of alkaline earth, transition and heavy metal cations by hydrous oxide gels of iron and aluminum. Soil Sci. Soc. Am. J. 40:796–799.

Kirk, G.J.D., and J.B. Bajita. 1995. Root-induced iron oxidation, pH changes and zinc solubilization in the rhizosphere of lowland rice. New Phytol. 131:129–137.

Kirk, G.J.D., and S. Staunton. 1989. On predicting the fate of radioactive caesium in soil beneath grassland. J. Soil Sci. 40:71–84.

Kochian, L.V. 1993. Zinc absorption from hydroponic solutions by plant roots. p. 45–57. *In* A.D. Robson (ed.) Zinc in soils and plants. Kluwer, Dordrecht, the Netherlands.

Kuo, S., and B.L. McNeal. 1984. Effects of pH and phosphate on cadmium sorption by a hydrous ferric oxide. Soil Sci. Soc. Am. J. 48:1040–1044.

Kwong, Ng Kee, K.F., and P.M. Huang. 1979. Surface reactivity of aluminum hydroxides precipitated in the presence of low molecular weight organic acids. Soil Sci. Soc. Am. J. 43:1107–1113.

Ladd, J.N., and J.H.A. Butler, 1975. Humus-enzyme systems and synthetic, organic polymers-enzyme analogs. p. 143–194. *In* E.A. Paul and A.D. McLaren (ed.) Soil biochemistry. Marcel Dekker, New York.

Langford, C.H., and D.W. Gutzman. 1992. Kinetic studies of metal ion speciation. Anal. Chim. Acta 256:183–201.

Laurie, S.H., N. Tancock, S.P. McGrath, and J.R. Sanders. 1991. Influence of complexation on the uptake by plants of iron, manganese, copper and zinc: II Effect of DTPA in a multi-metal and computer simulation study. J. Exp. Bot. 42:509–513.

Lavigne, J.A., C.H. Langford, and M.K.S. Mak. 1987. Kinetic study of the speciation of Ni (II) bound to a fulvic acid. Anal. Chem. 59:2616–2620.

Lepp, N. 1992. Uptake and accumulation of metals in bacteria and fungi. p. 277–298. *In* D.C. Adriano (ed.) Biogeochemistry of trace metals. Lewis Publ., Boca Raton.

Levy, R., and C.W. Francis. 1976. Adsorption and desorption of cadmium by synthetic and natural organo-clay complexes. Geoderma 15:361–370.

Li, X.L., E. Georg, and H. Marschner. 1991. Acquisition of phosphorus and copper by VA-mycorrhizal hyphae and root to shoot transport in white clover. Plant Soil 136:49–57.

Li, Y.-M., R.L. Chaney, and A.A. Schneiter. 1994. Effect of soil chloride level on cadmium concentration in sunflower kernels. Plant Soil 167:275–280.

Lindsay, W.L. 1974. Role of chelation in micronutrient availability. p. 507–524. *In* E.W. Carson (ed.) The plant root and its environment. Univ. of Virginia Press, Charlottesville.

Lindsay, W.L. 1979. Chemical equilibria in soils. Wiley-Interscience, New York.

Lindsay, W.L., and W.A. Norvell. 1969. Equilibrium relationships of Zn^{2+}, Fe^{3+}, Ca^{2+} and H^+ with EDTA and DTPA in soils. Soil Sci. Soc. Am. Proc. 33:62–65.

Linehan, D.J., A.H. Sinclair, and M.C. Mitchell. 1989. Seasonal changes in Cu, Mn, Zn and Co concentrations in soil in the root zone of barley (*Hordeum vulgare* L.). J. Soil Sci. 40:103–115.

Lion, L.W., R.S. Altmann, and J.O. Leckie. 1982. Trace-metal adsorption characteristics of estuarine particulate matter: Evaluation of contributions of Fe/Mn oxide and organic surface coatings. Environ. Sci. Technol. 16:660–666.

Loganathan, P., and R.G. Burau. 1973. Sorption of heavy metal ions by a hydrous manganese oxide. Geochem. Cosmochim. Acta 37:1277–1293.

Lorenz, S.E., R.E. Hamon, and S.P. McGrath. 1994. Difference between soil solutions obtained from rhizosphere and non-rhizosphere soils by water displacement and soil centrifugation. Eur. J. Soil Sci. 45:431–438.

Lumsden, D.G., and L.J. Evans. 1995. Predicting chemical speciation and computer simulation. p. 86–134. *In* A.M. Ure and C.M. Davidson (ed.) Chemical speciation in the environment. Blackie Academic & Professional, London.

Lund, W. 1990. Speciation analysis–why and how? Fresenius'. J. Anal. Chem. 337:557–564.

Lynch, J.M. 1983. Soil biotechnology. Microbial factors in crop productivity. Blackwell Scientific Publ., Oxford.

Ma, Q., and W.L. Lindsay. 1990. Divalent Zn activity in arid zone soils obtained by chelation. Soil Sci. Soc. Am. J. 54:719–722.

Margerum, D.W., G.R. Cayley, D.C. Weatherburn, and G.K. Pagenkopf. 1978. Kinetics and mechanisms of complex formation and ligand exchange. p. 1–220. *In* A.E. Martell (ed.) Coordination chemistry. Vol. 2. ACS Monogr. 174. Am. Chem. Soc., Washington, DC.

Marschner, H. 1995. Mineral nutrition of higher plants. Academic Press, London.

Marschner, H., and V. Römheld. 1983. In vivo measurement of root-induced pH changes at the soil–root interface: Effects of the plant species and nitrogen source. Z. Pflanzenernähr. Bodenk. 111:241–251.

Marschner, H, V. Römheld, W.J. Horst, and P. Martin. 1986. Root-induced changes in the rhizosphere: Importance for the mineral nutrition of plants. Z. Pflanzenernähr. Bodenk. 149:441–456.

Marschner, H., M. Treeby, and V. Römheld. 1989. Role of root-induced changes in the rhizosphere for iron acquisition in higher plants. Z. Pflanzenernähr. Bodenk. 152:197–204.

Masschelyn, P.H., and W.H. Patrick, Jr. 1994. Selenium, arsenic and chromium redox chemistry in wetland soils and sediments. p. 615–625. *In* D.C. Adriano et al. (ed.) Biogeochemistry of trace elements. Science & Technology Letters, Northwood.

McBride, M.B. 1989. Reactions controlling heavy metal solubility in soil. Adv. Soil Sci. 10:1–56.

McBride, M.B. 1991. Processes of heavy and transition metal sorption by soil minerals. p. 149–174. *In* G.H. Bolt et al. (ed.) Interactions at the soil colloid–soil solution interface. Kluwer Academic Publ., Dordrecht, the Netherlands.

McKenna, I.M., R.L. Chaney, and F.M. Williams. 1993. The effects of cadmium and zinc interactions on the accumulation and tissue distribution of zinc and cadmium in lettuce and spinach. Environ. Pollut. 79:113–120.

McLaren, R.G., J.G. Williams, and R.S. Swift. 1983. Some observations on the desorption and distribution behavior of copper with soil components. J. Soil Sci. 34:325–331.

McLaughlin, M.J., and K.G. Tiller. 1994. Chloro-complexation of cadmium in soil solutions of saline/sodic soils increases phyto-availability of cadmium. p. 195–196. *In* Trans. of the World Congress of Soil Science, 15th, Acapulco, Mexico. 1994. ISSS, Wageningen, the Netherlands.

McLaughlin, M.J., K.G. Tiller, T.A. Beech, and M.K. Smart. 1994. Soil salinity causes elevated cadmium concentrations in field grown potato tubers. J. Environ. Qual. 23:1013–1018.

McLaughlin, M.J., K.G. Tiller, R. Naidu, and D.P. Stevens. 1996. Review: The behavior and environmental impact of contaminants in fertilizers. Aust. J. Soil. Res. 34:1–54.

Mench, M., and S. Fargues. 1994. Metal uptake by iron-efficient and inefficient oats. Plant Soil 165:227–233.

Mench, M., and E. Martin. 1991. Mobilization of cadmium and other metals from two soils by root exudates of *Zea mays* L., *Nicotiana tabacum* L. and *Nicotiana rustica* L. Plant Soil 132:187–196.

Mench, M., J.L. Morel, and A. Guckert. 1987. Metal binding properties of high molecular weight soluble exudates from maize (*Zea mays* L.) roots. Biol. Fertil. Soils 3:165–169.

Mench, M., J.L. Morel, A. Guckert, and B. Guillet. 1988. Metal binding with root exudates of low molecular weight. J. Soil Sci. 39:521–527.

Merckx, R., J.H. van Ginkel, J. Sinnaeve, and A. Cremers. 1986. Plant-induced changes in the rhizosphere of maize and wheat: II. Complexation of cobalt, zinc and manganese in the rhizosphere of maize and wheat. Plant Soil 96:95–107.

Milberg, R.P., D.L. Brower, and J.V. Lagerwerff. 1978. Exchange adsorption of trace quantities of cadmium in soils treated with calcium and sodium: A reappraisal. Soil Sci. Soc. Am. J. 36:734–737.

Morel, J.L., M. Mench, and A. Guckert. 1986. Measurement of Pb^{2+}, Cu^{2+} and Cd^{2+} binding with mucilage exudates from maize (*Zea mays* L.) roots. Biol. Fertil. Soils 2:29–34.

Mück, K. 1995. Long-term reduction of caesium concentration in milk after nuclear fallout. Sci. Total Environ. 162:63–74.

Mullins, G.L., and L.E. Sommers. 1986. Characterization of cadmium and zinc in four soils treated with sewage sludge. J. Environ. Qual. 15:382–387.

Mullins, G.L., L.E. Sommers, and S.A. Barber. 1986. Modeling the plant uptake of cadmium and zinc from soils treated with sewage sludge. Soil Sci. Soc. Am. J. 50:1245–1250.

Naas, C.N., and N.D. Horowitz. 1986. Adsorption of cadmium to kaolinite in the presence of organic material. Water Air Soil Pollut. 27:131–140.

Navrot, J., A. Singer, and A. Banin. 1978. Adsorption of cadmium and its exchange characteristics in some Israeli soils. J. Soil Sci. 29:505–511.

Neal, R.H., and G Sposito. 1986. Effects of soluble organic matter and sewage sludge amendments on cadmium sorption by soils at low cadmium concentrations. Soil Sci. 142:164–172.

Nielsen, N.E. 1976. The effect of plants on the copper concentration in the soil solution. Plant Soil 45:679–687.

Norberg, A., and S. Rydin. 1984. Development of a continuous process for metal recovery by *Zooglea ramigera*. Biotechnol. Bioeng. 26:265–268.

Nye, P.H. 1984. On estimating the uptake of nutrients solubilized near roots or other surfaces. J. Soil Sci. 35:439–446.

Nye, P.H., and P.B. Tinker. 1977. Solute movement in the soil–root system. Stud. Ecol. 4. Blackwell Scientific Publ., Oxford.

O'Neill, P. 1995. Arsenic. p. 105–121. *In* B.J. Alloway (ed.) Heavy metals in soils. Blackie Academic & Professional, London.

Oscarson, D.W., P.M. Huang, U.T. Hammer, and W.K. Liaw. 1993. Oxidation and sorption of arsenite by manganese dioxide as influenced by surface coatings of iron and aluminum oxides and calcium carbonate. Water Air Soil Pollut. 20:233–244.

Otte, M.L., M.J. Dekkers, J. Rozema, and R.A. Broekman. 1991. Uptake of arsenic by *Aster tripolium* in relation to rhizosphere oxidation. Can. J. Bot. 69:2670–2677.

Otte, M.L., C.C. Kearns, and M.O. Doyle. 1995. Accumulation of arsenic and zinc in the rhizosphere of wetland plants. Bull. Environ. Contam. Toxicol. 55:154–161.

Otte, M.L., J. Rozem, L. Koster, M.S. Haarsma, and R.A. Broekman. 1989. Iron plaque on roots of *Aster tripolium* L.: Interaction with Zn uptake. New Phytol. 111:309–317.

Papadopoulos, P., and D.L. Rowell. 1988. The reactions of cadmium with calcium carbonate surfaces. J. Soil Sci. 39:23–36.

Parker, D.R., R.L. Chaney, and W.A. Norvell. 1995. Chemical equilibrium models: Applications to plant nutrition. p. 163–196. *In* R.H. Loeppert et al. (ed.) Soil chemical equilibrium and reaction models. SSSA Spec. Publ. 42. ASA and SSSA, Madison, WI.

Phinney, J.T., and K.W. Bruland. 1994. Uptake of lipophilic organic Cu, Cd and Pb complexes in the coastal diatom *Thallassiosira weissflogii*. Environ. Sci. Technol. 28:1781–1790.

Rao, D.N., and D.S. Mikkelsen. 1977. Effect of rice straw additions on production of organic acids in a flooded soil. Plant Soil 47:303–311.

Rengel, Z. 1993. Mechanistic simulation models of nutrient uptake: A review. Plant Soil 152:161–173.

Rengel, Z., and D.L. Robinson. 1990. Modeling magnesium uptake from an acid soil: I. Nutrient relationships at the soil–root interface. Soil Sci. Soc. Am. J. 54:785–791.

Reuss, J.O., H.L. Dooley, and W. Griffis. 1978. Uptake of cadmium from phosphate fertilizers by peas, radishes and lettuce. J. Environ. Qual. 7:128–133.

Ritchie, G.S.P., and G. Sposito. 1995. Speciation in soils. p. 201–233. *In* A.M. Ure and C.M. Davidson (ed.) Chemical speciation in the environment. Blackie Academic & Professional, London.

Robert, M., and J. Berthelin. 1986. Role of biological and biochemical factors in soil mineral weathering. p. 453–495. *In* P.M. Huang and M. Schnitzer (ed.) Interactions of soil minerals with natural organics and microbes. SSSA, Madison, WI.

Rogers, R.D., and S.E. Williams. 1986. Vesicular–arbuscular mycorrhiza: Influence on plant uptake of cesium and cobalt. Soil Biol. Biochem. 18:371–376.

Römheld, V., and H. Marschner. 1981. Effect of Fe stress on utilization of Fe chelates by efficient and inefficient plant species. J. Plant Nutr. 3:1–4.

Rovira, A.D. 1965. Plant root exudates and their influence upon soil microorganisms. p. 107–186. *In* K.F. Baker and W.C. Snyder (ed.) Ecology of soil-borne plant pathogens. Univ. of California Press, Berkeley.

Rovira, A.D., and C.B. Davey. 1974. Biology of the rhizosphere. p. 153–204. *In* E.W. Cursova (ed.) The plant root and its environment. Univ. of Virginia Press, Charlottesville.

Rovira, A.D., and B.M. McDougal. 1967. Microbial and biochemical aspects of the rhizosphere. p. 417–463. *In* A.D. McLaren and G.H. Peterson (ed.) Soil biochemistry. Vol. 1. Marcel Dekker, New York.

Saleque, M.A., and G.A.D. Kirk. 1995. Root-induced solubilization of phosphate in the rhizosphere of lowland rice. New Phytol. 129:325–336.

Santillan-Medrano, J., and J.J. Jurinak. 1975. The chemistry of lead and cadmium in soil: Solid phase formation. Soil Sci. Soc. Am. Proc. 39:851–856.

Sawhney, B.L. 1972. Selective sorption and fixation by clay minerals: A review. Clays Clay Mineral. 20:93–100.

Schindler, P.W., and G. Sposito. 1991. Surface complexation at (hydr)oxide surfaces. p. 115–145. *In* G.H. Bolt et al. (ed.) Interactions at the soil colloid–soil solution interface. Kluwer Academic Publ., Dordrecht, the Netherlands.

Schreiner, O., and M.X. Sullivan. 1910. Studies in soil oxidation. USDA Bureau Soils Bull. 73:1–57.

Shaw, G., R. Hewamanna, J. Lillywhite, and J.N.B. Bell. 1992. Radiocaesium uptake and translocation in wheat with reference to the transfer factor concept and ion competition effects. J. Environ. Radioactivity 16:167–180.

Shorey, E.C. 1913. Some organic soil constituents. USDA Bur. Soils Bull. 88:5–41.

Silberbush, M., S. Sorek, and A. Yakirevich. 1993. K^+ uptake by root systems grown in soil under salinity: I. A mathematical model. Transp. Porous Media 11:101–116.

Smith, W.H. 1976. Character and significance of forest tree root exudates. Ecology 57:324–331.

Smolders, E. 1993. Kinetic aspects of the soil-to-plant transfer of nitrate. Dissertationes de Agricultura, No. 230. Katholieke Universiteit Leuven, Belgium.

Smolders, E., L. Kiebooms, J. Buysse, and R. Merckx. 1996a. ^{137}Cs uptake in spring wheat (*Triticum aestivum* L. cv. Tonic) at different K supply: 1. The effect in solution culture. Plant Soil 181:205–209.

Smolders, E., L. Kiebooms, J. Buysse, and R. Merckx. 1996b. ^{137}Cs uptake in spring wheat (*Triticum aestivum* L. cv. Tonic) at different K supply: 2. A potted soil experiment. Plant Soil 181:211–220.

Smolders, E., and M.J. McLaughlin. 1996a. Effect of Cl on Cd uptake by Swiss chard in unbuffered and chelator buffered nutrient solutions. Plant Soil 179:57–64.

Smolders, E., and M.J. McLaughlin. 1996b. Influence of chloride on Cd availability to Swiss chard: A resin buffered solution culture system. Soil Sci. Soc. Am. J. 60:1443–1447.

Smolders, E., and G. Shaw. 1995. Changes in radiocaesium uptake and distribution in wheat during plant development: A solution culture study. Plant Soil 176:1–6.

Sparks, D.L., and D.L. Suarez. 1991. Rates of soil chemical processes. SSSA, Madison, WI.

Sposito, G. 1989. The chemistry of soils. Oxford Univ. Press, New York.

Stevenson, F.J. 1967. Organic acids in soil. p. 119–146. *In* A.D. McLaren and G.H. Peterson (ed.) Soil biochemistry. Vol. 1. Marcel Dekker, New York.

Stevenson, F.J. 1982. Humus chemistry: Genesis, composition, reactions. John Wiley & Sons, New York.

Stevenson, F.J., and M.S. Ardakani. 1972. Organic matter reactions involving micronutrients. p. 29–58. *In* J.J. Mordvedt et al. (ed.) Micronutrients in agriculture. SSSA, Madison, WI.

Ström, L., T. Olsson, and G. Tyler. 1994. Differences between calcifuge and acidifuge plants in root exudation of low-molecular organic acids. Plant Soil 167:239–245.

Surowitz, K.G., J.A. Titus, and M. Pfister. 1984. Effects of cadmium accumulation on growth and respiration of a cadmium-sensitive strain of *Bacillus subtilis* and a selected cadmium resistant mutant. Arch. Microbiol. 140:107–122.

Sweeck, L., J. Wauters, E. Valcke, and A. Cremers. 1994. The specific interception potential for radiocaesium. p. 249–258. *In* G. Desmet et al. (ed.) Transfer of radionuclides in natural and seminatural environments. Elsevier Applied Science, London.

Swift, R.S., and R.G. McLaren. 1991. Micronutrient adsorption by soils and soil colloids. p. 257–292. *In* G.H. Bolt et al. (ed.) Interactions at the soil colloid-soil solution interface. Kluwer Academic Publ., Dordrecht, the Netherlands.

Taylor, G.J., and C.D. Foy. 1985. Differential uptake and toxicity of ionic and chelated copper in *Triticum aestivum*. Can. J. Bot. 63:1271–1275.

Tiller, K.G. 1996. Soil contamination issues: Past, present and future, a personal perspective. p. 1–27. *In* R. Naidu et al. (ed.) Contaminants and the soil environment in the Australasia-Pacific region. Kluwer Academic Publ., Dordrecht, the Netherlands.

Tiller, K.G., J. Gerth, and G. Brummer. 1984. The relative affinities of Cd, Ni and Zn for different clay fractions and goethite. Geoderma 34:17–35.

Tiller, K.G., and J.F. Hodgson. 1962. The specific sorption of cobalt and zinc by layer silicates. Clays Clay Mineral. 9:393–403.

Tiller, K.G., J.F. Hodgson, and M. Peech. 1963. Specific sorption of cobalt by soil clays. Soil Sci. 95:392–399.

Tills, A.R., and B.J. Alloway. 1983. The use of liquid chromatography in the study of cadmium speciation in soil solutions from polluted soils. J. Soil Sci. 34:769–781.

Treeby, M., M. Marschner, and V. Römheld. 1989. Mobilization of iron and other micronutrient cations from a calcareous soil by plant-borne, microbial and synthetic metal chelators. Plant Soil 114:217–226.

Uren, N.C. 1989. Rhizosphere reactions of aluminum and manganese. J. Plant Nutr. 12:173–185.

Uren, N.C., and H.M. Reisenauer. 1988. The role of root exudates in nutrient acquisition. Adv. Plant. Nutr. 3:79–114.

Valcke, E., and A. Cremers. 1994. Sorption-desorption dynamics of radiocaesium in organic matter soils. Sci. Total Environ. 157:275–283.

Vancura, V. 1964. Root exudates of plants: 1. Analysis of root exudates of barley and wheat in their initial phases of growth. Plant Soil 21:231–248.

Wang, T.S.C., T.K. Tang, and T.T. Chuang. 1967. Soil phenolic acids as plant growth inhibitors. Soil Sci. 103:239–246.

Warncke, D.D., and S.A. Barber. 1973. Diffusion of zinc in soils: III. Relation to zinc adsorption isotherms. Soil Sci. Soc. Am. Proc. 37:355–358.

Wauters, J. 1994. Radiocesium in aquatic sediments: Sorption, remobilization and fixation. Dissertationes de Agricultura, 246. Katholieke Universiteit Leuven, Belgium.

Wauters, J., L. Sweeck, E. Valcke, A. Elsen, and A. Cremers. 1994. Availability of radiocaesium in soils: A new methodology. Sci. Total Environ. 157:239–248.

Webley, D.M., E.K. Henderson, and F. Taylor. 1963. The microbiology of rocks and weathered stones. J. Soil Sci. 14:102–112.

Wei, L.C., W.R. Ocumpaugh, and R.H. Loeppert. 1994. Plant growth and nutrient uptake characteristics of Fe-deficiency chlorosis susceptible and resistant subclovers. Plant Soil 165:235–240.

Whipps, J.M. 1990. Carbon economy. p. 59–97. *In* J.M. Lynch (ed.) The rhizosphere. John Wiley & Sons, Chichester, England.

Whitehead, D.C. 1964. Identification of p-hydroxybenzoic, vanillic, p-coumaric, and ferulic acids in soils. Nature (London) 202:417–418.

Willaert, G., and M. Verloo. 1992. Effects of various nitrogen fertilizers on the chemical and biological activity of major and trace elements in a cadmium contaminated soil. Pedologie 43:83–91.

Williams, C.H., and D.J. David. 1976. The accumulation in soil of cadmium residues from phosphate fertilizers and their effects on the cadmium content of plants. Soil Sci. 121:861–893.

Xue, J., and P.M. Huang. 1995. Zinc-adsorption-desorption on short-range ordered iron oxide as influenced by citric acid during its formation. Geoderma 64:343–356.

Zachara, J.M., S.C. Smith, C.T. Resch, and C.E. Cowan. 1992. Cadmium sorption to soil separates containing layer silicates and iron and aluminum oxides. Soil Sci. Soc. Am. J. 56:1074–1084.

Zhang, F., V. Römheld, and H. Marschner. 1989. Effect of zinc deficiency in wheat on the release of zinc and iron mobilizing root exudates. Z. Pflanzenernähr. Bodenk. 152:205–210.

Zhang, P.-C., and D.L. Sparks. 1989. Kinetics and mechanisms of molybdate adsorption–desorption at the goethite/water interface using pressure jump relaxation. Soil Sci. Soc. Am. J. 53:1028–1034.

10 Soil–Root Interface: Ecosystem Health and Human Food-Chain Protection

Rufus L. Chaney and Sally L. Brown

USDA-ARS Environmenal Chemistry Laboratory
Beltsville, Maryland

J. Scott Angle

University of Maryland
College Park, Maryland

10–1 INTRODUCTION

10–1.1 Natural History of Food Chain Element Poisoning

During human history, certain soils were found to be unproductive, or even dangerous to livestock because of geological enrichment with trace elements. Food-chain poisoning of livestock was observed at many locations where Mo or Se mineralization occurred in alkaline soils. Contamination of forages by industrial emission also has poisoned livestock in numerous cases. This historical evidence of the potential for food-chain toxicity formed the basis for modern research, development of improved management methods, and of regulations to limit soil contaminants in order to protect the environment. Several reviews have been especially instructive on the history of soil contamination, documenting adverse effects on wildlife, humans, or the environment (Allaway, 1968, 1977; Baker & Chesnin, 1976; Chaney, 1980, 1983a,b; Hansen & Chaney, 1984; Lisk, 1972; Reid & Horvath, 1980; Underwood,1977).

These subjects have become of greater public concern because humans were injured by contaminants in the general environment. Humans were poisoned where grain treated with Hg-fungicides was consumed directly, or indirectly through human consumption of livestock fed with the treated grain; excessive intake of organic Hg caused severe Hg health effects in developing children. Widespread use of persistent chlorinated hydrocarbons caused dispersal of these semivolatile compounds around the globe. Much later the biomagnification of DDT (1,1,1-trichloro-2-bis[*p*-chlorophenyl]ethane) in aquatic food-chains, and bioaccumulation of DDT by earthworms, caused injuries that threatened extinction of whole species. Industrial Hg contamination of Minamata Bay (Japan) caused severe health effects in consumers of fish and shellfish from the area, and led to identification of methyl-mercury as a food-chain poison. Methyl-mercury is now recognized as an important source of human risk when sediments are contaminated by inorganic Hg, and aquatic food-chains biomagnify trophic concentrations of these lipophilic compounds. And *Itai-itai* (ouch-ouch) disease caused a severe osteomalacia in subsistence farm families who consumed rice (*Oryza sativa* L.) grain that was home-grown in paddies Zn- and Cd-contaminated by industrial discharges from mining and smelting industries far upstream of the paddies (Kobayashi, 1978; Tsuchiya, 1978; Takijima & Katsumi, 1973). Although Se deficiency is a widespread disease in lesser-developed nations, Se has poisoned humans as well as livestock where ash from seleniferous coals were used to lime fields, and a change in weather caused a change in crop from rice to wheat (*Triticum aestivum* L.), which accumulated much higher levels of Se (Yang et al., 1983). The extensive injury to wildlife at Kesterson Reservoir in California showed again the remarkable biomagnification possible with Se in aquatic food-chains that can incorporate Se into proteins and cause severe teratogenic effects in birds and mammals at higher trophic levels (Ohlendorf et al., 1986).

These examples of soil contaminants causing adverse environmental effects provided support for research to understand processes that affect plant uptake, food-chain transfer, food-web biomagnification, and mechanisms of toxicity of contaminants. This knowledge is required to develop regulatory controls to prevent future excessive contamination that could cause environmental risk, and to identify soils that have become so contaminated that remediation may be required, or access restricted in order to reduce environmental harm.

10–1.2 Increasing Amounts of Organic Residues and Contaminants

Age-old methods of disposing human and livestock wastes are being strained by population growth. And the comingling of industrial wastes with domestic wastes in modern sewage systems causes contamination of some sludge materials at high levels (e.g., Sommers, 1980; Chaney & Giordano, 1977). With low population density, and recycling of wastes to the fields that produced the crops, food-chain transfer of soil contaminants did not cause health risks. But with higher populations, public health protection required sewage treatment to prevent water borne disease. Residues of sewage treatment (sludges or biosolids) needed to be safely disposed, and the high nutrient levels favored use on cropland. Even within agriculture, additives to livestock feeds can produce manures that carry higher levels of some trace elements (Cu, Zn, As, P, Cd) than levels resulting from crops grown on the farm (e.g., Chaney & Oliver, 1996).

By the 1970s, increased recognition of potential environmental disease from soil contaminants, and increased desire to beneficially use biosolids and manures on cropland, required deeper consideration of the safety of these practices and appropriate regulations. A 1972 conference on "Recycling Municipal Sludges and Effluents on Land" was sponsored by the U.S. Department of Agriculture (USDA), the U.S. Food and Drug Administration, the U.S. Environmental Protection Agency (USEPA), and the National Association of State Universities and Land Grant Colleges to identify the research necessary to develop management practices and regulatory controls needed to protect soil fertility, the food-chain and the environment from contaminants in municipal sludges and effluents (e.g., Chaney, 1973).

Since the early 1970s, a huge amount of research has been conducted to evaluate environmental risks from soil contaminants coming from geological sources, industrial sources, municipal sources, or agricultural sources. This body of research has identified soil, plant, and animal processes that affect the potential for transfer of soil contaminants thru agricultural food chains, and that have become the basis for regulatory controls in many nations.

Further research is needed to better define some of these processes for contaminant transfer and toxicity. Alternative industrial processes or consumer products that minimize soil contamination can reduce contaminant release and dispersal. Research is especially needed to better understand the limits for chemical forms of the contaminants that are needed to protect higher trophic levels of wildlife as well as humans. And research is needed to find methods to minimize the risks where soils have already been contaminated by natural or anthropogenic sources. Although research conducted to date has been largely successful in providing the information needed to protect humans and agricultural ecosystems, wildlife and natural ecosystems have received less study, and research is strongly needed.

10–2 PATHWAY APPROACH TO RISK ASSESSMENT FOR CONTAMINATED SOILS

The focus of this chapter is the current status of risk assessment for soil contaminants in the context of protecting soil fertility and food-chain safety. Two rather different philosophical approaches to soil protection are recognized. One approach says "add no contaminants" (or the similar "Do not allow contaminant concentrations to rise above background, or the 95th percentile of background"). An alternative approach, favored by the authors, says do the research and evaluation to "Assure that addition of contaminants does not comprise risk to even highly exposed organisms with lifetime exposure to the contaminated soil."

Davies (1992) suggested that the word *polluted* be distinguished from the word *contaminated* in scientific and public discussion about environmental contamination to help bring more precision into the public debate about addition of contaminants to soils. Elements are naturally present in all geological materials. On the one hand, all soils on Earth have become somewhat contaminated by industrial emissions of volatile organic and inorganic contaminants. Contaminants are found in arctic ice cores, and all cropland if one uses analytical methods with low enough detection limits. On the other hand, some soils contain such high levels of a contaminant that they should be labeled *polluted* because the contaminants may harm organisms. Risk assessment is the tool that modern science has developed to identify the level at which soils should be labeled polluted.

10–2.1 Food-Chain Transfer of Soil Contaminants

Over several decades, scientists in many disciplines have worked to develop methods for risk assessment of contaminants in soils to protect food chain crops, soil fertility, and the environment. In particular, work conducted to develop the Clean Water Act Part 503 Rule (now legally 40 CFR [Code of Federal Regulations] 503; referred to as USEPA 503 Rule) has received extensive review and is recognized as an improvement on previous approaches (USEPA, 1989a; National Research Council, 1996). Research had identified a few fundamental pathways for transfer of soil contaminants or applied biosolids to food-chains: (i) plant uptake and translocation to edible plant tissues; (ii) direct ingestion of soil by grazing livestock, children, or wildlife that consume earthworms; and (iii) ingestion by livestock of adhering soil splashed on the plants or fluid biosolids spray-applied on forages (Chaney & Lloyd, 1979).

The first fundamental pathway (food crops, Pathways 1 and 2) is the traditional food-chain transfer, while Pathway 3 (ingestion of soil and dust) is now recognized as a natural process that cannot be avoided, and which circumvents the normal Soil–Plant Barrier protections of the food-chain. The third fundamental pathway is similar to contamination of forages by stack emissions in that concentrations of contaminants can be present in the forage as consumed, which could not reach the aboveground plant parts by normal uptake processes. For this reason, spray application of biosolids on standing forages is prohibited under the USEPA 503 Rule (U.S. Environmental Protection Agency, 1993), and waiting periods are required before grazing to allow sufficient regrowth of the forage crop that exposures are low. Injection or incorporation of fluid biosolids is encouraged to prevent the adherence pathway entirely.

The Risk Assessment conducted for the USEPA 503 Rule on land application of biosolids integrated known environmental transfers of contaminants in biosolids into 14 Pathways (Table 10–1) that could potentially cause risk to highly exposed individuals (HEIs). Calculated limits for the different pathways are designed to protect the HEIs from adverse effects of contaminants that might be transferred to humans by each pathway if biosolids were applied to a soil at the maximum permitted cumulative application. To impose a strong measure of conservatism, the USEPA 503 rule was developed with the assumption that 1000 t ha^{-1} of biosolids would be applied over centuries as fertilizer or soil conditioner. Pathway 3 represents soil or dust ingestion by children, which occurs during children's exploration of their environment during infancy. Limits for Pathways 4 and 5 protect livestock, while limits for Pathways 6 and 7 protect humans who consume meats and organ meats from livestock that were maximally exposed to contaminants from biosolids, whether the livestock consumed only the forage materials, or also ingested soil or biosolids on the soil while grazing in pastures. Pathway 8 limits protect plants (phytotoxicity). Pathway 9 limits protect soil organisms (e.g., earthworms, bacteria, fungi), and Pathway 10 limits protect predators of soil organisms as examples of the wildlife species most exposed to soil contaminants. The remaining Pathways (11–14) involve human exposure through surface and groundwater, air, and suspended dust.

Table 10–1. Pathways for risk assessment for potential transfer of biosolids-applied trace contaminants to humans, livestock, or the environment, and the highly exposed individuals to be protected by a regulation based on the pathway analysis (U.S. Environmental Protection Agency, 1989a, 1993; Chaney & Ryan, 1994). Each Pathway presumes 1000 t dry biosolids ha^{-1} and/or annual application of biosolids as N fertilizer.

	Pathway	Highly Exposed Individuals
1	Biosolids→Soil→Plant→Human	Individuals with 2.5% of all food produced on amended soils.
2	Biosolids→Soil→Plant→Human	Home gardeners with 1000 t ha^{-1}; 60% garden foods for lifetime.
3	Biosolids→Human	Ingested biosolids product; 200 mg d^{-1}.
4	Biosolids→Soil→Plant→Animal→Human	Farms; 45% of homegrown meat.
5	Biosolids→Soil→Animal→Human	Farms; 45% of homegrown meat.
6	Biosolids→Soil→Plant→Animal	Livestock feeds; 100% on amended land.
7	Biosolids→Soil→Animal	Grazing Livestock; 1.5% biosolids in diet.
8	Biosolids→Soil→Plant	Crops; strongly acidic amended soil, but with limestone to prevent natural Al and Mn toxicity.
9	Biosolids→Soil→Soil Biota	Earthworms, microbes, in amended soil.
10	Biosolids→Soil→Soil Biota→Predator	Shrews; 33% earthworms diet, living on site.
11	Biosolids→Soil→Airborne Dust→Human	Tractor operator.
12	Biosolids→Soil→Surface Water→Human	Subsistence fishers.
13	Biosolids→Soil→Air→Human	Farm households.
14	Biosolids→Soil→Groundwater→Human	Well water on farms; 100% of supply.

10–2.2 Highly Exposed Individuals

Each pathway was constructed to estimate risk to HEIs; i.e., humans, plants, or animals at the 95th to 98th percentile of projected chronic lifetime exposure. These limits are more protective than it may seem at first glance. Instead of the 95th to 98th percentile of the whole U.S. population, the pathway estimates exposure among the subset of the population that was actually significantly exposed to contaminants from biosolids (assuming a long-term cumulative application of 1000 t ha^{-1}). In reality, only a small part of the population eat crops grown on a farm or garden that has high regular biosolids application or other sources of soil contamination, and perhaps no individual ever meets the full HEI definition; however, the need to assure low risk in the long term requires conservative definitions of the HEIs.

10–2.3 Soil–Plant Barrier as a Limit on Soil Contaminant Risks

Soil chemical processes strongly affect food-chain transfer of soil contaminants. Most elements in soils are relatively insoluble or they would have leached to groundwater long ago. Added contaminants are precipitated, coprecipitated, specifically adsorbed on hydrous Fe and Mn oxides, chelated by organic matter or otherwise strongly bound by soils. Some elements in soil are so insoluble that plant uptake is prevented (does not practically occur), and the elements are not absorbed by or toxic to animals even when element-rich soils are ingested (e.g., Cr, Sn, Si, Ti, Au, Ag). Another group of elements can be absorbed by livestock that ingest the soil, but the element is either not accumulated to an environmentally significant level by plants, or reach phytotoxic levels before reaching levels toxic to animals and humans. For these elements, soil ingestion is the pathway for risk transfer (e.g., F, Pb, As, Fe, Cu, Hg, Ni, Ba). Other elements are readily absorbed from soils, at least under conditions of soil pH that favor uptake and transport to shoots. For some of these elements, another natural food-chain protection exists in that the elements harm the plant before the concentration in the plant would harm livestock or humans who chronically consumed the plants. Most of the elements known for causing phytotoxicity are in this group (e.g., Zn, Mn, Ni, and B are absorbed to levels toxic to shoots, while Cu, As, and Al are toxic to roots with only small increases in shoots when phytotoxicity occurs). The plant tolerance for these elements is lower than the animal tolerance. This leaves us with a short list of elements that can move from contaminated soils to edible plant parts at high enough levels to comprise risk to the consumer of the plant before phytotoxicity is apparent (Mo, Se, Cd, and possibly Co). These hazards to the food chain include the well known toxic elements Se and Mo, and the more recently discovered food-chain hazards from toxic Cd. Cobalt is theoretically able to poison susceptible ruminants at plant concentrations below phytotoxic thresholds (Chaney, 1983b; Chaney & Ryan, 1994), but this has not been observed in the environment.

We noted above the cases in Japan (Kobayashi, 1978) and China (Cai et al., 1990) where subsistence farm families were harmed by soil Cd because they consumed locally grown rice for their lifetime. Rice is now recognized as uniquely able to transfer soil Cd into bioavailable forms in grain because of a combination of crop production practices, soil chemistry in flooded soils, the biochemistry of Zn and Cd in rice plants, and the low total and low bioavailability of Fe, Zn, and Ca in polished rice (Chaney et al., 1996a). The paddies that produced the rice that caused human disease were contaminated with Zn and Cd from mine wastes and smelter fumes, in the usual geological ratio of 1 μg Cd:100 μg Zn. But in flooded soils, ZnS and CdS are formed, but CdS is more rapidly oxidized upon drainage, releasing Cd for plant uptake. Oxidation of the soil also lowers soil pH, which promotes Cd uptake by rice. Further, Cd absorbed by roots during grain filling was found to be able to move directly to the grain (Chino, 1981; Chino & Baba, 1981). Thus, rice grain from these contaminated soils was increased in Cd up to 200-fold without a corresponding increase in grain Zn. The toxicity to humans was worsened by the fact that rice grain, especially polished rice grain, is so low in Fe, Zn, and Ca that human nutritional deficiencies are common in subsistence rice consumers. Fox (1988) has summarized the literature on Cd absorption by animals, and low or deficient levels of Fe, Zn, and Ca greatly increased absorption and retention of Cd by animals. Together, these factors led to human Cd disease in paddy soils that generally contained 2 to 10 mg Cd kg^{-1} (and 100–1200 mg Zn kg^{-1}).

In contrast, exposure to soil Cd has not caused detrimental effects in other situations. For example, housing was constructed on mine wastes highly contaminated with Zn and Cd after World War II, or in another location smelting contaminated the gardens with high levels of Zn and Cd. People grew and consumed garden foods, for decades, with no evidence of human risk from the soil-Cd or -

Zn. The combination of Zn inhibition of Cd uptake by plants, Zn phytotoxicity to crops if soil pH is allowed to fall (Baker & Bowers, 1988), Zn inhibition of Cd absorption in the intestine (Fox, 1988), and perhaps other factors not yet fully understood, prevented people from having many foods rich in Cd. Either the crops were low in Cd, or yield was depressed by Zn so that no food was produced to eat. Long-term residents of Shipham, England (Strehlow & Barltrop, 1988), Stolberg, Germany (Ewers et al., 1993), and Palmerton, PA (Sarasua et al., 1995), who consumed garden foods grown locally, and were exposed to Cd- and Zn-contaminated dusts, experienced no harm from the high soil-Zn and -Cd. Similarly, long-term consumers of oysters (*Ostrea lutaria*) at Bluff, New Zealand, had adequate bioavailable Zn, Fe, and Ca in their diets because of their high consumption of oysters with high levels of Cd but rich in Zn and Fe; these oyster consumers had no adverse effects from consuming high levels of Cd for most of their lives (Sharma et al., 1983; McKenzie-Parnell & Eynon, 1987; McKenzie-Parnell et al., 1988).

Table 10–2 shows the limits established for heavy metals by the final USEPA 503 Rule, and the Pathway that was most limiting (Column 2) for each metal using the final USEPA calculations (U.S. Environmental Protection Agency, 1993). Column 3 is the ceiling limit. A ceiling concentration limit for metals in biosolids was established for several reasons: (i) if only cumulative loading limits were imposed, pretreatment effectiveness might be threatened; (ii) if more highly contaminated biosolids were land applied, because the metals are more phytoavailable the higher the total metal concentration in a biosolid, phytotoxicity might occur at lower cumulative metals applications than estimated using data obtained from studies with better quality biosolids. If no ceiling were regulated, lower cumulative limits would have been reached by including data from highly contaminated biosolids and metals-salt field studies. Column 4 shows the cumulative application limit for the metals, in kilograms per hectare. These are the outcome of the risk assessment pathway calculations.

In order to develop a new method of biosolids regulation [the Alternative Pollutant Limit, APL], the cumulative application limit was assumed to be applied by 1000 t ha^{-1} of biosolids, and expressed on a milligrams per kilograms basis. Because limiting the concentration of metals in a biosolid can provide greater protection than simply limiting the cumulative application of metals in biosolids, the Expert Workgroup (Page et al., 1989) recommended that USEPA provide a regulatory mechanism that reduced the regulatory burden for higher quality biosolids that met the limits in Column 4. These APL biosolids may be marketed for general use without cumulative site loadings for the regulated metals, if the pathogen levels in the product are reliably reduced to nondetectable levels by heat and time. USEPA has further labeled "Exceptional Quality Biosolids" an APL quality biosolid that has been stabilized and received an effective pathogen reduction treatment. Industrial pretreatment has substantially reduced contaminant concentrations in municipal biosolids (U.S. Environmental Protection Agency, 1990). Treatment of drinking water to reduce corrosion of pipes has further reduced Pb, Zn, Cu, and Cd in biosolids in numerous cities.

Column 6 is the so-called No Observed Adverse Effect Level (NOAEL) quality of biosolid as recommended by Chaney and Ryan (1994). For a number

Table 10–2. Comparison of the ceiling (Column 3, based on the lower of the Pathway Limit and the 99th percentile) and Cumulative (Column 4, Pathway Risk Assessment based) limits for biosolids contaminants under the Final USEPA 503 Rule Limits (Feb. 19, 1993) vs. NOAEL limits (Column 6) estimated by Chaney and Ryan (1994). Biosolids that meet the Alternative Pollutant Limit (APL) or No Observed Adverse Effect Level (NOAEL) quality limits could be applied at up to 1000 t ha^{-1} before reaching the Cumulative Limit, and still protect Highly Exposed Individuals according to the Technical Support Document for the USEPA 503 Rule. For the APL and NOAEL biosolids, adsorption of contaminants by biosolids constituents lowers the potential for risk sufficiently to allow general marketing and continuing use in sustainable agriculture. The percentile of the NOAEL(93) concentration in the National Sewage Sludge Survey is shown for comparison (Column 7). Column 8 shows the metal limits that we believe attainable by Publicly Owned Treatment Works that enforce industrial pretreatment standards; the corrosivity of drinking water may need to be controlled in some cities to achieve NOAEL levels of Pb, Cu, Zn, and Cd.

	Limits under the USEPA 503 Rule			Limits under the NOAEL Approach			
Element	Limiting pathway	Ceiling 99th %	Cumulative, kg ha^{-1} = APL, mg kg−1	Limiting pathway	Limit	Percentile of NOAEL	Attainable quality
1	2	3	4	5	6	7	8
		mg kg^{-1}			mg kg^{-1}		mg kg^{-1}
As	3	75	41	3	54	98	<25
Cd	3	85	39	2	21	91	<5–10
Cd–Zn†	--	--	--	2	0.015	87	0.010
Pb	3	840	300	3	300	90	<100
Hg‡	3	57	17	12	17	93	<5
Mo§	6	75	35	6	54	98	<50
Se	6	100	36	6	28	98	<15
Cr	8	3000	1300	--	--	--	--
Cu	8	4300	1500	8	1500	89	<500–750
Ni	8	420	420	8	290	98	<100
Zn	8	7500	2800	8	2800	91	<1500–2000

† The ratio of Cd to Zn in such products strongly affects the potential of Cd to cause food-chain risk; although the USEPA did not regulate Cd/Zn ratio in the 40CFR503 Rule, the NOAEL and Attainable Quality include this characteristic.

‡ Valid for all biosolids uses except mushroom production.

§ Molybdenum limit corrected by adding appropriate data omitted by USEPA contractor, and omitting data from a biosolids containing 1500 mg Mo kg^{-1}.

of the elements, USDA review of the final USEPA Rule indicated that policy decisions had led to limits or limiting Pathways that USDA concluded were inappropriate (Administrator Finney of the USDA-ARS, 1993, personal communication). For several elements, the use of presumed 100% bioavailability of soil- or biosolid-metals ingested by children resulted in limits more restrictive than needed to protect Highly Exposed Individual children (e.g., As, Cd). In addition, USEPA used the 99th percentile of their National Survey (U.S. Environmental Protection Agency, 1990) as the ceiling limit instead of the 98th percentile used for the NOAEL ceiling limits. Using an experimentally-based biosolid Cd bioavailability to pigs (*Sus scrofa domesticus*) made soil ingestion no longer the most limiting Pathway for Cd. In the NOAEL list, Cr has been deleted because Cr(III) in biosolids has not been found to cause adverse effects in the environment, and thus did not require a limit (Chaney et al., 1998).

Ryan and Chaney (1998) provide a detailed discussion of the impact of protecting HEIs on the level of protection actually achieved by the complex algo-

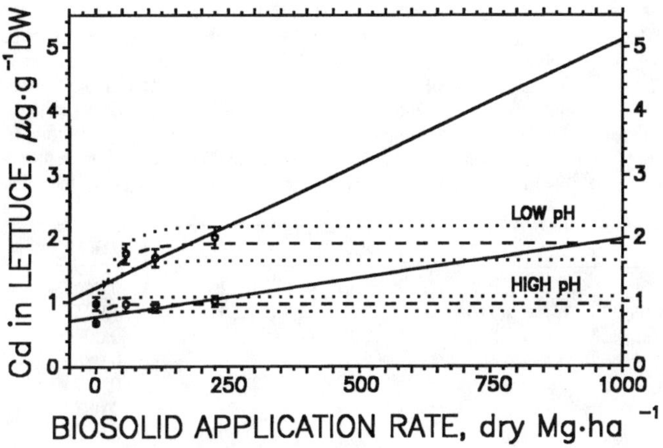

Fig. 10–1. Linear vs. plateau regression analysis of lettuce uptake of Cd from Christiana fine sandy loam (clayey, kaolinitic, mesic Typic Paleudult) amended with 0, 56, 112, or 224 Mg dry heat-treated sludge ha^{-1}, and pH adjusted to 6.2 to 6.5 with limestone (high pH) or uncontrolled (≤5.5 in 1983; low pH). Predicted responses extrapolated to 1000 Mg ha^{-1} to show implications of the model used. Results are average for 1976 to 1983. Data points shown are arithmetic means ± one standard error; plateau regressions show predicted (dashed lines) with ±95% confidence interval (dotted lines). Equations for linear regressions (solid lines) are: lettuce Cd = 1.22 + 0.00390 × rate (low pH); lettuce Cd = 0.774 + 0.00121 × rate (high pH). Sludge applied in 1976 contained 13.4 μg Cd, 1330 μg Zn, and 83 mg Fe g^{-1} dry weight (data originally reported in Chaney et al., 1982).

rithms of the USEPA 503 Rule risk assessment methodology (U.S. Environmental Protection Agency, 1989a). In the garden pathway, many factors are multiplied together to make the final calculation. The highly exposed population for this Pathway is those individuals who consume food grown on soil in home gardens amended with 1000 t ha^{-1} of biosolids. These individuals are assumed to be exposed for 70 yr for cancer risks, and 40 to 50 yr for Cd injury to the kidney (kidney Cd concentration naturally declines after about age 50). The soil is presumed to be acidic, pH ≤6.0 for the whole period. Transfer of Cd from the soil is estimated by the linear regression uptake slope for the crop and soil multiplied by the amount of a food ingested per day (lifetime average), multiplied by the fraction of diet presumed to be produced on the biosolid-amended home garden (39–60% of lifetime garden foods consumption). The uptake slope used in the USEPA 503 Rule calculation of food-chain transfer is not the increment reached at the plateau in the usually observed long-term relationship between soil-Cd and crop-Cd (Chaney & Ryan, 1994; Corey et al., 1987), but the linear regression for the data. But the linear regression approach gives a much higher predicted plant Cd concentration at 1000 t biosolids ha^{-1} than observed in long-term field studies (see Fig. 10–1). The smaller the cumulative application rate for the actual data used in the regression, the larger the error of over-prediction. And the reference dose (RfD) which may not be exceeded (e.g., for Cd, 1 μg Cd kg^{-1} body weight day^{-1}) is a conservative estimate of the intake of Cd, which over a lifetime causes the first sign of mild kidney disease (see Ryan et al., 1982; Chaney & Ryan, 1994).

Table 10–3. Increased Cd in garden food groups due to biosolids use calculated according to the USDA (Administrator Finney, USDA, 1993, personal communication) recommendation; arithmetic means of Cd uptake slopes were used rather than geometric means; mean for leafy vegetables calculated only for acid soils (pH < 6) and biosolids with Cd < 150 mg kg^{-1}. RP_C is the maximum allowable cumulative Cd concentration applied from biosolids; UC_i is the linear regression uptake slope for an element in food group (i); DC_i is the daily intake for food group (g dry weight d^{-1}; i) ,and FC_i is the fraction of annual intake of food group (i) that is grown on the soil presumed to have received 1000 t biosolids ha^{-1}. The lower panel shows the full detail of the calculation algorithm.

Food group	UC_i	DC_i	FC_i	$UC_i \bullet DC_i \bullet FC_i$	%
Potatoes	0.008	15.60	0.37	0.0462	1.8
Leafy vegetables	1.719	1.97	0.59	1.995	79.9
Nondry Legumes	0.004	3.22	0.59	0.0076	0.3
Root vegetables	0.094	1.60	0.59	0.0885	3.5
Garden fruits	0.113	4.15	0.59	0.277	11.1
Sweetcorn	0.097	1.60	0.59	0.0814	3.3
Grains and cereals	--	89.08	0.0043	0.0069	?
All garden foods				2.496	100

Calculation algorithm:

$$RP_C = \frac{\underset{\text{WHO limit}}{70\ \mu g\ d^{-1}} - \underset{\text{Background intake}}{16.1\ \mu g\ d^{-1}}}{\underset{\Sigma(UC_i \bullet DC_i \bullet FC_i)}{2.496}} = \frac{\underset{\text{Allowed increase}}{53.9\ \mu g\ d^{-1}}}{2.496} = \underset{\text{Pathway 2 limit}}{21.5\ kg\ ha^{-1}}$$

So the original algorithms (U.S. Environmental Protection Agency, 1989a) were revised to include some calculation factors based on central-tendency rather than worst case. The USEPA calculation of the mean uptake slope for valid field data used the geometric mean of all data, not just the soil with pH ≤6.0. Because this geometric mean increased the allowed soil-Cd for garden soils to 120 kg ha^{-1}, and acidic soils are well known to favor Cd uptake by plants, the USDA (1993) advised USEPA that they should use the arithmetic mean of the valid field data for plant uptake slopes for soils with pH ≤6.0. USDA argued that the dataset was biased by having more data about alkaline than acidic soils, and for cabbage (*Brassica oleraceae* L.) type plants with lower Cd uptake rather than lettuce- (*Lactuca sativa* L.) or spinach-like (*Spinacia oleracea* L.) plants with higher Cd uptake rates. If the arithmetic mean of the acidic soils data set were used, the maximum soil-Cd allowed would have been 12 kg ha^{-1}. In the real data for field studies with leafy vegetables grown on biosolids amended soils, a few studies stand out as having much higher slopes than most. In general, the higher the biosolid Cd concentration, the higher the uptake slope regardless of the cumulative applied Cd [shown in controlled studies by Jing & Logan (1992)]. Because the Rule would be limiting maximum biosolid Cd to relatively lower levels than used for a number of the studies which provided field data with high Cd uptake slopes, USDA reasoned that the uptake slopes from highly contaminated biosolids should not be used in the USEPA calculations. In fact, one-half of the total sum of slopes for all studies came from a single study with a highly contaminated biosolid on a very strongly acidic soil (see Chaney & Ryan, 1994). When the data from highly contaminated biosolids were omitted from the dataset used to make the calculation, the estimated allowed cumulative application of biosolids Cd was 21 kg ha^{-1} (see Table 10–3).

Even with the revised algorithms of the final USEPA 503 Rule, it is more likely that the Rule errs on the conservative side (making lower than necessary estimates of allowed cumulative contaminant applications) rather than on the high side. Because individuals cannot grow a mixture of high uptake slope leafy vegetables (lettuce, spinach, and others) on a single garden for the whole year due to climate limitations on crop growth, they cannot practically ingest 60% of their annual intake of leafy vegetables grown on a garden with 1000 t ha^{-1} of cumulative biosolids application. Thus, the estimate is higher than actual Cd intakes. Especially so for lettuce and spinach type leafy vegetables that accumulate higher Cd levels than cabbage, kale (*B. oleraceae*), broccoli (*B. oleraceae*) , and collards (*B. oleraceae*) type leafy vegetables. Use of linear regression slopes for the uptake of metals to edible portions of crops is a high estimate of the increase when the plateau is reached (see below). Multiplying the combination of central tendency and worst case variables together, one may still be estimating exposures far beyond the most highly exposed individual for their lifetime, and thus imposing lower than necessary allowed cumulative loadings. Many of the most limiting pathways for a contaminant were those which calculated a lower estimate than needed to provide full lifetime protection to the HEIs. We cannot estimate the actual percentile of the HEI in the final rule due to the lack of data on measured intake of vegetables grown on biosolid amended soils, and the cumulative application on these soils.

Although it is scientifically valid to criticize the final pathway analysis as being very conservative, some scientists have concluded that the pathway calculations are not protective enough (e.g., McBride, 1995). McBride challenged some of the methods and conclusions of the expert workgroup that assisted USEPA with this work, with considerable emphasis on the importance of organic matter in applied biosolids on the phytoavailability of soil-metals. This hypothesis was repeatedly posed by researchers concerned about biosolids-applied metals: "What happens to metal phytoavailability when the organic matter is biodegraded." In England, it was called the *Time-Bomb Model* for the worst case risk from biosolids metals (Beckett et al., 1979). They felt that organic matter must comprise the most important material adsorbing metals in biosolids, and thus in biosolids-amended soils, and because the added organic matter will eventually be oxidized to the level appropriate for the climate and texture of the soil in question, the added metals would become more plant available over time and eventually poison plants (thus the *Time-Bomb* of added metals). Many researchers had this concern in the 1970s, before the extensive research programs were conducted in several nations. McBride (1995) also challenged the concept of the plateau response (as illustrated by Chaney et al., 1982; Fig. 10–1), which results from the biosolid-applied adsorbent materials (hydrous oxides of Fe, Mn, Al, and others) that persist in biosolid-amended soils (see review in Corey et al., 1987).

10–2.4 Patterns of Soil→Plant Transfer of Soil-Cadmium

Many factors have been found to affect the soil→plant transfer of soil-Cd. Crops differ greatly in Cd accumulation from the same soil (e.g., Brown et al.,

1996). Lower soil pH generally increases Cd accumulation. Lower levels of soil components that adsorb or chelate Cd favor Cd accumulation by plants.

Increased soil Cl^- is now recognized to play an important role in increasing Cd uptake by plants. Bingham et al. (1984) noted that added Cl^- increased Cd uptake, while added sulfate had little effect on crop Cd uptake (Bingham et al., 1986). This difference between Cl^- and sulfate was quite unexpected because these anions form Cd complexes with approximately equally strength (see Parker et al., 1995). For a number of years, no field observations were available to illustrate the practical significance in the field of the laboratory controlled studies of Bingham and colleagues. But in 1994, McLaughlin et al. found that increasing soil Cl^- significantly increased Cd in potato (*Solanum tuberosum* L.) tubers and shoots. Their observation that soil Cl^- increased Cd uptake was confirmed for soil-Cd uptake to nonoilseed sunflower (*Helianthus annuus* L.) kernels even in calcareous soils (Li et al., 1995a). After a number of innovative experiments, Smolders and McLaughlin (1996) used a chelating resin-buffered recirculating nutrient solution system to show that the fundamental basis of the Cl^- effect was direct uptake or leakage of the $CdCl^{1+}$ complex into plant roots, not just increasing convection or mass flow of Cd into the root with the transpiration stream because more Cd was dissolved in the soil solution in the presence of high soil Cl^-.

The importance of Cl^- increasing food Cd concentrations has implications for Cd risk assessment in areas where soil Cl^- is normally high. When Cl^- is high in irrigation water, or when pedogenic Cl^- continues to be released over decades or centuries, crop uptake of Cd can be increased significantly. This process clearly indicates that breeding for salinity tolerant plants (to allow crop production under conditions now considered salt toxic) will increase Cd levels in the foods produced unless efforts are made to reduce Cd concentrations in the edible parts of the plants during the breeding program. Because Cl^- forms complexes with Cd but not so strongly with Zn, crop Cd in selectively increased compared with Zn. Because this increases the Cd/Zn ratio, the Cd in these crops has relatively higher bioavailability to consumers than the Cd in normal crops grown on soils with backgound levels of Cl^-. In comparison with the effect of flooded soil preventing increase in Zn in rice grain while allowing grain Cd to be increased, the effect of Cl^- is smaller. But the potential for Cl^- to increase crop Cd without any symptoms in the crops means that institutional programs are needed to make sure that excessive bioavailable Cd is not reached in salinity tolerant cultivars of major food crops. Comparison of the increase in potato tuber Cd level on relatively low Cd soils in the study of McLaughlin et al. (1994) with the lack of increase in tuber Cd in the field study of Harris et al. (1981) on a long-term sludge farm with about 19 mg Cd kg^{-1}, illustrates that the normal physiological processes that prevent increase of Cd in potato tubers are not operative in the case of soils rich in Cl^-.

Besides the importance of this understanding for management of soil- and crop-Cd, these findings illustrate what actually happened in one of the few illustrations of apparent synergistic interactions of metals added to soils. Hassett et al. (1976) and Miller et al. (1977) grew corn (*Zea mays* L.) in a loamy sand amended with a factorial combination of rates of Cd and Pb, and found that increasing Pb increased Cd concentrations in the roots and shoots significantly and substantially. Interelement competition for strong binding sites could allow patterns such as this to occur as suggested by McBride (1995). Soil acidification resulting from the large additions of $PbCl_2$ could have increased soil-Cd phytoavailability. But in this case, the combination of added Pb decreasing soil pH, and the large addition of chloride in the $PbCl_2$ causing increased solubility and plant uptake of Cd can explain these results without invoking the speculative claim of an important synergistic interaction between Pb and Cd. The experiment that was the basis for the suggestion of synergistic interactions had design features now recognized to be nonoptimal, including: (i) using soluble metal-salts at phytotoxic levels; (ii) using repeatedly air-dried soil; (iii) lack of added fertilizers needed to support plant growth such that low phosphate could allow higher Pb translocation to shoots; (iv) no adjustment of soil pH to the same initial pH level (to correct for the displacement of adsorbed protons by the added Pb and Cd); and (v) plants were grown in pots in a greenhouse that favors uptake to shoots. Together, these design characteristics enhance the likelihood that the experimental outcome of this study may be an artifact rather than a synergistic interaction between added metals.

Extrapolating from pot studies in the greenhouse or growth chamber to the field is one of the most misleading aspects of metal uptake by plants. de Vries and Tiller (1978) found very significantly higher Cd, Zn, and other element concentrations in lettuce and onion (*Allium cepa* L.) grown in pots compared to plants grown in microcosms or in the field. They believed that the exaggerated uptake resulted from (i) the high concentrations of soluble fertilizer salts added to small volumes of soil in pots studies, (ii) differences in soil temperatures in pots vs. fields, and (iii) differences in evapotranspiration in laboratory studies compared with the field. Other workers have confirmed this outcome.

Another source of error in uptake studies is caused by rapid biodegradation of the organic matter in manures or biosolids. Biodegradation by-products temporarily increase metal solubility in the soil and uptake by plants. An experiment by Sheaffer et al. (1981) using inadequately stabilized biosolids in heated soils caused very high uptake of Cu, Zn, and some other elements, and induced Cu phytotoxicity in radish (*Raphanus sativus* L.); however, the next crop season, after the organic C was stabilized, this unusual uptake was no longer observed, and biosolids stimulated yield of radishes.

The shape of the response curves of plant-Cd vs. soil-Cd are very important in assessment of soil-Cd risk to humans. Generally speaking, addition of soluble metal-salts to soils gives a linear response of increased plant concentration until the plant concentration approaches phytotoxic levels. But if Cd is added with metal adsorbing materials (e.g., hydrous Fe oxides), the extent of increase is reduced (Singh, 1981; Kuo, 1986).

Figure 1–2 shows the response patterns for uptake of Cd by soybeans [*Glycine max* L. (Merr.)] grown on two soils amended with a combination of soluble salts of Zn and Cd (White & Chaney, 1980). These metals were added together in the study because of their nearly universal geological co-occurrence. The Sassafras soil (fine-loamy, mixed, mesic Typic Hapludult) had smaller ability to adsorb metals; as the Zn plus Cd addition increased, it is believed that Zn filled the metal adsorption sites, causing plant uptake of Cd to have increasing

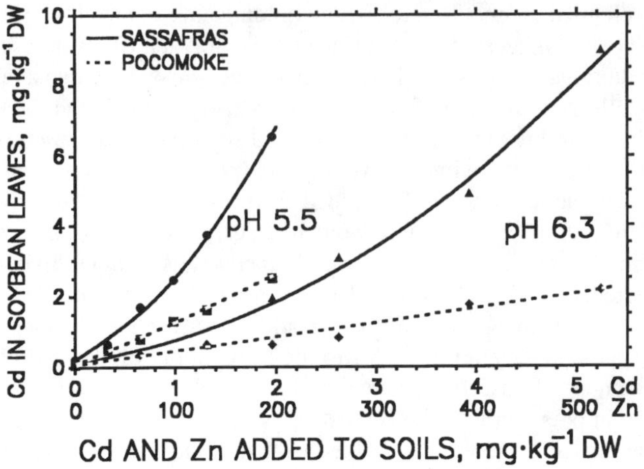

Fig. 10–2. Influence of Zn + Cd (at ratio of 3 µg Cd per 100 µg Zn) addition to Sassafras and Pocomoke sandy loam soils (fine-loamy, mixed, thermic Typic Umbraqualt), and soil pH (adjusted to 5.5 or 6.3) on the Cd concentration in soybean trifoliolate leaves (White & Chaney, 1980).

slope with increasing Cd plus Zn application (increasing slope, plateauing toward the *Y* axis). The Pokomoke soil had higher organic matter, and higher specific metal adsorption capacity; the plant uptake of Cd response pattern remained linear with increasing rate of Cd plus Zn-salt application. These patterns of soil-plant response are commonly found in pot experiments with additions of soluble metal-salts (if pH is maintained across treatments; otherwise pH drops as metal-salt addition increases, and uptake slope increases with increasing rate of metal application due to the acidification).

Figure 10–3 shows our view of the model patterns of plant uptake of metals in relation to soil-metal concentrations found in studies of long-term biosolids

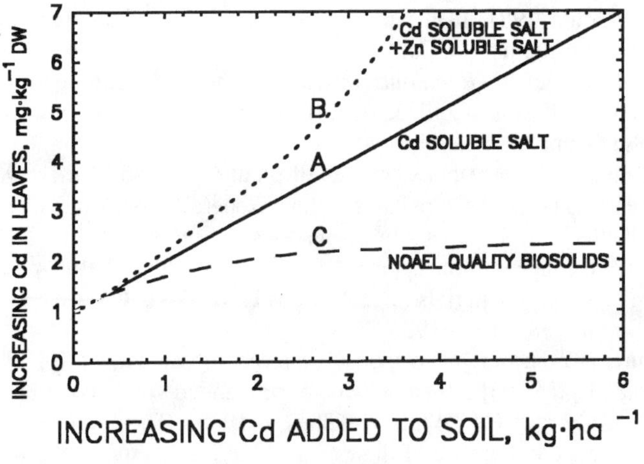

Fig. 10–3. Hypothetical models of increasing plant Cd concentration in response to increasing total soil Cd concentration: (A) From addition of a soluble Cd-salt; (B) From addition of a soluble Cd-salt with 100 times more Zn as a soluble Zn-salt; and (C) From addition of NOAEL quality biosolids, after organic matter stabilization to background levels.

application compared with those for metal-salt treated soils. In Fig. 10–3, all lines start at the linear slope usually found for added Cd-salts, and represent equal Cd additions in different forms, to one soil. Curve A represents the linear response to small additions of salt-Cd found in nearly all studies in the literature. In Curve B, the pattern is of increasing slope at increasing Cd applications because Zn also is added, at 100-times the Cd additions, and the added Zn competes for the stronger adsorption sites in the soil. These first two patterns have been repeatedly observed in many studies, and are illustrated well by the data in Fig. 10–2. In contrast, the model slope C in Fig. 10–3 is for biosolid applied Cd, which causes decreasing slope toward a plateau with the *X* axis.

Figure 10–1 shows the results for lettuce uptake of Cd on long-term biosolids field plots at Beltsville, MD, and shows the plateau response compared with linear regression to estimate the uptake slope for the plateau data (Chaney et al., 1982). As noted by Corey et al., 1987, biosolids contain about 50% inorganic matter with significant levels of hydrous metal oxides. Thus, one is not adding metal-salts, but metals that are equilibrated, adsorbed, or precipitated (or coprecipitated) with persistent metal oxide adsorption surfaces. Some of the reaction products appear more similar to those of solid solutions than coprecipitates. The addition of metal adsorbing materials other than organic matter provides persistent metal adsorption proportional to the metal additions, and prevents the *Time Bomb* from detonating. When Beckett's research group finally got to test the Time Bomb hypothesis in a field study, they found no evidence to support the model, and no evidence of synergistic interactions of metals in phytotoxicity or uptake (Johnson et al., 1983). A similar lack of additive or synergistic effects of adding Zn, Cu, and Ni in various combinations was obtained in a pot study of biosolids-applied metals by Davis and Carlton-Smith (1984).

Such a study as that of Davis and Carlton-Smith (1984) remains difficult to interpret because as sludge metal concentrations increase, the metals in the sludge–soil mixture have greater phytoavailability. As noted by Logan and Chaney (1983), nearly every experiment that examined high loadings of metals from biosolids sources used highly contaminated biosolids to be able to achieve the high loadings. The very nature of the experimental design prevented a meaningful answer because highly contaminated biosolids were used in the experiments. These many factors (pot studies, metal-salt studies, or studies on soils with freshly applied biosolids) that caused higher metal uptake or phytotoxicity than found if the quality of biosolids recommended for beneficial use are applied, made the data from those studies not valid predictors of food-chain transfer or phytotoxicity of metals. The only cases in which these high quality biosolids have caused metal phytotoxicity were when soil pH was allowed to drop to levels well below 5.0, where the combination of soil-Al and -Mn, and biosolids-Zn, -Cu, and -Ni reduced yields of soybeans (e.g., Lutrick et al., 1982). Biosolid-applied metals causing phytotoxicity under recommended soil pH management conditions has only occurred when highly contaminated biosolids were applied (e.g., Webber et al., 1981; Marks et al., 1980).

Chaney, Ryan, and other scientists have examined plant uptake of Cd from soils in farmers' fields that had long-term biosolids applications (e.g., Chaney & Hornick, 1978; Mahler & Ryan, 1988a). A very important set of studies by

Mahler et al. (1987) and Mahler and Ryan (1988a,b) involved additions of Cd-salts to soils from high cumulative biosolids application fields and from adjacent untreated fields of the same soil series. They grew Swiss chard (*Beta vulgaris* L.), a spinach-like vegetable with high Cd uptake slope. Chard is very tolerant of high Cd (Mahler et al., 1978), so that it has a very wide range of linear response to soil-available-Cd. For each Cd-salt amended soil, the response to the additional 5 or 10 mg Cd kg^{-1} gave a linear response with a high R^2. A careful examination of the full data of Mahler et al. (1978) shows that the uptake slope for the soils without biosolids are in general higher than the slope for the biosolid-amended soils. When the biosolid-untreated and nontreated soils were at the same pH, or taken to the same pH by addition of limestone, the slope was lower for the treated soil. This was especially true for those soils with high cumulative loadings of biosolids (Fig. 10–4).

In many long-term field studies of biosolids application, added organic C has declined with time to, or near to, the level of the control soil. Thus, other biosolids components must have provided increased metal adsorption capacity to the soil or the response pattern should approach that of a soluble Cd-salt added to the soil at equal cumulative rate. Analysis of data from low Cd biosolids does not allow testing this hypothesis because there is often no increase in plant Cd at any time. Fortunately, a relatively high Cd (210 mg Cd kg^{-1} dry weight) biosolid was applied (at 50 and 100 t ha^{-1}, applying 11.5 and 21 kg Cd ha^{-1}) to field plots at Beltsville in 1978, and equal Cd-salt (21 kg Cd ha^{-1}) was added to comparison plots. Low and high soil pH was maintained during the nearly two-decade long experiment. Analysis of the pattern of Cd phytoavailability in these plots has

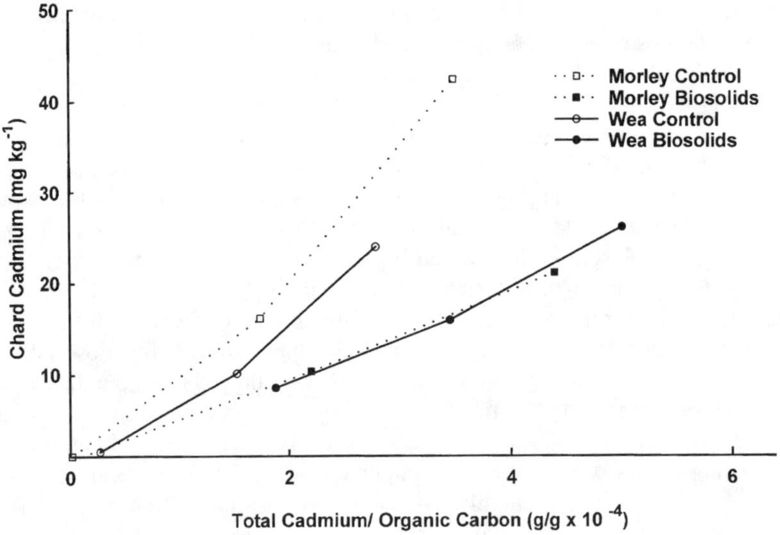

Fig. 10–4. Linear response of chard Cd concentration to added salt-Cd on long-term biosolids amended or nonamended soils. Carbon levels in the biosolid-amended soils are no longer above levels in the nonamended soils, and soil pH levels were made equivalent by making all soils calcareous (based on data from Mahler et al., 1987).

shown that as the organic carbon was lost, the lettuce Cd vs. soil Cd/OC slope declined somewhat instead of rising toward the slope of the Cd salt treatment (Fig. 10–5). The results of this experiment fail to show increasing phytoavailability of biosolid-applied Cd with time. This is true whether evaluated in the traditional plots of crop-Cd (Y) vs. soil-Cd (X) plots, or the plots of crop-Cd (Y) vs. soil-Cd:soil-OC recommended by McBride (1995). Thus, these field data with quite Cd-contaminated biosolids provide empirical refutation to the *Time Bomb* hypothesis.

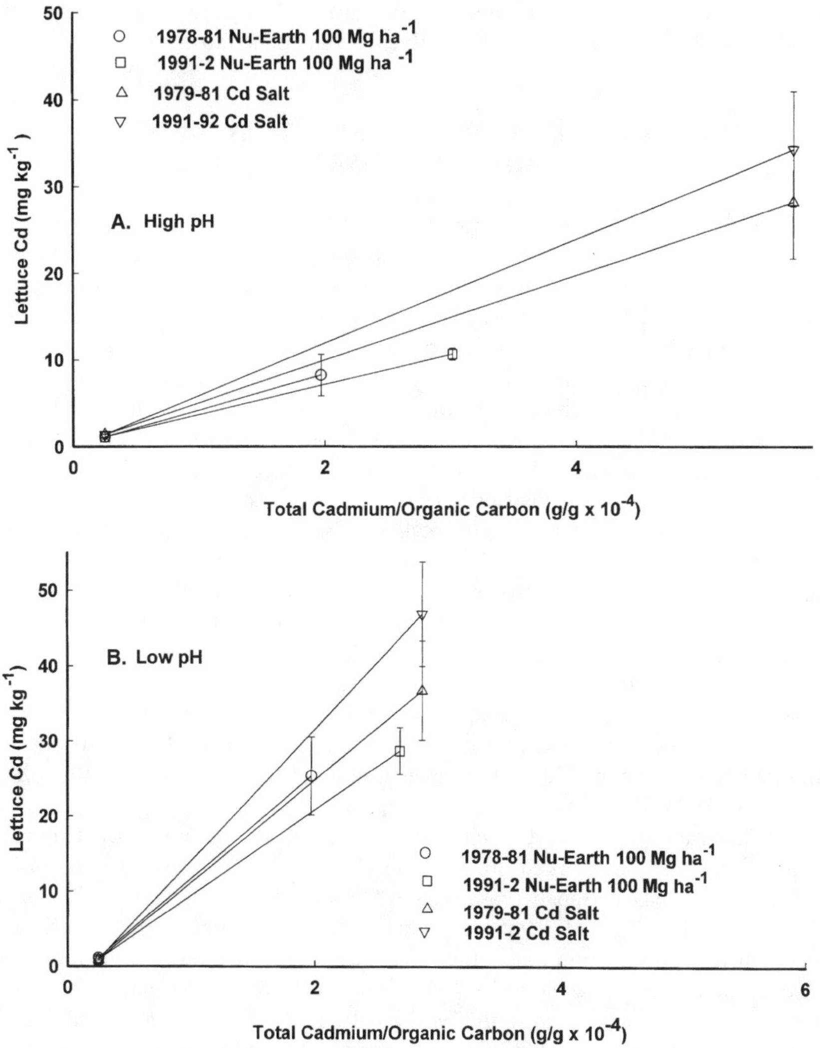

Fig. 10–5. Comparison of the relationship between plant Cd concentration and the ratio between total soil-Cd and soil organic-C for lettuce grown on long-term biosolids and Cd-salt amended soils maintained at (A) high soil pH and (B) low soil pH. Data points represent values collected during the 1991 and 1992 growing seasons.

Although this pattern has been observed in many long-term field studies of biosolids-applied metals, the mechanism that causes these results has not yet been fully demonstrated. Some evidence supports the role of hydrous oxides and phosphate in the biosolid as the persistent adsorption ability of biosolids, and higher Fe biosolids have been found to allow lower uptake of Cd (all other experimental variables unchanged). Additional research is needed to clarify the mechanisms involved—both those that affect adsorption on particle surfaces of the soil–biosolids mixture, and those that interfere with uptake of Cd (e.g., Zn). It is clear that Cd persists in soils over millennia based on the profile distribution of Cd in geogenic-Cd enriched soils in California (Burau et al., 1981), but plant uptake was far lower than seen for soluble Cd salts added to such soils.

10–3 NUTRITIONAL INTERACTIONS AND CONTAMINANT BIOAVAILABILITY

A further aspect of the agronomy and human nutrition of rice that is involved in the higher risk from rice Cd in subsistence rice eating populations was deduced by Chaney et al. (1996a). As shown in Fig. 10–6, rice (and corn) have much lower concentrations of Fe and Zn than other major staple grain foods (based on data in Wolnik et al., 1983a,b, 1985). As we have noted before, not only are the total concentrations of these nutrients in rice and corn low, but milling to prepare white rice or corn removes much of the Fe and Zn present. Rice is well known to supply inadequate Fe and Zn to support animal life (e.g., Pedersen & Eggum, 1983; Hallberg et al., 1974). Nutritionists have examined the likelihood of Zn or Fe deficiency in subsistence populations who consume rice or corn compared with populations with higher intake of Zn and Fe from other foods especially meat. Each year many children living on rice or corn diets in developing counties become moderately or severely deficient in Fe and Zn (United Nations ACC/SCN, 1992). The deficiencies are severe enough that the immune system of the children become deficient, and they suffer severe disease and death from mild microbiological infections. In contrast, these diseases harm few children in the developed countries that have much lower incidence or severity of Fe or Zn deficiency. These Fe and Zn deficiencies also would cause much higher absorption and retention of dietary Cd (e.g., Flanagan et al., 1978; Shaikh & Smith, 1980; Fox et al., 1984).

Thus agronomy of Cd can overwhelm the toxicology of Cd risk. Epidemiologists who worked diligently to characterize the medical basis of Cd risk to individuals in Japan (e.g., Nogawa et al., 1987; Tsuchiya, 1978) or in Europe (e.g., Friberg et al., 1985, 1986) had little appreciation for the soil science, plant nutrition, food-chain transfer, nor for the human nutrition aspects of Cd disease. It now seems clear how they were led to the conclusion that soil-Cd was dangerous to humans, because that is the pattern observed in Japan and China, where rice provides the bulk of the Cd exposure and is simultaneously the reason for high Cd bioavailability. But lettuce, wheat, and potatoes are much more important sources of exposure to Cd from contaminated soils in the West, and soil-Cd comprises far lower risk through these foods than through rice. As summarized fully in Chaney and Ryan (1994), Zn phytotoxicity provides a very strong limit on Cd in edible plant tissues, and the home garden cannot achieve excessive bioavailable dietary Cd when soil total Cd/Zn is 0.015 or lower. Cd and Zn in mine wastes and smelter emissions usually occur at the geological ratio, about 0.005 to 0.01 Cd/Zn. In situations where commercial Zn contaminates soils (e.g., from corrosion of galvanized steel, or burned tires), so little Cd accompanies the Zn that even lettuce is not increased in Cd at the point where Zn causes phytotoxity (Jones, 1983). Zinc in the plant competes with Cd uptake by intestinal cells therby reducing the net absorption of Cd by all species tested. Examples of the Zn inhibition of food Cd absorption are shown in Chaney et al. (1978) and McKenna et al (1994). Addition of salt Zn to purified diets also reduced absorption of salt Cd in the studies of Jacobs et al. (1983) and Fox et al. (1984).

However, Cd from plating wastes, wastes from the manufacture of Cd–Ni batteries, Cd pigments, or Cd-stabilizers for polyvinylchloride plastics have little accompanying Zn. These latter sources allow much greater potential for flow of

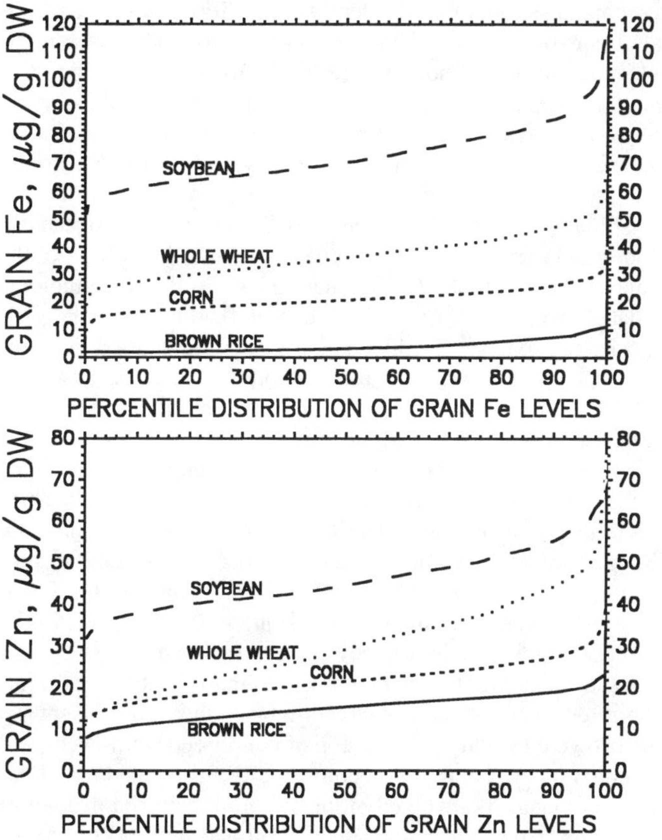

Fig. 10–6. Statistical distribution of Zn and Fe concentrations in the grain of rice, corn, wheat, and soybeans harvested from U.S. fields representing the major soil series and regions where these crops are normally grown (based on Wolnik et al., 1983a,b, 1985). The data are expressed on a dry weight basis.

Cd into the food chain. These Cd without Zn sources can cause potentially high Cd bioavailability to humans who consume crops grown on soils strongly contaminated by these sources, especially so if the soils are acidic. The case of high Cl^- allowing Cd to enter plants without increasing plant Zn, requires more evaluation to determine under what circumstances human or environmental risk could occur. In some nations, phosphate fertilizers have applied much higher amounts of Cd than in most of the USA, and the fertilizer Cd was accompanied by much lower Zn than occurs in other geologic Zn sources (Cd is usually ≥10% of Zn in P fertilizers vs. 0.5% Cd of Zn in Zn ore sources that contaminate soils).

As agronomists who have worked for several decades to set limits for Cd in biosolids, composts, soils, and crops, we find it ironic that agronomy has proved so important in soil-Cd risk. Toxicologists usually study the contaminant they are examining in a pure form, such as soluble Cd-salts without accompanying Zn or other materials in Zn ores. They provide the Cd-salts to animals who are often minimally adequate or even deficient in Fe or Zn in order to observe maximal Cd retention and adverse health effects, often supplying the Cd by injection rather than ingestion routes. Under these conditions, some of the adverse toxicological effects of Cd result from Cd-induced Zn deficiency, but the environmental relevance of these findings is restricted to sites where high Cd contamination is not accompanied by Zn at normal environmental Cd/Zn ratios.

Improved understandings about soil-Cd risk are of increasing importance today because some northern European nations want to impose limits on Cd in food grains far lower than needed to protect even the exposed farm families in Japan and China from rice-Cd. Sweden and other northern European nations have recommended to international food regulatory bodies that a limit of 0.10 mg Cd kg^{-1} be imposed on all grains. This regulation ignores the epidemiologic research in Japan that showed that if rice did not exceed 0.40 to 0.50 mg kg^{-1} for a lifetime of subsistence exposure, individuals were at no increased risk from soil-Cd. Society is harmed if otherwise wholesome food is rejected because of unscientific fear of Cd by individuals who have little appreciation of the role of agronomy or bioavailability in soil-Cd risk to humans.

As it happens, some crops naturally have higher Cd than do most other crops (Chaney et al., 1993). In particular, nonoilseed sunflower has higher Cd by about 10-fold than rice and bread wheat grown on the same uncontaminated soil. In the USA, the soils where nonoilseed sunflower is grown are alkaline, and are not contaminated by industrial Cd sources (Chaney et al., 1993). Because of the tendency of sunflower to accumulate Cd from soils, the nonoilseed sunflower industry supported research by Chaney and coworkers to search for low Cd sunflower germplasm, and for production practices that might lower Cd in sunflower kernels. This very successful cooperative research and development program has significantly lowered Cd in the USA crop of nonoilseed sunflowers, and new low Cd hybrid cultivars are being used commercially (Li et al., 1994, 1995b,c). The researchers and managers involved in this program believed making this investment to assure that Cd levels in marketed nonoilseed sunflower kernels are below the import limits was a prudent action, even if there is presently no indication that individuals will consume sufficient quantities of these kernels during their lifetime to comprise risk. The present World Health Organization (WHO) and USA limit for Cd ingestion, 70 μg d^{-1}, is for chronic lifetime exposure, and includes safety factors. Such prudence is indicated because of our incomplete understanding of effects of Cd on humans at presently subclinical exposures. Thus it is responsible to minimize the level of Cd in farm products where this is practicable. A few other crops may require similar attention to limiting grain Cd levels by breeding lower grain Cd genotypes or shifting soil series or cultural practices for production. Similar concerns indicate that the development of Al-tolerant or salinity-tolerant crop cultivars should include examination of the effect of highly acidic or highly saline conditions on the Cd levels in the crop (Chaney et al., 1996a).

10–4 ESTIMATING RISK FROM PERSISTENT XENOBIOTICS IN SOILS AND BIOSOLIDS

Chaney et al. (1996b) have recently summarized the USEPA 503 Rule risk assessment process for polychlorinated biphenyls (PCBs) in land-applied biosolids. Because of the long half-life of PCBs in soils, these compounds represent a model for food-chain transfer of persistent xenobiotics under the pathway approach for risk assessment. Table 10–4 provides a summary of the pathway limitations estimated.

Table 10–4. Summary of polychlorinated biphenyls (PCBs) application limits for each pathway from the USEPA 503 Proposed Rule (U.S. Environmental Protection Agency, 1989b) with the corrected versions based on USEPA (1993) and Chaney et al. (1996b). Units are changed in some corrected versions. Error in actual compound residue in plant per unit soil residue was corrected compared to the proposed USEPA 503 Rule (U.S. Environmental Protection Agency, 1989b). The last column shows the PCBs concentrations that would have been required to meet each pathway annual application rate for PCBs if biosolids were applied at 10 dry t ha^{-1} yr^{-1}.

Pathway	Limit units	Limit value	Concentration (Annual)
			mg kg^{-1} biosolids dry weight
1†	mg kg^{-1} soil maximum	290	
	kg $(ha\ yr)^{-1}$	37	3700
2†	mg kg^{-1} soil maximum	67.1‡	
	kg $(ha\ yr)^{-1}$	8.5	
2§	mg kg^{-1} soil maximum	17.2	
	kg $(ha\ yr)^{-1}$	2.3	231
3§	mg kg^{-1} biosolids dry weight	14	14
4†	kg $(ha\ yr)^{-1}$	4.6	460
4§	mg kg^{-1} soil maximum	18	
	kg $(ha\ yr)^{-1}$	2.4	246
5†	mg kg^{-1} biosolids dry weight	4.6	4.6
5-surface§	mg kg^{-1} biosolids dry weight	2.2	2.23
5-mixed§	mg kg^{-1} soil maximum	2.2	
	kg $(ha\ yr)^{-1}$	0.299	29.9
10§	mg kg^{-1} soil maximum	4.06	
	kg $(ha\ yr)^{-1}$	0.545	54.5

† Calculated limit according to U.S. Environmental Protection Agency (1993).

‡ U.S. Environmental Protection Agency used default assumption for their calculation which differed from those we judge to be most valid; carrot carries most PCB into garden foods in human diets, not potatoes as calculated by US-EPA.

§ Alternative calculation shows the corrected estimate for the limit based on Chaney et al. (1996b).

Several key points should be considered regarding soil-xenobiotics. For compounds such as the PCBs that are strongly adsorbed by organic matter, the vapor pressure and phytoavailability of biosolids-applied PCBs are much lower than equal amounts of pure PCBs added to soil. O'Connor et al. (1990) found that spiking soils or applied biosolids with PCBs also gave much higher transfer of PCBs to carrot (*Dacaus carota* L.) peels than did equivalent PCBs that had equilibrated in biosolids over many years. Insignificant transfer of soil-PCBs to forage crops were observed in field studies (Gan & Berthouex, 1994). Because of the low rates of transfer of PCBs from soils to crops or the air, the only significant food-chain hazard is from soil ingestion by livestock (Fries, 1996) or children, or from consumption of earthworms that bioaccumulate PCBs in soils. Similar biosolids binding of polycyclic aromatic hydrocarbons (PAHs) limits transfer of these xenobiotics to carrot roots (Wild & Jones, 1992).

The wildlife protection pathway for PCBs is interesting because it involves bioaccumulation of PCBs by earthworms. Like the waxy carrot peel, the surface of earthworms is lipophilic, and collects compounds such as PCBs from soils by vapor transfer. Because of the bioaccumulation of PCBs on the earthworms, the soil within the earthworm is a less important source of PCB exposure than is the earthworm body. A few wildlife species consume earthworms as an appreciable fraction of their diets during part of the year. Although earthworm-consuming bird species are well known models for DDT risk assessment, most birds have a wide range compared with small mammals such as shrews (*Sorex araneus*) or moles (*Talpa europea* L.) that might live their entire life within a contaminated site.

Beyer and Stafford (1993) and Eisler (1986) have discussed wildlife exposures to PCBs, and sensitivity of different wildlife species to PCBs. They identified small mammals such as shrews and moles are the HEIs for exposure through the earthworm pathway. Other small mammals might live their life within a contaminated field, but they consume plant leaves or seeds, or aboveground insects that do not have the extent of PCB bioaccumulation found for earthworms (see Forsyth & Peterle, 1984; Davis et al., 1981).

The tolerable concentration of PCBs in the diet of small mammals was obtained from the literature (Eisler, 1986; Peakall, 1986; Beyer & Stafford, 1993). The lowest no observed adverse effect level [NOAEL] for PCB toxicity to mammals and birds is 5 mg kg^{-1} diet in chickens (*Gallus domesticus*; Lillie et al., 1975). Although shrews and moles have not been studied, this NOAEL is considered appropriate for earthworm-consuming wildlife because chickens are more sensitive to dietary PCBs than mice (*Mus domesticus*) or rats (*Rattus norvegicus*).

As in the human and livestock Pathways, Chaney et al. (1996b) needed to make the pathway calculations of limits as realistic as available data allow. Thus, in contrast with the original USEPA pathway for wildlife in which ducks (*Anus* sp.) were presumed to consume earthworms as 100% of their diet (U.S. Environmental Protection Agency, 1989b), the exposed wildlife species is no longer presumed to consume only earthworms for their lifetime. Ingestion of earthworms as 100% of diet might be appropriate for consideration of acute exposure of wildlife species, but not for chronic exposure. After consideration of maximum chronic consumption of earthworms by wildlife (see review by MacDonald, 1983), 33% of diet was selected as the maximum chronic fraction of dietary earthworms. It should be recognized that the 45% soil content of earthworms as ingested in the field makes them have a much higher dry matter percentage than other soil biota and insects consumed by species such as shrews; thus, earthworms comprise a higher fraction of the diet wet matter than of the diet dry matter upon that these calculations are based. Available information indicates that earthworms comprising 33% of the lifetime diet dry matter of shrews is a high-end estimate. Thus, the maximum allowed diet PCB level (5 μg PCB g^{-1} diet dry weight) is multiplied by three to calculate the 15 mg PCB kg^{-1} dry weight allowed in whole earthworms, which comprise 33% of the diet of HEI wildlife.

Next, the bioaccumulation of PCB or similar compounds in earthworms, compared with soil in which they live, must be obtained from field studies with soil–biosolids PCBs. This transfer coefficient is available from studies by Beyer and Stafford (1993), Beyer and Krynitsky (1989), Diercxsens et al. (1985), Kreis et al. (1987), Marquenie et al. (1987), and Tarradellas et al. (1982).

For earthworms with soil in their digestive system (as they would normally be consumed in the field), the bioaccumulation factor is about 3.69 mg PCB kg^{-1} earthworm-dry weight (nonpurged) per 1 mg PCB kg^{-1} dry soil (the geometric mean for 1.8, 3.4, 3.3, 4.9, 5.0, and 5.1 [the arithmetic mean for these slopes is 3.9]) from the above noted studies.

Thus 15 mg PCB kg^{-1} worm fresh weight ÷ 3.69 = 4.06 mg PCB kg^{-1} soil dry weight = 8.12 kg PCB ha^{-1}.

For organics, the biodegradation or other dissipation must be considered in translating the maximum soil PCB concentration limit into annual application limits. The allowed loading must be considered the maximum concentration that may be reached at any time rather than the cumulative application. Equal annual applications are usually estimated based on the half-life of the organic compound in soil, although the regulation is met as long as the maximum soil concentration needed to protect wildlife is not exceeded.

For biosolids mixed with soil, the annual application of PCB that will not exceed the maximum allowed 4.06 mg PCB kg^{-1} DW soil [when equilibrium is reached between annual application and annual dissipation; this requires 81 yr for PCBs with the assumed 10 yr half-life] is calculated using the result from Eq. [7] of Chaney et al. (1996b):

Allowed annual biosolids-PCB application =

$$= (4.06 \text{ mg PCB kg}^{-1} \text{ soil-biosolids mixture dry weight}) \times 2000/1000 \times [14.9]^{-1}$$

$$= 0.545 \text{ kg PCB (ha} \times \text{yr)}^{-1}.$$

If one assumes 10 t of a biosolids would be applied per year as a N fertilizer, the concentration of PCB in that biosolids could be:

$$\frac{0.545 \text{ kg PCB}}{\text{ha} \times \text{yr}} \times \frac{1 \text{ ha} \times \text{yr}}{10 \text{ t biosolids dry weight}} = 54.5 \text{ mg PCB kg}^{-1} \text{ biosolids dry weight}$$

Because the concentration of PCBs in modern biosolids is generally <0.10 mg kg^{-1} dry weight, wildlife are highly protected from PCBs in biosolids applied to land.

10–5 OTHER SOIL CONTAMINANTS

10–5.1 Soil-Lead Risks to Children through Soil Ingestion

Readers can find detailed discussion of soil-Pb risk to children who ingest soil or dust in Chaney and Ryan (1994), Freeman et al. (1992), Ruby et al. (1993), or Ryan and Chaney (1998). Although urban garden soils and houseside soils have become highly Pb contaminated from paint and automotive emissions (Chaney et al., 1984), greater risk arises from soil ingestion rather than through plant uptake by garden foods (Sterrett et al., 1996). Soil adsorption of Pb and phosphate precipitation of Pb reduce soil-Pb bioavailability, perhaps in a manner similar to the way normal components of foods reduce absorption of dietary soluble Pb (James et al., 1985). Phytate in whole grain breads is highly effective in reducing bioavailability of dietary Pb (James et al., 1985).

Although some houseside soils contain up to 5% Pb, studies of soil replacement in Boston showed that current exposure to paint-Pb remains a much more important risk than soil-Pb (Weitzman et al., 1993). Young children commonly chew on painted surfaces and mouth dust-covered fingers and toys. A low percentage of children eat nonfood items, a behavior called *pica*. Questions still need to be raised about the high Pb residues in apple (*Malus domesticus* L.) orchard and certain other agricultural soils that became highly Pb contaminated from pesticide sprays before the advent of DDT (Chaney & Oliver, 1996).

Remediation of Pb-contaminated soils to which children have exposure has usually required removal of the contaminated soil and replacement with new topsoil. Such remediation costs on the order of $2.5 million per ha-30 cm, and alternatives are needed so that more of the Pb-contaminated urban soils are remediated to protect children. Formation of highly insoluble Pb phosphates such as chloropyromorphite may change the speciation and bioavailability of soil-Pb remarkably (Ma et al., 1993). Berti et al. (1997) found that addition of a combination of hydrous ferric oxides and phosphate were more effective than phosphate alone to reduce leachability and in vitro bioaccessibility of soil Pb. Chaney and Ryan (1994) summarized the evidence that incorporation of high Fe biosolid composts remediates soil Pb and corrects many of the phytotoxicity and infertility problems of smelter and urban metal contaminated soils. These in situ remediation methods may be applicable to moderately contaminated sites, or more highly Pb-contaminated soils where children will not have access.

10–5.2 Soil-Chromium Risks

Several recent reviews of the chemical species of Cr in soils discuss the potential of soil-Cr to be oxidized and leach as Cr(VI) to groundwater. As noted by Chaney et al. (1996b), serpentine-derived soils often contain up to 1% Cr in highly insoluble Cr(III) forms. The kinetics of dissolution or release of Cr(III) to react with Mn oxides in soil is too slow to allow formation of Cr(VI) in dangerous amounts. No evidence of potential for environmental risk has been demonstrated in study of soils amended with Cr-rich biosolids or leather-based organic fertilizers applied at fertilizer rates. The USEPA deleted Cr from the USEPA 503 Rule in response to a legal challenge from the U.S. leather industry. Indeed, in situ reduction and acidification of alkaline chromate smelting wastes rich in Cr(VI) may allow inexpensive remediation that protects humans and the environment (James, 1996).

10–5.3 Copper and Zinc in Livestock Manures

In many production regions, high levels of Cu and Zn are added to swine feeds to increase gain rates; and high levels of Cu, Zn, and As are added to poultry feeds. Such manures often contain higher Cu, Zn, and As levels than modern municipal biosolids. Although Cu toxicity to ruminants was a concern, it has not been found in experiments with season long grazing of pastures amended with high Cu manure or biosolids (Bremner, 1981; Poole, 1981). Copper in manures and biosolids is not equivalent to Cu-salts, but is adsorbed to the residues, and co-contaminating Zn and other elements reduce Cu bioavailability under practical grazing exposures (Suttle et al., 1975, 1984). When soil pH was managed reasonably, continuing fertilizer applications of these livestock manures have not caused phytotoxicity to corn (e.g., Zhu et al., 1991), but few tests have been conducted with the more Zn- and Cu-sensitive vegetable crops. The effect of pH mismanagement of amended soils has been observed in phytotoxicity of Zn sensitive crops such as peanut (*Arachis hypogea* L.) grown at pH 4.5. It is hard to be sympathetic to a farmer who would be growing peanut at pH 4.5. Addition of hydrous Fe oxides to manures might be an inexpensive way to reduce the potential risk of future Zn or Cu phytotoxicity under poorer pH management.

10–6 SUMMARY AND CONCLUSIONS

Research to characterize food-chain transfer of soil elements or xenobiotics has established a number of relationships. Much of the information on soil→plant→food-chain transfer is summarized by the soil–plant barrier model. Most elements and persistent xenobiotics in soil are insoluble or strongly adsorbed under usual and customary modern management practices, or sufficiently phytotoxic, that chronic lifetime consumption of foods or feeds grown on the soil have not been observed to cause adverse effects. A few elements are easily translocated to edible crop tissues and have caused livestock and even human poisoning (Se, Mo). Other elements are a risk in ingested soil, but typically not through the soil→plant→food-chain, and these include several important soil contaminants (Pb, As; persistent lipophilic xenobiotics such as PCBs).

Theoretically, Cd should be able to cause harm to consumers because it is so food-chain mobile. But in nearly every case where Cd contamination of soils occurs, the Cd is accompanied by 100 to 200-fold more Zn. The Zn can reduce crop yields before crop Cd levels become important to Cd exposure, Zn inhibits Cd uptake by plants, Zn inhibits Cd movement to plant storage tissues, and Zn inhibits Cd absorption in the intestine of the consumer. To date, only the flooded soil rice production system has been found to allow Cd to move into grain without corresponding increase in grain Zn. The recently identified effect of Cl^- increasing food-chain transfer of Cd but not Zn must raise a cautionary flag about subsistence agriculture in regions with soils rich in Cl^- from any source. If present research programs to increase salinity tolerance of crops are increasingly successful, such that an appreciable portion of foods are produced on such soils, it may become necessary to find methods (plant breeding or other) to keep the absorbed Cd from being moved into the edible crop tissues such as grains, fruits, and tubers.

Bioavailability of elements in ingested foods or soils has been shown to play a very important role in reducing toxicity to children, livestock, or wildlife that consume soil. Zinc-induced Cu deficiency in ruminant livestock and horses (*Equus caballus*) has occured when aerosol Zn deposition circumvents the phytotoxicity limit on maximum crop Zn levels. But crop Zn absorbed from soils has not been found to poison livestock that consume the crop for their lifetime.

The greater understanding about food-chain risk from Cd that has developed during the last 25 yr or more is an example of the complications in analysis of food-chain risk from soil-metals. Although subsistence rice farmers were seriously harmed by osteomalacia and renal tubular dysfunction due to excessive Cd absorption, Fe, Zn, and Ca deficiency in these subsistence consumers of polished rice makes this case highly different from western scenarios for soil→food-chain risk from Cd. Vegetables and staple grains other than rice reject Cd compared with Zn when both Cd and Zn contaminate the soil. These foods also provide higher levels of bioavailable Fe and Zn than does polished rice.

Although the information summarized in this chapter appears to integrate the existing research findings about food-chain Cd transfer, direct experimental testing of this model is now necessary. Experiments should be designed and carried out to clarify the relative importance of each of the deficient nutrients of rice grain vs. other grains in increasing the risk of soil-Cd to humans. This information is needed so that efforts to prevent this human Cd disease can be focused on the causal factors besides soil Cd contaminaton. Public policy about dietary and soil Cd cannot be significantly altered until these tests have been reported. It should be recognized that the overall pattern of human risk from Cd in rice vs. garden foods or oysters is well supported by epidemiologic studies. For other crops under good management practices, Zn phytotoxicity and Zn inhibition of Cd movement from soil to kidney prevent human Cd disease from occurring, and thereby protect the food-chain even for subsistence consumers. For the general population that consumes foods grown in many different locations, there is no evidence of lifetime dietary Cd risk from commercial foods.

Risks from soil-Cd cannot be estimated without knowledge of soil and water characteristics and management practices, the level of co-contaminating Zn, and the nature of the foods that will be grown on the contaminated soil. Some foods have been shown to have different Cd bioavailability than average foods (e.g., spinach-Cd has low bioavailability due to oxalate, and wheat and sunflower kernel Cd due to phytate; while lettuce has no chemical factor other than Zn that inhibits Cd absorption by animals); thus, different commodities should have different Cd limits rather than a single limit be imposed on all crops. And Cd limits should consider soil characteristics (e.g., high soil acidity), quality of irrigation waters (e.g., high Cl^-), and management practices, as well as the presence of Zn that reduces Cd absorption and risk to the food-chain.

It is important to recognize the remarkable difference between aquatic and terrestrial food-chains. In aquatic food chains, some elements and xenobiotics are strongly biomagnified between trophic levels (PCBs, methyl-mercury, Se), so that higher level predators can suffer health effects from eating apparently healthy fish. Terrestrial food-chains generally *biominify* rather than *biomagnify* contaminants between trophic levels (see Beyer & Stafford, 1993), except for lipophilic materials that are efficiently retained by storage in body fat, and efficiently absorbed when ingested by higher trophic level predators. For terrestrial food-chains, soil ingestion is the greater pathway for transfer of persistent lipophilic xenobiotics such as PCBs because the soil→plant transfer is very inefficient. Thus, the important food-chain risk situations usually involve aquatic food-chains with multiple trophic levels, poor soil and crop management practices, or problem soils, or the Se and Mo that are readily transferred through the terrestrial food chain when soils are rich in Se or Mo and pH is high. These principles of food-chain risk assessment are now well established by field experimental data.

Ingestion of soil by grazing livestock, wildlife, and children can circumvent the normal protections of the soil–plant barrier. The predominant pathway for risk for most toxic metals is through soil ingestion rather than plant uptake and food-chain transport. Human food-chains are often protected even when livestock may be injured by absorption of elements or xenobiotics from the ingested soils. For persistent xenobiotic soil contaminants, soil ingestion is the principal route of transfer to food webs. But for nonlipophilic elements, retention in higher trophic levels is so low that the element is biominified rather than bio-magnified. Excepting Se and Mo, and Cd contamination without Zn, terrestrial ecosystems are thus more likely to be affected by potential phytotoxicity from toxic elements limiting primary plant production, than from transfer of toxicants from soil to consumers.

Knowledge summarized in this chapter shows the contribution of basic and applied research to reduction in the uncertainty about ecosystem risks from contaminants in soils. Many important advances were attained because soil, plant, and animal scientists were concerned that contaminants would cause adverse effects in agriculture, to humans, or to wildlife or ecosystems. Industrial discharges to air and water have injured ecosystems because of the direct deposition on plants before the contaminants have time to reach the soil and enter soil equilibria. Contaminated soils must be examined as the complex chemical and biological system they are now known to be, and cocontaminants and normal diet components that counteract or promote food-chain risk of contaminants must be considered along with the contaminant. Ecotoxicology is very different than traditional toxicology in which individual chemicals are supplied to test species in soluble forms, to examine possible fate and effects of the contaminant. Ecotoxicology deals with soil and plant chemistry and biology, interactions among cocontaminants, bioavailability, biomagnification, and other phenomena that can reverse or induce adverse effects. Improved basic understandings of these soil, plant, and animal processes that affect ecosystem risk allow development of required protections of the environment, with limitations based on real risks from the contaminant dispersed in the environment rather than from injected soluble elements and xenobiotics. The hypothetical need for zero tolerance of contaminants is seldom confirmed by research on existing contaminants when all ecosystem processes are considered in the risk analysis. Although it is usually prudent and desirable to minimize dispersal of contaminants into the environment, elements are present in all soils and plants. Zero tolerance would prevent recycling of many beneficial materials in agricultural ecosystems. Scientific analysis of food-chain transfer processes and interactions can be used to characterize the potential for risk from contaminants in soils, and assist society in decisions about contaminants in soil and ecosystems.

REFERENCES

Allaway, W.H. 1968. Agronomic controls over the environmental cycling of trace elements. Adv. Agron. 20:235–274

Allaway, W.H. 1977. Food chain aspects of the use of organic residues. p. 282–298. *In* L.F. Elliott and F.J. Stevenson (ed.) Soils for management of organic wastes and wastewaters. ASA, SSSA, CSSA, Madison, WI.

Baker, D.E., and M.E. Bowers. 1988. Health effects of cadmium predicted from growth and composition of lettuce grown in gardens contaminated by emissions from zinc smelters. Trace Subst. Environ. Health 22:281–295.

Baker, D.E., and L. Chesnin. 1976. Chemical monitoring of soils for environmental quality and animal and human health. Adv. Agron. 27:305–374.

Beckett, P.H.T., R.D. Davis, and P. Brindley. 1979. The disposal of sewage sludge onto farmland: The scope of the problems of toxic elements. Water Pollut. Contr. 78:419–445.

Berti, W.R., S.D. Cunningham, and L.W. Jacobs. 1997. Sequential chemical extraction of trace elements: Development and use in remediating contaminated soils. p. 121–132. *In* M. Prost (ed.) Proc. Int. Conf. on the Biogeochemistry of Trace Elements, Paris. 15–19 May 1995. Colloque 85. INRA Editions, Paris.

Beyer, W.N., and A.J. Krynitsky. 1989. Long-term persistence of dieldrin, DDT, and heptachlor epoxide in earthworms. Ambio 18:271–273.

Beyer, W.N., and C. Stafford. 1993. Survey and evaluation of contaminants in earthworms and in soils derived from dredged material at confined disposal facilities in the Great Lakes region. Environ. Monitor. Assess. 24:151–165.

Bingham, F.T., G. Sposito, and J.E. Strong. 1984. The effect of chloride on the availability of cadmium. J. Environ. Qual. 13:71–74.

Bingham, F.T., G. Sposito, and J.E. Strong. 1986. The effect of sulfate on the availability of cadmium. Soil Sci. 141:172–177.

Bremner, I. 1981. Effects of the disposal of copper-rich slurry on the health of grazing animals. p. 245–260. *In* P. L'Hermite and J. Dehandtschutter (ed.) Copper in animal wastes and sewage sludge. Reidel Publ., Boston.

Brown, S.L., R.L. Chaney, C.A. Lloyd, J.S. Angle, and J.A. Ryan. 1996. Relative uptake of cadmium by garden vegetables and fruits grown on long-term sewage sludge amended soils. Environ. Sci. Technol. 30:3508–3511.

Burau, R.G., W.F. Jopling, C.V. Martin, and G.F. Snow. 1981. Monterey Basin pilot monitoring project. Vol. 2. Appendix L. Market basket survey of cadmium and zinc in the soils and produce of Salinas Valley. Appendix M. Cadmium, zinc, and phosphorus in Salinas Valley Soils. California Dep. of Health, Sacremento.

Cai, S., Y. Lin, H. Zhineng, Z. Xianzu, Y. Zhaolu, X. Huidong, L. Yuanrong, J. Rongdi, Z. Wenhau, and Z. Fangyuan. 1990. Cadmium exposure and health effects among residents in an irrigation area with ore dressing wastewater. Sci. Total Environ. 90:67–73.

Chaney, R.L. 1973. Crop and food chain effects of toxic elements in sludges and effluents. p. 120–141. *In* Proc. Joint Conf. on Recycling Municipal Sludges and Effluents on Land, Champaign, IL. 9–13 July 1973.. Natl. Assoc. of State Univ. and Land Grant Colleges, Washington, DC.

Chaney, R.L. 1980. Health risks associated with toxic metals in municipal sludge. p. 59–83. *In* G. Bitton et al. (ed.) Sludge—Health risks of land application. Ann Arbor Sci. Publ., Ann Arbor, MI.

Chaney, R.L. 1983a. Plant uptake of inorganic waste constituents. p. 50–76. *In* J.F. Parr et al. (ed.) Land treatment of hazardous wastes. Noyes Data Corp., Park Ridge, NJ.

Chaney, R.L. 1983b. Potential effects of waste constituents on the food chain. p. 152–240. *In* J.F. Parr et al. (ed.) Land treatment of hazardous wastes. Noyes Data Corp., Park Ridge, NJ.

Chaney, R.L., and P.M. Giordano. 1977. Microelements as related to plant deficiencies and toxicities. p. 234–279. *In* L.F. Elliott and F.J. Stevenson (ed.). Soils for management of organic wastes and waste waters. ASA, SSA, and CSSA, Madison, WI.

Chaney, R.L., and S.B. Hornick. 1978. Accumulation and effects of cadmium on crops. p. 125–140. *In* Proc. Int. Cadmium Conf., 1st, San Francisco. 31 Jan. –2 Feb. 1977. Metals Bulletin, London.

Chaney, R.L., Y.-M. Li, A.A. Schneiter, C.E. Green, J.F. Miller, and D.G. Hopkins. 1993. Progress in developing technologies to produce low Cd concentration sunflower kernels. p. 8092. *In* Proc. Sunflower Research Workshop, 15th, Fargo, ND. 14–15 Jan. 1993. Natl. Sunflower Assoc., Bismark, ND.

Chaney, R.L., and C.A. Lloyd. 1979. Adherence of spray-applied liquid sewage sludge to tall fescue. J. Environ. Qual. 8:407–411.

Chaney, R.L., and D.P. Oliver. 1996. Sources, potential adverse effects of and remediation of agricultural soil contaminants. p. 323–359. *In* R. Naidu et al. (ed.) Contaminants and the soil environment in the Australasia-Pacific region. Kluwer Academic Publ., Dordrecht, the Netherlands.

Chaney, R.L., and J.A. Ryan. 1994. Risk based standards for arsenic, lead and cadmium in urban soils. DECHEMA, Frankfurt.

Chaney, R.L., J.A. Ryan, and S.L. Brown. 1998. Development of the US-EPA limits for chromium in land-applied biosolids and applicability of these limits to tannery by-product derived fertilizers and other Cr-rich soil amendments. p. 229–295. *In* S. Canali et al. (ed.) Proc. chromium environmental issues. Franco Angeli, Milano, Italy.

Chaney, R.L., J.A. Ryan, Y.-M. Li, R.M. Welch, P.G. Reeves, S.L. Brown, and C.E. Green. 1996a. Phyto-availability and bio-availability in risk assessment of Cd in agricultural environments. p. 49–78. *In* OECD Proc. Sources of Cadmium in the Environment, Stockholm, Sweden. 15–22 Oct. 1995. OECD, Paris.

Chaney, R.L., J.A. Ryan, and G.A. O'Connor. 1996b. Organic contaminants in municipal biosolids: Risk assessment, quantitative pathways analysis, and current research priorities. Sci. Total Environ. 185:187–216.

Chaney, R.L., S.B. Sterrett, and H.W. Mielke. 1984. The potential for heavy metal exposure from urban gardens and soils. p. 37–84. *In* J.R. Preer (ed.) Proc. Symp. Heavy Metals in Urban Gardens, Washington, DC. 29 Apr. 1982. Univ. of the District of Columbia Ext. Serv., Washington, DC.

Chaney, R.L., S.B. Sterrett, M.C. Morella, and C.A. Lloyd. 1982. Effect of sludge quality and rate, soil pH, and time on heavy metal residues in leafy vegetables. p. 444–458. *In* Proc. Annual Madison Conf. on Applied Research Practices on Municipal Industrial Waste, 5th, Madison, WI. 22–24 Sept. 1982. Univ. of Wisconsin Ext., Madison.

Chaney, R.L., G.S. Stoewsand, A.K. Furr, C.A. Bache, and D.J. Lisk. 1978. Elemental content of tissues of Guinea pigs fed Swiss chard grown on municipal sewage sludge-amended soil. J. Agric. Food Chem. 26:944–997.

Chino, M. 1981. Uptake-transport of toxic metals in rice plants. p. 81–94. *In* K. Kitagishi and I. Yamane (ed.) Heavy metal pollution of soils of Japan. Japan Scientific Soc. Press, Tokyo.

Chino, M., and A. Baba. 1981. The effects of some environmental factors on the partitioning of zinc and cadmium between roots and tops of rice plants. J. Plant Nutr. 3:203–214.

Corey, R.B., L.D. King, C. Lue-Hing, D.S. Fanning, J.J. Street, and J.M. Walker. 1987. Effects of sludge properties on accumulation of trace elements by crops. p. 25–51. *In* A.L. Page et al. (ed.) Land application of sludge: Food chain implications. Lewis Publ., Chelsea, MI.

Davies, B.E. 1992. Trace metals in the environment: Retrospect and prospect. p. 1–17. *In* D.C. Adriano (ed.) Biogeochemistry of trace metals. Lewis Publ., Boca Raton, FL.

Davis, R.D., and C.H. Carlton-Smith. 1984. An investigation into the phytotoxicity of zinc, copper and nickel using sewage sludge of controlled metal content. Environ. Pollut. B8:163–185.

Davis, T.S., J.L. Pyle, J.H. Skillings, and N.D. Danielson. 1981. Uptake of polychlorobiphenyls present in trace amounts from dried municipal sewage sludge through an old field ecosystem. Bull. Environ. Contam. Toxicol. 27:689–694.

deVries, M.P.C., and K.G. Tiller. 1978. Sewage sludge as a soil amendment, with special reference to Cd, Cu, Mn, Ni, Pb, and Zn: Comparison of results from experiments conducted inside and outside a greenhouse. Environ. Pollut. 16:231–240.

Diercxsens, P., D. deWeck, N. Borsinger, B. Rosset, and J. Tarradellas. 1985. Earthworm contamination by PCBs and heavy metals. Chemosphere 14:511–522.

Eisler, R. 1986. Polychlorinated biphenyl hazards to fish, wildlife, and invertebrates: A synoptic review. U.S. Fish Wildlife Serv. Biol. Rep. 85(1.7). U.S. Dep. of Interior, Washington, DC.

Ewers, U., I. Freier, M. Turfeld, A. Brockhaus, I. Hofstetter, W. König, J. Leisner-Saaber, and T. Delschen. 1993. Heavy metals in garden soil and vegetables from private gardens located in lead/zinc smelter area and exposure of gardeners to lead and cadmium. Gesundheitswesen 55:318–325 (in German).

Flanagan, P.R., J.S. McLellan, J. Haist, M.G. Cherian, M.J. Chamberlain, and L.S. Valberg. 1978. Increased dietary cadmium absorption in mice and human subjects with iron deficiency. Gastroenterol. 74:841–846.

Forsyth, D.J., and T.J. Peterle. 1984. Species and age differences in accumulation of ^{36}Cl-DDT by voles and shrews in the field. Environ. Pollut. A33:327–340.

Fox, M.R.S. 1988. Nutritional factors that may influence bioavailability of cadmium. J. Environ. Qual. 17:175–180.

Fox, M.R.S., S.H. Tao, C.L. Stone, and B.E. Fry, Jr. 1984. Effects of zinc, iron, and copper deficiencies on cadmium in tissues of Japanese quail. Environ. Health Perspect. 54:57–65.

Freeman, G.B., J.D. Johnson, J.M. Killinger, S.C. Liao, P.I. Feder, A.O. Davis, M.V. Ruby, R.L. Chaney, S.C. Lovre, and P.D. Bergstrom. 1992. Relative bioavailability of lead from mining waste soil in rats. Fundam. Appl. Toxicol. 19:388–398.

Friberg, L., C.G. Elinder, T. Kjellstrom, and G.F. Nordberg (ed.). 1985. Cadmium and health: A toxicological and epidemiological appraisal. Vol. 1. Exposure, dose, and metabolism. CRC Press, Boca Raton, FL.

Friberg, L., C.G. Elinder, T. Kjellstrom, and G.F. Nordberg (ed.). 1986. Cadmium and health: A toxicological and epidemiological appraisal. Vol. II. Effects and response. CRC Press, Boca Raton, FL.

Fries, G.F. 1996. Ingestion of sludge applied organic chemicals by animals. Sci. Total Environ. 185:93–108.

Gan, D.R., and P.M. Berthouex. 1994. Disappearance and crop uptake of PCBs from sludge-amended farmland. Water Environ. Res. 66:54–69.

Hallberg, L., L. Garby, R. Suwanik, and E. Bjorn-Rasmussen. 1974. Iron absorption from southeast Asian diets. Am. J. Clin. Nutr. 27:826–836.

Hansen, L.G., and R.L. Chaney. 1984. Environmental and food chain effects of the agricultural use of sewage sludges. Rev. Environ. Toxicol. 1:103–172.

Harris, M.R., S.J. Harrison, N.J. Wilson, and N.W. Lepp. 1981. Varietal differences in trace metal partioning by six potato cultivars grown on contaminated soil. p. 399–402. *In* Proc. Int. Conf. Heavy Metals in the Environment, Amsterdam. Sept. 1981. CEP Consultants, Edinburgh, England.

Hassett, J.J., J.E. Miller, and D.E. Koeppe. 1976. Interactions of lead and cadmium on maize root growth and uptake of lead and cadmium by roots. Environ. Pollut. 11:297–302.

Jacobs, R.M., A.O.L. Jones, M.R.S. Fox, and J. Lener. 1983. Effects of dietary zinc, manganese, and copper on tissue accumulation of cadmium by Japanese quail. Proc. Soc. Exp. Biol. Med. 172:34–38.

James, B.R. 1996. The challenge of remediating chromium-contaminated soils. Environ. Sci. Technol. 30:248A–251A.

James, H.M., M.E. Hilburn, and J.A. Blair. 1985. Effects of meals and meal times on uptake of lead from the gastrointestinal tract in humans. Human Toxicol. 4:401–407.

Jing, J., and T. Logan. 1992. Effects of sewage sludge cadmium concentration on chemical extractability and plant uptake. J. Environ. Qual. 21:73–81.

Johnson, N.B., P.H.T. Beckett, and C.J. Waters. 1983. Limits of zinc and copper toxicity from digested sludge applied to agricultural land. p. 75–81. *In* R.D. Davis et al. (ed.) Environmental effects of organic and inorganic contaminants in sewage sludge. D. Reidel Publ., Dordrecht, the Netherlands.

Jones, R. 1983. Zinc and cadmium in lettuce and radish grown in soils collected near electrical transmission (hydro) towers. Water Air Soil Pollut. 19:389–395.

Kobayashi, J. 1978. Pollution by cadmium and the itai-itai disease in Japan. p. 199–260. *In* F.W. Oehme (ed.) Toxicity of heavy metals in the environment. Marcel Dekker, New York.

Kreis, B., P. Edwards, G. Cuendet, and J. Tarradellas. 1987. The dynamics of PCBs between earthworm populations and agricultural soils. Pedobiol. 30:379–388.

Kuo, S. 1986. Concurrent sorption of phosphate and zinc, cadmium, or calcium by a hydrous ferric oxide. Soil Sci. Soc. Am. J. 50:1412–1419.

Li, Y.-M., R.L. Chaney, and A.A. Schneiter. 1995a. Effect of soil chloride level on cadmium concentration in sunflower kernels. Plant Soil 167:275–280.

Li, Y.-M., R.L. Chaney, A.A. Schneiter, and J.F. Miller. 1995b. Genotypic variation in kernel cadmium concentration in sunflower germplasm under varying soil conditions. Crop Sci. 35:137–141.

Li, Y.-M., R.L. Chaney, A.A. Schneiter, and J.F. Miller. 1995c. Combining ability and heterosis estimates for kernel cadmium level in sunflower. Crop Sci. 35:1015–1019.

Li, Y.-M., R.L. Chaney, A.A. Schneiter, J.F. Miller, and C.E. Green. 1994. Progress in developing low cadmium nonoilseed sunflower genotypes. p. 113–121. *In* A.A. Schneiter et al. (ed.) Proc. 16th Sunflower Research Workshop, Fargo, ND. 13–14 Jan. 1994. Natl. Sunflower Assoc., Bismark, ND.

Lillie, R.J., H.C. Cecil, J. Bitman, and G.F. Fries. 1975. Toxicity of certain polychlorinated and polybrominated biphenyls on reproductive efficiency in caged chickens. Poult. Sci. 54:1550–1555.

Lisk, D.J. 1972. Trace metals in soils, plants, and animals. Adv. Agron. 24:267–325.

Logan, T.J., and R.L. Chaney. 1983. Utilization of municipal wastewater and sludges on land: Metals. p. 235–323. *In* A.L. Page et al. (eds.) Proc. Workshop on Utilization of Municipal Wastewater and Sludge on Land, Denver, CO. 23–25 Feb. 1983. Univ. of California, Riverside.

Lutrick, M.C., W.K. Robertson, and J.A. Cornell. 1982. Heavy applications of liquid-digested sludge on three ultisols: II. Effects on mineral uptake and crop yield. J. Environ. Qual. 11:283–287.

Ma, Q.Y., S.J. Traina, T.J. Logan, and J.A. Ryan. 1993. In situ lead immobilization by apatite. Environ. Sci. Technol. 27:1803–1810.

MacDonald, D.W. 1983. Predation on earthworms by terrestrial vertebrates. p. 393–414. *In* J.E. Satchell (ed.) Earthworm ecology: From Darwin to vermiculture. Chapman & Hall, London.

Mahler, R.J., F.T. Bingham, and A.L. Page. 1978. Cadmium-enriched sewage sludge application to acid and calcareous soils: Effect on yield and cadmium uptake by lettuce and chard. J. Environ. Qual. 7:274–280.

Mahler, R.J., and J.A. Ryan. 1988a. Cadmium sulfate application to sludge-amended soils: II. Extraction of Cd, Zn, and Mn from solid phases. Commun. Soil Sci. Plant Anal. 19:1747–1770.

Mahler, R.J., and J.A. Ryan. 1988b. Cadmium sulfate application to sludge-amended soils: III. Relationship between treatment and plant available cadmium, zinc, and manganese. Commun. Soil Sci. Plant Anal. 19:1771–1794.

Mahler, R.J., J.A. Ryan, and T. Reed. 1987. Cadmium sulfate application to sludge-amended soils: I. Effect on yield and cadmium availability to plants. Sci. Total Environ. 67:117–131.

Marks, M.J., J.H. Williams, and C.G. Chumbley. 1980. Field experiments testing the effects of metal-contaminated sewage sludges on some vegetable crops. p. 235–251. *In* Inorganic pollution and agriculture. Min. of Agric. Food, and Fish. Ref. Book 326. HMSO, London.

Marquenie, J.M., J.W. Simmers, and S.H. Kay. 1987. Preliminary assessment of bioaccumulation of metals and organic contaminants at the Times Beach confined disposal site, Buffalo, NY. U.S. Army Corps Eng. Waterways Exp. Stn. Misc. Pap. EL-87-6. U.S. Army, Vicksburg, MS.

McBride, M.B. 1995. Toxic metal accumulation from agricultural use of sludge: Are US-EPA regulations protective. J. Environ. Qual. 24:5–18.

McKenna, I.M., R. L. Chaney, S.H. Tao, R.M. Leach, Jr., and F.M. Williams. 1990. Interactions of plant zinc and plant species on the bioavailability of plant cadmium to Japanese quail fed lettuce and spinach. Environ. Res. 57:73–87.

McKenzie-Parnell, J.M., and G. Eynon. 1987. Effect on New Zealand adults consuming large amounts of cadmium in oysters. Trace Subst. Environ. Health 21:420–430.

McKenzie-Parnell, J.M., T.E. Kjellstrom, R.P. Sharma, and M.F. Robinson. 1988. Unusually high intake and fecal output of cadmium, and fecal output of other trace elements in New Zealand adults consuming dredge oysters. Environ. Res. 46:1–14.

McLaughlin, M.J., L.T. Palmer, K.G. Tiller, T.W. Beech, and M.K. Smart. 1994. Increasing soil salinity causes elevated cadmium concentrations in field-grown potato tubers. J. Environ. Qual. 23:1013–1018.

Miller, J.E., J.J. Hassett, and D.E. Koeppe. 1977. Interactions of lead and cadmium on metal uptake and growth of corn plants. J. Environ. Qual. 6:18–20.

National Research Council. 1996. Use of reclaimed water and sludge in food crop production. Natl. Academy Press, Washington, DC.

Nogawa, K., R. Honda, T. Kido, I. Tsuritani, and Y. Yamada. 1987. Limits to protect people eating cadmium in rice, based on epidemiological studies. Trace Subst. Environ. Health 21:431–439.

O'Connor, G.A., D. Kiehl, G.A. Eiceman, and J.A. Ryan. 1990. Plant uptake of sludge-borne PCBs. J. Environ. Qual. 19:113–118.

Ohlendorf, H.M., J.E. Oldfield, M.K. Sarka, and T.W. Aldrich. 1986. Embryonic mortality and abnormalities of aquatic birds apparent impacts by selenium from irrigation drain water. Sci. Total Environ. 52:49–63.

Page, A.L., T.J. Logan, and J.A. Ryan (ed.) 1989. W-170 Peer Review Committee analysis of the Proposed 503 Rule on sewage sludge. CSRS Technical Committee W-170, Univ. of California, Riverside.

Parker, D.R., R.L. Chaney, and W.A. Norvell. 1995. Equilibrium computer models: Applications to plant nutrition research. p. 163–200. *In* R.H. Loeppert et al. (ed.). Chemical equilibrium and reaction models. SSSA Spec. Publ. 42. SSSA, Madison, WI.

Peakall, D.B. 1986. Accumulation and effects on birds. p. 31–47. *In* J.S. Waid (ed.) PCBs and the Environment. CRC Press, Boca Raton, FL.

Pedersen, B., and B.O. Eggum. 1983. The influence of milling on the nutritive value of flour from cereal grains. 4. Rice. Qual. Plant Foods Human Nutr. 33:267–278.

Poole, D.B.R. 1981. Implications of applying copper rich pig slurry to grassland: Effects on the health of grazing sheep. p. 273–286. *In* P. L'Hermite and J. Dehandtschutter (ed.) Copper in animal wastes and sewage sludge. Reidel Publ., Boston, MA.

Reid, R.L., and D.J. Horvath, 1980. Soil chemistry and mineral problems in farm livestock: A review. Anim. Reed Sci. Technol. 5:95–167.

Ruby, M.V., A. Davis, T.E. Link, R. Schoof, R.L. Chaney, G.B. Freeman, and P.D. Bergstrom. 1993. Development of an *in vitro* screening test to evaluate the *in vivo* solubility of ingested mine-waste lead. Environ. Sci. Technol. 27:2870–2877.

Ryan, J.A., and R.L. Chaney. 1998. Issues of risk assessment and its utility in development of soil standards: The 503 methodology as an example. p. 393–414. *In* R. Prost (ed.) Proc. Int. Symp. on Biogeochemistry of Trace Elements, Paris. 15–19 May 1995. Colloque 85. INRA Editions, Paris.

Ryan, J.A., H.R. Pahren, and J.B. Lucas. 1982. Controlling cadmium in the human food chain: A rationale based on health effects. Environ. Res. 28:241–302.

Sarasua, S.M., M.A. McGeehin, F.L. Stallings, G.J. Terracciano, R.W. Amler, J.N. Logue, and J.M. Fox. 1995. Final Rep. Technical Assistance to the Pennsylvania Dep. of Health. Biologic indicators of exposure to cadmium and lead. Palmerton, PA. Part II. May 1995. Agency for Toxic Substances and Disease Registry, U.S. Dep. of Health and Human Services, Atlanta, GA.

Shaikh, Z.A., and J.C. Smith. 1980. Metabolism of orally ingested cadmium in humans. p. 569–574. *In* B. Holmstedt et al. (ed.) Mechanisms of toxicity and hazard evaluation. Elsevier/North-Holland Biomedical Press, Amsterdam.

Sheaffer, C.C., A.M. Decker, R.L. Chaney, G.C. Stanton, and D.C. Wolf. 1981. Soil temperature and sewage sludge effects on plant and soil properties. EPA-600/S2-81-069. Natl. Tech. Inf. Serv., Springfield, VA (NTIS:PB 81-191,199).

Sharma, R.P., T. Kjellstrom, and J.M. McKenzie. 1983. Cadmium in blood and urine among smokers and non-smokers with high cadmium intake via food. Toxicology 29:163–171.

Singh, S.S. 1981. Uptake of cadmium by lettuce as influenced by its addition to a soil as inorganic forms or in sewage sludge. Can. J. Soil Sci. 61:19–28.

Smolders, E., and M.J. McLaughlin. 1996. Chloride increases Cd uptake in Swiss chard in a resin-buffered nutrient solution. Soil Sci. Soc. Am. J. 60:1443–1447.

Sommers, L.E. 1980. Toxic metals in agricultural crops. p. 105–140. *In* G. Bitton et al. (ed.) Sludge: Health risks of land application. Ann Arbor Science Publ., Ann Arbor, MI.

Spurgeon, D.J., and S.P. Hopkin. 1996. Risk assessment of the threat of secondary poisoning by metals to predators of earthworms in the vicinity of a primary smelting works. Sci. Total Environ. 187:167–183.

Sterrett, S.B., R.L. Chaney, C.E. Hirsch, and H.W. Mielke. 1996. Influence of amendments on yield and heavy metal accumulation of lettuce grown in urban garden soils. Environ. Geochem. Health 18:135–142.

Strehlow, C.D., and D. Barltrop. 1988. The Shipham report: An investigation into cadmium concentrations and its implications for human health: 6. Health studies. Sci. Total Environ. 75:101–133.

Suttle, N.F., P. Abrahams, and I. Thornton. 1984. The role of a soil × dietary sulphur interaction in the impairment of copper absorption by ingested soil in sheep. J. Agric. Sci. 103:81–86.

Suttle, N.F., B.J. Alloway, and I. Thornton. 1975. An effect of soil ingestion on the utilization of dietary copper by sheep. J. Agric. Sci. 84:249–254.

Takijima, Y., and F. Katsumi. 1973. Cadmium contamination of soils and rice plants caused by zinc mining: l. Production of high-cadmium rice on the paddy fields in lower reaches of the mine station. Soil Sci. Plant Nutr. 19:29–38.

Tarradellas, J., P. Diercxsens, and M.B. Bouche. 1982. Methods of extraction and analysis of PCBs from earthworms. Int. J. Environ. Anal. Chem. 13:55–67.

Tsuchiya, K. (ed). 1978. Cadmium studies in Japan: A review. Elsevier/North-Holland Biomedical Press, New York.

Underwood, E.J. 1977. Trace elements in human and animal nutrition. 4th ed. Academic Press, New York.

United Nations ACC/SCN. 1992. Second report on the World Nutrition Situation. Vol. 1. Global and Regional Results. United Nations, Geneva, Switzerland.

U.S. Environmental Protection Agency. 1989a. Development of risk assessment methodology for land application and distribution and marketing of municipal sludge. EPA/600/6-89/001. U.S. EPA, Washington, DC.

U.S. Environmental Protection Agency. 1989b. Standards for the disposal of sewage sludge. Fed. Reg. 54(23):5746–5902.

U.S. Environmental Protection Agency. 1990. National sewage sludge survey. Availability of information and data, and expected impacts on proposed regulations: Proposed rule. Fed. Reg. 55(218):472448–47283.

U.S. Environmental Protection Agency. 1993. 40 CFR Part 257 et al. Standards for the use or disposal of sewage sludge: Final rules. Fed. Reg. 58(32):9248–9415.

Webber, M.D., Y.K. Soon, T.E. Bates, and A.V. Haq. 1981. Copper toxicity resulting from land application of sewage sludge. p. 117–135. *In* P. L'Hermite and J. Dehandtschutter (ed.) Copper in animal wastes and sewage sludge. Reidel Publ., Boston.

Weitzman, M., A. Aschengrau, D. Bellinger, R. Jones, J.S. Hamlin, and A. Beiser. 1993. Lead-contaminated soil abatement and urban children's blood lead levels. J. Am. Med. Assoc. 269:1647–1654.

White, M.C., and R.L. Chaney. 1980. Zinc, Cd and Mn uptake by soybean from two Zn- and Cd-amended coastal plain soils. Soil Sci. Soc. Am. J. 44:308–313.

Wild, S.R., and K.C. Jones. 1992. Polynuclear aromatic hydrocarbon uptake by carrots grown in sludge amended soils. J. Environ. Qual. 21:217–225.

Wolnik, K.A., F.L. Fricke, S.G. Capar, G.L. Braude, M.W. Meyer, R.D. Satzger, and E. Bonnin. 1983a. Elements in major raw agricultural crops in the United States: l. Cadmium and lead in lettuce, peanuts, potatoes, soybeans, sweet corn, and wheat. J. Agric. Food Chem. 31:1240–1244.

Wolnik, K.A., F.L. Fricke, S.G. Capar, G.L. Braude, M.W. Meyer, R.D. Satzger, and R.W. Kuennen. 1983b. Elements in major raw agricultural crops in the United States: 2. Other elements in lettuce, peanuts, potatoes, soybeans, sweet corn, and wheat. J. Agric. Food Chem. 31:1244–1249.

Wolnik, K.A., F.L. Fricke, S.G. Capar, M.W. Meyer, R.D. Satzger, E. Bonnin, and C.M. Gaston. 1985. Elements in raw agricultural crops in the United States: 3. Cadmium, lead, and eleven other elements in carrots, field corn, onions, rice, spinach, and tomatoes. J. Agric. Food Chem. 33:807–811.

Yang, G., S. Wang, R. Zhou, and S. Sun. 1983. Endemic selenium intoxication of humans in China. Am. J. Clin. Nutr. 37:872–881.

Zhu, Y.M., D.F. Berry, and D.C. Martens. 1991. Copper availability in two soils amended with eleven annual applications of copper-enriched hog manure. Commun. Soil Sci. Plant Anal. 22:769–783.

11 Nontarget Ecological Effects of Plant, Microbial, and Chemical Introductions to Terrestrial Systems[1]

Lidia S. Watrud and Ramon J. Seidler

U.S. Environmental Protection Agency
National Health and Ecological Effects Research Laboratory
Corvallis, Oregon

11–1 INTRODUCTION

There is a long published history of the nontarget effects of intentional introductions of plants, chemicals, and microbes to natural, agronomic, and forest ecosystems (Williams, 1980; Williamson & Brown, 1986; Miller, 1990; Edwards et al., 1992; Pell et al., 1992; Pipe, 1992; Addison, 1993; Visser et al., 1994). In this chapter, an overview is presented that focuses primarily on reported nontarget ecological effects of diverse nonengineered and engineered plants and microbes and chemical agents on soil biota. Effects discussed include changes in the numbers, diversity and activities of soil foodweb components such as bacteria, fungi, nematodes, and litter microarthropods. One of the unique discoveries that is noted in this review, is that both chemicals and biological products may induce similar effects on ecological processes and species composition.

The presentation of the reported nontarget effects of biologicals or chemicals on aboveground terrestrial ecosystem components is limited to effects of microbial insecticides on nontarget insects (Miller, 1990; James et al., 1993) and to the effects of agricultural chemicals on plant community composition (Gange & Brown, 1991). Escapes of selected horticultural and agronomic species are presented as examples of scenarios in which intentionally introduced plants have become invasive weeds (Williams, 1980; Mooney et al., 1986; Rissler & Mellon, 1993). Several case histories, highlighting the potential effects of engineered plants (Donegan et al., 1995, 1996, 1997; Vierheilig et al., 1995) and genetically engineered microbes (Bej et al., 1991; Seidler, 1992; Crawford et al., 1993; Doyle et al., 1995) on soil biota, also serve to illustrate the types of methodologies one might use to identify either biological or chemical effects. In contrast to the bio-

[1] This document has been subjected to the Agency's peer and administrative review and has been approved for publication. Mention of trade names or commercial products does not constitute endorsement or recommendation for use.

logicals, many of which are prototype products, the agricultural chemical examples include an array of commercial herbicides, fungicides, and insecticides that have been in long-term use. Findings and questions raised from the reported nontarget ecological effects of both the biologicals and chemicals highlight a number of research needs. These include development and use of molecular diagnostic, community, and trophic level testing methods and an increased emphasis on longer term field studies.

11–2 NONTARGET EFFECTS OF PLANT INTRODUCTIONS

The effects of introducing both nonengineered plants and engineered plants are discussed in the following section. Table 11–1 in Section 11–2.1 summarizes examples of escapes of both agronomic and horticultural species, which due in large part to their invasive nature, have altered community composition by reducing native plant and animal diversity in the affected habitats. Effects of genetically engineered plants on soil biota are summarized in Table 11–2 in Section 11–2.5.

11–2.1 Nonengineered Agronomic Species

Many of the world's major staple food crops such as corn (*Zea mays* L.), wheat (*Triticum aestivum* L.), potatoes (*Solanum tuberosum* L.) and barley (*Hordeum vulgare* L.) have been successfully introduced and contained in locales distant from their evolutionary centers of origin (Kloppenburg, 1988). Agronomic escapes or hybrids formed between crop and weed or native species, have however, occasionally become problematic or may be anticipated to become so (Dale, 1994; De Wet & Harlan, 1975; Williamson & Brown, 1986; Ellstrand, 1988; Keeler, 1989; Ellstrand & Hoffman, 1990; Darmency, 1992, 1994; Till-Bottraud et al., 1992; Raybould & Gray, 1993, 1994; Gliddon, 1994; Kareiva et al., 1994; Purrington & Bergelson, 1995). Among the reasons often cited for the apparent containment of agronomic cultivars of crops are physical requirements for water, fertilizer, and pest control. Biological constraints such as limited viability of pollen, incompatibility with other species, and a lack of viability or fertility of hybrid seeds also have generally limited the spread and persistence of *escapes* of either the introduced agronomic cultivars or their hybrids. Within

Table 11–1. Intentional nonengineered plant introductions that have become weeds.†

Plant	Use–effects
Johnson grass	Introduced as forage crop; widespread weed
Bermuda grass	Introduced for use as forage, turfgrass; widespread weed
Kudzu	Introduced as forage crop; serious weed in southeastern USA
Purple loosestrife	Horticultural escape; invasive–destructive in wetlands
Water hyacinth	Horticultural escape; clogs waterways and lakes
Melaleuca	Horticultural introduction from Australia; noxious, invasive pest in Everglades; no natural enemies

† Sources: Williams, 1980; Rissler & Mellon, 1993; U.S. Congress, Office of Technology Assessment, 1993.

agronomic situations, normal mechanical and chemical cultural practices for weed control have further acted to reduce the potential impact of escapes or volunteer plants. In the continental USA, hybrid progeny formation is considered more of a potential concern between cultivated and wild species of plants such as squash (*Cucurbita* sp.), sunflower (*Helianthus* sp.), or radish (*Raphanus* sp.), which have native or weedy relatives, than crops such as corn, soybeans [*Glycine max* (L.) Merr.], potatoes or tomatoes (*Lycopersicon lycopersicum* (L.) Karsten, which do not have close wild relatives with which they are likely to hybridize (Kirkpatrick & Wilson, 1988; Klinger et al.,1991, 1992; Klinger & Ellstrand, 1994; Arias & Rieseberg, 1994). In the USA, hybrids between cultivated rice (*Oryza sativa* L.) and wild rice (*Zizania aquatica* L.) have been reported (Langevin et al., 1990). In France, hybrids between sugar beets (*Beta vulgaris* L.) and wild beets (*Beta* sp.) have been reported (Boudry et. al., 1993). A number of plants intended as intentional agronomic introductions as forage crops, for example, kudzu [*Pueraria lobata* (Willd.) Owhl], Johnson grass (*Sorghum halepense* L.), Bermuda grass [*Cynodon dactylon* (L.) Pers], crab grass [*Digitaria sanguinalis* (L.) Scop.], and reed canary grass (*Phalaris arundinacea* L.), have had widespread adverse effects in diverse habitats ranging from agronomic fields, managed grasslands or residential turf to riparian areas (Williams, 1980; U.S. Congress, Office of Technology Assessment, 1993; Mooney et al., 1986; Rissler & Mellon, 1993). Johnson grass is itself the result of hybridization between cultivated sorghum [*S. bicolor* (L.) Moench] and a southeast Asian relative [*S. propinquam* Kunth (Hitchcock); Paterson et al., 1995]. In managed agricultural systems, the presence of escapes or volunteers has usually translated into a need for more chemical or mechanical weed control measures to limit yield losses due to competition between the weeds and the crop plants. In typically less intensively managed or nonagronomic situations, such as natural areas, parks, highway or railroad rights of way, residential, and commercial areas, the escapes or their hybrids with existing weeds may invade and displace more desirable native or introduced plantings; there too as in the agronomic situations, mechanical or chemical weed control measures may become necessary to keep the escapes or their hybrid progeny in check.

An emerging area of interest for both nonengineered and genetically engineered plants, is phytoremediation, the use of plants to clean up polluted soils (Anderson et al., 1993; Entry et al., 1995; Salt et al., 1995; Schnoor et al., 1995). In addition to the potential advantage of lower costs associated with in situ soil remediation with plant species selected or developed for phytoremediation, potential risks to herbivores of above- and belowground plant parts and to the soil biota that comprise soil foodwebs also need to be considered.

11–2.2 Nonengineered Horticultural Species

As with the escapes for agronomic introductions in terrestrial habitats cited above, a disturbing number of horticultural species have become invasive pests of wetlands, rangelands, and other unique habitats. Among the most notorious examples of such horticultural escapes into wetlands and waterways are purple loosestrife (*Lythrum salicaria* L.) and water hyacinth [*Eichornia crassipes*

(Mart.) Solms]; each has the capability to quickly take over and reduce the diversity of aquatic habitats. Melaleuca (*Melaleuca quinqenervia* S.T. Blake), an ornamental shrub introduced into Florida from Australia, is rapidly invading large areas of the Everglades. Estimates of its rate of spread are as high as 50 acres d^{-1} (Morganthaler, 1993). Factors cited in its successful invasion are the absence of natural disease or insect pests to keep it in check, and its ability to out compete native plants. Mechanical measures to control the spread of the plants are complicated by the allergenic responses the shrub may elicit in susceptible individuals upon touch or upon breathing fumes released in efforts to control it by fire. Ingestion of castor bean (*Riccinus communis* L.) escapes, a cultivated oil crop and ornamental, can cause poisoning in people and animals (Whitson, 1992). Dalmatian toadflax (*Linaria genistifolia* L.), Wyeth and silver lupines [*Lupinus wyethi* (S. Wats.) and *L. argenteus* (Pursh)], creeping buttercup (*Ranunculus repens*), foxglove (*Digitalis purpurea* L.), and Tansy ragwort (*Tanacetum vulgare* L.), are examples of horticultural species that escaped from garden cultivation, and became serious and noxious weeds, particularly in the western USA, because they can poison cattle or other herbivores that may graze on them. Other plants that have escaped from horticultural cultivation in the west and that have displaced native species, include Scotch broom (*Cytisus scoparius* L.) and Baby's breath (*Gypsophila paniculata* L.). For comprehensive lists of intentional and unintentional plant introductions that have become weeds, readers may refer to several excellent reviews: Williams (1980); Williamson and Brown (1986); Mooney et al. (1986); Whitson (1992); U.S. Congress Office of Technology Assessment (1993); and Rissler and Mellon (1993).

11–2.3 Genetically Engineered Plants

In the examples given above, nongenetically engineered plants were cited to have caused adverse ecological effects outside of their intended habitats. As new generations of genetically engineered plants are being developed, field tested, and commercialized, much excitement is being generated at the prospects of improving crop protection and crop quality, with potential concurrent decreases in the use of chemicals (Gasser & Fraley, 1989; Fraley, 1992; Watrud et al., 1996); however, unlike their nonengineered counterparts that may have had built-in genetic constraints because of their susceptibility to various physical, chemical, or biological stresses, the new generation of engineered plants may have built in competitive or survival advantages, should they escape from cultivation. Depending on the nature of the engineered traits they now have, these plants may be more tolerant to one or more factors including herbicides or other pesticides, insect pests, plant diseases, cold temperatures, heavy metals, high salinity, drought, low or high pH, or to other environmental or anthropogenic stresses. If these plants have the capacity to exchange genetic information with noncrop species, such as weeds, the potential exists to create weeds even more difficult to control, since they may now enjoy an additional competitive advantage, such as pest or pesticide resistance. The potential adverse effects of escapes of engineered plants or hybridization of engineered genes into native or weedy plant populations include reduced diversity in plant community composition and increased needs for chemical or mechanical weed control (Keeler, 1989; Rissler & Mellon, 1993). Biological approaches and considerations aimed at reducing the risks of introducing nonindigenous species have been proposed recently (Ruesink et al., 1995).

Although >2000 field tests have now been performed world-wide for various crops expressing diverse genetically engineered traits (Krattiger, 1994), little information is available in the peer reviewed literature on the potential ecological effects of genetically engineered plants (Rissler & Mellon, 1993; Mellon & Rissler, 1995; Purrington & Bergelson, 1995). The major purpose of the field tests to date, has generally been to determine the efficacy of the particular trait introduced into the engineered plants. In this section, several case histories will be presented to illustrate methodologies that have been used to determine potential ecological effects of genetically engineered pesticidal plants on soil biota. For an overview of above- and belowground concerns that have been raised regarding the release of transgenic plants, readers are referred to the Symposium Issue of the *Journal of Molecular Ecology* (Seidler & Levin, 1994). Issues highlighted there include gene flow from the transgenic plants to wild relatives that could result in weediness and changes to plant community composition and potential nontarget effects of transgenic plants on soil biota. Concerns regarding the development of resistance in insect pest populations, to active ingredients in pesticidal plants and pesticidal microbes are discussed below in sections 2.5.1 and 3.1.1.

11–2.4 Methods to Compare Effects of Plant Introductions on Soil Biota

Microbiological, zoological, molecular, and biochemical methods can and have been used to determine the potential effects of engineered plants on soil biota. Many of the methods being used to identify potential effects of transgenic plants also have been used to look at the effects of chemicals and microbials. The types of methodologies often used to determine ecological effects of chemicals or biologicals are described in the following general references: for determinations of soil microbial populations (Bourquin & Seidler, 1986; Levin et al., 1992; Stotsky et al., 1993); for isolation of soil fauna such as nematodes, protozoans and litter microarthropods, soil respiration, enzymatic activities, estimates of microbial biomass, and determinations of symbiotic activity (Doran et al., 1994; Carter, 1993); and for extraction of soil deoxyribonucleic acid (DNA) and polymerase chain reaction (PCR) technology (Steffan et al., 1988; Picard et. al., 1992; Porteous & Armstrong, 1991, 1993; Tsai & Olson, 1991; Porteous et al., 1994). Several references providing good overviews of comparative methodologies for the assessment of chemical effects are those by Zelles et al. (1985); Edwards (1988); Edwards & Bater (1990); Schuster & Schröder (1990); Pipe (1992); Wardle et al. (1995).

As readers examine those references, commonalities in the methods and endpoints used to evaluate the effects of biological and chemical stressors will be readily apparent. More surprising, however, may be the frequency with which certain indicators such as numbers of bacteria, algae or Collembola, soil respiration, or the functioning of plant-microbial symbioses, are inhibited by a variety of different chemical and/or biological treatments. Common effects noted with a variety of different biological and chemical stressors will be discussed below.

11–2.5 Results of Transgenic Plant Studies

A general summary of some of the key nontarget ecological effects that have been noted with engineered pesticidal plants is presented in Table 11–2. Four examples of pesticidal plants are discussed below; three of them have been engineered for insect control, one of them has been engineered for plant disease control. The studies illustrate the use of methods and indicators (e.g., diversity and abundance of soil biota and establishment of symbioses), that may be useful to determine potential nontarget effects in soil ecosystems.

11–2.5.1 Cotton with *Bacillus thuringiensis* Subspecies *kurstaki* Gene

Using cotton (*Gossypium hirsutum* L.) expressing the *Bacillus thuringiensis* subsp. *kurstaki* delta-endotoxin gene (Perlak et al., 1990), a broad array of biological techniques including microbiological, invertebrate zoological, molecular, and biochemical, were used to study the effects of decomposing leaf litter on soil food web components in a soil microcosm (Donegan et al., 1995). Among the noteworthy results of that study were that transient increases in populations of bacteria and fungi were found to be associated with only two of three transgenic cotton lines that were tested, while purified microbial toxin did not produce any change in bacterial populations. These results suggest that rather than the engineered gene product (i.e., the pesticidal delta-endotoxin expressed in the engineered plants), somaclonal variation (variability associated with the processes of plant tissue culture, transformation, and regeneration), may have produced a change(s) in the cotton that resulted in the increases in microbial populations. Another possible explanation is that the site of insertion of the *Bacillus*

Table 11–2. Nontarget effects of genetically engineered plants on soil biota and plants

Plant–engineered trait	Use–effects	References
Cotton/*B.t. kurstaki*	Insecticidal cotton; controls lepidopteran pests Transient increases in numbers of soil bacteria and fungi Change in species composition of soil bacteria	Donegan et al., 1995
Tobacco–protease inhibitor	Insecticidal tobacco model Decreased numbers of Collembola Increased number of nematodes	Donegan et al., 1997
Potato–*B.t. tenebrionis*	Insecticidal potato; controls Colorado potato beetle Longer viability of plants associated with increased infection of potato tubers/ *Verticillium dahliae*	Donegan et al., 1996
Tobacco–pathogenesis related proteins	Disease resistant tobacco model Delay in onset of Arbuscular Mycorrhizal infection Decreased level of Arbuscular Mycorrhizal infection	Vierheilig et al., 1995

thuringiensis subsp. *kurstaki* DNA into the cotton DNA may have caused changes in the expression of other genes. A position or pleiotropic effect unrelated to expression of the insecticidal delta-endotoxin, may have caused changes in the cotton resulting in the transient increase in microbial populations. A lack of genetic homogeneity in the lines tested (i.e., the lines possibly may not have been completely isogenic), also remains a possibility. If that latter possibility is true, factor(s) produced by the two lines that brought about the increases in populations of both soil bacteria and fungi, may have directly or indirectly caused the observed increases. Perhaps even more significant than the transient increases in numbers of bacteria and fungi were the findings by Donegan et al. (1995), which indicated that changes in the species composition of the bacterial community remained even after the transient increase in population numbers leveled off. The changes in microbial diversity were established using selective growth media and by analysis of substrate usage patterns (Garland & Mills, 1991). Changes in microbial diversity in the soil microcosms were further documented by the use of molecular ecology methods. These included sequentially, extraction of soil DNA, polymerase chain reaction amplification using a 16S rRNA eubacterial specific primer sequence, restriction endonuclease digestion of the amplified rDNA fragments, and agarose gel electrophoresis. Using these techniques, different DNA fragment banding patterns or *fingerprints* were found to be exhibited by microbial populations following exposure to transgenic or parental plant litter fragments (Donegan et al., 1995). In addition to the transient changes in the microbial populations noted above for two transgenic lines of cotton, numerous agricultural chemicals also have been reported to produce transient changes in populations, biomass, or enzymatic activities of soil microbiota (Tables 11–2, 11–3, 11–4, 11–5, 11–6, 11–7, and 11–8). Questions that arise when such changes have been observed are: (i) whether such changes are persistent and (ii) whether there are potential downstream ecological effects of such changes, regardless of whether the initial changes are persistent or not. An additional concern that has been raised for engineered plants, particularly those containing pesticidal genes from *B. thuringiensis* subsp. *kurstaki*, is development of resistance in target insect populations. Strategies to delay resistance development are therefore of interest and have been proposed (USDA, 1992; Mc Gaughey & Whalon, 1992; Alstad & Andow, 1995).

11–2.5.2 Tobacco with Protease Inhibitor Gene

In studies using leaf litter from pesticidal tobacco (*Nicotiana tabacum* L.) engineered to express a protease inhibitor of plant origin (Narváez-Vásquez et al., 1992), changes in the numbers of two different types of soil biota were observed by Donegan et al. (1997). The numbers of Collembola, representative of a group of litter decomposing microarthropods, decreased in soil surrounding litterbags containing the pesticidal tobacco, whereas the number of microbial feeding nematodes increased. Whether such changes in numbers of detritovores (e.g., Collembola) or nematodes could impact the levels of nutrients in soil that are available to the soil biota or that may directly or indirectly affect plant nutrient status, are areas for future research.

11–2.5.3 Potato with *Bacillus thuringiensis* Supspecies *tenebrionis* Gene

In a third study of transgenic plants (Donegan et al., 1996), phylloplane and soil microorganisms and plant diseases were monitored in field plots of potatoes (*Solanum tuberosum*) that had been engineered to express the delta-endotoxin gene from *B. thuringiensis* subsp. *tenebrionis* to resist the Colorado potato beetle (*Leptinotarsa decimlineata*). In that study, no changes were observed in the numbers of total bacterial or fungi. Furthermore, no consistent changes were observed in the species diversity of phyllosphere microorganisms; however, a higher incidence of infection with a soil-borne fungus, *Verticillium dahliae* Kleb., was found in the vascular bundles of transgenic potato tubers. The increased incidence of infection in the transgenic tubers is perhaps explained by the longer viability and, consequently, longer period of exposure of the transgenic plants to *V. dahliae)*. Use of the pesticidal potato plants also resulted in an increase in the number of beneficial predatory insects (Gary Reed, 1996, personal communication). Data from longer-term field studies carried out in multiple locations over multiyear periods would be needed to clarify short-term and long-term potential ecological effects of these and other pest tolerant transgenic crops.

11–2.5.4 Tobacco with Pathogenesis Related Proteins

In studies utilizing tobacco as a model species to test genes for their potential utility in resisting fungal root pathogens, a nontarget effect of certain transformant lines was demonstrated on a fungal symbiont of higher plants (Vierheilig et al., 1995). Although the majority of transformant lines constitutively expressing various pathogenesis-related (PR) proteins derived from tobacco or cucumber (*Cucumis sativa*) did not affect colonization by the arbuscular mycorrhizal fungus *Glomus mosseae* (Nicol. & Gerd.) Gerd. & Trappe, one line constitutively expressing the acidic isoform of tobacco PR-2 did inhibit both the onset and degree of mycorrhizal colonization. As in the example cited above for a potential nontarget effect of two lines of cotton expressing the *B. thuringiensis* subsp. *kurstaki* gene on soil biota (Donegan et. al., 1995), caution needs to be exercised in generalizing observations for given types of engineered plants. That is, causes unrelated to the activity of an engineered gene may result in undesired ecological consequences. Potential causes of adverse effects that may need to be explored include somaclonal variation, position effects related to the site of insertion of the gene, or unexpected pleiotropic effects of introduced genes.

11–3 NONTARGET EFFECTS OF MICROBIAL INTRODUCTIONS

General summaries of the nontarget effects of both nonengineered and engineered microbes are provided in Tables 11–3 through 11–5, and in the reviews by Seidler (1992), Stotsky et al. (1993), and Doyle et al. (1995). Specific examples discussed below include three registered nonengineered bacterial or fungal pesticides; two engineered and one nonengineered model bacterial bioremediation agents; two engineered model bacterial biomass conversion agents and three nonengineered mycorrhizal fungi used as biofertilizers. Summaries of the key effects of nongenetically engineered bacteria and fungi respectively, are found in Tables 11–3 and 11–4. The effects of genetically engineered bacteria are summarized in Table 11–5.

11–3.1 Microbial Pest Control Agents

The cases below include examples of both registered and experimental bacterial or fungal pesticides designed for insect, weed, or disease control. Nontarget effects on beneficial insects, plants, and mycorrhizal symbionts are discussed.

11–3.1.1 *Bacillus thuringiensis* Subspecies *kurstaki*

Bacillus thuringiensis subsp. *kurstaki, (Btk),* registered as a microbial insecticide since 1961 (W. Schneider, 1995, personal communication), represents a model type of microbial pesticide. This microorganism works against its target pests, the lepidopteran insects, with a biological specificity that is generally much higher than most chemical insecticides; it is widely recognized as safer to humans, wildlife, plants, and traditional nontarget species such as earthworms (*Lumbricus* and *Eisenia* sp), honeybees (*Apis mellifera*), *Daphnia*, and others, than most chemical insecticides. *Btk* also has served as the source of genes that subsequently have been introduced into plant colonizing microorganisms (Watrud et al., 1985; Obukowicz et al., 1986a,b, 1987; Watrud, 1987; Fischhoff & Watrud, 1988), and into plants as diverse as tomato, cotton, and corn (Fischhoff et al., 1988; Perlak et al., 1990; Umbeck et al., 1992; Koziel et al., 1993). Used alone, or in conjunction with Integrated Pest Management (IPM) practices such as crop rotation, scouting, pheromone traps, minimal chemical useage, *Btk* microbial and plant products have attracted considerable commercial attention. In addition to agronomic use for the control of corn earworm (*Heliothis zea*), *Btk* is used to control garden vegetable pests such as cabbage loopers (*Trichoplusia ni*) and tomato hornworms (*Manduca quinquemaculata*). It also is used by local municipalities, states, and federal agencies to control infestations of gypsy moths (*Lymantria dispar*) and spruce budworms (*Choristoneura occidentalis).* In spite of its long-term and widespread use, some of the nontarget effects of *Btk* have yet to be fully recognized. For example, the biomass of some beneficial nontarget insects (i.e., predators of lepidopteran pests), was decreased for a period of up to 3 yr, following the spraying of Christmas tree plantations with *Btk*, to control spruce budworm infestations (Miller, 1990). Another beneficial nontarget, the cinnabar moth (*Tyria jacobaeae*), introduced to feed on tansy ragwort (*Senecio jacobea*), a weed noxious to cattle and horses, similarly was inhibited by applications of *Btk* intended to control forest insect pests such as the gypsy moth and spruce budworm (James, et al., 1993).

In addition to the nontarget effects noted above, a concern associated with projected increased uses of both microbial formulations of *B. thuringiensis* and pesticidal plants expressing the active agents derived from this and related microbials, is accelerated development of resistance in populations of target pests (Mc Gaughey, 1985; Mc Gaughey & Beeman, 1988; Mc Gaughey & Johnson, 1992; Gould et al., 1992; Tabashnik et al., 1990; Tabashnik, 1994). Strategies to delay

Table 11–3. Nontarget effects of nongenetically engineered bacteria.

Bacterium	Use–effects	References
Bacillus thuringensis subsp. *kurstaki*	Biocontrol agent for Lepidopteran insect pests	
	Inhibition of Cinnabar Moth, a beneficial biocontrol insect	James et al., 1993
	Decreased biomass in nontarget Lepidopterans and insect predators of Lepidopterans; multiyear effect	Miller, 1990
	Inhibition of ammonification and nitrification	Addison, 1993
	Enhanced cellulose decay	
	Increased substrate induced respiration	Visser et al., 1994
	Variable effects (+/–) on bacterial population numbers	
	Binding of *B.t.* toxin to clays on soil and persistence of bioactivity	Stotsky, 1986
	Resistance development	Mc Gaughey, 1985; Mc Gaughey & Beeman, 1988; Tabashnik et al., 1990, 1991; Gould et al., 1992; Mc Gaughey & Johnson, 1992; Tabashnik, 1994
Pseudomonas SR3	Demonstrated efficacy for biodegradation of Pentachlorophenol	
	Inhibition of nodule number and size in *Lotus corniculatus*	Pfender et al., 1995
	Inhibition of substrate induced respiration	

the development of resistance are therefore being developed to prolong the useful life of the biological pesticides and to minimize potential adverse ecological effects (Tabashnik et al., 1991; USDA, 1992; Mc Gaughey & Whalon, 1992; Alstad & Andow, 1995).

11–3.1.2 *Beauveria bassiana* (Bals.) Vuill.

The filamentous entomopathogenic fungus *Beauveria bassiana*, is used for biocontrol of insects in many parts of the world, was registered in the USA in 1995 (W. Schneider, 1995, personal communication); however, unlike *Btk*, which is largely specific to lepidopteran insects, *B. bassiana* has a much broader host range. It can kill insects belonging to a number of different orders, including homopterans (e.g., aphids), orthopterans (e.g., grasshoppers), and coleopterans (e.g., Colorado potato beetle, *Leptinotarsa decimlineata*; Goettel et al., 1990). In a recent study in which *B. bassiana* was being used to control aphid pests of plants, James et al. (1995), observed a nontarget inhibitory effect of the biocontrol agent, (i.e., mortality of a beneficial predatory insect, the lady-bird beetle, *Hippodamia convergens*). Nontarget effects such as these noted above suggest that, as with chemicals, the environmental and agronomic risks as well as the benefits of microbial pesticides may need to be evaluated.

11–3.1.3 *Colletotrichum gloeosporioides* (Penz.) Penz. & Sacc.

The filamentous fungus *Colletotrichum gloeosporioides*, a biocontrol agent registered in the USA since 1982, (W. Schneider, 1995, personal communication), controls joint vetch (*Aeschynomene virginica* L.), a weed found in rice (*Oryza sativa*) fields; however, the results of laboratory studies point to a potential source of concern for this registered mycoherbicide, as it was demonstrated that the fungus could mate and exchange genetic information with other isolates of *Colletotrichum* that are pathogens on soybean [*Glycine max* (L.) Merr.], pecan (*Carya* sp.), apples (*Malus* sp.), and other crops (Cisar et al., 1994). Although evidence for genetic exchange between the biocontrol agent and pathogenic isolates has not been observed in the field, the possibility of introduced biocontrol agents developing broader host ranges, e.g., for nontarget crop and native species is a concern. Similarly, increases in the competitiveness of indigenous pathogens that may acquire traits such as resistance to fungicides from introduced biocontrol agents, also is a concern. Safety measures that should accordingly be addressed with biocontrol agents include: how to mitigate their viability, persistence, or dissemination and how to limit genetic exchange between the introduced agent and compatible pathogenic isolates.

11–3.1.4 Effects of Microbial Pesticides on Plant Growth and Symbionts

Several microbes proposed for biocontrol of plant diseases or for stimulation of plant growth, may have deleterious effects on the symbioses of plants with mycorrhizal fungi or on plants per se (Linderman et al., 1991; reviewed in Seidler, 1992). For example, the fungi *Trichoderma harzianum* Rifai Wt-6, *Talaromyces flavus* (Kloecher) Stolk & Samson, and *Gliocladium virens* Miller, Giddens, & Foster inhibited ectomycorrhizal formation on Douglas fir [*Pseudotsuga menziesii* (Mirbel)] seedlings by the fungus *Rhizopogon vinicolor* Smith. Furthermore, *G. virens* decreased the dry weights of the roots of Douglas fir seedlings. In dose response experiments, in which the level of the ecto-mycorrhizal fungus *Hebeloma crustuliniforme* (Builliard ex St. Amans) Quelet was reduced to increase the potential for seeing a microbial pesticide effect on the Douglas fir seedlings, two bacterial strains, *Pseudomonas flourescens* Pf-5 and *Enterobacter cloacae* EcCt-501, were associated with decreased root and shoot dry weights in the Douglas fir; ectomycorrhizal formation by *H. crustuliniforme* also appeared to be inhibited by the *E. cloacae*. In studies that examined the effects of microbial pesticides on arbuscular mycorrhizal (AM) colonization of cucumber (*Cucumis sativa* L.) and onion (*Allium cepa* L.) roots, *P. flourescens* Pf-5 (JL3832), and *Streptomyces* sp. H68-4 inhibited AM infection in cucumber and onion respectively (reviewed in Seidler, 1992). In a study to determine antagonisms between *Gliocladium virens* and AM fungi, no antagonisms were found (Paulitz & Linderman, 1991); however, ericoid mycorrhizal formation was inhibited in cranberry (*Vaccinium oxycarpum* L.) by *G. virens* Gv-p and by *E. aerogenes* B-8 (reviewed in Seidler, 1992). Although some differential effects were noted in the particular studies described above, caution needs to be exercised both in executing and analyzing studies to determine the effects of introduced microorganisms on plant symbionts. As well as evaluating different strains of microbial pesticides, plant growth promoting candidates, or different plant species, assays to assess the potential ecological effects of the microbial pesticide inocula should ideally incorporate evaluation of the effects of different concentrations of test species in different soil types under a variety of defined environmental conditions. Additional factors to consider in looking at the potential effects of microbial pesticides (e.g., for disease control) on mycorrhizal establishment and functioning, are not only an evaluation of effects on the symbiont, but effects on disease control, plant growth and yield as well (Linderman, 1992). In laboratory and field studies with a model engineered florescent pseudomonad and its nonengineered parent, both strains were observed to have similar patterns of root colonization; neither strain affected nodulation by *Bradyrhizobium* sp or infection by arbuscular mycorrhizal fungi (Kleupfel & Tonkyn, 1991). Similarly, in earlier controlled environment studies with the same parental organism and engineered pesticidal derivatives containing chromosomal inserts of the *B. t. kurstaki* delta endotoxin gene, no differences in the kinetics of colonization or survival were noted between the parental and engineered strains (Watrud et al., 1985).

11–3.2 Biofertilizers (Arbuscular Mycorrhizal Fungi and Rhizobia)

Arbuscular mycorrhizal (AM) fungi, ubiquitous symbionts of most herbaceous plant species, have long been recognized for their potential benefits to their plant partners, particularly in less intensively managed agronomic situations, such as pastures, and in natural areas that normally are not routinely fertilized (Harley & Smith, 1983). AM fungi presumably bring to a plant partnership relatively immobile elements such as P and S, in exchange for carbohydrate sources from their plant hosts. In contrast to ectomycorrhizal fungi, many of which can be cultured in the laboratory and added exogenously to nursery stock of pines and other tree species, AM fungi are still limited largely to being produced in conjunction with host plants in greenhouses, either in soil, sand, or nutrient mist cultures. Inocula obtained from these sources can then be added as spores, infected root fragments, or soil. Although most greenhouse studies done with AM fungi typically show increases in biomass or vigor of mycorrhizal plants as compared with their nonmycorrhizal controls, several reports of deleterious effects of these symbionts have been made. As summarized in Table 11–4, these include: (i) a greater susceptibility to the tobacco stunt virus noted with *Glomus macrocarpum* (Tul & Tul.; Modjo & Hendrix, 1986); and (ii) inhibition of nodulation of legumes by *Glomus fasciculatum* [(Thaxter) Gerd. & Trappe] and *Glomus mosseae* (Roldan et al., 1992; Thatoi et al., 1993).

Nongenetically engineered strains of the symbiotic N-fixing genera *Rhizobium* and *Bradyrhizobium* have been used agronomically for almost a century, to inoculate legumes (Catroux & Armager, 1992). More recently, with the advent of genetic engineering techniques, interest in developing and releasing rhizobial strains with higher efficacy in N fixation also has increased (Paau, 1991; Hall, 1995). Concerns that have been raised regarding the release of engineered rhizobia include their potential for increased competitiveness or survival, transfer of genes for antibiotic or pesticide resistances and effects on nontarget species (Tiedje et al., 1989; McClung & Sayre, 1994).

Table 11–4. Nontarget effects of nongenetically engineered fungi.

Fungus	Use–effect	References
Colletotrichum gloeosporioides	Biocontrol agent for joint-vetch in rice fields	
	Capability to exchange genetic information with pathogenic *Colletotrichum* isolates from pecan, soybean, apple	Cisar et al., 1994
Arbuscular mycorrhizae: *Glomus macrocarpum*, *Glomus fasciculatum*, *Glomus mosseae*	Biofertilizers	
	Greater susceptibility to tobacco stunt virus	Modjo & Hendrix, 1986
	Decreased plant biomass	
	Decreased plant growth in *Hedysarum* and *Leucaena*	Roldan et al., 1992
	Reduced nodulation	
	Reduced numbers of rhizobia in the rhizosphere	Thatoi et al., 1993
	Reduced nodulation	

11–3.3 Biomass Conversion Agents

Availability of low cost processes to convert agricultural or silvicultural wastes into useful by-products (e.g., energy sources, fertilizers, or specialty products) is attractive both economically and environmentally. Examples are given below and in Table 11–5 of ecological effects noted with two prototype agents genetically engineered for biomass conversion.

11–3.3.1 *Streptomyces lividans*

Alternatives to physical and chemical procedures to dispose of wood wastes or to convert them to useful byproducts are attractive for both economic and environmental reasons. Engineered isolates of *Streptomyces lividans* were developed accordingly (Wang et al., 1990) and tested as a model system to be used for biopulping of wood wastes (e.g., at kraft paper mills). Isolates of *S. livi-*

Table 11–5. Nontarget effects of genetically engineered bacteria.

Bacterium	Use–effects	References
Klebsiella planticola SDF20*pf1*::Mupf7701/ (pZM15)	Model biomass conversion agent for ethanol production	
	Transient decrease in Arbuscular Mycorrhizal infection in wheat	
	Increased number of bacterial feeding nematodes	Holmes & Ingham, 1994
	Phytotoxicity to wheat	
Pseudomonas putida PPO301 (pRO103)	Degrader of 2,4-D herbicide	
	Decreased numbers of soil fungi	Short et al., 1991, Doyle et al., 1991
	Decreased soil respiration	
Pseudomonas cepacia AC1100	Degrader of 2,4,5 T	
	Change in taxonomic diversity of soil microbiota	Bej et al., 1991
Streptomyces lividans TK23-3651 (pSE5) and TK23.1(pIJ702)	Degrader of cellulose; model biopulping agent	
	Transient increase in CO_2 evolution upon addition of lignocellulose	Wang et l., 1989; Crawford et al., 1993

dans transformed with a plasmid (pSE5), which encodes genes for lignocellulose degradation, were tested in soil microcosms both for their ability to degrade lignocellulose and for their potential effects on soil respiration. After a period of weeks (Wang et al., 1989; Crawford et al., 1993), a transient increase in soil respiration was found. The long lag period possibly signals a need for the development of proper nutrient, inoculum, and possibly cross-feeding conditions. It also may suggest that microbial activity in soil may involve depletion of more readily available substrates before a more complex substrate such as lignocellulose can be degraded. This in turn, may require longer assay times to determine the potential ecological effects of introduced microbes—engineered or not—that are added to soil in laboratory or natural environments.

11–3.3.2 *Klebsiella planticola*

As with wastes from the wood processing industry, herbaceous plant wastes from agricultural and food processing sources (e.g., straws, stover, peelings, seed hulls, mash, and others), may present either environmental challenges or economic opportunities for specialty products or for agricultural recycling. For example, these wastes may be used as green manures for plants or as constituents of animal feeds. One type of specialty product, ethanol, used as a fuel or as an ingredient in beverages or pharmaceuticals, has been traditionally produced by enclosed fermentation processes that have employed yeasts or other fungi. Using genetic engineering techniques, a plasmid encoding alcohol dehydrogenase functions (pZM15) was transferred into *Klebsiella planticola* (Feldmann et al., 1989). In laboratory studies, in which wheat was planted into nonsterile soils amended with the engineered *K. planticola*, effects of the engineered microbe on the rate of development of AM fungal infection and on the numbers and types of nematodes in the treated soils were determined (Holmes & Ingham, 1994). The findings from those studies suggested that the engineered microbe inhibited the rate of mycorrhizal infection. This inhibition of mycorrhizal infection was not apparent immediately post-inoculation; i.e., it occurred only after a lag period of several weeks post-inoculation. Increases in bacterial and fungal feeding nematodes in the rhizosphere also were noted, as was a yellowing and apparent phytotoxic response of the wheat plants. Whether the plants turned yellow as a result of the transient decrease in mycorrhizal infection or possibly from ethanol production by the engineered microbe was unclear from these studies. Follow-up studies using known amounts of mycorrhizal and bacterial inocula in pasteurized and nonpasteurized soils may help clarify relationships between the inocula and the observed responses in wheat plants and soil biota.

11–3.4 Bioremediating Agents

Bioremediating agents and processes are attractive for their potential efficacy and cost effectiveness. The technical and economic feasibilities of in situ remediation of polluted soils, sediments, surface waters, and aquifers are of interest to the private, public, and academic sectors. The examples given below, of several prototype organisms designed or selected for their ability to degrade a herbicide or a wood preservative, serve to illustrate both their promise and potential ecological concerns.

11–3.4.1 *Pseudomonas putida*

A soil isolate of *Pseudomonas putida* engineered to express the 2, 4-dichlorophenoxyacetate monooxygenase gene *tfdA* encoded on plasmid pRO103 (Streber et al., 1987), was shown to degrade the herbicide 2,4-D [(2,4-dichlorophenoxy)acetic acid] in laboratory cultures (Short et al., 1990). In laboratory studies using nonsterile soil pretreated sequentially with the herbicide and the remediating microbe, higher rates of germination of seeds of radish (*Raphanus sativa*) were found to be correlated with longer exposure of the soil to the remediating organisms prior to sowing seeds and to higher levels of initial inoculum density (Short et al., 1991). In the same study, it also was shown that pretreating seeds, rather than soil, with the engineered inoculum, also could enhance rates of germination of the radish. In addition to looking at the efficacy of the bioremediation treatment, potential nontarget ecological effects of the engineered organism on the remediating process also were studied (Short et al., 1991; Doyle et al., 1991). Two different types of adverse ecological effects were noted. The first was a reduction in the number of soil fungi and the second, was a decrease in soil respiration. A metabolite of the herbicide degradation, 2, 4-dichlorophenol, was shown to be the cause of both inhibition of soil respiration and a reduction in the number of soil fungi (Short et al., 1991). These results suggest that as new bioremediating agents are developed, they should be evaluated for the production of metabolites that may cause adverse ecological effects.

11–3.4.2 *Pseudomonas* sp. SR3

In laboratory microcosms in which nonsterile Milican gravelly sandy loam (ashy, frigid Vitritorrandic Durixeroll; unpublished soil survey of Deschutes County, Oregon, USDA National Resources Conservation Service), was spiked with radiolabeled pentacholorphenol (PCP) and concurrently amended with a nonengineered *Pseudomonas* strain SR3 (Resnick & Chapman, 1994), >65% of the PCP was broken down within 4 wk (Pfender et al., 1995). Following the remediation treatment, seed germination, and root elongation of lettuce (*Lactuca sativa* L.); perennial ryegrass (*Lolium perenne* L.), trefoil (*Lotus corniculatus* L.), millet (*Panicum miliaceum* L.), and cabbage (*Brassica oleracea* L.), and survival of earthworms (*Eisenia foetida*), were each found to be equivalent to that of controls that had neither PCP nor inoculum additions; however, lignocellulose stimulated respiration was lower in the bioremediated soil than in nonpolluted soil 6-wk post-treatment, suggesting inhibition of soil fungi or other microbes capable of metabolizing lignocellulose. Potential inhibition of a rhizobial-legume symbiosis also was observed in these studies, as suggested by a decrease in nodule size in trefoil grown in bioremediated soil and a lack of an effect on nodulation on trefoil grown in non-PCP amended soil inoculated with the SR3 strain and *Rhizobium meliloti*. High performance liquid chromatographic analyses following organic extractions failed to show the presence of PCP, pentacholoroanisole or other known PCP metabolites in these studies. Determining whether the observed inhibitory effects on nodulation or respiration were due to nonextractable or to nondetectable levels of PCP or to PCP metabolites, merits further investigation.

11–3.4.3 *Pseudomonas cepacia*

Another engineered *Pseudomonas* strain, *P. cepacia* AC1100, engineered to degrade the herbicide 2,4,5-T (2,4,5-tricholorophenoxy-acetic acid), also has been reported to cause population shifts and an increase in the diversity of soil microbiota (Bej et al., 1991). The persistence of the changes, determined largely by use of molecular techniques based on microbial community DNA reannealing kinetics, has not been established. Significantly, the greatest change in diversity was noted when both the engineered microbe and its substrate, the herbicide 2,4,5-T, were present. The possibility exists that gene transfer from the plasmid in the engineered microbe to indigenous microbes may have played a role in the observed changes in microbial diversity (Bej et al., 1991). As more microbes are developed for environmental applications, such as in situ remediation of soils, the risks of enhanced gene transfer need to be considered along with the benefit of enhanced efficacy of released strains. The possibilities of using specific substrates, conditions, or genetic selection or engineering, to enhance or limit expression or gene transfer of the engineered trait, and viability of the organism itself, are each areas for potential research aimed at both efficacy and environmental safety of bioremediating agents.

11–4 NONTARGET EFFECTS OF SELECTED CHEMICAL INTRODUCTIONS ON SOIL BIOTA OR NONTARGET PLANTS

Unlike the biologicals described above, most of which are laboratory prototypes or very early generation commercial products, the agricultural chemicals discussed below and in Tables 11–6 through 11–8 are commercial products, generally having a long history of agronomic use. The major nontarget effects that are being addressed here are those that may affect the soil biota.

Registration of chemical pesticides in the USA requires approval under the Federal Insecticide, Fungicide, and Rodenticide Act (FIFRA). In addition to efficacy testing, registrants need to demonstrate acceptable levels of safety of the product to nontarget organisms including invertebrates such as earthworms, daphnids, and honeybees, and vertebrates such as fish, birds, and mammals. The types of results presented in Tables 11–6, 11–7, and 11–8 for herbicides, fungicides, and insecticides, respectively, represent nontarget effects on the numbers, diversity, and activities of soil biota. If available, effects of agricultural chemicals on plant symbionts such as mycorrhizae or *Rhizobium* species, or on plant community composition also are presented. Unless specified otherwise, it may be assumed that concentrations equivalent to recommended field rates were used in the examples given below.

11–4.1 Effects of Herbicides

In a comprehensive review Pipe (1992), made a systematic summary of the effects of commercial herbicides and insecticides on the growth of algae and cyanobacteria tested under in vivo and in vitro conditions, and fungicides tested almost exclusively under in vitro conditions. Alachlor (2-chloro-2′-6′-diethyl-N-(methoxymethyl)-acetanilide) and butachlor (2-chloro-2′-6′-diethyl-N-(butoxy methyl) acetanilide), examples of two preemergent broad-leaf acetanilide herbicides, have each been reported to inhibit algal growth (Table 11–6) Two post-emergent broad leaf diphenoxyacetic acid herbicides—2,4-D and 2,4,5-T—have each been reported to inhibit the growth of cyanobacteria. Additionally 2,4,5-T has been noted to inhibit the nitrogenase activity of cyanobacteria (Pipe, 1992). The widely used nonselective post-emergent herbicide glyphosate [isopropylamine salt of N-(phosphono-methyl) glycine], used to control both grasses and broad leaf weeds, and unlike the previously mentioned herbicides does not contain chlorine groups, has been reported to cause a variety of nontarget effects on soil biota. As summarized in Table 11–6, these effects include a transient decrease in the number of microbes in samples taken from soils treated with glyphosate (Chakravarty & Chatarpaul, 1990), and inhibition of nodulation in small seeded legumes treated at concentrations higher than the recommended field rate (Mårtensson, 1992). In laboratory studies, glyphosate was reported to inhibit nitrification by microbes isolated from soil (Hendricks & Rhodes, 1992). A stimulatory effect of glyphosate was found on microbial biomass in the forest litter (FL) layer, but not in the forest soil (FS); however, respiration was generally found to increase in both the FS and FL (Stratton & Stewart, 1992) . In laboratory assays in which the herbicide was used at 10 to 100 times recommended field rates, Stratton and Stewart (1992) reported no significant effects on respiration. The herbicide bromoxynil (3,5-dibromo-4-hydroxybenzonitrile), a brominated hydroxy benzonitrile, was reported to cause an increase in algal growth under laboratory conditions (Pipe, 1992). Several herbicides have been reported to result in the development of spontaneous resistance in weed or crop populations in the field; these include the sulfonyl ureas, paraquat(1,1′-dimethyl-4,4′-

Table 11–6. Nontarget effects of herbicides on soil biota and plants.

Herbicide	Effects	References
Alachlor	Inhibition of *Chlamydomonas* (alga)	Pipe, 1992
Bromoxynil	Stimulation of algal growth	Pipe, 1992
Butachlor	Inhibition of algal growth	Pipe, 1992
2,4-D	Inhibition of growth of cyanobacteria	Pipe, 1992
2,4,5-T	Decreased biomass of soil microbes	Schönborn & Dumpert, 1990
	Suppression of growth of cyanobacteria	Pipe, 1992
	Inhibition of nitrogenase in cyanobacteria	Pipe, 1992
Glyphosate	Transient decrease in number of soil microbes	Chakravarty & Chatarpaul, 1990
	Inhibition of ecto-mycorrhizal fungi	Chakravarty & Chatarpaul, 1990
	Inhibition of nodule development in small-seeded legumes at high concentration	Mårtensson, 1992
Sulfonyl Ureas, Paraquat, Triazine, Trifluralin	Herbicide resistance development	Warwick, 1991

bipyridinium ion), triazines, and trifuralins (α,α,α-trifluoro-2,6-dinitro-N,N-dipropyl-p-toluidine; Warwick, 1991).

11–4.2 Effects of Fungicides

As summarized in Table 11–7, a common effect of several fungicides, as measured in laboratory bioassays, was the reduction in the rate of denitrification. Effects of fungicides on populations of microarthropods were variable, as were effects on microbial populations. Before drawing any general conclusions, however, more systematic side-by side evaluations of the effects of given chemicals on given indicators need to be made. As with the herbicides, fungicides may act in a variety of ways. Some, such as maneb (Mn ethylenebisdithiocarbamate), mancozeb, and copper, contain or are heavy metals. Others such as propiconazole [1-[[2,4-dichlorophenyl)-4-propyl-1,3dioxolan-2-yl]methyl]-1H-1,2,4-triazole], are ergosterol biosynthesis inhibitors. The reported nontarget effects of fungicides on soil biota are diverse (Table 11–7). These nontarget effects include changes in the number of invertebrates, inhibition of enzymatic activities associated with N cycling and inhibition of rhizobial or mycorrhizal symbioses. Specific effects noted for benomyl [methyl 1-[(butylamino)carbonyl])-1H-benzimidazol-2-ylcarbamate] include an increased number of microarthropods, reduced AM fungal infection in field samples and variable stimulation of algal and cyanobacterial growth in laboratory bioassays (Krogh, 1991; West et al., 1993; Borowicz, 1992; Pipe, 1992). Copper also has been reported to have numerous effects in the field, including decreasing the number of earthworms, Collembola and microbial biomass, and inhibition of the growth of cyanobacte-

Table 11–7. Nontarget effects of fungicides on soil biota.

Fungicide	Effects	References
Benomyl	Increased number of microarthropods	Krogh, 1991
	Reduced vesicular arbuscular mycorrhizal fungal infection	West et al., 1993; Borowicz, 1992
	Variable stimulation of algae and cyanobacteria	Pipe, 1992
Bengard	Depressed N mineralization	Banerjee & Dey, 1992
	Transient initial decrease in rhizosphere microflora	
Captan	Algicidal	Pipe, 1992
Copper	Decreased numbers and diversity of Collembola	Filser et al., 1995
	Decreased microbial biomass	Baath, 1989; Yang et al., 1993
	Inhibition of cyanobacteria	Pipe, 1992
Dithane	Transient initial decrease in thiosulfate oxidation at 20 d, followed by transient increase at 40 d	Banerjee & Dey, 1992
	Increased N mineralization (ammonification and nitrification)	
Ioxynil	Reduced rate of dentrification	Pell et al., 1992
Maneb	Reduced rate of dentrification	Pell et al., 1992
Mancozeb	Reduced rate of dentrification	
	Inhibition of nodule development in small-seeded legumes	Mårtensson, 1992
	Inhibition of algae	Pipe, 1992
Propiconazole	Ergosterol biosysthesis inhibitor; inhibition of filamentous fungi in soil	Elmholt, 1991

ria in the laboratory (Baath, 1989; Reber, 1992; Hattori, 1992; Pipe, 1992; Yang et al., 1993; Filser et al., 1995). Several fungicides, including bengard[2-(methoxy-carbomylamino)-benzimidazole], Ioxynil (4-hydroxy-3,5-diodobenzonitrile), Maneb, and Mancozeb, depress N mineralization or denitrification based on the laboratory assays used (Banerjee & Dey, 1992; Pell et al., 1992, Mårtensson, 1992). Increased N mineralization has been reported for Dithane (Banerjee & Dey, 1992). Dithane also has been reported to cause successive transient decreases and increases in thisoulfate oxidation (Banerjee & Dey, 1992). Two fungicides, Captan (cis-N-trichloromethylthio-4-cyclohexene-1,2-dicarboximide) and Mancozeb, have been reported to inhibit algal growth (Pipe, 1992). The ergosterol biosynthesis inhibitor propiconazole, used as a foliar treatment for rust and mildew control in cereals crops, can inhibit the growth of *Penicillium* in soil (Elmholt, 1991). Elmholt (1991) also noted that fluctuations in fungal (*Penicillium*) populations due to the propiconazole treatment were smaller than those associated with normal seasonal fluctuations.

11–4.3 Effects of Insecticides

As with the herbicides and fungicides, systematic side by side evaluations of given insecticidal compounds, using identical methods, need to be conducted before any broad generalizations can be made. Specific effects that have been reported (Table 11–8), are an increase in plant parasitic nematodes in field soils treated with DDT (Kaushik et al., 1988), and a transient increase in soil microbial populations following application of Furadan (2,3-dihydro-2,2-dimethyl-7-benzofuranyl methylcarbamate) to soil (Edwards et al., 1992). In laboratory bioassays, carbaryl (1-naphthyl N-methylcarbamate) and endosulfan (6,7,8,9,10,10-hexachloro-1,5,5a,6,9,9a-hexahydro-6,9-methano-2,4,3-benzodioxathiepin-3-oxide) inhibited algae and cyanobacteria (Pipe, 1992). The wood preservative PCP has been reported to cause a decrease in microbial biomass, especially of

Table 11–8. Nontarget effects of chemical insecticides on soil biota and plants.

Insecticide	Effects	References
Carbaryl	Inhibited growth of algae and cyanobacteria	Pipe, 1992
DDT	Increased number of plant parasitic nematodes	Kaushik et al., 1988
Dursban 5G	Granules applied to soil correlated with increase in perennial forb and decrease in perennial grass population	Gange & Brown, 1991
Dimethoate-40	Foliar application correlated with increase in perennial grasses and decrease in perennial forb population	Gange & Brown, 1991
Endosulfan	Toxicity to sulfur oxidizing bacteria Inhibition of algae and cyanobacteria	Bezbaruah & Saikia, 1990
Furadan	Transient increase in soil microbial populations 7 to 15 d after treatment	Edwards et al., 1992
Isofenphos	Decreased numbers of microarthopods (Collembola, Acarina, and others)	Krogh, 1991
Pentachlorophenol	Decreased microbial biomass, especially of fungi Decreased respiration	Schönborn & Dumpert, 1990

Table 11–9. Summary of nontarget effects of biologicals and chemicals on soil biota and plant symbionts.

Effects	Agent	References
Inhibition of Algal Growth	Alachlor Butachlor Captan Mancozeb Carbaryl Endosulfan	Pipe, 1992
Inhibition of Cyanobacteria	2,4-D 2,4,5-T Copper Mancozeb Carbaryl Endosulfan	Pipe, 1992
Decrease in soil bacteria	Glyphosate	Chakravarty & Chatarpaul, 1990
	2,4,5-T	Schönborn & Dumpert, 1990
	Bengard	Banerjee & Day, 1992
	Copper	Baath, 1989; Reber, 1992; Hattori, 1992; Yang et al., 1993
Decrease in soil fungi	PCP	Schönborn & Dumpert, 1990
	P. putida PPO301(pRO103)	Doyle et al., 1991; Short et al., 1991
Increase in soil bacteria	Furadan	Edwards et al., 1992
	Bacillus thuringiensis kurstaki	Visser et al., 1994
	Cotton/*Bt*k gene	Donegan et al., 1995
Increase in bacterial feeding nematodes	*Klebsiella planticola*	Holmes & Ingham, 1994
	Tobacco–protease inhibitor	Donegan et al., 1997
Inhibition of soil respiration	*P. putida* pR0103	Doyle et al., 1991; Short et al., 1991
	PCP	Schönborn & Dumpert, 1990
	Soil remediated of PCP/Ps. SR3	Pfender et al., 1995
Inhibition of V.A. mycorrhizal infection	Benomyl	West et al., 1993; Borowicz, 1992
	Tobacco–acidic PR proteins	Vierheilig et al., 1995
	Klebsiella planticola SDF20*pf1*::Mupf7701/(pZM15)	Holmes & Ingham, 1994
Inhibition of nodule development	Glyphosate	Mårtensson, 1992
	Mancozeb	Mårtensson, 1992
	Soil remediated of PCP/Ps. SR3	Pfender et al., 1995
Inhibition of ectomycorrihizal fungi	Glyphosate	Chakravarty & Chatatpaul, 1990
Inhibition of enzymatic activities	*Bacillus thuringiensis* kurstaki (ammonification and nitrification)	Addison, 1993
	2,4,5-T (cyanobacterial nitrogenase)	Pipe, 1992
	Ioxynil (nitrification)	Pell et al., 1992
	Maneb (nitrification)	Pell et al., 1992
	Mancozeb (nitrification)	Pell et al., 1992
	Dithane (thiosulfate oxidation)	Banerjee & Dey, 1992
	Glyphosate (nitrification)	Hendricks & Rhodes, 1992
Decrease in microarthropods	Isofenphos	Krogh, 1991
	Copper	Filser et al., 1995
	Tobacco–protease inhibitor gene	Donegan et al., 1997
Increase in microarthropods	Benomyl	Krogh, 1991

fungi, and to inhibit respiration in treated soil (Schönborn & Dumpert, 1990). Isofenphos [1-methylethyl 2-[[ethoxy[(1-methylethyl)-amino] phosphinothioy] oxy]benzoate] can cause a decrease in the number of microarthropods such as Collembola and mites in field soils (Krogh, 1991). In studies in which foliar and soil applied insecticides were compared for their effects on plant community composition, the application of Dursban [O,O-diethyl O-(3,5,6-trichloro-2-pyridinyl)phosphorothioate] 5G granules applied to soil was correlated with an increase in perennial forb and a decrease in perennial grass populations. In contrast, a foliar application of Dimethoate-40 [O,O-dimethyl-S(N-methylcarbamoylmethyl) phosphorodithioate] was correlated with an increase in perennial grasses and a decrease in perennial forb populations (Gange & Brown, 1991). As pesticidal genes for insect, nematode, mite, and plant disease control are cloned into a variety of herbaceous and woody plant species, the fate and effects of the pesticidal products on the soil biota will be areas of continuing research interest for both short-term and long-term ecological effects.

11–5 SUMMARY AND CONCLUSIONS

Both biologicals and chemicals introduced to the environment may affect ecosystem composition and function. As summarized in Table 11–9 and detailed in Tables 11–6 through 11–8, similar types of ecological effects have been reported for diverse agricultural chemicals. For example, algae and cyanobacteria can be inhibited by herbicides (alachlor and butachlor); fungicides (Captan and Mancozeb) and insecticides (Carbaryl and Endosulfan; Pipe, 1992). Similarly, both chemicals and biologicals may produce a common effect; for example, rhizobial-legume nodule development can be inhibited by a herbicide (glyphosate), and a fungicide (Mancozeb; Mårtensson, 1992); by two AM fungi (*Glomus fasciculatum* and *Gl. mosseae*; Roldan et al., 1992; Thatoi et al., 1993), and possibly by metabolites resulting from biodegradation of PCP, or from residual PCP (Pfender et al., 1995; Schönborn & Dumpert, 1990). Decreases in litter microarthopods have been reported to result from applications of a fungicide (Cu; Filser et al., 1995), an insecticide (isofenphos; Krogh, 1991), and a transgenic pesticidal plant (tobacco expressing a plant derived protease inhibitor; Donegan et al., 1997). Increases in bacterial feeding nematodes have been noted in soil inoculated with a model engineered biomass conversion agent that was engineered to produce ethanol (*Klebsiella planticola*; Holmes & Ingham, 1994), and with genetically engineered tobacco containing a protease inhibitor gene (Donegan et al., 1997). AM fungal infections can be inhibited by a fungicide (Benoyml; West et al., 1993), by an organism engineered for biomass conversion (*Klebsiella planticola*; Holmes & Ingham, 1994), and by one type of transgenic tobacco that was engineered to express an acidic isoform of pathogenesis related (PR) proteins for resistance to plant fungal pathogens (Vierheilig et al., 1995). Respiration of soil has been reported to be inhibited by a 2,4-D bioremediating microbe (*Pseudomonas putida*; Doyle et al., 1991; Short et al., 1991); and by PCP (Schönborn & Dumpert, 1990). Decreases in the number or biomass of bacteria in soil has been reported for the herbicides glyphosate and 2,4,5-T (Chakravarty

& Charpaul, 1990) and for the fungicide Bengard (Banerjee & Dey, 1992). Decreases in the number of soil fungi have been reported following applications of PCP (Schönborn & Dumpert, 1990) and the 2,4-D degrading microbe *Pseudomonas putida* PPO301 (pRO103; Doyle et al., 1991; Short et al., 1991). Increases in bacterial counts have been noted upon addition of the insecticide Furadan (Edwards et al. , 1992); addition of cotton leaf litter expressing the *B. thuringiensis* subsp. *kurstaki* delta-endotoxin protein (Donegan et al. , 1995) and upon the addition of *B. thuringiensis* to soil (Visser et al., 1994).

In the examples cited above, a variety of methods, ranging from single species toxicity assays to community level assays were used to determine the ecological effects of both abiotic stressors such as chemicals and biotic stressors such as microbes or plants. Whereas mortality or survival estimates generally measure effects on given species, measures such as respiration, enzymatic activities or biomass may provide estimates of toxicity to populations or communities of organisms, which may contain multiple genera, species or taxa. Tables 11–6 through 11–9 provide examples of information derived by using each of these types of assays. Interest also is increasing in using trophic indices that reflect effects on given trophic groups within taxa, rather than on the entire taxon of, e.g., nematodes (Bongers, 1990; Yeates, 1994). Effects of a given treatment on specific trophic groups such as plant parasitic nematodes, bacterial feeding nematodes, fungal feeding nematodes, may thus be evaluated, rather than effects on the entire taxon of nematodes.

Relatively few references are available regarding the permanence of the ecological changes caused by biotic or abiotic treatments; many of the reports cited the apparent transient nature of the observed changes. Experimental variables were generally monitored for relatively short periods of days or weeks and very rarely, for year-to-year effects. Greater emphasis should be placed on identifying ecological effects at population or community levels, based on treating, testing, and sampling in the field over several seasons, rather than simply testing single species in the laboratory. Accordingly, longer assay or monitoring time frames may be needed to identify potential adverse effects that may directly or indirectly result in decreases in diversity of the soil biota and/or inhibition of nutrient cycling processes. Continued development of diagnostic, data analysis, and mitigation methods to facilitate early identification and effective management of complex biological and chemical interactions that may lead to potential adverse ecological effects is needed. Data collection both from simulated environments and from field sites, over longer periods of time than are now typically used in risk assessments is recommended. Use of multiple approaches ranging from molecular, organismal, community and ecosystem level characterizations of changes over time, especially in field situations, also are needed. Availability of data generated by these various approaches ultimately may permit us to better understand, predict, manage, and mitigate the effects of both planned and unplanned biological and chemical introductions to soil ecosystems.

REFERENCES

Addison, J.A. 1993. Persistence and nontarget effects of *Bacillus thuringiensis* in soil: A review. Can. J. Res. 23:2329–2342.

Alstad, D.N., and D.A. Andow. 1995. Managing the evolution of insect resistance to transgenic plants. Science (Washington, DC) 268:1894–1895.

Anderson, T.A., E.A. Guthrie, and B.T. Walton. 1993. Bioremediation in the rhizosphere. Environ. Sci. Technol. 27:2630–2636.

Arias, D.M., and L.H. Rieseberg. 1994. Gene flow between cultivated and wild sunflower. Theor. Appl. Genet. 89:655–660.

Baath, E. 1989. Effects of heavy metals in soil on microbial processes and populations. Water Air Soil Pollut. 47:335–379

Banerjee, M.R., and B.K. Dey. 1992. Effects of different pesticides on microbial populations, nitrogen mineralization, and thiosulfate oxidation in the rhizosphere of jute (*Corchorus capsularis* L. cv.). Biol. Fertil. Soils 14:213–218.

Bej, A.K., M. Perlin, and R.M. Atlas. 1991. Effect of introducing genetically engineered microorganisms on soil microbial community diversity. FEMS Microbiol. Ecol. 86(2):169–176.

Bezbaruah, B., and N. Saikia. 1990. Pesticide influence on sulphur oxidation in soil and bacterial isolates. Indian J. Agric. Sci. 60(6):406–412.

Bongers, T. 1990. The maturity index: An ecological measure of environmental disturbance based on nematode species composition. Oecologia 83:14–19.

Borowicz, V.A. 1992. Effects of benomyl, clipping, and competition on growth of prereproductive *Lotus corniculatus*. Can J. Bot. 71:1169–1175.

Boudry, P., M. Morchen, P. Sanmitou-Laprade, P. Vernet, and H. Van Dijk. 1993. The origin and evolution of weed beets: Consequences for the breeding and release of herbicide-resistant transgenic sugar beets. Theor. Appl. Genet. 87:471–478.

Bourquin, A., and R.J. Seidler. 1986. Research plan for test methods development for risk assessment of novel microbes released into terrestrial and aquatic ecosystems. p. 18–61. *In* G.S. Omen (ed.) Biotechnology and the environment: Research need. Noyes Data Corp, Park Ridge, NJ.

Carter, M.R. (ed.). 1993. Soil sampling and methods of analysis. Canadian Society of Soil Science. Lewis Publ., Boca Raton, FL.

Catroux, G., and N. Armager.1992. Rhizobia as soil inoculants in agriculture. p. 178. *In* J.C. Fry and M.J. Day (ed.) Release of genetically engineered and other microorganisms. Cambridge Univ. Press, Cambridge, England.

Chakravarty, P., and L. Chatarpaul. 1990. Non-target effect of herbicides: I. Effect of glyphosate and hexazinone on soil microbial activity. Microbial population, and in-vitro growth of ectomycorrhizal fungi. Pestic. Sci. 28:233–241.

Cisar, C.R., F.W. Spiegel, D.O. TeBeest, and C. Trout. 1994. Evidence for mating between isolates of *Colletotrichum gloeosporioides* with different host specificities. Curr. Genet. 25:330–335.

Crawford, D.L., J.D. Doyle, Z. Wang, C.W. Hendricks, S.A. Bentjen, H. Bolton, Jr., J.K. Fredrickson, and B.H. Bleakley. 1993. Effects of a lignin peroxidase-expressing recombinant *Streptomyces lividans* TK23.1 on biogeochemical cycling and microbial numbers and activities in soil microcosms. Appl. Environ. Microbiol. 59(2):508–518.

Dale, P.J. 1994. The impact of hybrids between genetically modified crop plants and their related species: General considerations. Molecular Ecol. 3:31–36.

Darmency, H. 1992. Outcrossing and hybridization in wild and cultivated foxtail millets: Consequences for the release of transgenic crops. Theor. Appl. Genet. 83:940–946.

Darmency, H. 1994. The impact of hybrids between genetically modified crop species and their related species: Introgression and weediness. Molec. Ecol. 3:37–40.

DeWet, J.M.K., and J.R. Harlan. 1975. Weeds and domesticated evolution in the man-made habitat. Econ. Bot. 29:99.

Donegan, K.K., C.J. Palm, V.J. Fieland, L.A. Porteous, L.M. Ganio, D.L. Schaller, L.Q. Bucao, and R.J. Seidler. 1995. Changes in levels, species and DNA fingerprints of soil microorganisms associated with cotton expressing the *Bacillus thuringiensis* var. *kurstaki* endotoxin. Appl. Soil Ecol. 2:111–124.

Donegan, K.K., D.L. Schaller, J.K. Stone, L.M. Ganio, G. Reed, P.B. Hamm, and R.J. Seidler. 1996. Microbial populations, fungal species diversity and plant pathogen levels in field plots of potato plants expressing the *Bacillus thuringiensis* var. *tenebrionis* endotoxin. Transgenic Res.5:25–35.

Donegan, K.K., R.J. Seidler, V.J. Fieland, D. L. Schaller, C.J. Palm, L.M. Ganio, D.M. Cardwell, and Y. Steinberger. 1997. Decomposition of genetically engineered tobacco under field conditions: Persistence of the proteinase inhibitor I product and effects on soil microbial respiration and protozoa, nematode and microarthropod populations. J. Appl. Ecol. 34:767–777.

Doran, J.W., D.C. Coleman, D.F. Bezdicek, and B.A. Stewart (ed.). 1994. Defining soil quality for a sustainable environment. SSSA Spec. Publ. 35. SSSA and ASA, Madison WI.

Doyle, J.D., K.A. Short, G. Stotzky, R.J. King, R.J. Seidler, and R.H. Olsen. 1991. Ecologically significant effects of *Pseudomonas putida* PP0301 (pRO103), genetically engineered to degrade 2,4-dichlorophenoxyacetate, on microbial populations and processes in soil. Can. J. Microbiol. 37:682–691.

Doyle, J.D., G. Stotzky, G. McClung, and C.W. Hendricks. 1995. Effects of genetically engineered microorganisms on microbial populations and processes in natural habitats. Adv. Appl. Microb. 40:237–287.

Edwards, C.A. 1988. The use of key indicator processes for assessment of the effects of pesticides on soil ecosystems. p. 739–746. *In* Brighton Crop Protection Conf., Pests and Diseases, Brighton, England. November 1988. British Crop Protection Council, Surrey, England..

Edwards, C.A., and J.E. Bater. 1990. An evaluation of laboratory and field studies for the assessment of the environmental effects of pesticides. p. 963–968. *In* Brighton Crop Protection Conf., Pests and Diseases, Brighton, England. 19–22 Nov. 1990. British Crop Protection Council, Surrey, England.

Edwards, D.E., R.J. Kremer, and A.J. Keaster. 1992. Characterization and growth response of bacteria in soil following application of carbofuran. J. Environ. Sci. Health B27(2):139–154.

Ellstrand, N. 1988. Pollen as a vehicle for the escape of engineered genes? Trends Ecol. Evol./ Trends Biotechnol. 3/6(4):S30–S32.

Ellstrand, N.C., and C.A. Hoffman. 1990. Hybridization as an avenue of escape for engineered genes. BioScience 40:438–442.

Elmholt, S. 1991. Side effects of propiconazole tilt 250 EC™ on non-target soil fungi in a field trial compared with natural stress effects. Microb. Ecol. 22:99–108.

Entry, J.A., N.C. Vance, M.A. Hamilton, D. Zabowski, L.S. Watrud, and D.C. Adriano. 1995. Phytoremediation of soil contaminated with low concentrations of radionucldies. Water Air Soil Pollut. 88:167–176.

Feldman, N.S., G.A. Sprenger, and H. Sahm. 1989. Ethanol production from xylose with a pyruvate-formate-lyase mutant of *Klebsiella planticola* carrying a pyruvate-decarboxylase gene from *Zymomonas mobilis*. Appl. Microb. Biotech. 31:152–157.

Filser, J., H. Fromm, R.F. Nagel, and K. Winter. 1995. Effects of previous intensive agricultural management on microorganisms and the biodiversity of soil fauna. Plant Soil 170:123–129.

Fischhoff, D.A., K.S. Bowdish, F.J. Perlak, P.G. Marrone, S.M. McCormick, J.G. Niedermeyer, D.A. Dean, K. Kusano-Kretzmer, E.J. Mayer, D.E. Rochester, S.G. Rogers, and R.T. Fraley. 1988. Insect tolerant transgenic tomato plants. Bio/Technology 5:807–813.

Fischhoff, D.A., and L.S. Watrud. 1988. Microbes as a source of genes for agricultural biotechnology. p. 65–71. *In* P. DeForest et al (ed.) Biotechnology: Professional issues and social concerns. Am. Assoc. for Adv. of Science., Washington, DC.

Fraley, R. 1992. Sustaining the supply. Bio/Technology 10:40–43.

Gange, A.C., and V.K. Brown. 1991. Effects of insecticide application on weed and pasture plant communities. p. 901–910. *In* Brighton Crop Protection Conf., Weeds, Brighton, England. 18–21 Nov. 1991. British Crop Protection Council, Surrey, England.

Garland, J.L., and A.L. Mills. 1991. Classification and characterization of heterotrophic microbial communities on the basis of patterns of community-level sole-carbon-source. Appl. Environ. Microb. 57:2351–2359.

Gasser, C.S., and R.T. Fraley. 1989. Genetically engineering plants for crop protection. Science (Washington, DC) 244:1293–1299.

Gliddon, A. 1994. The impact of hybrids between genetically modified crop plants and their related species: Biological models and theoretical perspectives. Molec. Ecol. 3:41–44.

Goettel, M.S., T.J. Poprawski, J.D. Vandenberg, Z. Li, and D.W. Roberts. 1990. Safety to non-target invertebrates of fungal biocontrol agents. p. 209–231. *In* M. Laird et al. (ed.) Safety of microbial insecticides. CRC Press, Boca Raton, FL.

Gould, F., A. Martinez-Ramirez, A. Anderson, J. Ferre, F.J. Silva, and W.J. Moar. 1992. Broad spectrum resistance to *Bacillus thuringiensis* toxins in *Heliothis virescens*. Proc. Natl. Acad. Sci. (USA) 89(17):7986–7990.

Hall, G. 1995. Environmental release of genetically modified rhizobia and mycorrhizas. p. 64–92. *In* G.T. Tzotozos (ed.) Genetically modified organisms. Biddles, Guildford, England.

Harley, J.L., and S.E. Smith. 1983. Mycorrhizal symbiosis. Academic Press, London.

Hattori, H. 1992. Influence of heavy metals on soil microbial activities. Soil. Sci. Plant Nutr. 38(1):93–100.

Hendricks, C.W., and A.N. Rhodes. 1992. Effect of glyphosate and nitrapyrin on selected bacterial populations in continuous-flow culture. Bull. Environ. Contam. Toxicol. 49:417–424.

Holmes, M.T., and E.R. Ingham. 1994. The effects of genetically engineered microorganisms on soil foodwebs. p. 97. *In* Program and abstracts. Ecol. Soc. of Am., Tempe, AZ.

James, R.R., J.C. Miller, and B. Lighthart. 1993. *Bacillus thuringiensis* var. *kurstaki* affects a beneficial insect, the Cinnabar moth (Lepidoptera: Arctiidae). J. Econ. Entom. 86(2):334–339.

James, R.R., B.T. Shaffer, B.A. Croft, and B. Lighthart. 1995. Field evaluation of *Beauveria bassiana*: Its persistence and effects on pea aphid and a non-target coccinellid in alfalfa. Biocontrol Sci. Technol. 5:425–437.

Kareiva, P., W. Morris, and C.M. Jacobi. 1994. Studying and managing the risk of cross-fertilization between transgenic crops and wild relatives. Molec. Ecol. 3:15–22.

Kaushik, C.P., M.K.K. Pillai, M.L. Chawla, and H.C. Agarwal. 1988. Impact of HCH and DDT residues on soil nematodes. J. Ent. Res. 12(2):142–148.

Keeler, K. 1989. Can genetically engineered weeds become crops? Bio/Technology 7:1134–1139.

Kirkpatrick, K.J., and H.D. Wilson. 1988. Interspecific gene flow in *Cucurbita: C. texana* vs. *C. pepo.* Am. J. Bot. 75:519–527.

Klinger, T., P.E. Arriola, and N.C. Ellstrand. 1992. Crop-weed hybridization in radish (*Raphanus sativus* L.): Effects of distance and population size. Am. J. Bot. 79:1431–1435.

Klinger, T., D.R. Elam, and N.C. Ellstrand. 1991. Radish as a model system for the study of engineered gene escape rates via crop-weed mating. Conserv. Biol. 5:531–535.

Klinger, T., and N.C. Ellstrand. 1994. Engineered genes in wild populations: Fitness of weed-crop hybrids of *Raphanus sativus*. Ecol. Appl. 4:117–120.

Kloppenburg, J.R. 1988. First the Seed: The political economy of plant biotechnology 1492–2000. Cambridge Univ. Press, London.

Kleupfel, D.A., and D.W. Tonkyn. 1991. Release of soil-borne bacteria: Biosafety implications from contained experiments. p. 55–65. *In* D.R. MacKenzie and S.C. Henry (ed.) Proc. of the Kiawah Island Conf., South Carolina. 27–30 Nov. 1990. Agric. Res. Inst., Bethesda, MD.

Koziel, M.G., G.L. Beland, C. Bowman, N.B. Carozzi, R. Crenshaw, L. Crossland, J. Dawson, N. Desai, M. Hill, S. Kadwell, K. Launis, K. Lewis, D. Maddox, K. McPherson, E. Meghji, M.R. Merlin, R. Rhodes, G.W. Warren, M. Wright, and S.V. Evola. 1993. Field performance of elite transgenic maize plants expressing an insecticidal protein derived from *Bacillus thuringiensis*. Bio/Technology 11:194–200.

Krattiger, A.F. 1994. The field testing and commercialization of genetically modified plants: A review of worldwide data (1986 to 1993/94). p. 247–266. *In* A.F. Krattiger and A. Rosemarin (ed.) Biosafety for a sustainable agriculture: Sharing biotechnology regulatory experiences of the western hemisphere. ISAAA, Ithaca, NY.

Krogh, P.H. 1991. Perturbation of the soil microarthropod community with the pesticides benomyl and isofenphos: I. Population changes. Pedobiologia 35:71–88.

Langevin, S.A., K. Clay, and J. Grace. 1990. The incidence and effects of hybridization between cultivated rice and its related weed rice (*Oryza sativa* L.). Evolution 44:1000–1008.

Levin, M.A., R.J. Seidler, and M. Rogul. 1992. Microbial ecology. McGraw-Hill, New York.

Linderman, R.G. 1992. Vesicular–arbuscular mycorrhizae and soil microbial interactions. p. 45–70. *In* G.J. Bethlenfalvay, and R.G. Linderman (ed.) Mycorrhizae in sustainable agriculture. ASA Spec. Publ. 54. ASA, Madison, WI.

Linderman, R.G., T.C. Paulitz, N.J. Mosier, R.P. Griffiths, J.E. Loper, B.A. Caldwell, and M.E. Henkels. 1991. Evaluation of the effects of biocontrol agents on mycorrhizal fungi. p. 379. *In* D.L. Keister and P.G. Cragan (ed.) The rhizosphere and plant growth. Kluwer Academic Publ., Dordrecht, the Netherlands.

Mårtensson, A.M. 1992. Effects of agrochemicals and heavy metals on fast-growing rhizobia and their symbiosis with small-seeded legumes. Soil Biol. Biochem. 24(5):435–335.

McClung, G., and P. Sayre. 1994. A review of ecological assessment case studies from a risk assessment perspective. Vol. II. Section Two. EPA/630/R-94/003 U.S. Environ. Protection Agency, Office of Res. and Dev., Washington, DC.

Mc Gaughey, W.H. 1985. Insect resistance to the biological insecticide *Bacillus thuringiensis*. Science (Washington, DC) 229:193–194.

Mc Gaughey, W.H., and R.W. Beeman. 1988. Resistance to *Bacillus thuringiensis* in colonies of Indian meal moth and almond moth (Lepidoptera: Pyralidae). J. Econ. Entom. 81:28–33.

Mc Gaughey, W.H., and D.E. Johnson. 1992. Indianmeal moth (Lepidoptera: Pyralidae) resistance to different strains and mixtures of *Bacillus thuringiensis*. J. Econ. Entom. 85(5):1594–1600.

Mc Gaughey, W.H., and M.E. Whalon. 1992. Managing insect resistance to *Bacillus thuringiensis* toxins. Science (Washington, DC) 258:1451–1455.

Mellon, M., and J. Rissler. 1995. Transgenic crops: USDA data on small-scale tests contribute little to commercial risk assessment. Bio/Technology 13:96.

Miller, J.C. 1990. Field assessment of the effects of a microbial pest control agent on nontarget lepidoptera. Am. Entom. 36:135–139.

Modjo, H.S., and J.W. Hendrix. 1986. The mycorrhizal fungus *Glomus macrocarpum* as a cause of tobacco stunt disease. Phytopathology 76(7):688–691.

Mooney, H., S. Hamburg, and J. Drake. 1986. The invasions of plants and animals into California. p. 250–269. *In* J.A. Mooney and J.A. Drake (ed.) Ecology of biological invasions of North America and Hawaii. Springer-Verlag, New York.

Morganthaler, E. 1993. What's Florida to do with an explosion of Melaleuca trees? *Wall Street Journal*, 8 Feb. 1993.

Narváez-Vásquez, J., M.L. Orozco-Cardénas, and C.A. Ryan. 1992. Differential expression of a chimeric CaMV-tomato proteinase inhibitor I gene in leaves of transformed nightshade, tobacco and alfalfa plants. Plant Molec. Biol. 20:1149–1157.

Obukowicz, M.G., F.J. Perlak, S.L. Bolten, K. Kusano-Kretzmer, E.J. Mayer, and L.S. Watrud. 1987. IS50L as a non-self transposable vector used to integrate the *Bacillus thuringiensis* delta endotoxin gene into the chromosome of root-colonizing pseudomonads. Gene 51(1):91–96.

Obukowicz, M.G., F.J. Perlak, K. Kusano-Kretzmer, E.J. Mayer, and L.S. Watrud. 1986a. Integration of the delta endotoxin gene of *Bacillus thuringiensis* into the chromosome of root- colonizing strains of pseudomonads using Tn5. Gene 45(3):327–332.

Obukowicz, M.G., F.J. Perlak, K. Kusano-Kretzmer, E.J. Mayer, and L.S. Watrud. 1986b. Tn5-mediated integration of the delta endotoxin gene from *Bacillus thuringiensis* into the chromosome of root-colonizing pseudomonads. J. Bacteriol. 168(2):982–989.

Paau, A.S. 1991. Improvements of *Rhizobium* inoculants by mutation, genetic engineering and formulation. Biotech. Adv. 9:173–184.

Paterson, A.H., K.F. Schertz, and Y.-R. Linn. 1995. The weediness of wild plants: Molecular analysis of genes influencing dispersal and persistence of Johnsongrass, *Sorghum halepense* (L.) Pers. Proc. Natl. Acad. Sci. (USA) 92:6127–6231.

Paulitz, T.C., and R.G. Linderman. 1991. Lack of antagonism between the biocontrol agent *Gliocladium virens* and vesicular arbuscular mycorrhizal fungi. New Phytol. 117:303–308.

Pell, M., B. Stenberg, J. Stenström, and L. Torstensson. 1992. Dentrification potential assay as a tool for evaluating effects of pesticides on soil biota. p. 128. *In* J.P.E. Anderson et al. (ed.) Proc. Int. Symp. Environmental Aspects of Pesticide Microbiology, Sigtuna, Sweden. 17–21 Aug. 1992. Swedish Univ. of Agric. Sci., Uppsala.

Perlak, F.J., R.W. Deaton, T.A. Armstrong, R.L. Fuchs, S.R. Sims, J.T. Greenplate, and D.A. Fischhoff. 1990. Insect resistant cotton plants. Bio/Technology 8:939–943.

Pfender, W.F., S.P. Maggard, and L.S. Watrud. 1995. Soil microbial activity and plant/microbe symbioses as indicators for ecological effects of bioremediation biotechnology. p. 269–279. *In* Proc. Biotech. Risk Assessment Symp., College Park, MD. 22–24 June 1994. Publ. 1001. Univ. of Maryland Biotech. Inst., College Park.

Picard, C., C. Ponsonnet, E. Paget, X. Nesme, and P. Simonet. 1992. Detection and enumeration of bacteria in soil by direct DNA extraction and polymerase chain reaction. Appl. Environ. Microb. 58:2717–2722.

Pipe, A.E. 1992. Pesticide effects on soil algae and cyanobacteria. Rev. Environ. Contam. Toxicol. 127:95–171.

Porteous, L.A., and J.L. Armstrong. 1991. Recovery of bulk DNA from soil by a rapid, small-scale extraction method. Curr. Microb. 22:345–348.

Porteous, L.A., and J.L. Armstrong. 1993. A simple mini-method to extract DNA directly from soil for use with polymerase chain reaction amplification. Curr. Microb. 27:115–118.

Porteous, L.A., J.L. Armstrong, R.J. Seidler, and L.S. Watrud. 1994. An effective method to extract DNA from environmental samples for polymerase chain reaction amplification and DNA fingerprint analysis. Curr. Microb. 29:301–307.

Purrington, C.B., and J. Bergelson. 1995. Assessing weediness of transgenic crops: Industry plays plant ecologist. Trends Ecol. Evol. 10(8):340–342.

Raybould, A.F., and A.J. Gray. 1993. Genetically modified crops and hybridization with wild relatives: A UK perspective. J. Appl. Ecol. 30:199–219.

Raybould, A.F., and A.J. Gray. 1994. Will hybrids of genetically modified crops invade natural communities? Trends Ecol. Evol. 9:85–89.

Reber, H.H. 1992. Simultaneous estimates of the diversity and the degradative capability of heavy-metal-affected soil bacterial communities. Biol. Fertil. Soils 13:181–186.

Resnick, S.M., and P.J. Chapman. 1994. Physiological properties and substrate specificity of a pentachlorophenol-degrading *Pseudomonas* species. Biodegradation 5(1):47–54.

Rissler, J., and M. Mellon. 1993. Perils amidst the promise, ecological risks of transgenic crops in a global market. Union of Concerned Scientists, Cambridge, MA.

Roldan, A., G. Diaz, and J. Albaladejo. 1992. Effect of VAM-fungal inoculation on growth and phosphorus uptake of two *Hedyesarum* species in a xeric torriorthent soil from southeast Spain. Arid Soil Res. Rehabil. 6:33–39.

Ruesink, J.L., I.M. Parker, M.J. Groom, and P.M. Kareiva. 1995. Reducing the risks of non-indigenous species introductions. Bioscience 45:465–477.

Salt, D.E., M. Blaylock, N.P.B.A. Kumar, V. Dushenkov, B.D. Ensley, I. Chet, and I. Raskin. 1995. Phytoremediation: A novel strategy for the removal of toxic metals from the environment using plants. Biotechnology 13:468–474.

Schnoor, J.L., L.A. Light, S.C. McCutcheon, N.L. Wolfe, and L.H. Carreira. 1995. Phytoremediation of organic and nutrient contaminants. Environ. Sci. Technol. 29:318–323.

Schönborn, W., and K. Dumpert. 1990. Effects of pentachlorophenol and 2,4,5-trichlorophenoxyacetic acid on the microflora of the soil in a beech wood. Biol. Fertil. Soils 9:292–300.

Schuster, E., and D. Schröder. 1990. Side-effects of sequentially-applied pesticides on non-target soil microorganisms: Field experiments. Soil Biol. Biochem. 22:367–373.

Seidler, R.J. 1992. Evaluation of methods for detecting ecological effects from genetically engineered microorganisms and microbial pest control agents in terrestrial systems. Biotech. Adv. 10:149–178.

Seidler, R.J., and M. Levin. 1994. Potential ecological and non-target effects of transgenic plant gene products on agriculture, silviculture and natural ecosystems: General introduction. Molec. Ecol. 3:1–3.

Short, K.A., J.D. Doyle, R.J. King, and R.J. Seidler. 1991. Effects of 2,4-dichlorophenol, a metabolite of a genetically engineered bacterium, and 2,4-dichlorophenoxyacetate on some microorganism-mediated ecological processes in soil. Appl. Environ. Microb. 57(2):412–418.

Short, K.A., R.J. Seidler, and R.H. Olsen. 1990. Survival and degradative capacity of *Pseudomonas putida* induced or constitutively expressing plasmid-mediated degradation of 2,4-dichlorophenoxyacetate in soil. Can. J. Microb. 36:821–826.

Steffan, R.J., J. Goksoyr, A.K. Bej, and R.M. Atlas. 1988. Recovery of DNA from soils and sediments. Appl. Environ. Microb. 54:2908–2915.

Stotzky, G. 1986. Influence of soil mineral colloids on metabolic processes, growth, adhesion, and ecology of microbes and viruses. p. 305–428. *In* P.M. Huang and M. Schnitzer (ed.) Interactions of soil minerals with natural organics and microbes. SSSA, Madison, WI.

Stotzky, G., M.W. Broder, J.D. Doyle, and R.A. Jones. 1993. Selected methods for the detection and assessment of ecological effects resulting from the release of genetically engineered microorganisms to the terrestrial environment. Adv. Appl. Microb. 38:1–98.

Stratton, G.W., and K.E. Stewart. 1992. Glyphosate effects on microbial biomass in a coniferous forest soil. Environ. Toxicol. Water Qual. 7:223–236.

Streber, W.R., T.N. Timmis, and M.H. Zenk. 1987. Analysis, cloning, and high-level expression of 2,4-dichlorophenoxyacetate monooxygenase gene *tfdA* of *Alcaligenes eutrophus* JMP134. J. Bacteriol. 169:2950–2955.

Tabashnik, B.E. 1994. Evolution of resistance to *Bacillus thuringiensis*. Annu. Rev. Entomol. 39:47–79.

Tabashnik, B.E., N.L. Cushing, N. Finson, and M.W. Johnson. 1990. Field development of resistance to *Bacillus thuringiensis* in diamondback moth (Lepidoptera:Plutellidae). J. Econ. Entomol. 83(5)1671–76.

Tabashnik, B.E., N. Finson, and M.W. Johnson. 1991. Managing resistance to *Bacillus thuringiensis*: Lessons from the diamondback month (Lepidoptera:Plutellidae). J. Econ. Entomol. 84:49–55.

Thatoi, H.N., S. Sahu, A.K. Misra, and G.S. Padhi. 1993. Comparative effect of VAM inoculation on growth, modulation and rhizobium population of subabul (*Leucaena leucocephala* (Lam.) de Wit.) grown in iron mine waste soil. Indian For. 119(6):481–489.

Tiedje, J.M., R.K. Colwell, Y.L. Grossman, R.E. Hodson, R.E. Lenski, R.N. Mack, and P.J. Regal. 1989. The planned introduction of genetically engineered organisms: Ecological considerations and recommendation. Ecology 70(2):298–315.

Till-Bottraud, I., X. Rebould, P. Brabant, M. Lefranc, B. Rherissi, F. Vederl, and H. Darmency. 1992. Outcrossing and hybridization in wild and cultivated foxtail millets: Consequences for the release of transgenic crops. Theor. Appl. Genet. 83:940–946.

Tsai, Y.-L., and B.H. Olson. 1991. Rapid method for direct extraction of DNA from soil and sediments. Appl. Environ. Microb. 57:1070–1074.

Umbeck, P.F., K.A. Barton, E.V. Nordheim, J.C. McCarty, W.L. Parrott, and J.N. Jenkins. 1992. Degree of pollen dispersal by insects from a field test of genetically engineered cotton. J. Econ. Entomol. 84:1943–1950.

U.S. Congress, Office of Technology Assessment. 1993. Harmful non-indigenous species in the United States. U.S. Congress, Washington, DC.

U.S. Department of Agriculture. 1992. Scientific evaluation of the potential for pest resistance to the *Bacillus thuringiensis* (Bt) delta-endotoxins. *In* A Conf. to Explore Resistance Management Strategies, Beltsville, MD. 21–23 Jan. 1992. USDA, Washington, DC.

Vierheilig, H., M. Alt, J. Lange, M. Gut-Rella, A. Wiemken, and T. Boller. 1995. Colonization of transgenic tobacco constitutively expressing pathogenesis-related proteins by the vesicular–arbuscular mycorrhizal fungus *Glomus mosseae*. Appl. Environ. Microb. 8(61):3031–3034.

Visser, S., J.A. Addison, and S.B. Holmes. 1994. Effects of DiPel® 176, a *Bacillus thuringiensis* subsp. *kurstaki* (B.t.k.) on the soil microflora and the fate of B.t.k. in an acid forest soil: A laboratory study. Can. J. Res. 24:462–471.

Wang, Z., B.H. Bleakley, D.L. Crawford, G. Hertel, and F. Rafii. 1990. Cloning and expression of a lignin peroxidase gene from *Streptomyces viridosporus* in *Streptomyces lividans*. J. Biotech. 13:131–144.

Wang, Z., D.L. Crawford, A.L. Pometto, III, and F. Rafii. 1989. Survival and effects of wild-type, mutant, and recombinant *Streptomyces* in a soil ecosystem. Can. J. Microb. 35:535–543.

Wardle, D.A., G.W. Yeates, R.N. Watson, and K.S. Nicholson. 1995. The detritus food-web and the diversity of soil fauna as indicators of disturbance regimes in agro-ecosystems. Plant Soil 170:35–43.

Warwick, S.I. 1991. Herbicide resistance in weedy plants: physiology and population biology. Annu. Rev. Ecol. Syst. 22:95–114.

Watrud, L.S. 1987. Developing biotechnology products for agricultural pest control. p. 323–328. *In* D.D. Hemphill (ed.) Trace substances in environmental health. Univ. of Missouri, Columbia.

Watrud, L.S., S.G. Metz, and D.A.. Fischhoff. 1996. Engineered plants in the environment. p. 165–189. *In* E. Israeli and M. Levin (ed.) Engineered organisms in environmental settings. Biotechnological and Agricultural Applications. CRC Press, Boca Raton, FL.

Watrud, L.S., F.J. Perlak, M.T. Tran, K. Kusano, M.A. Miller-Wideman, M.G. Obukowicz, D.R. Nelson, J. Kreitinger, and R.J. Kaufman. 1985. Cloning of the *Bacillus thuringiensis* subsp. Delta-Endotoxin Gene into *Pseudomonas fluorescens*. Molecular biology and ecology of an engineered microbial pesticide. p. 40–46. *In* H.O. Halvorson et al. (ed.) Engineered organisms in the environment: Scientific issues. Am. Soc. for Microbiol., Washington, DC.

West, H.M., A.H. Fitter, and A.R. Watkinson. 1993. The influence of three biocides on the fungal associates of the roots of *Vulpia ciliata* ssp. *ambigua* under natural conditions. J. Ecol. 81:345–350.

Whitson, T.D. 1992. Weeds of the West. Western Soc. of Weed Science and the Western U.S. Land Grant Univ. Coop. Ext. Serv., Jackson, WY.

Williams, M.C. 1980. Purposefully introduced plants that have become noxious or poisonous weeds. Weed Sci. 28:300–305.

Williamson, M.H., and K.C. Brown. 1986. The analysis and modelling of British invasions. Phil. Trans. R. Soc. London B 314:505–522.

Yang, C.-H., J. Menge, and D.A. Cooksey. 1993. A role of copper resistance in competitive survival of *Pseudomonas fluorescens* in soil. Appl. Environ. Microb. 59(2):580–584.

Yeates, G.W. 1994. Modification and qualification of the nematode maturity index. Pedobiologia 38:97–101.

Zelles, L., I. Scheunert, and F. Korte. 1985. Side effects of some pesticides on non-target soil microorganisms. J. Environ. Sci. Health B 20(5):457–488.

12 Ecosystem Health and Its Relationship to the Health of the Soil Subsystem: A Conceptual and Management Perspective

David J. Rapport

Faculty of Environmental Science
University of Guelph
Guelph, Ontario
and
Department of Pharmacology and Toxicology
University of Western Ontario
London, Ontario

Connie Gaudet

Soil and Sediment Quality Section
Environment Canada
Ottawa, Ontario

John McCullum and Murray Miller

Department of Land Resource Science
University of Guelph
Guelph, Ontario

12–1 INTRODUCTION

Soil degradation is increasingly recognized as an urgent environmental issue with ramifications at the local, regional, and global scales. The 1995 announcement for the 9th Conference of the International Soil Conservation Organization (ISCO) emphasized that "*...continuous soil degradation destroys the basis of life of future generations. This constitutes an environmental threat comparable to global climate change and therefore demands the same attention.*" This realization is not new—the 1984 report: "*Soil at Risk: Canada's Eroding Future*" (Standing Committee on Agriculture, Fisheries, and Forestry, 1984), prefaced their discussions with a quote by Lester Brown (World Watch Institute) that "*civilization as we know it cannot survive the continuing loss of topsoil at current rates.*" These two declarations, more than a decade apart, capture the fact that soil health is a central component in sustaining the world's ecosystems and the myriad of natural and socioeconomic systems they support.

To deal effectively with soil health in this broader context, we need a fundamental shift in the way we view soil—from a medium that supports human activities to a dynamic, multifunctional ecological component of the larger biophysical and socioeconomic environment. Soil is a vital component in the functioning of ecosystems, accounting for virtually all decomposition processes and a significant proportion of energy flow, a critical link in sequestering C and mitigating global climate change, and a vital economic resource required for agriculture, silviculture, many raw materials and indirectly as a platform for most infrastructure development (Blum & Sandelises, 1994). An ecosystem health approach to the evaluation and management of the soil subsystem incorporates the associated ecological and socioeconomic values into a useful assessment or diagnostic framework, reconciling the apparent conflict between what have traditionally been seen as competing values.

12–2 ECOSYSTEM HEALTH AS AN EMERGING INTEGRATIVE SCIENCE

An ecosystem approach does not embody any one definition or course of action but is both a way of doing things and way of thinking (Royal Commission on the Future of Toronto's Waterfront, 1992). Attention is focused on interrelationships among ecosystem components, integrated multisectoral management, and the role of culture, values, and socioeconomic systems (Berry & Marmorek, 1992, unpublished data). Two concepts have emerged as part of the ecosystem approach—ecosystem health and ecosystem integrity. Ecosystem integrity is a term used to describe the functional and structural attributes of an ecosystem in terms of resilience, biodiversity, and freedom from human impact. Ecosystem health is more commonly used in a broader sense as an integrative science drawing together ecology, geography, ethics, environmental management, and health sciences. It encompasses and builds on the concept of ecosystem integrity. By definition, whether an ecosystem is healthy or not becomes a social judgement as well as a scientific exercise.

Definitions and understanding of ecosystem health vary but there are at least three fundamental attributes or criteria of a healthy ecosystem:

- The ecosystem should be free of "ecosystem distress syndrome (EDS)" (Rapport et al., 1985; Rapport, 1989a,b; Costanza, 1992; Hildén & Rapport, 1993; Rapport, 1995a);
- The ecosystem should be self-sustaining with minimal subsidy, e.g., applications of fertilizer to sustain yield in agricultural systems; and
- The ecosystem should not adversely affect or degrade surrounding systems. For example, a healthy managed forest ecosystem should not cause injury to the fisheries in the drainage basin; a healthy agroecosystem should not add pollution or nutrient burden to its drainage.

Associated with each of these criteria are evolving methods for assessment (Cairns & Pratt, 1995; Hansen, 1995; Karr, 1995; Rapport, 1995a,b,c; Smol,

1995). One of the challenges is to develop indicators that have been shown in practice to be highly reliable and informative about the nature of dysfunctions, probable causes, or probable remedies. While much attention has been given to selection of indicators of ecosystem condition, and various schemes have been devised to classify indicators, few of these have been validated (Rapport, 1995a,b). It is apparent, however, that single indicators are insufficient: it is syndromes not individual signs or symptoms that can relied upon for the evaluation of the health of ecosystems (Hildén & Rapport, 1993; Rapport, 1989b; Whitford, 1995).

12–3 SOCIETAL VALUES IN THE CONTEXT OF AN ECOSYSTEM MANAGEMENT FRAMEWORK

A discussion of soil health in an ecosystem health context must include a consideration of values, since the two are difficult to separate (O'Neill et al., 1992). Evaluation of ecosystem health is meaningful in relation to societal values that must be assessed through broad public consultation, though this does not mean that any environmental situation should be considered healthy, simply because it has sufficient public support (Rapport, 1995a). In agroecosystems, for example, the objectives for evaluating health will vary with different perspectives. For the farmer, explicit objectives for agroecosystems and their soils relate largely to quantity and quality of yield. Policy-makers may be interested in production levels of various commodities as well as soil loss, tillage systems or other soil-related public policy issues or related environmental policy issues such as water quality and wildlife habitat. Regulators will largely be interested in compliance with environmental standards and guidelines. The public may be interested in their exposure to contaminants in the run off from agricultural soils and from the produce grown on these soils; also in recreational opportunities and in the cost of produce to the consumer. The strength of an ecosystem health approach is that it seeks to integrate these seemingly isolated values and to place them within the context of an integrated, interrelated ecosystem. This requires that science and values are reconciled in defining a healthy state that not only produces the *ecological goods and services* recognized by society, but maintains the functional and structural integrity necessary to sustain this healthy and productive state over the long term.

In general, three activities appear critical to the success of ecosystem-based management plans (Canadian Council of Ministers of the Environment, 1996a,b): (i) a clear articulation of what is desirable with respect to that ecosystem; (ii) an evaluation mechanism to indicate whether desired conditions have been achieved; and (iii) a reporting mechanism to convey information to stakeholders and decision makers.

Ecosystem goals and objectives, based on societal values and the inherent potential of the ecosystem, as described below, can meet the first requirement of a management plan. Indicators, such as measures of the quality or health of the soil subsystem, can serve the latter two requirements. Several ecosystem management frameworks have been developed, based on the relationship between societal values and indicators of ecosystem health, two of which are discussed briefly below.

The U.S. Environmental Monitoring and Assessment Program (EMAP), explicitly incorporates societal values for major ecosystems upon which *assessment endpoints* are based. These assessment endpoints relate to the measurable aspects of the environment that can be used to evaluate and monitor the health of the system (Heck et al., 1993). These endpoints are intended to be unambiguous, have social or biological relevance, and be quantifiable. The quantifiable measurements associated with the assessment endpoints are EMAP's indicators and would include many of the measures of soil quality or health described in the discussion of soil health indicators that follow. In such a framework, measures of soil health do not represent a laundry list of unknown or unclear relevance and importance, but are linked to the health of the *ecosystem properties* and the value that an ecologically literate society places on that ecosystem.

An ecosystem management framework incorporating societal values also has been endorsed in Canada (Canadian Council of Ministers for the Environment, 1996a). This framework includes ecosystem goals, objectives, and indicators as the basic template for management of ecosystems. In this framework, *ecosystem goals* reflect broad societal values for a system that are clear, attainable, sustainable, and supported by ecological, economic, and social reasoning. An example of an ecosystem goal for Lake Ontario is that "the presence of contaminants shall not limit the use of fish, wildlife and waters of the Lake Ontario basin by humans and shall not cause adverse health effects in plants and animals" (Ecosystem Objectives Working Group, 1992, unpublished data). Goals are ecosystem specific and in an agroecosystem may include such goals as sustainable production of agricultural produce that does not negatively impact related aquatic and terrestrial systems and their ability to support wildlife. Such broad societal goals are agreed upon by all stakeholders and are then translated into more specific *ecosystem objectives* that describe the desired conditions for a given ecosystem, which relate to the broader goals. It is here that the condition of the soil subsystem is placed in the context of broader ecosystem goals. For example, an ecosystem objective may be to maintain the productive capacity of the soil subsystem with minimal fertilizer subsidy, which in turn will limit impact on wildlife and water quality. As in the EMAP framework, *ecosystem indicators* are the quantifiable measurements that can be used to evaluate progress towards meeting each objective. For the Lake Ontario or the agroecosystem examples above, indicators might include such things as "levels of contaminants in surface water do not exceed the Canadian Water Quality Guidelines (Canadian Council of Resource and Environmental Ministers, 1987), or fertilizer subsidy per unit yield shall not exceed *x*; or there will be no net loss of soil organic content from the system, and others.

Ecosystems naturally depend upon the well-functioning of subsystems such as the soil subsystem. The number of possible soil health indicators is large, and the choice daunting. What is important is that these indicators are unambiguous, measurable, and have relevance to broader ecosystem goals and objectives. For example, most agroecosystem objectives will rely on soil and soil properties to be fully or partially achieved. A less explicit, and less precise, objective for a farm family might relate to their quality of life. Quality of life includes a number of components, but is definitely affected by income. Income is partially, but not completely determined, by productivity and crop yield, which is in turn determined by soil. From an ecosystem health perspective, it is important to identify objectives for the entire system early on, prior to defining what specific soil-related objectives and indicators might be. The frameworks discussed above provide a unifying template for such considerations.

Within the context of such general management frameworks, there are several fundamental considerations that must be kept in mind in an ecosystem health approach to soils. The principles of ecosystem health, and particularly EDS, show that suites of indicators are the most reliable approach to assessing ecosystem health, because single indicators or indices can be misleading. This maxim does not make the selection of indicators easier though. As discussed above, in general we seek indicators that are mechanistically linked to human-set objectives for the system and/or its ecological health (Bernstein, 1992; Canadian Council of Ministers for the Environment, 1996a). It is at this stage that the critical connective links can be established between indicators of subsystem function and indicators of ecosystem health bearing in mind that some subsystem functions are more vital than others for the larger system, and that relationships may be scale-dependent. Emergent properties are pervasive in nature (Funtowicz & Ravetz, 1994) and clearly, in dealing at the scale of ecosystem health, it must be remembered that the whole is more than the sum of the parts.

Objectives for soils must be realistic, taking into account the inherent soil properties that affect the potential of the system, as well as the properties that can be affected by management. Soil characteristics such as texture, topography, depth, drainage, and mineral composition change little over years and decades (although they can be altered with sufficient investment and/or inputs), and so set limits on objectives that management can reasonably expect to achieve. On the other hand, fertility characteristics can be modified with chemical and organic inputs over a period of days. Some economically nonviable soils can become viable and highly productive with judicious application of lime and/or particular nutrients. When setting objectives for the system, managers must have a clear understanding of limitations posed by inherent characteristics of the system in question as well as the range of management practices, and associated costs, available to modify the system.

Finally, there is some ambiguity surrounding the boundaries of the ecosystem for which we are defining indicators and objectives. There are a number of suitable demarcations for specific types of analyses. For example, drainage basins and subbasins are appropriate for analysis of many aquatic and terrestrial ecosystems. Other classifications based on topography, climate, land use, and dominant vegetation have been constructed and used in environmental assessments and management at regional, national, and international levels (Friend & Rapport, 1991; United Nations Statistical Office, 1991). These varying geographies suggest that the choice of boundaries, although to some extent arbitrary, is feasible on a pragmatic basis. The guiding principle is one of suiting the geography to the problem being addressed.

12–4 SOIL QUALITY OR SOIL HEALTH?

Much of the discussion related to ecosystem health assessment and management deals with the importance of indicators as measures of ecosystem health. It is at the indicator level, that much of the attention on measures of soil health or quality is focused. At this point, it is important to define the difference between soil quality and soil health; terms that are used interchangeably in the scientific literature (Harris & Bezdicek, 1994). Soil quality relates to the capability of soil for production or provision of other services beneficial to humans such as pollution attenuation (Doran & Parkin, 1994). Soil quality indicators, such as those used by the U.S. Environmental Protection Agency's Environmental Monitoring and Assessment Program's (EMAP) Agroecosystems component, generally include a range of site-specific measures such as pH, texture, organic matter content, or microbial biomass with a strong focus on biophysical features (Meyer et al., 1992). They provide a framework within which to assess the ability of the soil to meet human objectives. Soil quality, then, includes inherent soil characteristics that may assist or constrain the soil's ability to support the achievement of human objectives. Such inherent characteristics could be considered somewhat equivalent to genetic potential and are not included in soil health. Soil health strictly includes only those characteristics that can be affected by management at scales relevant to managers, e.g., over years, not centuries. The fact that a soil is too steep, or too stony, to be economically farmed cannot be sufficient reason to assess that soil as unhealthy. Nor is it correct to say that soil health is simply a subset of soil quality, as is explained below; soil health includes factors that may be unrelated to the achievement of human objectives. Though the term soil *quality* is used in Canada (Canadian Council of Ministers of the Environment, 1991) to refer to levels of toxic substances in soil that are considered *unhealthy* and unsustainable, using the above argument, these should be more appropriately considered as measures of soil *health*.

Another area of emphasis in the soil quality literature relates to the development of indices of soil quality (Parr et al., 1992). The use of indices has been considered in the ecosystem health literature and the ability of aggregate indexes, i.e., formulaic or algorithmic calculations that condense values of a number of indicators into one overall value, to accurately measure ecosystem health or detect changes has been questioned and/or treated skeptically by many investigators, although the value of suites of indicators is emphasized (Milne, 1992). Karr (1992), who developed the much respected Index of Biotic Integrity for aquatic ecosystems, has advocated a multimetric approach, but warned that each test of which an index is composed must have a sound theoretical and empirical basis, and that aggregation of test results into a single measure results in a loss of information about specific system attributes (Karr, 1992). Rapport et al. (1985) emphasize that detection of EDS requires a number of different metrics.

12–5 SOIL INDICATORS IN AN AGROECOSYSTEM CONTEXT

Soil quality and soil health cannot necessarily be equated with agroecosystem health, or even with health of the soil sub-system. Agroecosystem health requires: (i) freedom from EDS; (ii) minimal, nonincreasing subsidy per unit yield, and (iii) prevention of negative impacts to surrounding systems. Soil quality indicators address the second consideration and, in some cases, aspects of the third, but have little direct relation to the first. There is another set of soil ecological indicators that addresses these gaps, and that intersects the set of soil quality indicators. Further, the soil subsystem is directly linked to the rest of the agroecosystem, so an assessment of its health cannot be made focusing solely on soil-related indicators or biophysical indicators. For example, there can be human health impacts from nitrate in groundwater; there is a relationship with soil health but it is an indirect one. The assessment of agroecosystem health is incomplete without the socioeconomic component. We now turn to examine the two sets of indicators mentioned above, namely indicators relating to economic objectives and indicators relating to ecological objectives; then we consider how these two elements of soil health relate to agroecosystem health.

There is a broad range of quality indicators for agricultural soils. In terms of productivity, perhaps the best set of general screening indicators relate to yield trends over time under a consistent management system, i.e., rotation, tillage type, fertilizer regime, and others (Powlson & Johnston, 1994), with a decrease indicating a loss of health. Large year-to-year variations in yield have been shown to be a prelude to complete crop failure (Woodward, 1993). Yield integrates an array of factors, some related to soil, so it is important to consider the impact of nonsoil factors, particularly climatic factors, on yield changes before drawing conclusions regarding soil health. There are a number of other objectives that managers also may seek to assess, biodiversity for instance, for which indicators other than yield would be more appropriate.

Indicators of soil ecological health relate to ecological functioning within the soil rather than directly to management objectives such as productivity. There has been a great deal of interest recently in indicators of soil ecological health, particularly bioindicators (e.g., Doran & Parkin, 1994). Anderson (1994) stated that management practices that reduce complex biological interactions in soils are not inherently unstable, but depend on the ecological and agronomic context. It appears that much more work is required before there is an immediate understanding of the relationship between the various bioindicators currently being investigated and soil ecological health. For example, it is not enough to say that microbial activity is critical to soil health when levels of such activity vary by orders of magnitude throughout a growing season in response to climatic factors and seasonal plant growth patterns, or even fertilizer application (Grace et al., 1994; Hassink et al., 1991). Some form of standard would be ideal, including ranges of healthy levels and thresholds, to allow the assessment of a measurement or series of measurements; however, it must be recognized that the development of absolute standards for assessment of ecosystem health may not be scientifically possible, underlining the need for suites of indicators that, taken together, can provide the required information. It also is necessary to monitor indicators over time to assist in determining whether a healthy or unhealthy trend exists.

Once a trend of declining health has been identified, the difficult work of diagnosing the cause begins. Diagnosing decreasing productivity often begins with an analysis of the availability of water and nutrients in conjunction with pH, and there may be specific symptoms in plants that indicate such problems. Toxic effects or soil-borne pathogens also may affect productivity. Changes in management practices, such as tillage type, can affect yield directly (Ball et al., 1994; Carter, 1994). Some of these causes–symptoms occur over short time-scales, certainly within a growing season, and can often be corrected with one-time management intervention. In some cases, the factors may interact. For example, aluminum toxicity is associated only with pH below 5.5. A difficulty in using these short-term changes as indicators is their transience and the possibility of other transient factors, notably climate, creating *noise*, which makes the cause–effect relationship to indicator response difficult to interpret.

Other changes in soil may be more chronic, responding slowly over years and decades to agricultural practices such as tillage, cropping and harvest, residue management, or agrochemical use and include: soil erosion, organic matter content, structure, in-soil species diversity and population sizes, and factors related to infiltration and water-holding capacity. For example, there is evidence that the amount of organic matter will decline when a previously uncultivated soil is cropped, eventually reaching a new equilibrium, depending on the amounts of C and N returned to it by the specific farming system (Swift, 1994). The new equilibrium is not necessarily quickly reached, however. In Rothamstead, England, it took 130 yr for a percentage of C to reach an equilibrium under a continual, specific farming system (Powlson & Johnston, 1994). Other work suggests that this kind of equilibrium may not be reached in all agricultural systems (Bird & Rapport, 1986). Battiston et al. (1987) have shown that soil erosion decreased corn (*Zea mays* L.) yield minimally until a particular soil depth threshold was crossed (which varied for differing soil types) at which point average yield losses were 59% (range from 16–80%) on these severely eroded soils. Soil structure and aggregation, particularly pore size distribution, continuity, and stability, may be the most critical determinant of productivity but is difficult to measure in a single, consistently meaningful way (Kay, 1990).

Finally, long-term changes in soil can occur over centuries or millennia. These include topographic changes due to geomorphological processes and, of particular importance for agriculture, loss of clay minerals and associated cation-exchange capacity due to natural weathering processes over many thousands of years (Chesworth, 1983). These latter processes result in decreasing fertility and ultimately barren-ness unless new sources of rock are added to the system. These processes also are accelerated by one or two orders of magnitude by agriculture (Chesworth, 1983). Such changes occur far too slowly to provide a useful indication of stress on the agroecosystem.

The characteristics of soil, discussed above, which respond over the medium term may be the most useful indicators of soil health, partly because they integrate many different processes affecting soil, including abiotic, biotic, and management and partly because they result from the combined effects of management over time. Soil aggregation and structure are affected by a combination of abiotic processes such as wet–dry cycles, biotic processes such as the formation of bacterial gels, and management practices such as tillage, and their subsequent interactions (Coleman et al., 1994; Foster, 1994). More transient responses, such as N concentration, change quickly and can easily be modified by management so that the symptom is masked. Nitrogen fertilization can have an immediate effect on N concentration in soil. Similarly, very slow processes, which result in what we have called *inherent* soil characteristics, are of little value in assessing the agroecosystem's health.

There are at least two intersecting sets of indicators of soil health: those related to productivity and other human-determined objectives and those related to the soil ecosystem. Economic-driven productivity can sometimes appear to be in conflict with ecology and for purposes of comparison among indicators, and to identify processes where the dual roles of soils apparently conflict, it will be most useful to concentrate on indicators that fall into the overlap area between the two sets. Our analysis suggests that measures related to erosion and soil structure are among those that measure both aspects and so may be the most useful in assessing soil health from an agroecosystem perspective.

12–6 ECOSYSTEM MANAGEMENT: NITRATE–PHOSPHATE PROBLEM IN AGROECOSYSTEMS

Agroecosystems provide an excellent example of the apparent conflict between the ecological and economic roles of soils. The farmer also has multiple roles. As producer, the farmer is dependent on sufficient production to maintain an adequate income and lifestyle, yet is also dependent on the satisfactory functioning of the agroecosystem, especially the soil subsystem, to maintain production (farmer as steward). Agricultural management can be seen to affect all aspects of ecosystem health as set out in the definition above. In many agricultural soils degradation occurs, such as erosion or loss of structure, that may ultimately lead to destabilization of the ecosystem (Bormann & Likens, 1979) and symptoms of EDS. Agricultural systems are necessarily subsidized with energy, nutrients and other agrochemicals (Gliessman, 1990). These subsidies, together with other aspects of agricultural management, affect the functioning of the agroecosystem and have the potential to affect surrounding systems. Yet the ills that may be wreaked on agricultural soils are often based on the intent to increase yield and economic output. An agroecosystem that forces farmers off the land because they can't make a living cannot be considered healthy no matter how good the ecological condition. Economic viability must include the cost of production and ecological viability must incorporate human needs. For example, maximum production from any crop variety cannot generally be achieved using the soil medium, but with hydroponics (Miller et al., 1989). If maximum production per plant were the only criterion for healthy agriculture, all crops would be grown hydroponically; however, the cost of hydroponic production is much greater than soil-based farming. The farmer effectively uses certain ecological properties of the soil to economically subsidize the cost of production. Agroecosystem health requires a balance among all these interrelated parts.

The complexity of evaluating ecosystem health, and the need for reconciling ecological functioning with economic production in agroecosystems is well-illustrated by the current debate over control of NO_3 leaching and associated potential for groundwater contamination, and phosphate run-off causing eutrophication. Nitrogen and P are essential crop nutrients. There is a strong argument that the addition of N and P fertilizer is a necessary component of many farming systems, at least if economic viability is to be maintained; however, fertilizers can have negative consequences on ecological processes and on humans and other species living within the agroecosystem, and in adjoining ecosystems. Since the early 1970s, there has been serious public concern over the effects of P pollution, such as eutrophication by phosphates, as evidenced by activities of the Pollution from Land Use Activities Reference Group (PLUARG), set up by the International Joint Commission on the Great Lakes (Coote et al., 1982). More recently, the issue of NO_3 contamination of groundwater with its attendant negative human health effects on those reliant on this water has attracted attention and public debate[1].

A number of possible responses to fertilizer impacts on the environment can be readily prescribed. The most obvious is to decrease the amount of fertilizer input to the system in order to reduce the amount entering the water system through runoff and leaching to the groundwater. In general, excessive leaching of NO_3 occurs only when N availability exceeds crop N requirement (Addiscott et al., 1991). In some cases, fertilizer inputs are much greater than those recommended for a healthy economic return and can easily be decreased without significantly affecting yield (Legg et al., 1989; Schepers et al. 1991). In other cases, recommended fertilizer application may result in groundwater contamination (Power & Broadbent, 1989). Here, both choices result in an unhealthy situation: loss of production due to below-minimum fertilizer inputs results in unhealthy agroecosystems from an economic perspective, while raising concentration of NO_3 in groundwater above safe levels creates a human health hazard and may disrupt ecological processes, by polluting local water systems.

The element of spatial and temporal scales add further complexity to the discussion. Should NO_3 leaching be minimized at the field scale or at a higher level of aggregation such as the watershed where a number of different rotations may individually be above or below a selected threshold but collectively meet the agreed standard? Temporally, it may be possible for a rotation over a period of years to meet an average annual NO_3 leaching standard while exceeding it in certain years. It may seem obvious that there is no single correct approach to the question of scale. From an ecosystem perspective, management at the watershed level makes sense, but this does not make the contamination of an individual farm well less likely, nor any less of a disaster for the farmer involved. In general, management occurs at the farm scale, while factors affecting management decisions occur at many scales: fertility at the field scale; agricultural policy at the regional scale, provincial and national scales; economic forces at the local to global scales. The NO_3–PO_3 issue exemplifies the need for a holistic approach that incorporates values, such as the maximum acceptable NO_3–PO_3 concentration in groundwater, in determining ecosystem health[2].

[1] Note that NO_3contamination can result from synthetic fertilizers and from organic forms such as manure or legume cover crops.

[2] The maximum acceptable NO_3 concentration can never be determined solely on a scientific basis, since different concentrations will result in different outcomes, depending on exposure, individual health, genetic history, and other factors.

A decision on the acceptable level of risk for any hazard must be based on all the associated costs and benefits including: health, economic, policy and social values (Gentile & Slimak, 1992).

Finally, it should be obvious that the role of NO_3 and other fertilizers in agroecosystem health is but one component of a much larger set of factors that determine the system's total health picture. There is a great deal of scope for interaction of factors in different and complex ways and it is this scope that makes ecosystem health assessment challenging, controversial, and absolutely critical to the future of humanity.

12–7 INDICATORS OF SOIL QUALITY FROM A CONTAMINATED LAND PERSPECTIVE

Traditionally, soil health has focused on the structure and function of the soil as related to crop productivity. During the past decade, soil contamination has emerged as a key environmental issue not only in Canada, but worldwide (Cairns, 1991; Canadian Council of Ministers for the Environment, 1991; Sheppard et al., 1992). Soil contamination places both ecological and human health at risk. To deal effectively with ecosystem health as it relates to the health of the soil subsystem, our understanding of soil quality and soil health must be expanded to include contaminant levels and their effects. The following definition captures a broader, ecosystem-based goal for soil health with respect to contaminants, in that soil of good quality or health must "pose no harm to any normal use by humans, plants or animals, not adversely affect natural cycles or functions and not contaminate other components of the environment" (Moen, 1988). This definition, which formed the basis of soil policy in the Netherlands, clearly challenges the scientists, managers, and regulators to place soil health in a dynamic, ecosystem context.

Our understanding of contaminants in soil is still evolving, and this has presented considerable scientific and regulatory challenges. Indicators of soil *quality* (here used interchangeably with the term *health*), with respect to contaminants must be practical as well as scientifically defensible. Since the restoration of contaminated soil to pristine levels is rarely feasible (Cairns, 1989), soil quality standards and guidelines usually aim to achieve some defined level of ecological and human health protection, but what this level should be is far from resolved (Sheppard et al., 1992). Increasingly, these levels are based on the emerging principles of ecological and human health risk assessment.

Risk is generally characterized as the potential for a substance in the environment to adversely affect a valued ecosystem component (VEC) or receptor. The evaluation of risk depends not only on identifying ecosystem components that are related to societal values and ecosystem integrity, but also in identifying and evaluating the environmental pathways by which a contaminant in soil may reach these ecosystem components. The evaluation of soil health from this perspective represents a challenge not only to soil scientists, but to biologists, chemists, physical scientists, toxicologists, and human health experts. The range of contaminants to be addressed, from heavy metals to excess nutrients and inhalable fibers and the inherent variation in their behavior and effects are only some of the aspects that must be considered. The impact of contaminated soil on the environment is not always direct. Humans eat plants grown on soil, drink water that passes through soil, and inhale air that exchanges with air in the soil. In addition, there is a complex web of political, social, and legal concerns as well as technological limitations in dealing with contaminated soil.

The Canadian program for the development of national soil quality guidelines (standards) is illustrative of an ecosystem management framework for soil quality with respect to contaminants. The broad goals for the development of Canadian Soil Quality Guidelines is to protect, sustain, and enhance the major beneficial uses of the Canadian environment with equal emphasis on the protection of the environment and human health (Canadian Council of Ministers for the Environment, 1991). This program was initiated in 1989 in response to an increasing concern about the human health and environmental risk of contaminated land and the constraints this places on sustainable land use in Canada (Gaudet et al., 1992). Soil quality guidelines or indicators developed under this program are a very specialized subset of the indicators of soil health described above and should not be considered in isolation.

The ecosystem management framework described earlier in this paper incorporates as its central tenet, societal values. The development of national soil quality guidelines, uses as its basis, the major socioeconomic value of land (Canadian Council of Ministers for the Environment, in press). The major land-use categories so far considered include: (i) agricultural; (ii) residential–parkland; (iii) commercial; and (iv) industrial.

The Canadian protocol for development of soil quality guidelines describes the receptors or VECs to be protected in order to sustain the defined land uses in a *healthy* state according to the definition advanced by Moen (1988; Table 12–1). The final recommended soil quality guidelines or standards are expressed simply as numerical limits or levels of contaminants in soil that should not be exceeded in order to protect the environment and human health with respect to a specified land use. They are actually integrated indicators of soil-subsystem health that are clearly linked to the health and sustainability of the overlying terrestrial and aquatic ecosystems and their societal value.

In developing Canadian soil quality guidelines, the soil subsystem is placed in the context of the receptors or VECs directly or indirectly exposed to, and affected by, soil contaminants (e.g., through direct contact, ingestion of contaminated crops, or infiltration to groundwater). Both human and ecological components are considered. For example, under an ecological scenario, the receptors identified for protection in an agricultural setting include grazing wildlife and livestock, crops and native plants, soil invertebrates, microbial function (e.g., respiration, C fixation), and aquatic life. Figure 12–1 illustrates potential uptake scenarios in a terrestrial ecosystem, considering only ecological components. In complex multi-use scenarios, humans, crops and livestock also would be included as part of the food chain or direct exposure pathways. From the human health perspective, there is only one receptor, but exposure will vary widely with land use and site conditions. In an agricultural setting, direct soil ingestion, inhalation of vapors from volatile contaminants in soil, as well as indirect exposure to soil

Table 12–1. Framework for development of Canadian soil quality guidelines. These guidelines recognize the multifunctional use of soil as an integral part of both aquatic and terrestrial ecosystems. They are based on potential risk to key human and environmental receptors and consider both direct and indirect exposure to soil contaminants (Moen, 1988).

Land use	Agricultural	Residential–parkland	Commercial	Industrial
Receptors of concern	—humans (child most sensitive) —livestock —crops —invertebrates —native plants and animals —microbial function	—humans (child most sensitive) —native plants and animals —microbial function	—humans (child most sensitive) —native plants and animals —microbial function	—humans (adult) —native and ornamental plants —microbial function
Exposure pathways of concern	Humans (24 h exposure) *Direct:* —soil ingestion, dermal, inhalation *Indirect:* —contaminated drinking water —soil vapours in basements —consumption of produce (meat, dairy, crops)	Humans (24 h exposure) *Direct:* —soil ingestion, dermal, inhalation *Indirect:* —contaminated drinking water —soil vapours into basements —consumption of backyard garden produce	Humans (10 h exposure) *Direct:* —soil ingestion, dermal, inhalation *Indirect:* —contaminated drinking water —soil vapours into basements	Humans (8 h exposure) *Direct:* —soil ingestion, dermal, inhalation *Indirect:* —contaminated drinking water —contamination of adjacent residential properties
	Environmental *Direct:* —soil contact (plants, microbial processes, invertebrates) —soil ingestion (wildlife and livestock) *Indirect:* —ingestion of contaminated forage —contaminated surface water (aquatic life)	Environmental *Direct:* —soil contact (plants, microbial processes, invertebrates) *Indirect:* —contaminated surface water (aquatic life)	Environmental *Direct:* —soil contact (plants, microbial processes, invertebrates) *Indirect:* —contaminated surface water (aquatic life)	Enviroмmental *Direct:* —soil contact (plants, microbial processes, invertebrates) *Indirect:* —contaminated surface water (aquatic life)

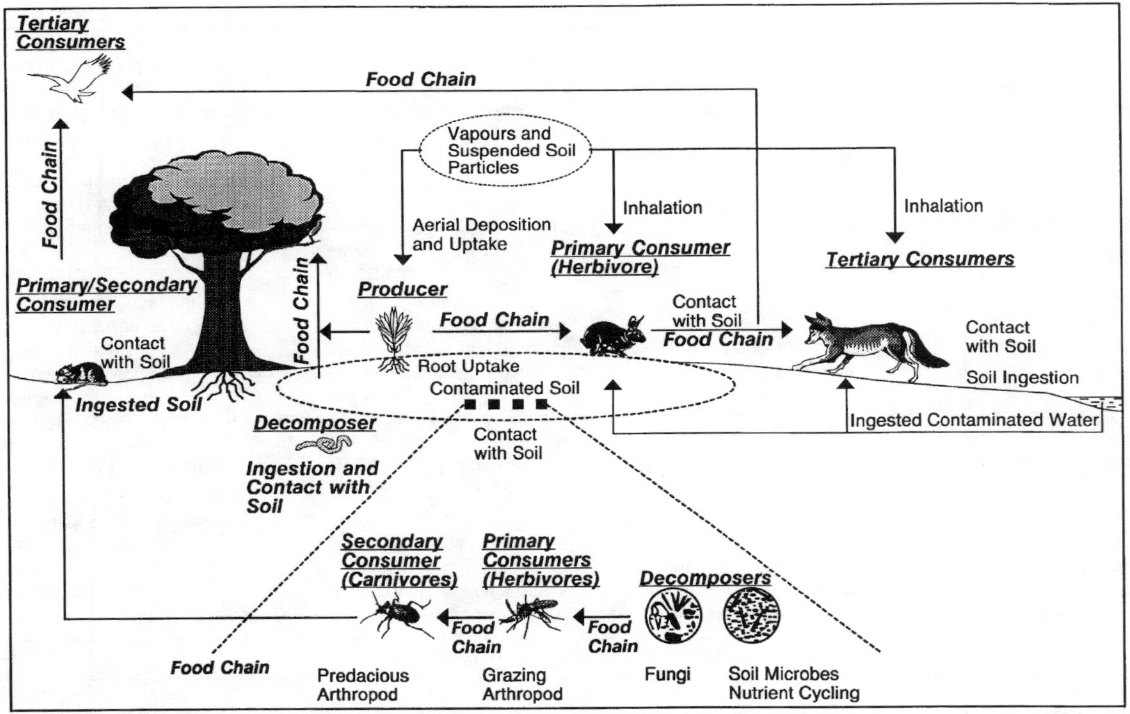

Fig. 12–1. The impact of contaminated soil on the terrestrial ecosystem for a simplified natural scenario. In complex, multi-use settings such as agricultural lands, these receptors and pathways would include humans, crops, and livestock and such factors as infiltration of soil vapors into basements.

contaminants through ingestion of produce (fruit–vegetable, meat, and dairy products) grown on the soil are considered. In addition, the potential contamination of groundwater is evaluated. Though produce grown on contaminated soil is important in an agricultural scenario, soil quality for industrial lands will emphasize, from the human health perspective, occupational exposure on site, as well as potential off-site impacts such as infiltration of vapors into basements and the off-site migration of contamination to surrounding lands. From an ecological perspective, valued ecosystem components related to soil health under an industrial scenario may exclude grazing, livestock, and crops but would include protection of ornamental and native plants and of the invertebrate populations and microbial function. The latter two considerations are regarded as fundamental to sustaining the capacity of soil in a healthy productive state over the long term and therefore are not considered *expendable* even for industrial land uses.

A series of calculations described in the Canadian Council of Ministers for the Environment protocol are used to evaluate toxicity and exposure information in terms of the potential risk to the environment and human health. These calculations are used to define maximum levels of contaminants in soil to sustain healthy terrestrial and associated aquatic ecosystems relative to the specified land use. These indicators are clear, unambiguous, science-based indicators of the health of the soil subsystem that can be linked to the health of the larger ecosystem and to socioeconomic values (Table 12–2). Importantly, they are integrated, quantitative measures that can be easily evaluated, interpreted, monitored, and communicated to the public and stakeholders.

12–8 SUMMARY AND CONCLUSIONS

In this chapter we have suggested a framework for ecosystem-based management incorporating societal values as the basis for development of goals, objectives, and indicators of ecosystem health. To implement such a framework, research and management emphasis is urgently needed for the development of quantifiable indicators of the health of the soil subsystem that are clearly linked to the broader ecosystem goals and objectives. Though many such indicators have been proposed, they require validation with respect to their relevance and importance as indicators of health. In this process two elements are important: (i) that there is the need for a group of indicators—syndromes, not symptoms (more technically signs) are to be sought; and (ii) that indicators fall into a number of categories. Some of these categories are useful for diagnostic purposes, identifying the potential causes of particular dysfunctions, and some are useful for general screening purposes, that is overall assessments of health. Some are also useful for risk assessment, that is to evaluate potential losses that may not yet have surfaced, as a result of particular types of stresses impacting the environment. In this discussion, we have placed emphasis on the importance of both biophysical and socioeconomic conditions to provide a spectrum of indicators for assessing ecosystem health. Presently, statistical methods that employ classical statistics, are inadequate for situations wherein replicates are not possible. New methodologies will need to be developed whereby one can confirm the utility of such

Table 12–2. The relationship among ecosystem health goals, objectives, and indicators using an agroecosystem example. Measures of soil health are generally specified at the indicator level and may be the kinds of quantified measures traditionally used for evaluating soil quality; however, to be valid as indicators of ecosystem health, they must be clearly linked to, and measures of, broader goals and objectives as in the table below. See Table 12–1 for further details on Canadian soil quality guidelines.

	Goals	Objectives	Indicators
Main characteristics	—define broad societal goals for the ecosystem —negotiated with all stakeholders on a region or site-specific basis	—apply to a particular aspect of the desired ecosystem goal such as wildlife habitat or crop productivity —usually narrative	—quantifiable measures of progress towards objectives —numerical or narrative —based on best scientific information and measurements available
Soil-related examples for agroecosystems	—the agroecosystem should be maintained and as necessary restored to support diverse biological communities, sustainable production of crops and livestock and healthy human communities	—levels of contaminants in soil will not adversely affect human or livestock health through direct soil exposure or ingestion of produce grown on the soil —levels of contaminants in soil will not adversely affect adjacent ground water and surface water systems and their uses, including aquatic life, drinking water, irrigation and recreation	—levels of substance *x* in soil should not exceed the Canadian soil quality guidelines for protection of: —plants and soil invertebrates —grazing livestock and wildlife —soil microbial processes —aquatic life form —groundwater recharge into surface water —human health (through exposure from direct soil ingestion; inhalation of soil vapours, drinking ground water; ingestion of farm produce)

groups of indicators as suggested here. These methods would need to establish that where indicators indicate poor ecosystem health, then in the absence of interventions, there are further declines in health and these can be directly related to the loss of ecosystem services and management options (Rapport, 1995a,b,c). These challenges are not only in the areas of soil health, but are general challenges as the entire area of ecosystem health advances to a firmer footing in practice and in application.

ACKNOWLEDGMENTS

This work was sponsored by the Eco-Research Chair Program in Ecosystem Health with funding from the three research councils of Canada (Medical, Social, and Natural Sciences) and partner Agencies (Environment Canada, Statistics Canada, Canadian Forest Service, and Ontario Ministry of Agriculture, Food and Rural Affairs).

REFERENCES

Addiscott, T.M., A.P. Whitmore, and D.S. Powlson. 1991. Farming, fertilizers and the nitrate problem. CAB Int., Wallingford, Oxon, England.

Anderson, J.M. 1994. Functional attributes of biodiversity in land use systems. p. 267–290. *In* D.J. Greenland and I. Szabolcs (ed.) Soil resilience and sustainable land use. CAB Int., Wallingford, England.

Ball, B.C., R.W. Lang, E.A.G. Robertson, and M.F. Franklin. 1994. Crop performance and soil conditions on imperfectly drained loams after 20–25 years of conventional tillage and direct drilling. Soil Tillage Res. 31:97–118.

Battiston, L.A., M.H. Miller, and I.J. Shelton. 1987. Soil erosion and corn yield in Ontario: 1. Field evaluation. Can. J. Soil Sci. 67:731–745.

Bernstein, B.B. 1992. A framework for trend detection: Coupling ecological and management perspectives p. 1101–1114. *In* D.H. McKenzie et al. (ed.) Ecological indicators. Vol. 2. Elsevier, London.

Bird, P.M., and D.J. Rapport. 1986. State of the environment report for Canada. Can. Gov. Publ. Ctr., Ottawa.

Blum, W.E.H., and A.A. Sandelises. 1994. A concept of sustainability and resilience based on soil functions: The role of ISSS in promoting sustainable land use. p. 535–542. *In* D.J. Greenland and I. Szabolcs. (ed.) Soil resilience and sustainable land use. CAB Int., Wallingford, England.

Bormann, F.H., and G.E. Likens. 1979. Pattern and process in a forested ecosystem. Springer-Verlag, New York.

Cairns, J.. Jr. 1991. Restoration ecology: A major opportunity for ecotoxicologists. Environ. Toxicol. Chem. 10:429–432.

Cairns, J., and J.R. Pratt. 1995. The relationship between ecosystem health and delivery of ecosystem services. p. 63–76. *In* D.J. Rapport, C. Gaudet, and P. Calow. (ed.) Evaluating and monitoring the health of large-scale ecosystems. Springer-Verlag, Heidelberg.

Canadian Council of Ministers of the Environment. 1991. Interim Canadian environmental quality criteria for contaminated sites. CCME EPC-CS34. CCME, Winnipeg, MB.

Canadian Council of Ministers of the Environment. 1996a. A framework for developing goals, objectives and indicators of ecosystem health. Draft Rep. Water Quality Guidelines Task Group of the CCME, Winnepeq, MB.

Canadian Council of Ministers of the Environment. 1996b. A protocol for the derivation of soil quality guidelines for protection of environmental and human health. CCME EPC. CCME, Winnipeg, MB.

Canadian Council of Resource and Environment Ministers 1987. Canadian water quality guidelines. Task Force on Water Quality guidelines of the CCREM, Ottawa, ON.

Carter, M.R. 1994. A review of conservation tillage strategies for humid temperate regions. Soil Tillage Res.. 31:289–301.

Chesworth, W., F. Macias-Vazquez, D. Acquaye, and E. Thompson. 1983. Agricultural alchemy: Stones into bread. Episodes 1:3–7.

Coleman, D.C., P.F. Hendrix, M.H. Beare, D.A. Crossley, Jr., S. Hu, and P.C.J. van Vliet. 1994. The impacts of management and biota on nutrient dynamics and soil structure in sub-tropical agroecosystems: Impacts on detritus food webs. p. 133–143. *In* C.E. Pankhurst et al. (ed.) Soil biota: Management in sustainable farming systems. CSIRO, East Melbourne, Australia.

Coote, D.R., E.M. MacDonald, W.T. Dickinson, R.C. Ostry, and R. Frank. 1982. Agriculture and water quality in the Canadian Great Lakes Basin: 1. Representative agricultural watersheds. J. Environ. Qual. 11:473–481.

Costanza, R. 1992. Toward an operational definition of ecosystem health. p. 239–256. *In* R. Costanza et al. (ed.) Ecosystem health: New goals for environmental management. Island Press, Washington, DC.

Doran, J.W., and T.B. Parkin. 1994. Defining and assessing soil quality. p. 3–21. *In* J.W. Doran et al. (ed.) Defining soil quality for a sustainable environment. SSSA Spec. Publ. 35. SSSA, Madison, WI.

Foster, R.C. 1994. Microorganisms and soil aggregates. p. 144–155. *In* C.E. Pankhurst et al. (ed.) Soil biota: Management in sustainable farming systems. CSIRO, East Melbourne, Australia.

Friend, A.M., and D.J. Rapport. 1991. Evolution of macro-information systems for sustainable development. Ecol. Econ. 3:59–76.

Funtowicz, S., and J.R. Ravetz. 1994. Emergent complex systems. Futures. 26(6):568–582.

Gaudet, C., A. Brady, M. Bonnell, and M.P. Wong. 1992. Canadian approaches to establishing cleanup levels for contaminated sites. p. 49–65. *In* Hydrocarbon contaminated soils and groundwater. Lewis Publ. Chelsea, MI.

Gentile, J.H., and M.W. Slimak. 1992. Endpoints and indicators in ecological risk. p. 1385–1397. *In* D.H. McKenzie et al. (ed.) Ecological indicators. Vol. 2. Elsevier, London.

Gliessman, S.R. 1990. Agroecology: Researching the ecological basis for sustainable agriculture. p. 3–10. *In* S.R. Gliessman (ed.) Agroecology: Researching the ecological basis for sustainable agriculture. Springer-Verlag, New York.

Grace, P.R., J.N. Ladd, and J.O Skjemstad. 1994. The effect of management practices on soil organic matter dynamics. p. 162–171. *In* C.E. Pankhurst et al. (ed.) Soil biota: Management in sustainable farming systems. CSIRO, East Melbourne, Australia.

Hansen, P.D. 1995. Assessment of ecosystem health: Development of tools and approaches. p. 195–217. *In* D.J. Rapport et al. (ed.) Evaluating and monitoring the health of large-scale ecosystems. Springer-Verlag, Heidelberg.

Harris, R.F., and D.F. Bezdicek. 1994. Descriptive aspects of soil quality/health. p. 23–35. *In* J.W. Doran et al. (ed.) Defining soil quality for a sustainable environment. SSSA Spec. Publ. 35. SSSA, Madison, WI.

Hassink, J., J.H. Oude Voshaar, E.H. Nijhuis, and J.A. Van Neen. 1991. Dynamics of the microbial populations of a reclaimed-polder soil under a conventional and reduced-input farming system. Soil Biol. Biochem. 23(6):515–524.

Heck, W.W., C.L. Campbell, D.A. Neher, and M.J. Munster. 1993. An agroecosystem monitoring and assessment program for sustainable agriculture. *In* 12th Annual Organic Agriculture Conf., Univ. of Guelph, Guelph, ON. 29 Jan. 1993.

Hildén, M., and D.J. Rapport. 1993. Four centuries of cumulative impacts on a Finnish river and its estuary: An ecosystem health approach. J. Aquatic Ecosyst. Health. 2:261–275.

Karr, J.R. 1992. Ecological integrity: protecting earth's life support systems. p. 223–238. *In* R. Costanza, B.G. Norton, and B.D. Haskell. (ed.) Ecosystem health: New goals for environmental management. Island Press, Washington, DC.

Karr, J.R. 1995. Using biological criteria to protect ecological health. p. 137–152. *In* D.J. Rapport et al. (ed.) Evaluating and monitoring the health of large-scale ecosystems. Springer-Verlag, Heidelberg.

Kay, B.D. 1990. Rates of change of soil structure under different cropping systems. Adv. Soil Sci. 12:3–50.

Legg, T.D., J.J. Fletcher, and K.W. Easter. 1989. Nitrogen budgets and economic efficiency: A case study of southeastern Minnesota. J. Prod. Agric. 2:110–116.

Meyer, J.R., C.L. Campbell, T.J. Moser, G.R. Hess, J.O. Rawlings, S. Peck, and W.W. Heck. 1992. Indicators of the ecological status of agroecosystems. p. 629–658. *In* D.H. McKenzie et al. (ed.) Ecological indicators. Vol.1. Elsevier, London.

Miller, M.H., G.K. Walker, M. Tollenaar, and K.G. Alexander. 1989. Growth and yield of maize (*Zea mays*) grown outdoors hydroponically and in soil. Can. J. Soil Sci. 69:295–302.

Milne, B.T. 1992. Indicators of landscape condition at many scales. p. 883–895. *In* D.H. McKenzie (ed.) Ecological indicators. Vol. 2. Elsevier, London.

Moen, J.E.T. 1988. Soil protection in the Netherlands. p. 1495–1503. *In* K. Wolf et al. (ed.) Contaminated soil 88. Kluwer Academic Publ., Dordrecht, the Netherlands.

O'Neill, R.V., C.T. Hunsaker, and D.A. Levine. 1992. Monitoring challenges and innovative ideas. p. 1443–1460. *In* D.H. McKenzie et al. (ed.) Ecological indicators. Vol. 2. Elsevier, London.

Parr, J.F., R.I. Papendick, S.B. Hornick, and R.E. Meyer. 1992. Soil quality: Attributes and relationship to alternative and sustainable agriculture. Am. J. Altern. Agric. 7:5–11.

Power, J.F., and F.E. Broadbent. 1989. Proper accounting for N in cropping systems. p. 159–161. *In* R.F. Follett (ed.) Nitrogen management and ground water protection. Elsevier, Amsterdam.

Powlson, D.S., and A.E. Johnston. 1994. Long-term field experiments: Their importance in understanding sustainable land use. p. 367–394. *In* D.J. Greenland and I. Szabolcs. (ed.) Soil resilience and sustainable land use. CAB Int., Wallingford, England.

Rapport, D.J. 1989a. Symptoms of pathology in the Gulf of Bothnia (Baltic Sea): Ecosystem response to stress from human activity. Biol. J. Linnean Soc. 37:33–49.

Rapport, D.J. 1989b. What constitutes ecosystem health? Perspect. Biol. Med. 33:120-132.

Rapport, D.J. 1995a. Ecosystem health: An emerging integrative science. p. 5–31. *In* D.J. Rapport et al. (ed.) Evaluating and monitoring the health of large-scale ecosystems. Springer-Verlag, Heidelberg.

Rapport, D.J. 1995b. Ecosystem health: Exploring the territory. Ecosyst. Health. 1:5–13.

Rapport, D.J. 1995c. Ecosystem services and management options as blanket indicators of ecosystem health. J. Aquatic Ecosyst. Health. 4:97–105.

Rapport, D.J., H.A. Regier, and T.A. Hutchinson. 1985. Ecosystem behavior under stress. Am. Nat. 125:617–640.

Royal Commission on the Future of Toronto's Waterfront. 1992. Regenerations: Toronto's waterfront and the sustainable city. Final report. Royal Commission on the Future of Toronto's Waterfront, Toronto.

Schepers, J.S., M.G. Moravek, E.E. Alberts, and K.D. Frank. 1991. Maize production impacts on groundwater quality. J. Environ. Qual. 20:12–16.

Sheppard, S.C., C. Gaaudet, M.I. Sheppart, P.M. Cureton, and M.P. Wong. 1992. The development of assessment and remediation guidelines for contaminated soils: A review of the science. Can. J. Soil Sci. 72:359–394.

Smol, J.P. 1995. Paleolimnological approaches to the evaluation and monitoring of ecosystem health: Providing a history for environmental damage and recovery. p. 301–318. *In* D.J. Rapport et al. (ed.) Evaluating and monitoring the health of large-scale ecosystems. Springer-Verlag, Heidelberg.

Standing Senate Committee on Agriculture, Fisheries and Forestry, 1984. Soil at risk: Canada's eroding future. Rep. to the Senate of Canada, Ottawa, Ontario.

Swift, M.J. 1994. Maintaining the biological status of soil: A key to sustainable land management. p. 235–247. *In* D.J. Greenland and I. Szabolcs. (ed.) Soil resilience and sustainable land use. CAB Int., Wallingford, England.

United Nations Statistical Office. 1991. Concepts and methods of environment statistics: Statistics of the natural environment. United Nations, New York.

Whitford, W.G. 1995. Decertification: Implications and limitations of the ecosystem health metaphor. p. 273–294. *In* D.J. Rapport et al. (ed.) Evaluating and monitoring the health of large-scale ecosystems. Springer-Verlag, Heidelberg.

Woodward, F.I. 1993. How many species are required for a functional ecosystem. p. 271–291. *In* E.D. Schulze and H.A. Mooney. (ed.) Biodiversity and ecosystem function. Springer-Verlag, Berlin.

13 Role of Soil Chemistry in Soil Remediation and Ecosystem Conservation

Domy C. Adriano and Anna Chlopecka

University of Georgia
Savannah River Ecology Laboratory
Aiken, South Carolina

Daniel I. Kaplan

Pacific Northwest National Laboratory
Richland, Washington

13–1 INTRODUCTION

In densely populated and industrialized countries, soil resources are under severe exploitation for the production of food and fiber and energy-related biomass. The ever increasing world population means more land, including those of marginal quality, needs to be converted for man's use. There is growing awareness of the importance of a sustainable environment in natural resource conservation. This was a theme of the 1992 United Nations Conference on Development and the Environment held in Rio de Janeiro, Brazil. In essence the importance of sustainable environments to sustainable development and economy was emphasized.

While the role of physically degraded soils, i.e., soils eroded by wind and water, on soil productivity is well known, much less is known about the impact of adverse biological and chemical processes on the soil's well-being (Lal & Pierce, 1991; Logan, 1992; Adriano et al., 1997). Thus, biochemical processes leading to soil contamination[1] need to be elucidated. A knowledge of these processes is needed to enable scientists and engineers to develop and design methods of mitigating the adverse effects of polluted[2] soils.

In this workshop it has become apparent that a multidisciplinary but holistic approach is necessary to sustain ecosystem health. In addition, the unique role

[1] Contamination occurs when an ecosystem's chemical composition deviates from a normal composition of an environment; contaminants may not be classified as pollutants unless they have some detrimental effect on the biota.

[2] Pollution occurs when a substance is present in greater than natural concentration (or background values) as a result of human activity and having a net detrimental effect upon the environment and its components.

 Soil Chemistry and Ecosystem Health. Special Publication no. 52.

of the soil as the ecosystem organizer casts soil scientists as key players in sustaining ecosystem health (Coleman et al., 1998, this publication). It is therefore essential that the role of the soil in our society and the various physical, biological, and chemical processes in the soil that may affect its quality be discussed. It is a main objective of this chapter to clarify these processes but with more emphasis on the chemical ones.

13–2 DEFINITION AND FUNCTIONS OF SOIL

The international definition of soil follows: "soil is the upper layer of the earth's crust composed of mineral particles, organic matter, water, air, and organisms" (Bachmann-Erdt, 1993). For the purpose of soil protection, due regard must be directed to the topsoil, subsoil, and deeper layers and to the associated mineral deposits in groundwater. Because soil pollution decreases the availability of resources, it is imperative that soils be treated carefully in order to ensure sustainable ecosystems for future generations. To appreciate the role of soils in ecosystems it is necessary to enumerate their various functions for the society and environment; the most important include (Blum & Santelises, 1994):

- As a controller of element and energy cycles in ecosystems
- As a supporter of plants, animals, and man
- As a base for construction of buildings and various structures
- As a base for the production of food and fiber and other biomass
- As a bearer of groundwater aquifer and mineral deposit
- As a genetic reservoir (i.e., biodiversity)
- As an archive of archaeology and natural history

In addition, society uses soils for the disposal of materials of anthropogenic origin, including hazardous waste and airborne and waterborne contaminants.

13–3 DEFINITION OF SOIL QUALITY

Because of society's increasing awareness of the important role of soil on the health of plants, animals, and human, soil quality has been defined in a number of ways. The following are the more commonly used definitions for soil quality, which are interchangeably used for soil health (soil health is preferred by the farming community, whereas soil quality is preferred by the science community).

Larson and Pierce (1991) define "soil quality as the capacity of a soil to function both within its ecosystem boundaries (e.g., soil map unit boundaries) and within the environment external to that ecosystem." They proposed *fitness for use* as a simple operational definition for soil quality and stressed the major functions of soil as a medium for plant growth, in partitioning and regulating the flow of water in the environment, and as an environmental buffer (the concept of environmental buffer is further discussed later in this chapter). Parr et al. (1992) define "soil quality as the capability of a soil to produce safe and nutritious crops in a sustained manner over the long-term, and to enhance human and animal

Table 13–1. Indicators of changes in soil quality and their relationship to soil function (National Research Council, 1993).

	Soil functions			
Soil quality indicator	Promote plant growth	Promote biodiversity	Regulate water flow	Buffer Environmental changes
Nutrient availability	Direct	Direct	Indirect	Direct
Organic matter	Indirect	Direct	Direct	Direct
Infiltration	Direct	Direct	Direct	Indirect
Aggregation	Direct	Direct	Direct	Indirect
pH	Direct	Direct	Direct	Indirect
Soil fauna	Indirect	Direct	Indirect	Indirect
Bulk density	Direct	Indirect	Direct	Indirect
Topsoil depth	Direct	Direct	Indirect	Indirect
Salinity	Direct	Direct	Direct	Indirect
CEC	Indirect	Indirect	Indirect	Indirect
Water holding capacity	Direct	Direct	Direct	Indirect
Soil enzymes	Indirect	Indirect	Indirect	Indirect
Soil flora	Indirect	Direct	Indirect	Indirect
Heavy metal availability	Direct	Direct	Indirect	Direct

health without impairing the natural resource base or harming the environment." More recently, Karlen et al. (1997) define "soil quality as the fitness of a specific kind of soil to function within its capacity and within natural or managed ecosystem boundaries, to sustain plant and animal productivity, maintain or enhance water and air quality, and support human health and habitation." This definition was developed by a special committee on soil quality within the Soil Science Society of America. Thus, it appears that soil quality can best be defined in relation to the functions that soils perform in ecosystems (Table 13–1).

The quality of soil resources historically has been closely related to soil productivity. Indeed, in many cases soil quality and soil productivity have been used interchangeably (National Research Council, 1993). Other definitions relevant to soil quality include "healthy soil, an essential component of a healthy ecosystem, is the foundation upon which sustainable agriculture is built" and "soil health is the soil fitness to support crop growth without becoming degraded or otherwise harming the environment" (Acton & Gregorich, 1995).

Soil health deteriorates mainly by wind and runoff erosion, loss of organic matter, salinization, breakdown of soil structure, and chemical contamination. As such soil quality cannot be measured directly much less quantitatively but it can be deduced or estimated by measuring key indicators including physical, biological, and chemical properties such as pH, electrical conductivity, nutrient content, organic matter content, and other indirect indicators, such as crop performance and surface water and groundwater quality. Thus, soil quality reflect the composite (holistic) picture of soil physical, chemical, and biological properties and the processes that interact to determine its condition (Table 13–1).

In practice, soil quality is largely dependent on human activities (Fig. 13–1 and 13–2). The ways man uses soil affects soil quality. Soil erosion strips away fertile topsoils, leaving the soil less hospitable to plants. Heavy farm equipment can compact the soil and impede its capacity to accept and store water. Loss of organic matter due to erosion or poor farming practices can seriously hamper the

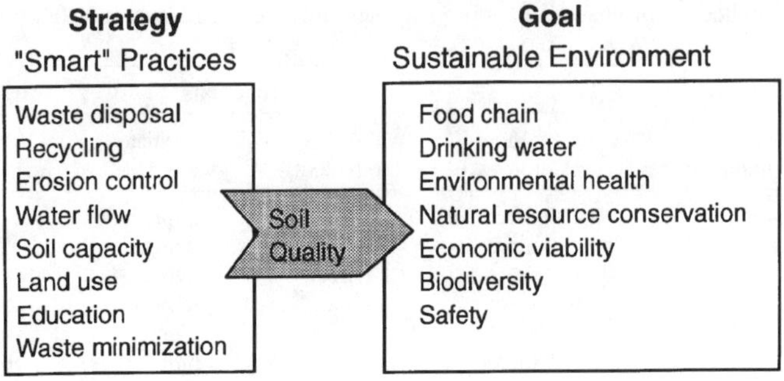

Fig. 13–1. Some key smart practices that may enhance soil quality necessary to achieve sustainable environment.

soil's ability to filter out pollutants (Table 13–1). This can be demonstrated in Fig. 13–1 where the kind and extent of man's activities can influence the quality of the environment, i.e., *smart* practices can enhance environmental quality. To achieve this goal, it is necessary to tailor land use to specific soil conditions, i.e., land use should be compatible with the soil fitness for that purpose.

13–4 SOIL AS A SINK FOR CONTAMINANTS

A paradigm,

- "Soil is an environmental crossroad" underlines the soil's role as an environmental buffer and central organizer of ecosystem.

Soils make it possible for plants to grow by mediating the biological, physical, and chemical processes that supply plants with nutrients, water, and other elements. The physical, biological, and chemical processes that occur in soils buffer environmental changes in air quality, water quality, and global climate (Lal & Pierce, 1991). The soil serves as a major incubation chamber for the decomposition of organic waste such as pesticides, sewage sludge, solid waste, and other organic materials. Depending on how they are managed, soils can be important sinks or sources of CO_2 and other gases that contribute to the greenhouse effect. Soils store, degrade, or immobilize nitrates, phosphates, metals, pesticides, radionuclides, and other substances that eventually can become air or water pollutants.

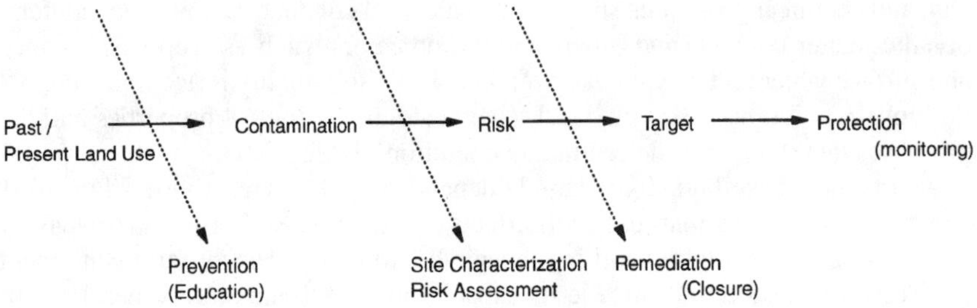

Fig. 13–2. Sequence of soil contamination-remediation indicating critical junctions for action.

In addition to intensifying exploitation of the land to meet the demands of the world's burgeoning population, the stress on soil will be exacerbated by the increasing trend to dispose of chemical waste on land, as a result of banning other disposal options. Concomitant with society's demand for more and improved chemical products is the ever-increasing chemical waste to be disposed of. Waste disposal technology has not caught up with production technology. More and more waste will accumulate on land, thereby endangering the quality of the soil, surface water, groundwater, and the atmosphere. Worldwide, there are >100 000 commercially available chemicals with 1000 new substances becoming available every year (Tolba & El-Kholy, 1992).

The following factors ensure greater chemical stress on the soil:

- Burgeoning population
- More chemicals being produced
- Implementation of more air and water pollution control measures
- More incentives to recycle renewable materials onto land
- Difficulty in incinerating and ocean dumping of hazardous waste
- Land as a cheap disposal medium
- Poor public awareness
- Expensive waste minimization technology and
- Expensive environmental restoration

In the USA, the Clean Air (1963) and the Clean Water (1977) Acts have shifted the disposal burden more onto the land, and the public's demand for cleaner air and water is tantamount to generating more waste. In the 1970s and 1980s, the Comprehensive Environmental Response Compensation and Liability Act (CERCLA), or Superfund Act, the Toxic Substances Control Act (TSCA), and the Resource Conservation and Recovery Act (RCRA) were enacted and caused a significant change in environmental policy in the USA. Implementation of RCRA may result in the accumulation of supposedly renewable materials, i.e., by-products of commercial operations that may provide some economic benefits (Travis & Cook, 1989; Andelman & Underhill, 1990; Brooks, 1991). Examples of these renewable resources are municipal sewage sludge, livestock waste, coal combustion residues, and phosphate and lime manufacturing by-products. Commercial operators tend to dispose of these waste by-products as close as possible to their facilities, as this offers a cheaper disposal option. There are >1200 contaminated sites in the USA designated as Superfund (or CERCLA) sites. Most of the contamination was caused by chemical manufacturing and operation of industrial and municipal landfills.

Because our soils have limited resiliency to contain and transform chemicals, these substances can become mobile and bioavailable to organisms (Runge et al., 1986; Bolt & Bruggenwert, 1976). Research has shown that polluted soils can lead to tainted foodstuff and drinking water (Winteringham, 1984; Barth & L'Hermite, 1987). The pathways of contaminants in the soil system is depicted in Fig. 13–3. Superfund sites in the USA are a good example that the soil's buffering capacity has been exceeded, resulting in the transport of pollutants to deep soil strata and the food chain. These sites are considered to pose the greatest ecological risks regarding the quality of groundwater and soil health (Andelman &

Underhill, 1990). Thus, we now have ample evidence that degraded soils may not only imperil our life support system but also may threaten our own well-being. Although the public has gained some awareness of potential health consequences of environmental pollution in general, society may still be largely uneducated as to the soil's role as a sink for contaminants.

The soil's capacity for contaminants, which influences the soil's resiliency to chemical stress, depends on the chemical properties of the contaminants, as well as the following soil properties (Adriano, 1986; Alloway, 1995).

Physical:
- Soil type
- Amount and type of clay
- Porosity (bulk density)
- Degree of saturation (i.e., water)

Biological:
- Organic matter
- Soil animals
- Bacteria
- Fungus

Chemical:
- pH
- Ion exchange capacity
- Redox potential
- Fe and Mn oxides
- Salinity
- Nutrients

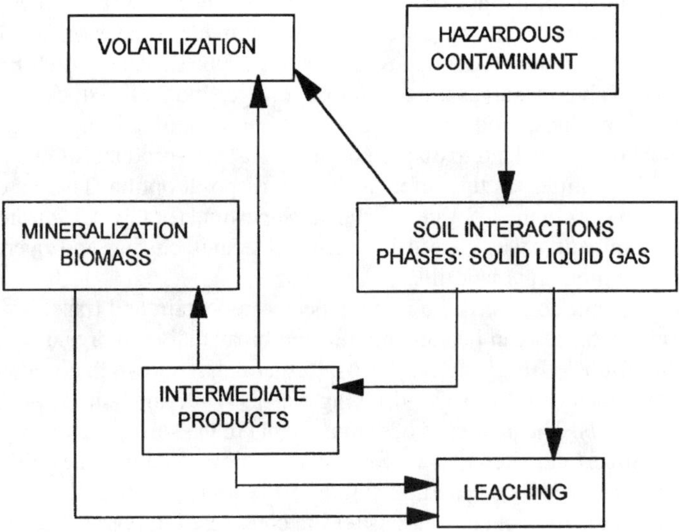

Fig. 13–3. Contaminant pathways in a terrestrial ecosystem.

In addition to these factors, the type of landscape could have some bearing on the soil's capacity as a sink. Soils that are vegetated could possess improved physical, biological, and chemical attributes as compared with those that are not. This is primarily due to the importance of plant roots in terms of enhancing soil structure, microbial activities, and soil conservation.

13–5 CHEMICAL DEGRADATION OF SOILS

Chemical degradation of soils is a consequence of the postindustrial revolution. It is an environmental price that society has to pay for increasing energy usage, modern agriculture, rapid population growth, and urbanization. Land contamination has been associated with the accumulation of unwanted waste by-products and major environmental perturbations resulting from accelerated human activities. Soil degradation through physical, biological, and chemical processes can reduce soil quality by changing soil attributes, such as nutrient status, C content, texture, available water holding capacity, structure, rooting depth, pH, and others (Fig. 13–4).

The following are major biological and chemical processes in soil that may influence its quality:

- Weathering
- Salinization
- Organic matter buildup and depletion
- Nutrient buildup and depletion
- Acidification
- Pollution

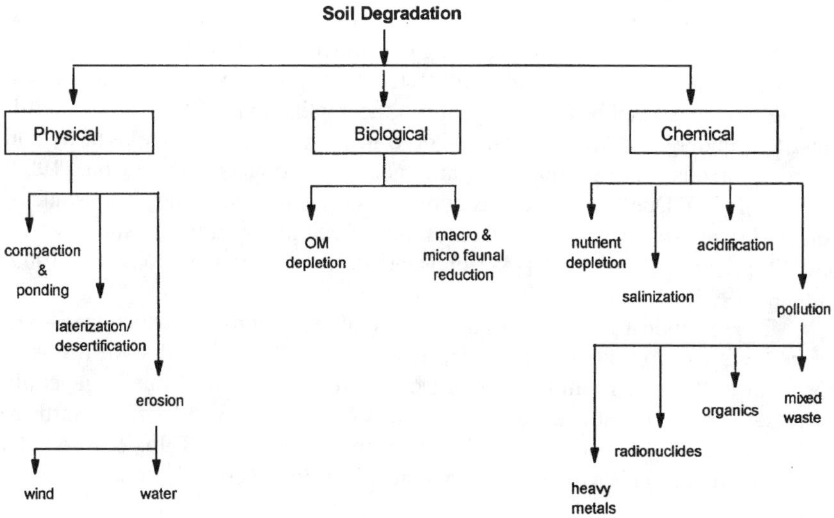

Fig. 13–4. Processes that influence the physical–biological–chemical degradation of soils.

Weathering in the pedosphere is a natural process that alters the chemical composition and structure of soil minerals, as well as the soil parent rocks (Dixon & Weed, 1989; Mortvedt et al., 1991). This results in the gradual loss of the alkaline earth cations (e.g., Ca^{2+} and Mg^{2+}) and the accumulation of insoluble compounds of Si, Al, and Fe. Over the long-term, this process usually results in the shifting of soil reaction to more acidic conditions. Young, unweathered soils (i.e., those formed from basic parent rocks) have a greater capacity to buffer against acid constituents than older, more weathered soils, such as those in the tropics and in the southeastern part of the USA. Weathering also may result in the mineralization of metal-bearing rocks, which at times cause anomalous levels of certain elements, such as Zn, Cu, Ni, Co, Se, and others.

Salinization often occurs under semiarid and arid conditions. It is frequently the result of poor irrigation management practices (U.S. Salinity Laboratory Staff, 1954; Tolba & El-Kholy, 1992). Major factors inducing salinization include excessively high salt contents of irrigation water and improper leaching and drainage conditions of irrigated fields. This problem is the net result of salt accumulation on the soil surface due to an imbalance in the evaporation and leaching of salts in the soil profile.

Organic matter and nutrients may build up or deplete in soil depending on farming practices (Stevenson, 1982). In intensive agriculture, such as in the midwestern U.S. Corn Belt, heavy fertilizer applications are required, especially in irrigated areas. This usually results in the leaching of excess nitrates into the groundwater (Williams, 1992; Powe, 1992). Organic matter from plant residues normally can build up in prairies and forest ecosystems, whereas loss of organic matter can occur when grasslands are burned or when forests are slashed and burned for shifting agriculture.

In soil, acidification can be caused by weathering of certain soil constituents and by certain cultural practices. Soil acidification is a natural weathering process in humid environments, but is greatly accelerated by certain factors. When soils are perturbed to the extent that pyrite-containing strata are exposed to more aerobic conditions, acidic-forming constituents, such as H_2SO_4, are generated. This is the case in mine spoils generated by mineral ore exploration and coal mining. Soils also can become acidic from the applications of nitrogenous fertilizers that undergo nitrification and when organic matter is decomposed by soil microorganisms that produce organic acid by-products (Stevenson, 1982; Sumner, 1991). Due to industrialization, acid rain can significantly contribute to soil acidification because of the direct input of acidic constituents, such as NO_x and SO_x compounds (Reuss & Johnson, 1986; Adriano & Johnson, 1989; Sumner, 1993).

Due to industrialization, many types of pollution are recognized—air, water, soil, and groundwater pollution from gaseous, liquid, and solid forms of chemical pollutants. Pollution varies from country to country but is generally more serious in developed countries with high population densities (Barth & L'Hermite, 1987; Harrison, 1990; Hansen & Jorgensen, 1991; Tolba & El-Kholy, 1992). Air pollution is a transboundary issue in North America, Europe, and other regions.

There are other chemical processes in soil that may influence degradation—ion exchange, complexation, oxidation–reduction, sorption–desorption, precipitation–dissolution, and others (Bolt & Bruggenwert, 1976; Sposito, 1989; McBride, 1994). It is important in the context of soil protection that soil possesses some capacity to resist (buffer) drastic changes in soil reaction and pollutant load. Soils that are healthy, such as typical prairie soil and agricultural soil, have high buffering capacities. When organic matter in soil decomposes it is humified and produces compounds that may influence the toxic effects of pollutants. These compounds can form complexes with pollutants, such as heavy metals, rendering them more or less mobile and bioavailable. The O_2 status of soil varies according to the presence of water and organic matter. Certain pollutants may become more mobile and toxic as they change their oxidation state in response to the redox potential. Toxicity of pollutants also depends on the chemical form in the soil–solution interface. Those in the precipitate or sorbed form are not mobile and bioavailable (Chlopecka et al., 1996).

The following are some of the more common practices that are conducive to chemical degradation of soil:

- Fertilizer–pesticide application
- Land disposal of waste
- Landfilling–storage of waste
- Mining–smelting of metal ores
- Combustion of fossil fuels
- Disposal–releases of radioactive waste
- Storage of petroleum products
- Dredging
- Deforestation

Repeated applications of pesticides and fertilizers are known to have contaminated groundwaters with pesticides and nitrates (Greenwood et al., 1990; Racke & Coats, 1990; Bar-Yosef et al., 1991; Williams, 1992; Powe, 1992). Once ingested, nitrates can be transformed to nitrites that become a health risk when converted to nitrosamine. Pesticide residues can degrade the quality of crops and the quality of drinking water from the aquifer (Ekstrom & Akerblom, 1990). Lately, land users are resorting to recycling municipal sewage sludge, livestock manure, and other resource materials, such as coal combustion residues (i.e., fly ash and flue gas desulfurization sludge). Potential problems with sewage sludge and livestock manure arise not only from their pathogen and excessive nutrient contents, especially N and P, but also from heavy metals (Adriano, 1986; Page et al., 1983). Cadmium is the metal of most concern from sewage sludge application on land; concerns with excessive Cu, Zn, and As may arise from swine and poultry manure application on land.

While modern landfills are engineered to contain chemical constituents by installing clay and plastic liners, old landfills often leak due to the absence of these barriers. Thus, old landfills are more prone to leak potentially hazardous constituents to underlying soil strata and groundwater (Fuller & Warrick, 1985; Andelman & Underhill, 1990; Suter et al., 1993).

Mining of metal ores and coal almost invariably results in drastically perturbed landscape (Davies, 1991). Deep geologic materials, as well as subsoils, are brought up to the surface along with pyrite-containing materials, resulting in an acidic soil environment. During the smelting and refining of metal ores, acidic substances and metal-bearing particulates escape into the atmosphere. Combustion of fossil fuels, such as coal and oil, for generating electric power results in the release of SO_x and NO_x compounds (Adriano & Johnson, 1989; Longhurst, 1990; Moldan & Schnoor, 1992). Most of these acid rain constituents, along with heavy metals, are deposited in nearby areas but some are transported over long distances.

The Chernobyl accident in the Ukraine in 1986 demonstrates that severe ecological consequences can arise from failure of nuclear reactors and releases of their waste (Eisenbud, 1987; Kryshev, 1992). Today, massive programs are being planned to restore some health to the ecosystems of the Chernobyl area and the neighboring areas in the Ukraine and Belarus.

Stockpiling of fossil fuel, especially coal high in S, produces acid leachates and runoff water from the oxidation of pyrite-type material in coal (Carlson & Adriano, 1993). Leachate and runoff waters have $pH < 3$. In addition, the runoff water is usually high in soluble constituents such as Fe, Mn, Ni, Zn, sulfate, and others (Anderson et al., 1992). These contaminants may adversely affect the quality of surface and groundwaters when they leach from the runoff storage basin. Leaking of petrol products, such as refined oil, from large storage tanks frequently occurs and can potentially contaminate the underlying groundwater (Cairney, 1993).

To render waterways commercially navigable, sediments have to be dredged. Dredged materials are often disposed of in adjoining areas that may overlie already contaminated land. Once exposed, certain chemical constituents in the sediment may undergo transformation, causing certain pollutants to become more toxic (Salomons & Forstner, 1988). For example, Cd in its nontoxic sulfide form in the sediment is transformed to the more toxic form once aerated upon spreading on land resulting in the formation of H_2SO_4.

Commercial logging and clear-cutting of forests may lead to the depletion of soil organic matter (Allen, 1985). The tree biomass is burned for rapid cleanup and for rapid mineralization of plant nutrients. Under this practice, the soil may only be productive for a few years, after which it becomes unproductive due to nutrient depletion and soil erosion (Sumner & Miller, 1992). This practice is considered a major reason for the controversial global change issue.

13–6 SOIL CONTAMINATION AND REMEDIATION

Contamination of soils from anthropogenic chemicals and their subsequent degradation has become a major concern because of the critical role of soil resources in promoting sustainable environment and economic development. Both inorganic and organic anthropogenic compounds in soils may not only adversely affect their production potential but also may compromise the quality of the food chain and the underlying groundwater. This scenario may require regulatory-driven risk assessment and evaluation of remedial technology to a specific soil condition.

In the USA, approximately 700 tons of hazardous waste are produced daily, equivalent to 250 million tons annually or nearly 1 ton per person per year (Iskandar & Adriano, 1997). Major concerns pertinent to land disposal of hazardous waste include the contamination of both surface and groundwaters, transport of pathogens and other toxins to man through such pathways as crops grown in waste-amended soils and the export of substances to nontarget ecosystems.

Land use, past and present, critically influences the extent and intensity of soil contamination (Fig. 13–2). The logical junctures where intervention measures could be employed also are indicated. For example, contamination prevention measures such as recycling, waste minimization, and others, are important tools in minimizing soil contamination. Furthermore, it is very important to educate the public about potential consequences of soil contamination.

Physical, biological, and chemical processes may contribute to soil degradation (Fig. 13–4). While much is known about physical processes of soil degradation, much less is known about the biological and chemical processes, especially pollution with heavy metals, radionuclides, organic chemicals, and mixed waste (i.e., inorganic and organic). Chemically degraded soils may result in an imbalance in the physical, biological, and chemical processes upon which soil productivity depends.

Although less acreage is involved compared to physically degraded soils, chemically degraded soils are becoming important in terms of food production and quality (Adriano et al., 1997). Crops cultivated on soils of marginal quality may not only suffer in yield but also in quality, making them less competitive in the world market. For example, certain cereal and vegetable crops from eastern Europe may not be permitted in the European common market because of questionable quality. Therefore, it has become apparent that polluted soils are not only a social and health issue but an economic issue as well (World Commission on Environment and Development, 1987).

The following are potential major consequences of chemically degraded soils:

- Poor physical–biochemical properties of soil
- Increased susceptibility of soil to erosion
- Diminished sustainability of the soil
- Tarnished quality of the food chain
- Diminished quality of surface and groundwater
- Human illness
- Economic loss

Severely chemically degraded soils could manifest one or more of these consequences. The lesson from the collapse of the former Soviet Bloc countries where the environment was compromised for industrialization and militarization attest to this statement.

The degree of soil degradation depends on numerous factors including soil factors (see section on Soil as a Sink for Contaminants), the type, amount, and persistence of chemicals in soil, and others. When adversely affected, the soil may be viewed as dysfunctional in the sense that biological activity is diminished. This and similar conditions may require remediation (otherwise known as restoration) in order to mitigate the adverse effects. Such remediation practices may restore the soil to some healthy condition. In addition to mitigating ecological and health risks associated with pollution, it also is important that the adverse effects do not spread to surrounding environments. The following major steps are considered in soil remediation:

- Site discovery, site inspection and preliminary assessment
- Site characterization, contaminant characterization, analyses of restoration alternatives, and selection of restoration technique
- Remediation and site closure
- Site compliance monitoring

Therefore, in restoring degraded soil, it is not only necessary to eliminate the source, movement and risk of chemicals, but also to manage the remediated soil so that similar problems do not reappear.

The following are some major factors that need to be considered in assessing the risks involved in contaminated sites:

- Site
  - Location
  - Extent of the contaminated area
  - Terrain–slope
  - Vicinity to population
  - Vicinity to surface water and groundwater
- Contaminant–Waste
  - Type of contaminants
  - Number of contaminants
  - Concentration of contaminants
- Soil
  - Type and texture
  - Depth
  - Chemical, biological, and physical properties
- Geological–hydrological factors
- Intended land use
  - Residential
  - Agriculture, forestry
  - Industrial, and others
- Meteorological–climatological factors
- Risks
  - Ecological
  - Human Health

The complexity in addressing soil cleanup is further exhibited by the extreme difficulty in establishing cleanup standards, i.e., how clean is clean? Should the soil in question be cleaned to its original condition, i.e., background level? The background itself is another issue. Should there be a universal background value? Another issue pertains to the soil conditions after remediation.

Certain remedial techniques employ drastic measures to contain or separate contaminants, such as using solvents to extract contaminants from the soil. Soil structure and chemical and mineralogical properties may be altered upon such drastic chemical *washing*. What to do with the soil depends on its future land use.

Most modern remedial techniques, no matter how innovative, are expensive and generally unaffordable in developing countries. The cost is variable and depends on the type of technique. More recent cost estimates of some of these technologies range from $5 to $50 per cubic meter for in situ soil venting of volatile organics, to as high as $100 to $500 per cubic meter for redisposal in sanitary landfill or by incineration. Vitrification may cost upward of $250 per cubic meter of affected soil.

Because of the dynamic nature of soil, contaminants, and microbiota, a promising option called natural remediation (also known as intrinsic remediation) is now being evaluated. Contaminants have the tendency to dissipate in the natural soil environment because they are transformed by various physical, biological, and chemical processes. Unassisted intrinsic remediation is considered as an emerging option that requires several years of basic research. This approach seems promising for areas where human exposure is not an issue or where there is insignificant ecological risk.

13–7 ROLE OF SOIL CHEMISTRY IN SOIL REMEDIATION

Soil cleanup requires an interdisciplinary approach because of its complex and heterogeneous nature. Scientists, engineers, policy makers, regulators, the industry, and the public will need to be involved in the decision making for soil cleanup. In general, there are three major categories for the treatment of polluted soils: physical, biological, and chemical. In some cases, a combination of treatments is necessary. Selection of cleanup technology depends on the desired level of cleanup, the length of time involved in the cleanup, the forms and amounts of contaminants, site characteristics, the cost involved, and others. In the USA, the initial regulatory guideline for cleanup was based on an extraction procedure referred to as TCLP (i.e., toxicity characteristic leaching procedure) which involves the leaching of a polluted soil or waste with weak acids, and represents a worst case scenario (U.S. Environmental Protection Agency, 1990a). This extraction does not mimic the natural process by which contaminants are released from soils or waste. In the absence of TCLP data, total contaminant concentrations could be used.

In the USA, development of environmental technologies related to soil and water issues has been advancing rapidly since 1986 when SARA (Superfund Amendments and Reauthorization Act) established a coordinated research and development program and promoted technology demonstration in the field. When the Superfund program was initiated about 15 yr ago, acceptable technologies for hazardous waste and soil cleanup included excavation, incineration, and land disposal. At that time, mainly constructed landfills with double liners were employed for land containment of hazardous waste. In 1984, the U.S. Environmental Protection Agency emphasized the use of in situ treatment of soil rather than removal and disposal or incineration. Recently, the U.S. Environmental Protection Agency formed a network with the industry whereby technologies are made conveniently available (U.S. Environmental Protection Agency, 1994a,b).

The U.S. Environmental Protection Agency through its national contingency plan established the following criteria for selecting remedial techniques:

- The overall protection of human health and the environment
- Compliance with applicable or relevant and appropriate requirements
- Long-term effectiveness and performance
- Reduction of toxicity, mobility, and volume through treatment
- Short-term effectiveness
- Implementation ability
- Cost
- State acceptance
- Community acceptance

These requirements for site cleanup demand vast amounts of financial resources, beginning with the initial site visit and assessment and ending with site closure and community acceptance. In some cases, some sort of litigation may ensue between the site custodian and the U.S. Environmental Protection Agency that may not only delay the cleanup but also escalate the total cost.

Although the soil cleanup issue needs an interdisciplinary solution, soil scientists can provide the central role of disseminating basic soil science principles pertinent to the problem. Crucial to this understanding are chemical processes that may influence the transport and toxicity of contaminants. A soil chemist should be able to integrate the importance of the source term, waste and soil characteristics, waste form and speciation, various soil chemical processes, and monitoring of the technology performance with time.

The role of soil scientists, especially soil chemists, in soil remediation can be demonstrated by citing some of the simpler approaches using soil ameliorants. These ameliorants are rather inexpensive and are readily available on a global basis. Application of limestone to elevate soil pH to about 6.5 is recommended when agricultural soils are amended with sewage sludge high in metals, especially cadmium (Adriano, 1986; Pierzynski et al. 1994; Page et al., 1983). The addition of certain clay minerals, such as vermiculite, to contaminated soils may be effective in immobilizing metals and radionuclides, such as ^{137}Cs and ^{90}Sr (Adriano et al., 1997; Mench et al., 1997). The application of organic matter may not only aid in immobilizing some contaminants but also may serve as an energy source for microbes to decompose–transform organics and metalloids (Frankenberger & Benson, 1994). Zeolites serve as molecular sieves and have been used in industry for treatment of waste waters. Their unique ion exchange, adsorption, dehydration, and catalytic properties have prompted their use in agricultural and industrial processes. Kesraoul-Oukl et al. (1993) found Pb and Cd removal efficiencies of >99% when conditioning natural zeolites, suggesting their potential in treating effluent contaminated with high levels of these metals. Addition of synthetic zeolite to soils contaminated with metals in Poland significantly reduced the concentrations of Cd and Pb in the roots and shoots of a num-

Table 13–2. Selected ameliorants that are adapted to metal contaminated soils.

Technique	Target contaminants	Soil processes involved	Constraints
Limestone	Metals, radionuclides	Precipitation, sorption	Ineffective for oxyanions; certain crops (lettuce, spinach, tuber, and others); short term
Zeolite	Metals, radionuclides	Ion exchange, sorption, fixation	Insufficient dat;, short term
Apatite	Metals	Sorption, precipitation, complexation	Selective; insufficient data
Clay mineral	Metals, radionuclides	Ion exchange, sorption fixation	Type of clay; short term

ber of pot-grown crops, again suggesting the potential applicability of this ameliorant in metal-contaminated soil cleanup (Gworek, 1992a,b). The use of P and K fertilizers may have some beneficial effect on certain contaminants. It is well known that phosphate forms a fairly stable complex with Pb in soils (Rabinowitz, 1997; Ma et al., 1993, Nriagu, 1974) and catapults this process as promising in in situ immobilization of Pb. In practice, hydroxyapatite ($Ca_5(PO_4)_3OH$) will dissolve and precipitate when mixed with lead as insoluble hydroxypyromorphite, $Pb_{10}(PO_4)_6(OH)_2$, (Ma et al., 1993). The use of K fertilizer is based on its similar geochemical and physiological behavior with Cs. As such, they compete for the same exchange sites in soil and the same absorption sites on plant roots, causing them to become antagonistic. The ameliorants in Table 13–2 may provide only interim solutions in stabilizing contaminants and have been applied in agricultural soils. Recently, Chlopecka and Adriano (1996) observed enhanced stabilization of metals in a cropped contaminated soil using clinoptilolite, hydroxyapatite, and Fe-oxide waste by-product. But their efficacy over the long term still need to be proven.

Several techniques that are considered innovative or emerging by the cleanup industry have parts of their basic principles related to soil science or its allied sciences. Two examples are described below:

13–8 SOIL WASHING

13–8.1 Process Description

Soil washing is an ex-situ process employing chemical and physical extraction and separation techniques to remove a broad range of organic, inorganic, and radioactive contaminants from soils (U.S. Environmental Protection Agency, 1989, 1990b; Everson, 1989; Anderson, 1994a). The process entails excavation of contaminated soil, mechanical screening to remove various oversized materials, separating coarse- and fine-grained fractions, treatment of these fractions, and management of the generated residuals. It is a separation and volume reduction process that is typically used in conjunction with other technologies. Concentrating the contaminants in a smaller volume for further treatment enables a more overall cost-effective treatment.

Surface-associated contaminants are removed through abrasive scouring and scrubbing using wash water that sometimes is augmented by surfactants or extractants. The soil is then separated from the spent washing fluid, which carries with it some of the contaminants. The recovered soils consist of a clean coarse fraction (sand and gravel textured soils, generally >50 μm), a contaminated fine fraction (silts and clays, generally <50 μm), and a contaminated organic–humic fraction (material that floats on the washing solution). In order for the process to be effective, essentially all of chemical contaminants must be associated with the fine grain fraction. The fine-grain material generally requires further treatment, such as stabilization–solidification (Anderson, 1994b), soil flushing (Anderson, 1994a), and prepared bed reactors (Lynch & Genes, 1989; Sims, 1990; Kim et al., 1991; U.S. Environmental Protection Agency, 1994b).

13–8.2 Potential Applications

Soil washing performance is closely tied to three key physical soil characteristics: particle size distribution, contaminant distribution among the different size particles, and how strongly the soil binds the contaminant. Soil washing is likely to be more successful when applied to soils containing a relatively high percentage of sand and gravel than to soils high in clay- and silt-sized particles. In general, soil washing is most appropriate for soils that contain at least 50% sand and gravel, such as coastal sandy soils and soils with glacial deposits (Westinghouse Hanford Company, 1994). Therefore, knowledge of the typical particle-size distribution that will be encountered throughout the contaminated soil can be valuable as an early indicator (or screening tool) of the potential effectiveness of soil washing in separating out contaminants. In order for soil washing to be effective, the contaminants must be associated with a relatively small range of particle sizes: the smaller the range of particle sizes that are contaminated, the greater the reduction potential in volume of contaminated soils (Serne et al., 1992). The distribution of contaminants among the different soil size fractions is primarily controlled by the process by which the contaminant becomes associated with the soil: adsorption and precipitation. If a contaminant adsorbs onto the soil, the finer soil fraction tends to contain higher concentrations of the contaminant than the coarser size fraction. This occurs because the finer particles possess a greater surface area per unit mass than the coarser particles. Zinc, Cd, and Cs are commonly associated with the solid phase through adsorption. If a contaminant precipitates onto the soil, the contaminant is associated with a wide range of particle sizes. Occasionally, contaminants also can exist as distinct particles. Precipitated and distinct particle contaminants have been separated from the uncontaminated portion of soil by two general methods: particle size fractionation–specific gravity separation or magnetic separation (U.S. Department of Energy-RL, 1994).

13–8.3 Process Evaluation

When used as a pretreatment step for other remediation processes, soil washing presents two key advantages. The first is its ability to substantially contribute to waste minimization by concentrating a large proportion of the nonvolatile and heavy metal contaminants into a residual soil product typically representing less than half the original soil volume. The second advantage lies in its potential cost-effectiveness. Lower remediation costs result for many sites through the reduction of sheer volume of contaminated soil that must be treated by more expensive methods (Gombert, 1992; Murray, 1993; Phillips et al., 1993; U.S. Department of Energy-RL, 1994; U.S. Environmental Protection Agency, 1994a).

The waste matrix may pose the most significant limitation. Complex mixtures of contaminants make it difficult to formulate a single suitable washing fluid and may require sequential washing steps with different extractants (Gombert, 1992). Further, frequent changes in the contaminants and their concentrations in the feed soil can disrupt the process requiring modification of the wash fluid formulation and the operating settings. Soil washing will usually not be cost-effective in treating soils having a high percentage of clay and silt (>30–50 %). Also, high humic content in the soil makes separation of contaminants very difficult (Rhoades, 1981). It is generally ineffective in treating soils contaminated with a high concentration of mineralized metals (metals held within the structure of minerals) or hydrophobic organics. Hydrophobic contaminants are usually difficult to separate from soil particles into an aqueous washing fluid.

Several full-scale soil washing projects are now being implemented in the USA, primarily using applications already proven in the Netherlands and Germany; a brief description of several case studies are presented by Anderson (1994a). An especially successful application of soil washing–flushing was recently reported for an ammunition plant in New Brighton, Minnesota (Fristad, 1995). More than 20 000 tons of soil were treated. The cleaned soil remained on-site, and the heavy metal contaminants were removed, recovered, and recycled. Eight heavy metals were removed from the contaminated soil achieving the stringent cleanup criteria of <175 mg kg^{-1} Pb and achieving background concentrations for Sb, Cd, Cr, Cu, Hg, and Ni. More than 98% of the contaminant metals were removed. No hazardous waste requiring landfill disposal was generated during the entire operation.

Several variations of soil washing are or have been tested in the Superfund Innovative Technology Evaluation Program (U.S. Environmental Protection Agency, 1994a,b). At 23 sites in this program, washing soils contaminated with semivolatile organics, pesticides, polynuclear aromatic hydrocarbons, and heavy metals gave removal efficiencies for residual metals and hydrocarbons of 90 to 98%. Removal of metals from the finer grain fractions required treatment with acids or chelating agents. Although studies have shown that soil washing can be effective in removing gasoline and diesel fuels from soils, thermal desorption, soil vacuum extraction, biodegradation, vapor extraction, or other processes may be more appropriate.

13–9 PHYTOREMEDIATION

13–9.1 Process Description

Plants have been described in engineering terms as solar driven pumps and filtering systems that extract and concentrate certain elements from the environment (Salt et al., 1995; Cunningham & Berti, 1993). Typically, the metals that may accumulate in plants are those that have nutritional value, such as Ca, Mg, Fe, Mn, Zn, Cu, and Mo. Certain plants also have the ability to accumulate metals that have no known biological function, such as Cd, Cr, Pb, Ni, Co, Se, and Hg (Baker & Brooks, 1989); however, excessive accumulation of these nonessential metals is toxic to most plants.

Recently the value of metal-accumulating plants for environmental remediation has been realized (Baker et al., 1988; Cunningham & Berti 1993; Wenzel et al., 1993; Banuelos et al., 1993; Pierzynski, 1997). There are four general processes by which plants can be used to remediate metal-contaminated soils (Salt et al., 1995):

1. Phytoextraction: process in which metal-accumulating plants are used to transport and concentrate metals from the soil into the harvestable parts of roots and aboveground shoots (Brooks, 1997; Brown et al., 1994, 1995; Kumar et al., 1995);
2. Plant-assisted bioremediation: process in which plant roots in conjunction with their rhizospheric microorganisms are used to remediate soils contaminated with organic compounds (Walton & Anderson, 1992; Anderson et al., 1993);
3. Rhizofiltration: process in which plant roots absorb, precipitate, and concentrate toxic metals from polluted wastewater streams (Dushenkov et al., 1995); and
4. Phytostabilization: process in which plants tolerant to contaminant metals are used to reduce the mobility of contaminant metals, thereby reducing the risk of further environmental degradation by leaching into the groundwater or by being airborne (Smith & Bradshaw, 1979; Losi et al., 1994).

Of these processes, only phytoextraction can be used to remove metals from contaminated soils. Plant-assisted bioremediation is used exclusively with organic compounds; rhizofiltration is a process to remove contaminants from an aqueous waste stream and not from soil; phytostabilization does not remove metals from the soil, rather, it attempts to immobilize the metals in the soil to limit migration.

13–9.2 Potential Application

The availability of contaminant metals for plant uptake is a limiting factor of phytoextraction of soils. One approach that has been taken to enhance phytoextraction is to use soil amendments to solubilize metals and bring them into the soil solution (Blaylock et al., 1995; Salt et al., 1995). For example, soil-applied chelating solutions increased shoot Cd concentrations of Indian mustard (*Brassica juncea* L.) from 150 to >875 mg kg^{-1} without significantly decreasing shoot dry weight (Blaylock et al., 1995). Applications of chelators to lead contaminated soils (1200 mg Pb kg^{-1}, pH 8.3) also produced substantial increases in shoot Pb concentrations. A cost analysis of this approach was not addressed by the authors.

Soil pH is another important factor controlling the solubility of metals in soils. Numerous studies have shown that lowering the pH of a soil will induce desorption of heavy metals and thus increase their concentration in soil solutions (Harter, 1983; Chlopecka & Adriano, 1996). By maintaining a moderately acid pH in the soil through the use of ammonium-containing fertilizers or soil acidifying amendments, it may be possible to increase metal bioavailability and hence plant uptake. The addition of ascorbic acid to soils high in manganese oxides amended with selenite increased the solubility of Se by enhancing Mn-mediated oxidation of selenite to selenate (Blaylock & James, 1994).

Another rather limited, yet apparently affective application of plants for remediation is to enhance the volatilization (i.e., phytovolatilization) of selenium from soil (Parker, 1995). This process has some potential to remediating soil contaminated with Hg and possibly other metals.

13–9.3 Process Evaluation

The ideal plant for the phytoextraction process must tolerate and accumulate high levels of metals in its harvestable parts, have a rapid growth rate, and produce a high biomass. Thus, to improve the annual removal rate, one may try to improve either the biomass yield or the plant uptake rate. Both approaches are being attempted. Improvements in biomass yield is accomplished through typical agricultural optimization experiments where soil texture and nutrient status is optimized for site conditions while at the same time maintaining contaminant uptake (Baker et al., 1994). The other approach is to use plants that accumulate exceptionally high concentrations of the metal. This is accomplished through phenotypic screening of plant species (Baker & Brooks, 1989) and by genetic engineering (Lefebvre et al., 1987; Misra & Gedamu, 1989; Maiti et al., 1991). Some successes have already been reported in the application of genetic engineering to hyperaccumulators. For example, genes encoding the Cd-binding protein metallothionein, have been expressed in plants in a seemingly successful attempt to increase Cd resistance (Lefebvre et al., 1987; Misra & Gedamu, 1989; Maiti et al., 1991). For initial attempt at phytoextraction, it is likely easier to increase the biomass yield than it is to increase plant uptake rate.

The length of time required for remediation (in years) can be estimated by dividing the total quantity of contaminant metal in the soil in excess of the allowable limits by the annual removal rate of metals on a hectare basis.

The contaminant concentration plays an important role in determining the length of time required for the phytoremediation of a site. This is generally not the case with other decontamination technologies. The economics of stabilization, landfilling, incineration, soil washing, and soil flushing do not change significantly if the metal content of a soil ranges over an order of magnitude. The effect of contaminant concentration on the feasibility of phytoremediation, however, is dramatic. Most researchers are attempting to develop plants to remove between 500 and 1000 kg of metal ha^{-1} yr^{-1} (Cunningham & Berti, 1993). Sites containing metals barely above regulatory limits might be cleaned up in 1 yr, whereas heavily contaminated sites might take hundreds of years to remediate by this technology. In one field experiment, it was determined that the best metal accumulator, *Thlaspi caerulescens* (belonging to the mustard family) would take 13 to 14 yr of continuous cultivation to clean a Ni and Zn contaminated soil (Baker et al., 1994).

Phytoextraction is an emerging technology that holds great potential. In order to realize this potential, it will be necessary to improve our understanding of the wide range of processes that are involved, including: plant biology, agricultural engineering, soil science, microbiology, and genetic engineering. Phytoextraction is most applicable to contaminants that: (i) are near the surface, (ii) are relatively nonleachable, (iii) pose little imminent risk to health or the environment, and (iv) cover large surface areas. Initial cost projections are favorable. Phytoextraction is not yet ready for full scale application.

13–10 SOIL REMEDIATION AND SUSTAINABLE ECOSYSTEM

The traditional definition of an ecosystem is: an interconnected community of living things, including humans, and the physical environment in which they interact. An operational definition of a healthy ecosystem is one that is free from distress syndrome, is stable, and is sustainable, i.e., that is active and maintains its organization and autonomy over time and is resilient to stress (Costanza et al., 1992); however, Lackey (1998) stated that the ability of ecosystems to respond to a variety of stresses, natural and manmade, may be limited. This implies that ecosystem approach is necessary in sustaining or restoring natural systems and their functions and values (Interagency Ecosystem Management Task Force, 1995). The goal of the ecosystem approach is to restore and sustain the health, productivity, and biological diversity of ecosystems and the overall quality of life through natural resource management that is fully integrated with social and economic goals. This again implies a need for a multidisciplinary approach to sustain or restore soil health. Perhaps a useful tool in sustainable agroecosystems or other ecosystems is to employ a fairly recent idea of adaptive management. It is a process of adjusting management actions and directions in view of new information about the ecosystem and about progress toward ecosystem goals (Fig. 13–5). In essence, adaptive management works as follows: restoration or management measures are implemented; monitoring is conducted; feedback is provided based on new information gain; and adjustments are made. There is a need to constantly review and revise environmental and other restoration and management approaches because of the dynamic nature of ecosystems. When new information becomes available, a decision is made whether or how to adjust the strategy and actions. Thus, management decisions are viewed as experiments subject to modifications rather than as fixed and final rulings. As we increase our understanding of ecosystem structure and function and their relationship to management actions we also need to adjust our actions accordingly.

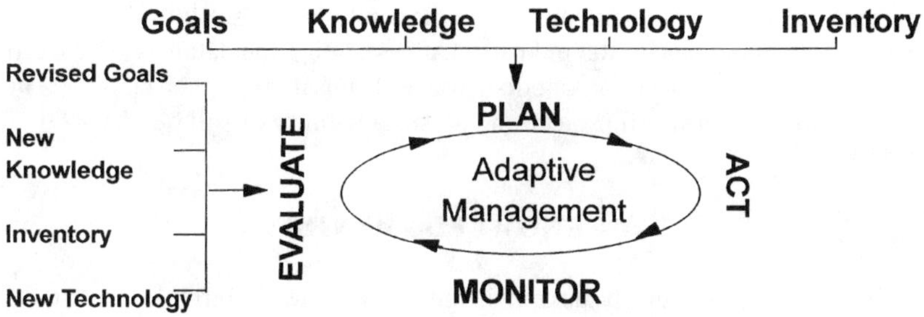

Fig. 13–5. An adaptive approach to ecosystem management.

13–11 SUMMARY AND CONCLUSIONS

Because of society's increasing demand for land-based products such as food, fiber, and energy related biomass, the soil is increasingly subjected to physical, biological, and chemical stresses. These stresses can be exacerbated by human activities that result in the generation of hazardous waste and their subsequent deposition on land. Under these conditions, soil quality becomes an issue. Man's need to exploit the soil for food production and at the same time use it as a repository for unwanted waste requires that soil quality be enhanced or sustained because soil quality, i.e., soil health, bears on the sustainability of ecosystems. Since the soil is regarded as the central organizer of the ecosystem, a healthy soil is a prerequisite to maintaining a healthy ecosystem. While our understanding of soil quality related to production agriculture is more advanced, our understanding of this concept with regard to chemically-stressed ecosystems is still in its infancy. Nevertheless, it is logical to define soil quality relative to intended land use.

The soil contamination issue needs an interdisciplinary approach. Central to this is basic soil science and the understanding of the fundamental physical, biological, and chemical processes in it. These processes influence the capacity of the soil to contain contaminants. Although knowledge of the physical and biological properties and processes in soils also is important, it is largely due to our understanding of the chemical properties and processes that enable us to determine the reactivity, mobility, and bioavailability of contaminants. These processes influence the extent by which plants, animals, and man are impacted by pollution. The ultimate consequences of soils that are severely degraded by chemicals include human illness and economic loss. Therefore, it is necessary to manage the risk arising from contaminated sites, but in a technologically and cost-effective manner. For agricultural soils needing remedial action with regard to excessive heavy metal uptake by food crops, some fairly abundant and inexpensive soil ameliorants are available. Examples are limestone, zeolite, hydroxyapatite, and other clay minerals. These soil amendments can aid in manipulating soil chemical properties and processes to assist in stabilizing the contaminants in place, rendering them immobile and nonbioavailable for plant uptake. Several remedial techniques widely used in the industry have their roots in soil science; examples of this are the soil washing and phytoremediation techniques. Both are considered innovative technologies by the industry. Understanding the nature and properties of soil with regard to food production, waste disposal, and other land uses may be crucial to long-term soil protection and sustainability of soil quality.

ACKNOWLEDGMENTS

The authors gratefully acknowledge the review of Terry Logan of Ohio State University and the editorial assistance of Caroline Sherony at SREL. This research was supported by Financial Assistance Award Number DE-FC09-96SR18546 from the U.S. Department of Energy to the University of Georgia Research Foundation.

REFERENCES

Acton, D.F., and L.J. Gregorich (ed.). 1995. The health of our soils: Toward sustainable agriculture in Canada. Ctr. for Land and Biol. Resour. Res., Agric. and Agri-Food Canada. Central Exp. Stn., Ottawa, ON.

Adriano, D.C. 1986. Trace elements in the terrestrial environment. Springer-Verlag, New York.

Adriano, D.C., J. Albright, F.W. Whicker, and I.K. Iskandar. 1997. Remediation of metal- and radionuclide-contaminated soil. p. 27–45. *In* I.K. Iskandar and D.C. Adriano (ed.) Remediation of metal-contaminated soils. Science Reviews, Northwood, England.

Adriano, D.C., and A.H. Johnson (ed.). 1989. Acid precipitation. Vol. 2. Springer-Verlag, New York.

Allen, J.C. 1985. Soil responses to forest clearing in the U.S. and the tropics: Geological and biological factors. Biotropica 17:15–27.

Alloway, B.J. 1995. Heavy metals in soil. Chapman & Hall, London.

Andelman, J.B., and D.W. Underhill. 1990. Health effects from hazardous waste sites. Lewis Publ., Chelsea, MI.

Anderson, M.A., P.M. Bertsch, and L.W. Zelazny. 1993. Multicomponent transport through soil subjected to coal pile runoff under steady saturated flow. p. 137–164. *In* R.F. Keefer and K.S. Sajwan (ed.) Trace elements in coal and coal combustion residues. Lewis Publ., Boca Raton, FL.

Anderson, T.A., E.A. Guthrie, and B.T. Walton. 1993. Bioremediation. Environ. Sci. Technol.27:2630–2636.

Anderson, W.C. 1994a. Innovative site remediation technology: Soil washing/soil flushing. Vol. 3. Water Environ. Fed., Alexandria, VA.

Anderson, W.C. 1994b. Innovative site remediation technology: Stabilization/solidification. Vol. 4. Water Environ. Fed., Alexandria, VA.

Bachmann-Erdt, G. 1993. Regulating soil cleanup and protection: An evaluation of policy options in Europe and the USA. Bd Wi-Verlag Warburg, Germany.

Baker, A.J.M., and R.R. Brooks. 1989. Terrestrial higher plants which hyperaccumulate metallic elements: A review of their distribution, ecology and phytochemistry. Biorecovery 1:81–126.

Baker, A., R. Brooks, and R.D. Reeves. 1988. Growing for gold... and copper... and zinc. New Sci. 10:44–48.

Baker, A., S.P. McGrath, C.M.D. Sidoli, and R.D. Reeves. 1994. The possibility of *in-situ* heavy metal decontamination of polluted soils using crops of metal-accumulating plants. Resour. Conserv. Recycling. 11:41–91.

Banuelos, G.S., G. Cardon, B. Mackey, J. Ben-Asher, L. Wu, P. Beuselinck, S. Akohoue, and S. Zambrzuski. 1993. Plant and environment interactions, boron and selenium removal in boron-laden soils by four sprinkler irrigated plant species. J. Environ. Qual. 22:786–792.

Bar-Yosef, B., N.J. Barrow, and J. Goldshmid (ed.) 1991. Inorganic contaminants in the vadose zone. Springer-Verlag, Berlin.

Barth, H., and P. L'Hermite. 1987. Scientific basis for soil protection in the European Community. Elsevier Applied Science Publ., Essex, England.

Blaylock, M.J., and B.R. James. 1994. Redox transformation and plant uptake of selenium resulting from root–soil interactions. Plant Soil. 158:1–12.

Blaylock, M.J., O. Zakharova, D.E. Salt, and I. Raskin. 1995. Increasing heavy metal uptake through soil amendments: The key to effective phytoremediation. p. 218. *In* Agronomy abstracts. ASA, Madison, WI.

Blum, W.E.H., and A.A. Santelises. 1994. A concept of sustainability and resiliency based on soil functions. p. 535–542. *In* D.J. Greenland and I. Szabolcs (ed.). Soil resilience and sustainable land use. CAB Int., Wallingford, Oxon, England.

Bolt, G.H., and M.G.M. Bruggenwert (ed.). 1976. Soil chemistry: A basic element. Elsevier, Amsterdam.

Brooks, C.S. 1991. Metal recovery from industrial waste. Lewis Publ., Chelsea, MI.

Brooks, R.R. 1997. Plant hyperaccumulators of metals and their role in mineral exploration, archaeology, and land reclamation. p. 123–133. *In* I.K. Iskandar and D.C. Adriano (ed.) Remediation of metal-contaminated soils. Science Reviews, Northwood, England.

Brown, K.W. 1997. Decontamination of polluted soils. p. 47–66. *In* I.K. Iskandar and D.C. Adriano (ed.) Remediation of metal-contaminated soils. Science Reviews, Northwood, England.

Brown, S.L., R.L. Chaney, J.S. Angle, and A.J.M. Baker. 1995. Zinc and cadmium uptake by hyperaccumulator *Thlaspi caerulescens* grown in nutrient solution. Soil Sci. Soc. Am. J. 59:125–133.

Brown, S.L., R.L. Chaney, J.S. Angle, and A.J.M. Baker. 1994. Phytoremediation potential of *Thlaspi caerulescens* and bladder campion for zinc and cadmium-contaminated soil. J. Environ. Qual. 23:1151–1157.

Bryda, L.D., and J.A. Simon. 1994. Recent developments in cleanup technologies. Remediation 5:137–148.

Cairney, T. (ed.). 1993. Contaminated land: Problems and solutions. Lewis Publ., Boca Raton, FL.

Carlson, C.L., and D.C. Adriano. 1993. Environmental impacts of coal combustion residues. J. Environ. Qual. 22:227–247.

Chlopecka, A., and D.C. Adriano. 1996. Mimicked *in situ* stabilization of metals in a cropped soil. Environ. Sci. Technol. 30:3294–3303.

Chlopecka, A., J.B. Bacon, J. Kay, and M.J. Wilson. 1996. Forms of cadmium, lead, and zinc in contaminated soils from southwestern Poland. J. Environ. Qual. 25:69–79.

Coleman, D.C., P.F. Hendrix, and E.P. Odum. Ecosystem health: An overview. p. 1–20. *In* P.M. Huang et al. (ed.) Soil chemistry and ecosystem health. SSSA Spec. Publ. 52. SSSA, Madison, WI (this publication).

Costanza, R., B.G. Norton, and B.D. Haskell. 1992. Ecosystem health . . . new goals for environmental management. Island Press, Washington, DC.

Cunningham, S.D., and W.R. Berti. 1993. Remediation of contaminated soils with green plants: An overview. In Vitro. Cell. Dev. Bio. 29P:207–212.

Davies, M.C.R. (ed.). 1991. Land reclamation. Elsevier Applied Science Publ. Essex, England.

Dixon, J.B., and S.B. Weed (ed.). 1989. Minerals in soil environments. 2nd ed., SSSA Book Ser. 1. SSSA, Madison, WI.

Dushenkov, V., P. Dumar, H. Motto, and I. Raskin. 1995. Rhizofiltration: The use of plants to remove heavy metals from aqueous streams. Environ. Sci. Technol. 29:1239–1245.

Eisenbud, M. 1987. Environmental radioactivity. Academic Press, Orlando, FL.

Ekstrom, G., and M. Akerblom. 1990. Pesticide management in fowl and water safety: International contributions and national approaches. Rev. Environ. Contam. Toxicol. 114:23–55.

Everson, F. 1989. Overview: Soil washing technologies, for comprehensive and liability act resource conservation and recovery act leaking underground storage tanks, site remediation. U.S. Environ. Protection Agency Rep. USEPA 1:440. USEPA, Washington, DC.

Fuller, W.H., and A.W. Warrick. 1985. Soils in waste treatment and utilization. Vol. 1 and 2. CRC Press, Boca Raton, FL.

Frankenberger, W.T., and S. Benson. 1994. Selenium in the environment. Marcel Dekker, New York.

Fristad, W.E. 1995. Case study: Using soil washing/leaching for the removal of heavy metal at the twin cities army ammunition plant. Remediation 5:61–72.

Gombert, D. 1992. Soil washing evaluation by sequential extraction for test reactor area warm waste pond. Westinghouse Idaho Nuclear Company, Idaho Falls.

Greenwood, D.J., P.H. Nye, and A. Walker. 1990. Quantitative theory in soil productivity and environmental pollution. The Royal Society, London, England.

Gworek, B. 1992a. Lead inactivation in soils by zeolites. Plant Soil 143:71–74.

Gworek. B. 1992b. Inactivation of cadmium in soils by synthetic zeolites. Environ. Pollut. 75:269–271.

Hansen, P.E., and S.E. Jorgensen (ed.). 1991. Introduction to environmental management. Developments in Environmental Modeling 18. Elsevier Science Publ., B.V., Amsterdam, Netherlands.

Harrison, R.M. (ed.). 1990. Pollution: Causes, effects, and control. The Royal Society of Chemistry, Cambridge, England.

Harter, R.D. 1983. Effect of soil pH on adsorption of lead, copper, zinc, and nickel. Soil Sci. Soc. Am. J. 47:47–51.

Iskandar, I.K., and D.C. Adriano. 1997. Remediation of soils contaminated with metals: A review of current practices in the USA. p. 1–26. *In* I.K. Iskandar and D.C. Adriano (ed.) Remediation of metal-contaminated soils. Science Reviews, Northwood, England.

Karlen, D.L., M.J. Mausbach, J.W. Doran, R.G. Cline, R.F. Harris, and G.E. Schuman. 1997. Soil quality: A concept, definition and framework for evaluation. Soil Sci. Soc. Am. J. 61:4–10.

Kesraoul-Oukl, S., C. Cheeseman, and R. Perry. 1993. Effects of conditioning and treatment of chabazite and chinoptilolite prior to lead and cadmium removal. Environ. Sci. Technol. 27:1108–1116.

Kim, B.J., C.S. Gee, J.T. Bandy, and C. Huang. 1991. Hazardous waste treatment technologies. Res. J. WPCF 63:501–509.

Kumar, P., V. Dushenkov, H. Motto, and I. Raskin. 1995. Phytoextration: The use of plants to remove heavy metals from soils. Environ. Sci. Technol. 29:1232–1238.

Kryshev, I.I. (ed.). 1992. Radioecological consequences of the Chernobyl accident. Nuclear Soc. Int., Moscow.

Lackey, R.T. 1998. Seven pillars of ecosystem management. The Environ. Prof. (in press).

Lal, R., and F.J. Pierce. 1991. The vanishing resource. p. 1–5. *In* R. Lal and F.J. Pierce (ed.) Soil management for sustainability. Soil Conserv. Soc. Am., Ankeny, IA.

Larson, W.E., and F.J. Pierce. 1991. Conservation and enhancement of soil quality. *In* Evaluation for sustainable land management in the developing world. Vol. 2 1BSRAM. Proc. 12 (2). Int. Board for Soil Res. and Manage., Bangkok, Thailand.

Lefebvre, K.K., B.L. Miki, and J.F. Laliberte. 1987. Mammalian metallothionein functions in plants. Biotechnology 5:1053–1056.

Logan, T.J., 1992. Reclamation of chemically degraded soils. Adv. Soil Sci. 17:13–35.

Longhurst, J.W.S. (ed.). 1990. Acid deposition: Origins, impacts and abatement strategies. Springer-Verlag, New York.

Losi, M.E., C. Amrhein, and W.T. Frankenberger, Jr. 1994. Bioremediation of chromate-contaminated groundwater by reduction and precipitation in surface soils. J. Environ. Qual. 23:1141–1150.

Lynch, J., and B.R. Genes. 1989. Land treatment of hydrocarbon contaminated soils. p. 163–181. *In* P.T. Kostecki and E.J. Calabrese (ed.) Petroleum contaminated soils. Vol. 1. Remediation techniques, environmental fate and risk assessment. Lewis Publ., Chelsea, MI.

Ma, Q.Y., T.J. Logan, J.A. Ryan, and S.J. Traina. 1993. *In situ* lead immobilization by apatite. Environ. Sci. Technol. 27:1803–1810.

Maiti, I.B., G.J. Wagner, and A.G. Hunt. 1991. Light inducible and tissue specific expression of a chimeric mouse metallothionein cDNA gene in tobacco. Plant Sci. 76:99–107.

McBride, M.B. 1994. Environmental chemistry of soils. Oxford Univ. Press, New York.

Mench, M., V. Amans, V. Sappin-Didier, S. Fargues, A. Gomez, M. Loffler, P. Masson, and D. Arrouays. 1997. A study of additives to reduce availability of Pb in soil to plants. p. 185–202. *In* I.K. Iskandar and D.C. Adriano (ed.) Remediation of metal-contaminated soils. Science Reviews., Northwood, England.

Moldan, B., and J.L. Schnoor. 1992. Czechoslovakia: Examining a critically ill environment. Environ. Sci. Technol. 26:14–21.

Mortvedt, J.J., F.R. Cox, L.M. Shuman, and R.M. Welch (ed.). 1991. Micronutrients in agriculture. SSSA Book Ser. 4. SSSA Madison, WI.

Misra, S., and L. Gedamu. 1989. Heavy metal tolerant *Brassica napus* L. and *Nicotiana tabacum* L. plants. Theor. Appl. Genet. 78:161–168.

Murray, J.R. 1993. Soil washing treatability testing of warm waste pond soils at INEL. *In* Proc. of Soil Decon '93. Oak Ridge Natl. Lab., Oak Ridge, TN.

National Research Council. 1993. Soil and water quality: An agenda for agriculture. Natl. Academy Press, Washington, DC.

Nriagu, J. 1974. Lead orthophosphates: IV. Formation and stability in the environment. Geochim. Cosmochim. Acta. 38:887–898.

Page, A.L., T.L. Gleason, J.E. Smith, I.K. Iskandar, and L.E. Sommers (ed.). 1983. Utilization of municipal wastewater and sludges on land. Univ. of California, Riverside.

Parker, D.R. 1995. Prospects for phytoremediation of selenium-contaminated soils in the western U.S. p. 213. *In* Agronomy abstracts. ASA, Madison, WI.

Parr, J.F., R.I. Papendick, S.B. Hornick, and R.E. Meyer. 1992. Soil quality: Attributes and relationship to alternative and sustainable agriculture. Am. J. Altern. Agric. 7:5–11.

Phillips, C.R., W.S. Richardson, C. Cox, and M.C. Eagle. 1993. A pilot plant for the remediation of radioactive contaminated soils using particle-size separation technology. *In* Proc. of Soil Decon '93. Oak Ridge Natl. Lab., Oak Ridge, TN.

Pierzynski, G.M. 1997. Strategies for remediating trace-element contaminated sites. p. 67–84. *In* I.K. Iskander and D.C. Adriano (ed.). Remediation of metal-contaminated soils. Science Reviews, Northwood, England.

Pierzynski, G.M., J.T. Sims, and G.F. Vance. 1994. Soils and environmental quality. Lewis Publ., Boca Raton, FL.

Powe, J.F. 1992. Integrated nutrient management in agricultural systems. p. 85–96. *In* M.S. Bajwa et al., (ed.) Proc. Int. Symp. Nutr. Manage. for Sust. Prod., Punjab Agric. Univ. Press, Ludiahana, India.

Rabinowitz, M.B. 1997. Phosphate application reduces the health hazard of lead-contaminated soil. p. 203–208. *In* I.K. Iskandar and D.C. Adriano (ed.). Remediation of metal-contaminated soils. Science Reviews, Northwood, England.

Racke, K.D., and J.R. Coats (ed.). 1990. Enhanced biodegradation of pesticides in the environment. ACS Symp. Ser. 426. Am. Chem. Soc., Washington, DC.

Reuss, J.O., and D.W. Johnson. 1986. Acid deposition and the acidification of soils and waters. Springer-Verlag, New York.

Rhoades, L.J. 1981. Cost guide for automatic finishing processes. Soc. of Manufacturing Eng., Dearborn, MI.

Runge, E.C.A., K.W. Brown, B.L. Carlile, R.H. Miller, and E.M. Rutledge (ed.). 1986. Utilization, treatment, and disposal of waste on land. SSSA, Madison, WI.

Salomons, W., and U. Forstner. 1988. Chemistry and biology of solid waste-dredged material and mine tailings. Springer-Verlag, Berlin.

Salt, D.E., M. Blaylock, N. Kumar, V. Dushenkov, B.D. Ensley, I. Chet, and I. Raskin. 1995. Phytoremediation: A novel strategy for the removal of toxic metals from the environment using plants. Biotechnology. 13:468–474.

Serne, R.J., C.W. Lindenmeier, P.K. Bhatia, and V.L. Legore. 1992. Contaminant concentration versus particle size for 300 Area North Process Pond sediments. WHC-SD-EN-TI-049, Rev. 0. Westinghouse Hanford Company, Richland, WA.

Sims, R.C. 1990. Soil remediation techniques at uncontrolled hazardous waste sites. J. Air Waste Manage. Assoc. 40:704–732.

Smith, R.A.H., and A.D. Bradshaw. 1979. The use of metal tolerant plant population for the reclamation of metalliferous wastes. J. Appl. Ecol. 16:595–612.

Sposito, G. 1989. The chemistry of soils. Oxford Univ. Press, New York.

Stevenson, F.J. 1982. Humus chemistry: Genesis, composition, reactions. Wiley & Sons, New York.

Sumner, M.E. 1991. Soil acidity control under the impact of industrial society. p. 517–541. *In* G.H. Bolt et al. (ed.) Interactions at the soil colloid–soil solution interface. Kluwer Academic Publ., the Netherlands.

Sumner, M.E. 1993. Gypsum and acid soils: The world scene. Adv. Agron. 51:1–32.

Sumner, M.E., and W.P. Miller. 1992. Soil crusting in relation to global soil degradation. Am. J. Altern. Agric. 7:56–62.

Suter, G.W., R.J. Luxmoore, and E.D. Smith. 1993. Compacted soil barriers at abandoned landfill sites are likely to fail in the long term. J. Environ. Qual. 22:217–226.

Tolba, M.K., and O.A. El-Kholy. 1992. The world environment, 1972–1992. Chapman & Hall, London.

Travis, C.C., and S.C. Cook. 1989. Hazardous waste incineration and human health. CRC Press, Boca Raton, FL.

U.S. Department of Energy-RL. 1994. 300-FF-1 Operable unit remedial investigation Phase II Report: Physical separation of soils treatability study. U.S.DOE/RL-93-96. REV. 0. U.S. Dep. of Energy, Richland Operations Office, Richland, WA.

U.S. Environmental Protection Agency. 1989. Innovative technology: Soil washing. Office of Solid Waste and Emergency Response Directive 9200.5-25PFS. USEPA, Washington DC.

U.S. Environmental Protection Agency. 1990a. Hazardous waste: Toxicity characteristic leaching produce. Fed. Register. 55(126):40 CFR. Parts 261–302.

U.S. Environmental Protection Agency. 1990b. Engineering bulletin: Soil washing treatment. EPA 540/2-90/017. Office of Emergency and Remedial Response, Washington, DC, Office of Research and Development, Cincinnati.

U.S. Environmental Protection Agency. 1994a. Innovative treatment technologies: Annual Status Rep. 6th ed. EPA/542-R-94-005. USEPA, Office of Solid Waste and Emergency Response, Washington, DC.

U.S. Environmental Protection Agency. 1994b. Superfund innovative technology evaluation program: Technology profiles. 7th ed. EPA/540R-94/526. Office of Research and Development, Washington, DC.

U.S. Salinity Laboratory Staff. 1954. Diagnosis and improvement of saline and alkali soils. USDA Handb. 60. USDA, Washington, DC.

Walton, B.T., and T.A. Anderson. 1992. Plant-microbe treatment systems for toxic waste. Current Opinion Biotech. 3:267–270.

Wenzel, W.W., H. Sattler, and F. Jockwer. 1993. Metal hyperaccumulator plants: A survey on species to be potentially used for soil remediation. p. 52. *In* Agronomy abstract. ASA, Madison, WI.

Westinghouse Hanford Company. 1994. 100 Area soil washing: Bench-scale tests on 116-F-4 Pluto Crib soil. WHC-SD-EN-TI-268. Westinghouse Hanford Company, Richland, WA.

Williams, P.H. 1992. The role of fertilizers in environmental pollution. p. 195–215. *In* Proc. Int. Symp. Nutr. Manage. for Sust. Prod., Punjab Agric. Univ. Press, Ludiahana, India.

Winteringham, F.P.W. (ed.). 1984. Environment and chemicals in agriculture. *In* Proc. of a symposium held in Dublin. 15–17 Oct. Elsevier Applied Science Publ., Essex, England.

World Commission on Environment and Development. 1987. Our common future. Oxford Univ. Press, Oxford, England.